Hydrogels in Regenerative Medicine and Other Biomedical Applications

Hydrogels in Regenerative Medicine and Other Biomedical Applications

Editor

Peter John Jervis

MDPI • Basel • Beijing • Wuhan • Barcelona • Belgrade • Manchester • Tokyo • Cluj • Tianjin

Editor
Peter John Jervis
Chemistry
University of Minho
Braga
Portugal

Editorial Office
MDPI
St. Alban-Anlage 66
4052 Basel, Switzerland

This is a reprint of articles from the Special Issue published online in the open access journal *International Journal of Molecular Sciences* (ISSN 1422-0067) (available at: www.mdpi.com/journal/ijms/special_issues/Hydrogels_Regenerative_Biomedical).

For citation purposes, cite each article independently as indicated on the article page online and as indicated below:

LastName, A.A.; LastName, B.B.; LastName, C.C. Article Title. *Journal Name* **Year**, *Volume Number*, Page Range.

ISBN 978-3-0365-7251-2 (Hbk)
ISBN 978-3-0365-7250-5 (PDF)

Contents

About the Editor

Peter John Jervis

Dr. Peter J. Jervis obtained his PhD in Organic Chemistry from the University of Birmingham (UK) in 2008, under the supervision of Dr. Liam Cox, working on the total synthesis of natural products. He then completed two postdoctoral positions in Birmingham, the first with Dr. Richard Grainger on organosulfur chemistry, and the second in the group of Prof. Gurdyal Besra, on a Medicinal Chemistry project developing compounds active against tuberculosis. He spent some years in industry, working for GlaxoSmithKline and then AstraZeneca on Medicinal Chemistry projects. In 2018, he returned to academia with a move to Portugal, to work on collaborative project between the University of Minho (Prof. Paula M.T. Ferreira and Prof. José A. Martins) and the University of Porto (Prof. David M. Pereira) to carry out a research project on supramolecular hydrogels.

Preface to "Hydrogels in Regenerative Medicine and Other Biomedical Applications"

Great strides have been made in the area of hydrogel science since the first hydrogels were described in the 1960s. Hydrogels usually consist of three-dimensional matrices of hydrophilic polymers, held together by chemical or physical crosslinks, or supramolecular assemblies of small amphiphilic molecules. The gelation process occurs in response to a physical or chemical stimulus, such as temperature, pH, electric or magnetic field, enzymatic modification, light, and others. Consisting of mainly water molecules, they represent a unique class of materials, with many applications such as cell therapeutics, cartilage/bone regeneration, sustained drug release and drug delivery systems, tissue engineering, and 3D bioprinting. Despite these great strides, there is still much more to discover in this area. This Special Issue is focused on the use of hydrogels in tissue and bone regeneration. Hydrogels are particularly suited for this purpose as their physical characteristics resemble that of the extracellular matrix; as such, they have found applications as an extracellular medium for cancer cells, stem cells, and neuronal cells. This Special Issue also includes research papers on the other biomedical applications of hydrogels.

Peter John Jervis
Editor

International Journal of *Molecular Sciences*

MDPI

Editorial

Hydrogels in Regenerative Medicine and Other Biomedical Applications

Peter J. Jervis

Centre of Chemistry, Campus de Gualtar, University of Minho, 4710-057 Braga, Portugal; peterjervis@quimica.uminho.pt

Citation: Jervis, P.J. Hydrogels in Regenerative Medicine and Other Biomedical Applications. *Int. J. Mol. Sci.* **2022**, *23*, 3270. https://doi.org/10.3390/ijms23063270

Received: 15 March 2022
Accepted: 16 March 2022
Published: 18 March 2022

It is my great pleasure to be part of this Special Issue in the *International Journal of Molecular Sciences*—"Hydrogels in Regenerative Medicine and Other Biomedical Applications". Hydrogels are made up of three-dimensional matrices of hydrophilic polymers, which are held together by chemical or physical crosslinks or by supramolecular assemblies of smaller amphiphilic molecules. The gelation process can be initiated by a range of stimuli, including temperature, pH, electric or magnetic field, enzymatic modification, light, and others. Remarkably, these materials mainly consist of water molecules, in some cases containing over one million water molecules for every molecule of gelator. This unique class of materials represents a vast research field, owing to their many biomedical applications. These include uses in tissue engineering, 3D scaffolds for cell culture, biosensors, contact lenses, vehicles for controlled drug delivery, and many others. The focus of this Special Issue is related to regenerative medicine, an area that has witnessed rapid developments in recent years. Although much has already been discovered in this area, we continue to see excellent novel research, as exemplified by the seventeen manuscripts presented in this Special Issue, fourteen of which are original research articles.

A common theme in this Special Issue is the use of naturally occurring compounds and polymers for tissue engineering and wound-healing applications. For example, the group of Kang have improved the properties of fibrogen hydrogels by creating conjugates with keratin obtained from human hair. These new materials showed improved porosity. The keratin–fibrogen hydrogel also improved the proliferation of encapsulated human gingival fibroblasts (HGFs), which displayed a diffuse morphology when encapsulated by the new hydrogel [1]. Moon et al. report the development of hydrogels of triclosan (TCS)-complexed beta-cyclodextrin(β-CD)-conjugated methacrylated glycol chitosan (MGC) as antibacterial tissue adhesives. The hydrogel was formed through irradiation with UV light, and the in vitro and in vivo results reveal that the new hydrogel is able to inhibit bacterial infection and improve wound healing, showing its potential as a tissue bio-adhesive [2]. Tamer et al. used marjoram essential oil and kaolin to improve the properties of polyvinyl alcohol (PVA) sponges and created materials suitable for treating irregular wounds. Pore size, swelling capacity, antibacterial and antioxidant properties, and non-toxicity to fibroblasts were all improved, indicating the potential of the PVA/marjoram/kaolin sponge for hemostatic and antibacterial wound treatment [3]. Lin and Tang have described nanocomposites based on hydroxylapatite (HA), a naturally occurring mineral form of calcium apatite with the formula $Ca_5(PO_4)_3(OH)$. Combining cobalt complexes of this material with PVA produced a hydrogel (PVA–CoHA) that improved the mechanical and antibacterial properties over hydrogels made up of solely PVA. Furthermore, the local release of cobalt ions improved cell growth, thus showing potential for wound healing applications in the context of diabetic foot ulcers [4]. The group of Brun produced chitosan conjugated to cell adhesion peptide motifs (RGD and HVP). Their chitosan–RGD and chitosan–HVP conjugates were evaluated for their potential in bone tissue engineering. The optimal biomimetic properties were obtained using a 1:1 chitosan/chitosan–HVP system, which was superior to chitosan alone and chitosan—RGD systems, preserving the antibacterial effect of chitosan and supporting osteoblast cell adhesion [5].

Another important application of hydrogels is controlled drug delivery. García-Couce et al. have developed a novel strategy for the controlled release of betamethasone from AlgNa-g-poly(QCL-co-HEMA) hydrogels, allowing a sustained drug release over 8 hours. Furthermore, the new material was biocompatible with human chondrocytes and fibroblasts, which could be cultured on the hydrogels [6]. Additionally, related to cell culture, Svozilová et al. have created superporous poly(2-hydroxyethyl methacrylate-co-2-aminoethyl methacrylate) (P(HEMA-AEMA)) hydrogel scaffolds for in vitro 3D culturing of leukemic B cells. Modification with the RGDS cell-adhesion motif improved the survival of B-CLLs in cell culture, by promoting the cell–cell and cell–surface interactions. This model has potential for improving the screening of leukemia drugs, supplementing the use of in vivo animal protocols through the early exclusion of toxic and poorly active candidates [7].

Kim et al. present a proof-of-concept study where a hydrogel formed from methacrylated gelatin co-polymerized with a synthetic heparin mimic (GelMA-PSS) was able to serve as an in vitro platform to recapitulate in vivo muscle atrophy-like phenotypes. This was confirmed by the downregulation of key signaling pathways in cells cultured on the heparin-mimicking matrix. This in vitro platform may serve as a promising model of muscle injury model for drug screening and toxicity testing [8]. Safakas et al. have developed a hydrogel system with potential cell transplantation potential. Their NaALG-g-P(NIPAM80-co-NtBAM20)/DMEM formulation showed excellent shear-induced injectability at room temperature and instantaneous thermo-induced gel stiffening at body temperature. The rheological properties of this formulation are similar to those previously reported as ideal for cell transplantation applications. The biocompatibility of PNIPAM-based grafted polysaccharides makes this hydrogel a good candidate for further investigation [9].

Hydrogels continue to be developed as scaffolds related to 3D-bioprinting. Lee et al. used dual-nozzle, three-dimensional printing to develop adipose tissue-derived decellurized extracellular matrix (dECM) hydrogels based on highly elastic poly (L-lactide-co-ε-caprolactone) (PLCL). The 3D-printing technique used for the scaffold fabrication can be used to produce patient-specific scaffolds for later clinical application. The highly elastic PLCL co-polymer used in this study possesses physical properties closer to those of native adipose tissue than other polymers studied. The adECM hydrogel promotes adipose tissue reconstruction by encouraging neovascularization and tissue formation. Taken together, these materials could represent an alternative method for adipose tissue engineering [10]. Setayeshmehr et al. report novel, cell-compatible, dual-component, biomaterial inks, and bio-inks based on PVA and solubilized decellularized cartilage matrix (SDCM) hydrogels that can be utilized for cartilage bioprinting. Functionalizing PVA with cis-5-norbornene-endo-2,3-dicarboxylic anhydride (Nb) gave better results than functionalizing PVA with amine groups. The results show that the incorporation of SDCM into PVA-Nb reduces the compression modulus, enhances cell viability and bioprintability, and modulates the swelling ratio of the resulting hydrogels. Overall, PVA-Nb hydrogels containing SDCM show promise as versatile bio-inks for cartilage bioprinting [11].

Jin and Neogi present ultrasound imaging using a thermally tunable solid-state phononic crystal lens (TSSL). The results demonstrate the premise of applying a hydrogel-filled tunable solid-state lens in frequency domain echo-intensity modes and temporal domain time-of-flight modes, for monostatic detections and mappings in simple experimental environments. This polymer-based TSSL system presented improved tuneability in its focal length and far-field detection capability, compared with the water-based focusing sonic lens (FSL) of the same shape. This work represents the first practical experimental study of a thermally tunable solid-state phononic crystal focusing lens prototype [12].

In a study related to fertility restoration in men following cancer treatment, Del Vento et al. have used hydrogels supplemented with vascular growth factor-loaded nanoparticles in the transplantation of testicular tissue, ultimately leading to accelerated and improved vascular maturity. Poly(D,L-lactide-co-glycolide)/poly(ethylene glycol (PLGA/PLGA–PEG) was used to form nanoparticles containing vascular growth factors VEGF or PDGF. Each molecule was encapsulated individually in separate nanoparticles that were then combined within the alginate hydrogel. Supplementation of alginate hydrogels with nanoparticles delivering PDGF induced improvement in terms of vascularization and vascular maturity in testicular tissue grafts compared to supplementation with VEGF alone [13].

Hsiao et al. have developed a novel entrapment neuropathy model induced by ultrasound-guided perineural hydrogel injections. In entrapment neuropathy, a nerve becomes compressed between two other structures in the body (e.g., ligament and bone). Repetitive motion causes the ligament and bone to press or rub against the nerve, ultimately leading to numbness, tingling, burning, and weakness in the extremities. Inducing the condition in rats provides a surgery-free, minimally invasive animal model of entrapment neuropathy that could serve as a versatile tool in the search for effective treatments of this disease [14].

This Special Issue also contains three review articles, which we hope will provide excellent information resources for the hydrogel research community. Maji and Lee have provided an overview of the use of hydrogels as materials for developing three-dimensional in vitro models [15], while Alven and Aderibigbe have reviewed the use of hydrogels based on chitosan and cellulose and their applications in wound management [16]. Both of these reviews outline the outstanding progress being made in these respective areas. Finally, I had the pleasure of contributing a review from my own research team, whose research focuses on supramolecular peptide hydrogels. Compiled in collaboration with my esteemed colleagues Paula MT Ferreira and Jose Alberto Martins, we have published the first review of the literature specifically related to use of hydrogelators constructed from dehydropeptides, which are showing great promise for biomedical applications [17].

Funding: This editorial received no external funding.

Conflicts of Interest: The author declares no conflict of interest.

References

1. Kang, H.J.; Ko, N.; Oh, S.J.; An, S.Y.; Hwang, Y.-S.; Kim, S.Y. Injectable Human Hair Keratin–Fibrinogen Hydrogels for Engineering 3D Microenvironments to Accelerate Oral Tissue Regeneration. *Int. J. Mol. Sci.* **2021**, *22*, 13269. [CrossRef] [PubMed]
2. Moon, Y.J.; Yoon, S.-J.; Koo, J.-H.; Yoon, Y.; Byun, H.J.; Kim, H.S.; Khang, G.; Chun, H.J.; Yang, D.H. β-Cyclodextrin/Triclosan Complex-Grafted Methacrylated Glycol Chitosan Hydrogel by Photocrosslinking via Visible Light Irradiation for a Tissue Bio-Adhesive. *Int. J. Mol. Sci.* **2021**, *22*, 700. [CrossRef] [PubMed]
3. Tamer, T.M.; Alsehli, M.H.; Omer, A.M.; Afifi, T.H.; Sabet, M.M.; Mohy-Eldin, M.S.; Hassan, M.A. Development of Polyvinyl Alcohol/Kaolin Sponges Stimulated by Marjoram as Hemostatic, Antibacterial, and Antioxidant Dressings for Wound Healing Promotion. *Int. J. Mol. Sci.* **2021**, *22*, 13050. [CrossRef] [PubMed]
4. Lin, W.-C.; Tang, C.-M. Evaluation of Polyvinyl Alcohol/Cobalt Substituted Hydroxyapatite Nanocomposite as a Potential Wound Dressing for Diabetic Foot Ulcers. *Int. J. Mol. Sci.* **2020**, *21*, 8831. [CrossRef] [PubMed]
5. Brun, P.; Zamuner, A.; Battocchio, C.; Cassari, L.; Todesco, M.; Graziani, V.; Iucci, G.; Marsotto, M.; Tortora, L.; Secchi, V.; et al. Bio-Functionalized Chitosan for Bone Tissue Engineering. *Int. J. Mol. Sci.* **2021**, *22*, 5916. [CrossRef] [PubMed]
6. García-Couce, J.; Vernhes, M.; Bada, N.; Agüero, L.; Valdés, O.; Alvarez-Barreto, J.; Fuentes, G.; Almirall, A.; Cruz, L.J. Synthesis and Evaluation of AlgNa-g-Poly(QCL-co-HEMA) Hydrogels as Platform for Chondrocyte Proliferation and Controlled Release of Betamethasone. *Int. J. Mol. Sci.* **2021**, *22*, 5730. [CrossRef] [PubMed]
7. Svozilová, H.; Plichta, Z.; Proks, V.; Studená, R.; Baloun, J.; Doubek, M.; Pospíšilová, Š.; Horák, D. RGDS-Modified Superporous Poly(2-Hydroxyethyl Methacrylate)-Based Scaffolds as 3D In Vitro Leukemia Model. *Int. J. Mol. Sci.* **2021**, *22*, 2376. [CrossRef] [PubMed]
8. Kim, H.; Jeong, J.H.; Fendereski, M.; Lee, H.-S.; Kang, D.Y.; Hur, S.S.; Amirian, J.; Kim, Y.; Pham, N.T.; Suh, N.; et al. Heparin-Mimicking Polymer-Based In Vitro Platform Recapitulates In Vivo Muscle Atrophy Phenotypes. *Int. J. Mol. Sci.* **2021**, *22*, 2488. [CrossRef] [PubMed]
9. Safakas, K.; Saravanou, S.-F.; Iatridi, Z.; Tsitsilianis, C. Alginate-*g*-PNIPAM-Based Thermo/Shear-Responsive Injectable Hydrogels: Tailoring the Rheological Properties by Adjusting the LCST of the Grafting Chains. *Int. J. Mol. Sci.* **2021**, *22*, 3824. [CrossRef] [PubMed]

10. Lee, S.; Lee, H.S.; Chung, J.J.; Kim, S.H.; Park, J.W.; Lee, K.; Jung, Y. Enhanced Regeneration of Vascularized Adipose Tissue with Dual 3D-Printed Elastic Polymer/dECM Hydrogel Complex. *Int. J. Mol. Sci.* **2021**, *22*, 2886. [CrossRef] [PubMed]
11. Setayeshmehr, M.; Hafeez, S.; van Blitterswijk, C.; Moroni, L.; Mota, C.; Baker, M.B. Bioprinting Via a Dual-Gel Bioink Based on Poly(Vinyl Alcohol) and Solubilized Extracellular Matrix towards Cartilage Engineering. *Int. J. Mol. Sci.* **2021**, *22*, 3901. [CrossRef] [PubMed]
12. Jin, Y.; Neogi, A. Ultrasound Imaging by Thermally Tunable Phononic Crystal Lens. *Int. J. Mol. Sci.* **2021**, *22*, 7966. [CrossRef] [PubMed]
13. Del Vento, F.; Poels, J.; Vermeulen, M.; Ucakar, B.; Giudice, M.G.; Kanbar, M.; des Rieux, A.; Wyns, C. Accelerated and Improved Vascular Maturity after Transplantation of Testicular Tissue in Hydrogels Supplemented with VEGF- and PDGF-Loaded Nanoparticles. *Int. J. Mol. Sci.* **2021**, *22*, 5779. [CrossRef]
14. Hsiao, M.-Y.; Wu, Y.-W.; Chen, W.-S.; Lin, Y.-L.; Kuo, P.-L.; Wu, C.-H. Pathogenic Hydrogel? A Novel-Entrapment Neuropathy Model Induced by Ultrasound-Guided Perineural Injections. *Int. J. Mol. Sci.* **2021**, *22*, 3494. [CrossRef] [PubMed]
15. Maji, S.; Lee, H. Engineering Hydrogels for the Development of Three-Dimensional In Vitro Models. *Int. J. Mol. Sci.* **2022**, *23*, 2662. [CrossRef] [PubMed]
16. Alven, S.; Aderibigbe, B.A. Chitosan and Cellulose-Based Hydrogels for Wound Management. *Int. J. Mol. Sci.* **2020**, *21*, 9656. [CrossRef] [PubMed]
17. Jervis, P.J.; Amorim, C.; Pereira, T.; Martins, J.A.; Ferreira, P.M.T. Dehydropeptide Supramolecular Hydrogels and Nanostructures as Potential Peptidomimetic Biomedical Materials. *Int. J. Mol. Sci.* **2021**, *22*, 2528. [CrossRef] [PubMed]

International Journal of *Molecular Sciences*

MDPI

Review

Engineering Hydrogels for the Development of Three-Dimensional In Vitro Models

Somnath Maji [1] and Hyungseok Lee [1,2,*]

1 Department of Mechanical and Biomedical Engineering, Kangwon National University (KNU), Chuncheon 24341, Korea; somnath.2812@gmail.com

2 Department of Smart Health Science and Technology, Kangwon National University (KNU), Chuncheon 24341, Korea

* Correspondence: ahl@kangwon.ac.kr

Abstract: The superiority of in vitro 3D cultures over conventional 2D cell cultures is well recognized by the scientific community for its relevance in mimicking the native tissue architecture and functionality. The recent paradigm shift in the field of tissue engineering toward the development of 3D in vitro models can be realized with its myriad of applications, including drug screening, developing alternative diagnostics, and regenerative medicine. Hydrogels are considered the most suitable biomaterial for developing an in vitro model owing to their similarity in features to the extracellular microenvironment of native tissue. In this review article, recent progress in the use of hydrogel-based biomaterial for the development of 3D in vitro biomimetic tissue models is highlighted. Discussions of hydrogel sources and the latest hybrid system with different combinations of biopolymers are also presented. The hydrogel crosslinking mechanism and design consideration are summarized, followed by different types of available hydrogel module systems along with recent microfabrication technologies. We also present the latest developments in engineering hydrogel-based 3D in vitro models targeting specific tissues. Finally, we discuss the challenges surrounding current in vitro platforms and 3D models in the light of future perspectives for an improved biomimetic in vitro organ system.

Keywords: in vitro model; hybrid hydrogel; extracellular matrix; microenvironment mimicking matrix

Citation: Maji, S.; Lee, H. Engineering Hydrogels for the Development of Three-Dimensional In Vitro Models. *Int. J. Mol. Sci.* **2022**, 23, 2662. https://doi.org/10.3390/ijms23052662

Academic Editor: Peter John Jervis

Received: 15 December 2021
Accepted: 26 February 2022
Published: 28 February 2022

1. Introduction

In a living body, cells are enclosed firmly in a three-dimensional (3D) mass of matrix, where they are constantly proliferating, migrating, differentiating, and communicating with each other and their immediate microenvironment. This area that surrounds the cells is an intricate network of multi-domain macromolecules and other biological factors organized in a cell- and tissue-specific manner, and is termed as extracellular matrix (ECM) [1]. Faithfully replicating such a complex microenvironment of any tissue in in vitro cell culture conditions is still a challenge for the scientific community. For a very long time, two-dimensional (2D) cell culture was commonly used for any kind of cell- and tissue-based assay required in biomedical applications [2]. However, researchers are well aware of the shortcomings of 2D culture, as it causes unnatural changes in the cell morphology and behavior, resulting in misleading and non-predictive data [3,4]. In contrast, 3D culture and in vitro tissue models have the advantages of providing a microenvironment to cells, which enables them to interact with the matrix and neighboring cells in more physiologically relevant conditions (Table 1) [5–7].

Table 1. Comparison of cell behavior under 2D and 3D culture conditions.

S. No.	Cell-Specific Features	2D Culture	3D Culture	Ref.
1	Morphology and architecture of cells	Flat and extended morphology with poor cell architecture	A rich architecture with round and contracted morphology	[8–10]
2	Migration of cells	Cells migration is fast and directional	Migration of cells in all directions with slow and restricted motility	[11,12]
3	Proliferation of cells	High proliferation rate	Relatively low proliferation compared to 2D culture	[11,13]
4	Interaction with surroundings	Limited interaction with cells and ECMs	Cells can interact with their surrounding in all directions	[14]
5	Polarization of cells	Partly polarized	Full polarization	[11]
6	Intracellular metabolism	High metabolic rate	Relatively low rate of intracellular metabolism	[12]
7	Diffusion of fluids and cell signaling	Limited perfusion of fluids with asymmetric metabolite diffusion and cell signaling	Three-dimensional fluid perfusion and symmetric diffusion of metabolites	[15,16]
8	ECM remodeling	Poor or near absent	Close to mimicking the in-vivo ECM remodeling	[13]
9	Cell viability against cytotoxic agents	High loss of cells	High cell survival rate	[17]
10	Cell death/Apoptosis	Induced apoptosis	Tissue-like apoptosis	[12,18]

The recent advances in the tissue engineering field have fueled the imagination of scientists towards the development of in vitro 3D models, which could superiorly mimic the structural and functional aspects of the native tissues and organs [19–21]. There are three stimuli for this major paradigm shift in biomedical research. First is the absence of proper genetically and physiologically applicable animal model that can recapitulate human conditions [22]. Although large-scale animal models are still in use for biomedical research, they are significantly inferior in faithfully recapitulating human conditions due to substantial species genetic variation [23]. This difference also leads to poor clinical outcomes, even after successful animal trials [24]. Second is the ethical and moral dilemmas present with animal testing. Additionally, animal testing is also expensive and time-consuming and is not readily available for many researchers [25]. Finally, owing to the current advances in cell culture research [26]. Improvement in the proficiency of isolation, expansion, preservation, and guiding the growth and differentiation of human primary and stem cells towards a particular lineage has provided the means and methods for the development of cell-based in vitro 3D tissue/organ models. Applying patients' cells for developing in vitro 3D models could not only accelerate our understanding of tissue development and genetic alteration in the disease state, but also revolutionize the disease screening, diagnosis, and treatment development. Although the superiority of three-dimensional in vitro models is not universal, it provides an edge over the 2D culture in better mimicking the structural and functional characteristics of native human tissue and organs [27].

In this regard, the goal is to utilize the combined knowledge of material science and life sciences in developing an amicable microenvironment that supports the growth and survivability of physiological features of tissue and organs on a long-term basis [28]. The essence of this supportive biological environment lies in the selection of appropriate biomaterials that can most closely resemble the physiological and structural architecture of the native matrix. From a material design perspective, among the various biomaterials, hydrogels are considered the most relevant option for mimicking the native extra-cellular matrix of tissue [29]. A significant growth in research interest of hydrogels has been seen in last three decades, partly due to the emergence of the tissue engineering field and the appealing applications of 3D culture (Figure 1). Composed of various polymeric materials (natural, synthetic, and composite), hydrogel forms an interconnected polymeric network in which it can hold a large amount of water, thus closely resembling a hydrated form of native ECM. The hydration and porosity of hydrogel also facilitate better exchange of nutrients and gases among the cells, as well as the removal of waste products. Hydrogels fabricated from natural polymers such as carbohydrates and proteins can present the biological active cues required for the growth and proliferation of cells [30,31], while hydrogel based on synthetic polymers provides tunable mechanical properties that can resemble the strength of the native tissues [32]. Since hydrogels fabricated purely from natural or synthetic polymers do not match the overall structural or functional aspects of the 3D tissues, a hybrid system was developed to overcome the shortcomings of the individual components. Hybrid hydrogels can be defined as having building blocks that are chemically, physiologically, and functionally distinctive at the microscopic or molecular level. This system might contain biologically active molecules such as protein, peptides, polysaccharides, or nano-/microstructure polymers that are connected via physical or chemical means. There are a few general strategies for fabricating hybrid hydrogels, such as in situ synthesis [33], physical blending of polymers [34], the formation of interpenetrating networks (IPNs) [35,36], and bio-conjugation [37].

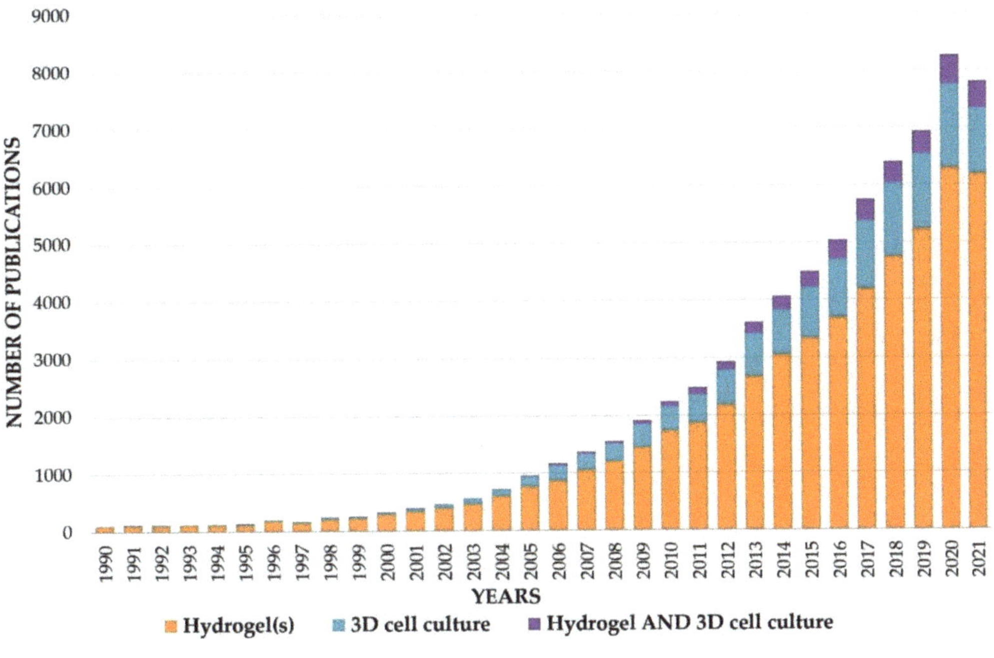

Figure 1. Number of publications related to "hydrogels" and "3D cell culture" from 1990 onward. The total number of studies was calculated based on data from PubMed (www.pubmed.ncbi.nlm.nih.gov (accessed on 20 November 2021)).

Further, in addition to conventional approaches, which include cell and spheroid laden bulk hydrogel constructs and cell-cultured porous and fibrous scaffolds, numerous micro-technologies have emerged, including microspheres, microfibers, sandwich systems, micro-patterned membranes, microfluidic systems, and 3D bioprinting platforms, and they have helped facilitate the development of hydrogel-based tissue/organ models with increased fidelity. In this review article, we highlight the recent progress in the development of a 3D in vitro tissue/organ model using hydrogel as the biomaterial. We start with a broad overview of different types of hydrogel systems based on their polymeric sources, followed by a review of the various gelation techniques. A summary of hydrogel design consideration, the development of various hydrogel modules, the fabrication technologies used, and the platform developed for establishing in vitro models is also provided. Furthermore, we present 3D in vitro models of various types of tissues, including skin, liver, intestine, bone, and cartilage, as well as cancer models. Despite being such a widely used biomaterial, hydrogel presents certain challenges, which are addressed in our final discussion of various limitations and future perspectives for engineering 3D in vitro modeling. A schematic of the various components and factors in the development of composite hydrogels and hydrogel-based in vitro 3D models is provided in Figure 2.

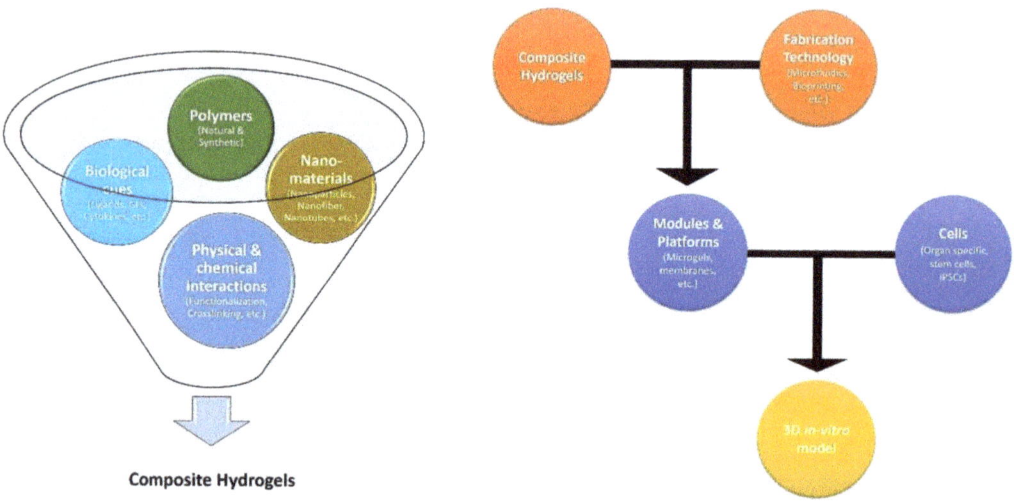

Figure 2. Schematic representation of various factors and components in the development of hydrogels and hydrogel-based in vitro 3D model.

2. Types of Hydrogels

Based on their composition, hydrogels have been commonly classified into natural, synthetic, and hybrid types.

2.1. Natural-Based Hydrogels

As these hydrogels are derived from natural polymers, they are biocompatible, interact favorably with cells, and usually show inherent bioactivity [38]. However, they have poor mechanical properties and relative instability due to their suboptimal polymer interactions and biodegradability, which affect their use in long-term cell culture [1,39]. Due to the presence of a large number of free functional groups, which include amines, carboxyl, and alcohols, hydrogels made from natural polymers can be customized synthetically to tune their properties [38,40]. Based on their chemical nature, natural hydrogels can be further classified into polysaccharide-, protein-, and peptide-based hydrogels.

2.1.1. Polysaccharide-Based Hydrogels

Polysaccharides are one of the most abundant classes of biopolymers and can be derived from animals, plants, microbes, and algae. In the past few decades, researchers have focused on this class of biopolymer for hydrogel development due to its multidimensional properties, such as high stability, non-toxicity, biodegradability, cytocompatibility, availability, and its relative cheapness [41]. Although a large number of polysaccharides (dextran, guar gum, xylan, etc.) have been studied for hydrogel development, this review focuses on alginate, hyaluronic acid (HA), and chitosan-based hydrogels due to their frequent use in 3D in vitro models [42,43].

Alginate polysaccharides are widely applied as a single polymer-based hydrogel system. Alginate is composed of two uronic acids and is mainly derived from bacteria or brown algae. Since it is not a component of mammalian ECM, it lacks cell attachment motifs. However, the presence of hydroxyl side groups allows for easy chemical modifications, making it a multi-purpose material in biomedical applications [44,45]. Gelation of alginate solution is achieved by treating it with calcium chloride to form a non-covalent electrostatic complex [46]. The degree of gelation can be controlled to tune the hydrogel's physical and chemical properties and generate more tissue mimetic viscoelastic behavior [47]. Scaglione et al. showed that by tuning the rheological behavior of alginate hydrogel, human intestine spheroids could be passaged and maintained for 90 days without significant reduction in the expression of some critical markers [48]. They also established a breast cancer model in alginate gel. However, the study also showed the need for cell adhesion motif and a more permissive environment by mixing alginate with Matrigel to demonstrate a model for invasive forms of cancer [49].

HA and chitosan are other widely used classes of polysaccharides for tissue engineering applications. Both HA and chitosan are types of glycosaminoglycan (GAG) composed of similar repeating units of disaccharides. GAGs are commonly used in combining biomaterials due to their involvement in cell signaling and communication [50]. Both polymers are biodegradable and non-toxic, albeit they have poor mechanical stability [51]. Several approaches have been used to enhance their modulus (methacrylation, hydrazide, or thiol functionalization) by exploiting the hydroxyl group of HA and the amine group of chitosan to modify their stability and degradability. In a study by Zhu et al., the stiffness of HA hydrogels was modified using bis-cystine containing matrix metalloproteinase degradable crosslinker [52]. In the same study, arginine-glycine-aspartate (RGD) peptides were used to increase the hydrogel adhesion property, after which mesenchymal stem cells (MSCs) were shown to form 3D cellular networks within seven days of culture. The same gel-based culture was used to embed noggin and BMP-2 morphogens. Under the influence of both the morphogens, the cells form a thicker and denser cellular network, with miscellaneous multi-cellular morphological features closely mimicking the morphogenesis of trabecular bone. Maji et al. used a frugal and combined method of high-speed stirring and freeze-drying to fabricate a macroporous scaffold combining chitosan with gelatin. Enhanced penetration of MSC spheroids into the construct was observed, which led to an increase in the cellular viability, migration, proliferation, and differentiation toward osteogenic lineage [53]. HA has also been extensively used in the development of cancer models for disease modeling and drug screening because it is a major component of tumor ECM. A study by Soker et al. demonstrates a successful establishment of cancer spheroid models of breast, liver, colon, and intestine using HA-based microbeads [54]. The models were demonstrated to be effective in cancer metastasis investigation, drug screening, and drug resistance evaluation.

2.1.2. Protein- and Peptide-Based Hydrogels

Protein- and peptide-based hydrogels are another big class of biomaterials and are popular among materials scientists due to their biocompatibility, biodegradability, and inherent possession of cell adhesion motifs [55]. Such intrinsic bioactive attributes, including proteins such as collagen, gelatin, fibronectin, and laminin, are used in combination with other natural or synthetic polymers to increase cell attachment and growth. Most of the proteins used in hydrogel preparation are structural proteins. In the following paragraph, we provide a brief overview of a few proteins used for the preparation of hydrogels.

Collagen I is a highly abundant structural protein present in the ECM of connective tissue types found in bone, cartilage, tendon, and ligament. It is a well-established solubilized fibrillary gel used in 3D cell culture of a vast range of cell types, including mesenchymal stem cells [56], fibroblasts [57], smooth muscle cells [58], adipocytes [59], and kidney cells [60], as well as in the culture of pancreatic epithelium [61], the growth and branching of mammary epithelium [62], cancer models (skin [63], breasts [64], colon [65], etc.), and in vascularization studies [66]. The strength of collagen-based gels can be easily tuned by crosslinking them with glutaraldehyde [67], gel compression [68], solvent evaporation [69], and other chemical methods [70]. Altering the collagen gel stiffness has resulted in enhancement in endothelial cell spreading and sprouting [71]. Jabaji et al. found that a long-term culture of human intestinal organoid in collagen type I gel resulted in the formation of fully elaborated intestinal epithelial tissue [72,73]. In another recent study, Wimmer et al. used a hydrogel blend of collagen I/Matrigel for the development of human blood vessels by direct differentiation of human embryonic cells and induced pluripotent stem cells (iPSCs) in a 96-well format [74]. Some other examples of collagen I hydrogel application in 3D culture are in vitro development of murine stomach tissue and murine and human colon tissue [73].

Gelatin is a hydrolyzed derivative of triple helix structured collagen. It is water solubilized, contains RGD adhesion motifs, and is less immunogenic than collagen [75]. Although highly biologically compatible, gelatin hydrogels have poor mechanical strength. Chemical and physical crosslinking methods such as glutaraldehyde, EDC-NHS, and ultraviolet have been used to strengthen gelatin's mechanical stability when used in in vitro cell culture [76]. Methacrylamide derivative of gelatin (GelMA) has been developed, which undergoes photopolymerization, thus enabling the tuning of the mechanical properties of the hydrogel [77]. The utilization of GelMA has tremendously increased in recent years due to it being biocompatible and non-toxic and having the potential to form vasculature networks [78]. The incorporation of nanomaterials, such as carbon-nanotube, graphene oxide (GO), and inorganic particles, has allowed for the engineering of GelMA-based hydrogels that mimic load-bearing tissues [79–81]. Recently, GelMA has also been applied in the bioprinting of tissue analogs [82].

Silk fibroin is another class of fiber-forming protein that has seen increasing use in tissue engineering applications [83]. Through the manipulation of some environmental factors, such as the pH, osmolarity, and shear stress, the large hydrophobic regions of the protein can be physically crosslinked to form β-sheets to form silk hydrogels [84]. Silk-based constructs have been applied to regenerate cartilage tissue by differentiating bone marrow-derived mesenchymal stem cells [85]. Self-assembling silk fibers have been used to develop minimally invasive therapy for brain-related biomedical conditions. In a study by the Kaplan group, silk/collagen-based fibrillated hydrogels were used to develop a compartmentalized structure of cortical brain tissues [86]. Skin reconstruction [87], the engineering of intestinal epithelium tissue [88], the mimicking of bone tissue [89], and nerve regeneration [90] are some of the bioengineering fields in which silk-based hydrogels and constructs have been used.

2.2. *Hydrogels from Synthetic Sources*

Synthetic polymers are usually termed inert or blank state as they do not have any biological active factors and do not promote cellular activity [45,91]. Usually, synthetic polymers are blended with natural, bioactive substances to enhance their cell compatibility [92]. Cell adhesion motifs, growth factors, or other biologically active small molecules can be covalently bonded to synthetic polymers in a controllable or anisotropic manner to guide directional cell growth and organization. The main advantage of this is that based on the various compositions, the mechanical condition (stiffness, elasticity, and durability) of the polymers can be easily defined, allowing regulation of cell growth, migration, and proliferation [93,94]. Some of the synthetic polymers commonly used in tissue engineering and in vitro cultures are polyethylene glycol (PEG), polylactic acid (PLA), poly (ε-caprolactone) (PCL), polyurethane (PU), and poly(lactic-co-glycolic acid) (PLGA) [95–97].

2.2.1. Polyethylene Glycol (PEG)

PEG is one of the most common and favorable synthetic polymers for tissue engineering due to its chain flexibility, hydrophilicity, and low non-specific interaction with the living tissue [98]. The PEG hydroxyl end group, presents in a broad range of structures and molecular weights, can be easily functionalized (dithiol, acrylate, NHS ester, etc.). Such multi-functionality allows the incorporation of bioactive moieties through a compatible condition such as photopolymerization, Michael-type addition, or chain polymerization [99,100]. Garcia et al. showed that the modulation of mechanical properties of maleimide-terminated PEG (PEG-4MAL) incorporated with RGD motifs strongly influences the epithelial morphogenesis of kidney cells [101]. They further demonstrated the influence of mechanically tunable hydrogel in guiding tissue morphogenesis of different origins (lungs and intestines) [102].

2.2.2. Other Synthetic Polymers

In comparison to natural polymers, synthetic hydrogels offer physicochemical properties that can be modulated according to the particular experimental need [103], thereby making them a preferable option in preparing ECM analogs. Examples of such synthetic polymers used in 3D modeling include polyvinyl alcohol, poly(hydroxyethyl methacrylate), poly(isopropylacrylamide) (PNIPAm), self-assembling polypeptides, and synthetic GAGs [104]. In recent years, synthetic GAG analog fabrication has been investigated due to its capability of binding growth factors, long-term availability, and control degradability. For example, in a study conducted by Chang et al., it was shown that poly(sodium-4-styrene sulfonate) (PSS) mimics synthetic heparin and has a strong binding affinity for fibroblast growth factor (bFGFs). A fabricated PSS functionalized PAm demonstrated maintenance and growth of iPSCs over 20 passages without losing its pluripotency or its morphological or karyotypic stability [105].

2.3. *Hybrid Hydrogels*

The inability of any single polymeric material to recapitulate the complexity of the in vivo microenvironment and natural ECM is well documented. The natural-based polymers contain bioactive cues for cellular growth and proliferation, but produce batch-to-batch discrepancies due to their poor mechanical and stability issues. In contrast, synthetic polymers provide excellent control over physical and chemical properties, but generally lack cell instructive motifs. Therefore, the hybridization of different compositions and types of hydrogels fabricated via various physical interactions or types of chemical polymerization provides an enticing solution. These hybrid hydrogels can be developed by blending between natural–natural, synthetic–synthetic, and natural–synthetic polymers. Natural–natural hybrid hydrogels are produced by the physical mixing of two or more different types of polymers derived from a natural source, whereas synthetic–synthetic hybrid polymers are fabricated by the copolymerization of two or more synthetic polymers to tune their biodegradability and physical properties [106,107]. Natural–synthetic hybrid hydrogels, however, are considered the most favorable for in vitro cell culture as they combine both the mechanical stability of synthetic polymers and the bioactivity of natural polymers into a single system. Further, bioactive moieties, as well as different types of nanomaterials, are integrated into the polymeric composition to impart specific properties on hybrid hydrogels [108]. In the following section, we discuss hybrid hydrogels, which we categorize based on their constituting material. A list of hybrid hydrogels along with their composition, specific features, and findings concerning in vitro cell culture applications is presented in Table 2.

Table 2. Survey of hybrid hydrogel composition for in vitro 3D cell culture.

Hydrogel Composition	Gelation/Fabrication Method Used	Distinctive Features	Findings	Reference
		Natural–Natural Polymers		
Alginate–Matrigel	Ionic crosslinking with Ca++ ions	Increased hydrogel stiffness	Progression of the normal mammary epithelium into malignant cells	[109]
Alginate–Marine Collagen–Agarose	Self-assembly	Increased cytocompatibility	High yield in multicellular spheroids	[110]
Alginate–Gelatin	Pre-crosslinking with calcium chloride	Similar mechanical properties as crosslinked alginate but superior cytocompatibility	Prospective bio-ink provides a good means for myoregenerative applications	[111]
Chitosan–Gelatin	Ionic interaction to form polyelectrolyte complexes	High shape fidelity and good biocompatibility	Bioprinting of skin fibroblasts	[112]
Gelatin–Silk Fibroin	Enzymatic and physical (sonication) crosslinking	Superior mechanical strength, tunable degradability, and improved multilineage differentiation ability	Site-specific bioprinting of progenitor cells and differentiation	[113]
GelMA–Gellan gum MA	Photo-crosslinking	Achieved similar viscoelasticity to native cartilage	Enhanced viability and growth for human articular chondrocytes	[114]
Thiolated Gelatin–Vinyl sulfonated HA	Click chemistry (-SH with C=C)	High viability and proliferation capability	Bone MSC in vitro differentiation toward chondrocytes	[115]
Amphiphilic peptides–keratin	Self-co-assembly	Integration of self-assembly and bioprinting. The process directed molecular assembly and nanomaterials into ordered structures with various sizes and geometries	Introduced a 3D bioprinting platform encapsulating cells in the pericellular environment	[116]
		Natural–Synthetic Polymers		
PEG–Keratin	thiol-norbornene "click" reaction	Highly tunable mechanical properties and long-term stability	Suitable for various microfabrication techniques (e.g., micropatterning, wet spinning) for fabrication of 3D cell-laden tissue constructs	[117]
PCL–RGD peptides	Acrylic acid (AAC) grafting by γ irradiation and crosslinking by 1-ethyl-3-(3-dimethylaminopropyl) carbodiimide/N-hydroxysuccinimide (EDC/NHS)	Provides a fibrous anchorage site in the 3D hydrogel environment	Mesenchymal stem cells show remarkable spreading properties with augmented viability and differentiation	[118]
PCL–Alginate	Interfacial bonding	Composite mimics the microarchitecture and mechanical properties of soft tissue	Subcutaneous implantation shows infiltration of pro-regenerative macrophages and leads to gradual remodeling and replacement of the composites with vascularized soft tissue	[119]
PCL–PEG–heparin	Melt-electrospinning writing combined with additive manufacturing	Fibrous networks exhibited mechanical anisotropy, viscoelasticity, and morphology similar to native cartilage tissue	In vitro neo-cartilage formation	[120]
PCL–GelMA–Alginate	Melt-electrospinning writing	Stiffness and elasticity were similar to those of the native articular cartilage tissue	Embedded human chondrocytes retain their round morphology and are responsive to in vitro physiological loading regime	[121]
Sodium alginate/PLGA microspheres–HSP27–TAT peptide	Crosslinking with calcium sulfate followed by physical incorporation of microspheres	Porous microsphere enabled sustained release of HSP27-TAT hybrid system for over two weeks	Sustained delivery of HSP27-TAT reduced the infarcted site and improved the end-systolic volume in the heart	[122]

Table 2. *Cont.*

Hydrogel Composition	Gelation/Fabrication Method Used	Distinctive Features	Findings	Reference
Chondroitin sulfate–PEG	Enzymatic crosslinking (Transglutaminase factor XIII)	Modular design allows the facile incorporation of additional signaling element	Tunable matrix with BMP2 binding and sustained release allows enhanced proliferation of MSCs and differentiation toward osteogenic lineage	[123]
Chitosan/HA–PLGA microsphere	Reversible Schiff's base reaction	The linking of vancomycin with PLGA microspheres enabled the hydrogel system to inhibit bacterial growth	Vascular endothelial growth factor (VEGF) encapsulation to the PLGA microspheres accelerated the growth and proliferation of endothelial cells and increased angiogenesis, thereby promoting management of non-healing wounds	[124]
Hydrogel linked with biological factors				
PEGDA–RGD peptide	Two-photon laser scanning photolithography	Generates microscale patterns with control over the spatial distribution of biomolecules	Human dermal fibroblasts cultured in fibrin clusters of hybrid hydrogel underwent a guided 3D migration	[125]
Star PEG–desulfated Heparin–VEGF	EDC/NHS crosslinking	The presence of heparin provides anticoagulant activity, while sustained release of VEGF ensures the growth of endothelial cells	In vitro tube formation of human umbilical vein endothelial cells (HUVEC) and promotion of wound healing in genetically diabetic mice	[126]
GelMA–VEGF	EDC/NHS crosslinking; Extrusion-based direct-writing bioprinting	Inside the bioprinted 3D construct, a central cylinder of GelMA was printed to allow the formation of perfusable blood vessels	Co-culture of MSCs enabled the formation and stabilization of endothelial cells. VEGF-linked hydrogel induced differentiation of MSCs toward osteogenesis	[127]
GelMA–BMP2–TGF-β1	Bioprinting; Photo-crosslinking	Established an anisotropic biomimetic fibrocartilage microenvironment by bioprinting a nanoliter droplet encapsulating MSCs, BMP-2, and TGF-β1	Genomic expression study findings indicate differentiation of MSCs and simultaneous upregulation of osteogenic and chondrogenic factors during the in vitro culture on the model 3D construct	[128]
Collagen–fibrin–VEGF	Culture plate coated with a nebulized layer of sodium bicarbonate; Bioprinting	An artificial neural tissue construct was fabricated by bioprinting collagen-laden neural stem cells and VEGF-embedded fibrin gel	A sustained release of VEGF was found from bioprinted fibrin gel, which enhanced the migration and proliferation of neural stem cells	[129]
Matrigel–Gelatin microparticles–VEGF	Glutaraldehyde crosslinking	The construct of gelatin microparticles is suitable to generate sustained release profiles of bioactive VEGF	In vitro study shows real-time migration of endothelial progenitor cells and reveals enhanced in vivo vasculogenic capacity	[130]
Hydrogels incorporated with nanomaterials				
Dopamine–Folic acid–ZnO–quantum dot	Crosslinking by transition metal ions to form metal–ligand coordination	The hydrogels provide greater antibacterial efficacy when illuminated at 660 and 808 nm (generate ROS and heat).	The hydrogels release zinc ions over two weeks, provide a sustained antimicrobial effect against *S. aureus* and *E. coli*, and promote fibroblast growth	[131]
Alginate-SiO_2 nanofibers	Ionic crosslinking through Al^{3+}	The method helps the nanofibrous hydrogels retain a large amount of water, which helps in producing desirable shapes at a larger scale	Results of zero Poisson's ratio, memory of shape, injectability, and conductivity provide insight into the development of future multifunctional hydrogels	[132]

Table 2. *Cont.*

Hydrogel Composition	Gelation/Fabrication Method Used	Distinctive Features	Findings	Reference
Poly hydroxyethyl-methacrylate (pHEMA)–multiwalled carbon nanotube (MWCNT)	Polymerization at 4 °C	With incorporated MWCNT, the construct has a more porous structure with better elastic modulus and electrical conductivity	The in vitro study shows the viability of neuroblastoma cells and that they help conduct electricity, indicating that the construct is more suitable as a nerve conduit	[133]
HA–dopamine–rGO	EDC/NHS crosslinking; Oxidative coupling of catechol groups by H_2O_2/HRP as the initiator	Multifunctional, including tissue adhesiveness, antibacterial and antioxidant ability, and good mechanical properties	Shows significant skin regeneration capacity with enhanced vascularization; promoted as an excellent wound-dressing hydrogel	[134]
Quaternized Chitosan–benzaldehyde terminated F127	Crosslinking between Schiff base bond and PF127 micelle	The dressings system showed good stretchability, similar mechanics to human native skin, and rapid self-healing	In an in vivo study with a full-thickness skin defect model, the hydrogel showed deposition of collagen with upregulated VEGF, which led to accelerated wound healing	[135]
Alginate–PLA nanofibers	Ultrasonication	Comparable compressive characteristics to native alginate hydrogels with better cytocompatibility	The nanofiber-blended bio-ink allowed the enhancement of adipose-derived stem cells proliferation, with the presence of collagen and proteoglycans indicating chondrogenic differentiation	[136]
Alginate–Nanofibrillated cellulose	90 mM $CaCl_2$	The bio-ink shear-thinning behaviors enable printing of 2D grid structures as well as complex 3D soft tissue-like constructs	Bioprinted human chondrocytes show over 80% viability 7 days post-culture, demonstrating the potential of the bio-ink for 3D bioprinting	[137]

2.3.1. Blend of Natural and Synthetic Polymers

Hybrid hydrogels composed of polymers from both natural and synthetic sources are the most researched and reported hydrogel type, as they provide the best features of both sources [138]. The mixing of polymers, which are termed double networks or IPNs, presents the advantages of both natural polymers (cell-adhesion ligands, hydrophilicity, fibrous architecture) and synthetic polymers (controlled degradability, tunable modulus, etc.) [139]. For example, different variants of PEG, such as Poly(ethylene glycol) diacrylate (PEGDA) and poly[oligo(ethylene glycol)] methacrylate (PEGMA), have been used to bioprint hydrogels with a combination of gelatin or alginate to increase their mechanical and biological properties [2,117]. Composite bio-inks were developed by using a blend of synthetic amphiphilic peptides with keratin. The potential of this self-assembled biomaterial was studied through the development of a 3D bioprinting platform, which encapsulated cells within the modular pericellular microenvironment [116]. The term soft network composites (SNC) was recently coined based on the architecture of soft tissue, in which fibrous polymers are embedded in a weak hydrogel system. To mimic this pattern, researchers have used fibrous synthetic polymers such as PCL and embedded them in a variety of synthetic and natural polymers such as PEG, HA, and alginate hydrogels [140]. In the first and an upgrade to the above model, Bas et al. introduced the method of melt electrospinning writing (MEW) to form a continuous interconnected fibrous network of PCL combined with PEG/heparin/fibrin hydrogel [120]. This composite hydrogel system not only exhibits better pore size and mechanical properties, but also provides a suitable environment for the growth and proliferation of the human chondrocytes forming neo-cartilage in vitro. In a similar study by Visser et al., PCL microfibers were introduced into a hydrogel composed of GelMA and alginate, and a significant increase (52-fold) in hydrogel stiffness was observed [121]. The technique of SNC is promising but still at the early development stage. Further studies targeting the interaction of fibrous synthetic materials with biodegradable hydrogels and remodeling of SNC over time are in progress to develop novel matrix designs and customizations.

2.3.2. Hybrid Hydrogels with Biological Factors

The ECM contains numerous biologically active macromolecules, such as cell adhesion moieties, growth factors, hormones, cytokines enzymes, and other signaling molecules, which control the spatio-temporal behavior and fate of cells [1]. PEG, PCL, PU, and other synthetic polymers as well as some natural polymers such as alginate and chitosan do not possess the biological motifs and cues of ECM. In this regard, much of the tissue engineering-based research has been focused on ornamenting biomacromolecules onto polymeric hydrogels to provide an in vivo ECM-imitating environment. One of the most common approaches is the tethering of integrin-binding peptides such as RGD and LDV in fibronectin to the synthetic polymer backbone to render the system biocompatible [141]. Further, with the advent of the third generation of biomaterials, there is a significant emphasis on the use of growth factors on biopolymers to enhance cell growth, proliferation, and in some cases differentiation or wound-healing applications. The vascular endothelial growth factor is one of the key players in the wound healing process, as it promotes angiogenesis and blood capillary formation, which are prerequisites for a successful tissue-engineered graft. In one such study, Byambaa et al. used the EDC/NHS coupling method to link the VEGF molecule with an injectable gelatin hydrogel to treat a non-union bone defect [127]. The study revealed that this hydrogel has better osteogenic and angiogenic potential, thus promoting the growth and differentiation of encapsulated cells. For certain stem cells, growth, and differentiation to a particular lineage require continuous external stimuli in the form of growth factors. In such a study, Gurkan et al. added TGF-b and FGF into the GelMA-based bio-ink [128]. They bioprinted this bio-ink with embedded mesenchymal stem cells in a growth factor gradient manner to mimic the fibro–cartilage transition in the bone–ligament interface. The study demonstrated the differentiation of MSCs into respective osteoblast and chondrocyte phenotypes in a spatially defined manner.

2.3.3. Hybrid Hydrogels Incorporated with Nanomaterials

The incorporation of nanomaterials into biomaterials is a popular, versatile, and useful technique to engineer a multi-functional construct for tissue engineering applications. Nanomaterial in the form of nanoparticles, nanofibers, and nanotubes has been used to ornament hydrogels, enhancing the mechanical, electrical, and optical properties as well as augmenting the bioactivity of the system [142,143]. In this section, we briefly discuss various types of nanomaterials including metal-based, carbon-based, and polymer nanofillers, which have been used to enhance the functional aspect of biomaterials. Among many metal-based polymers, silver nanoparticles have been extensively used in combination with many synthetic polymers such as PVA and PAm to provide antibacterial properties to the construct [144]. Based on its ability to avoid the issue of particle aggregation and ensure even distribution among the gel, the in situ synthesis approach has gained popularity among researchers [145]. This technique has also been applied with many natural-based polymers such as chitosan and gelatin to obtain biocompatible and biodegradable NPS/hydrogel composites [143]. Silver nanoparticle-based hydrogel composites have been applied in skin wound repair due to their antimicrobial properties, although their high cost has prevented their popularity as an effective hydrogel system. Other metal nanoparticles such as zinc oxide, iron oxide, nickel, and silica have been used in combination with different polymeric hydrogels to impart ferromagnetic and semi-conducting properties. Iron oxide is very commonly synthesized as a ferromagnetic material and has been used in combination with many hydrogels to form ferrogels. In one study, these ferrogels were applied as actuators to mimic human muscle movement [146]. Carbon-based nanomaterials have been extensively used in the fields of engineering and medicine, with applications including multi-walled carbon nanotubes (CNT), graphene, and GO [147]. Carbon nanotubes have seen wide application in hard tissue regeneration due to their ability to increase the overall strength of the structure as well as enhance the cell attachment process. In one study, a chitosan-based polymeric material was developed through combination with CNT, which shows higher blood cell and platelet adhesion than gelatin sponges and gauges, thus increasing the ability of the material to induce a blood clot [148]. When applied to the skin wound, the material was also found to promote healing better than the commercially available Tegaderm^{TM} film. GO is another attractive nanomaterial that has been widely used to impart multifunctionality in polymer-based composite hydrogels. Tai et al. developed a GO/PAA-based material that showed excellent swelling and electrical properties [149]. In another study, Liu et al. fabricated a GO/PAm-based material using in situ polymerization, and it showed a 4.5-fold increase in tensile strength compared to PAm [150]. Polymeric nano-fillers are a class of nanomaterials that have recently become popular among researchers in the bioengineering field and include micelles, dendrimers, polymeric nanofibers, nanocrystals, and hyper-branched polymers. In one study, Maji et al. fabricated a gelatin/chitosan-based generation 4 Polyamidoamine (PAMAM) dendrimer-based 3D construct for skin regeneration [151]. The crosslinked scaffold exhibited improved mechanical strength in terms of modulus and a tensile strength comparable to that of skin, along with increased keratinocyte migration and proliferation. Nanocellulose (CNF) is a natural-based polymeric nano-filler derived from plants and bacteria and is popularly used in the biomedical and energy fields [152,153]. CNF is compatible with a wide variety of natural and synthetic polymers. Its engraftment into polymers was done to improve their physicochemical properties, particularly to create a tough and flexible biomaterial for specific biomedical applications. There are other nano-materials, such as quantum dots [154], nanoclay [155], and calcium-phosphate nanocrystals (nano-hydroxyapatite) [156], whose incorporation into biomaterials has been found to be beneficial in terms of both physicochemical and biological attributes of the hydrogels and has been reviewed elsewhere in great detail [143,157,158].

2.4. *Animal Matrix-Derived Hydrogels*

Animal matrix-based hydrogels are usually cell-, tissue-, and organ-derived matrices and are composed of proteins (collagen, laminin, fibronectin, etc.), polysaccharides (including proteoglycans, glycoproteins, and GAGs), and other bioactive molecules (such as growth factors, cytokines, and hormones). The use of animal matrix for hydrogels for 3D cell culture has a clear advantage of better mimicking the architecture, composition, and microenvironment of native tissue. However, certain limitations such as time-consuming extraction and purification processes and batch-to-batch compositional variation have limited their popularity among researchers. In this section, we briefly discuss the application of animal matrix-based hydrogels in 3D in vitro modeling that were derived from the decellularization of tissues and organs and originated from the basement membranes of mouse sarcoma (commercially available as Matrigel).

2.4.1. Decellularized ECM (dECM)

The ECM is obtained by the process of decellularization, which removes inhabiting cells from the tissue/organ, preserving only the structural and functional macromolecules and other biologically active small molecules. The process was first reported in 1948 and was later applied in 1964 for the preparation of skin homograft [159,160]. In recent years, the steady rise of patients suffering from late and end-stage organ failure and the shortage of organ donors has increased the demand for decellularized organs, which include liver [161], heart [162], lungs [163], and kidney [164]. One of the main advantages of using a decellularized matrix in organ transplantation is the ease of reestablishing vascular networks at all hierarchical levels [165]. In recent years, as an alternate to tissue- or organ-based matrices, in vitro cell-cultured-based dECMs have attracted the interest of the scientific community as they are comparatively more homogenous and can be used as in vitro models in a more controlled way, leading to results that can easily be interpreted during analysis [166]. Various types of cells, including fibroblasts, mesenchymal stromal cells, and chondrocytes, have been used to establish matrices and been investigated for their ability in cell growth, proliferation, and guiding differentiation [167]. However, it is essential to carefully consider the choice of cells for ECM production. For instance, it was observed that fibroblast-derived matrices are unsuitable for imitating the lung microenvironment in composition and complexity [168]. Nevertheless, the idea of matrix precipitation and its solubilization into hydrogels has significantly increased the utility of ECM in both in vitro as well as in vivo applications. Despite losing its architectural and structural features, the solubilized form of ECM retains all the biochemical, growth-supportive, and cell-instructive properties of its native tissue, thereby affecting the metabolic activity and morphology of cultured cells. For example, rodent and human islet cells showed an increased level of insulin secretion when cultured long-term on bladder- and pancreas-derived ECM [169]. Another study by French et al. demonstrated the use of dECM hydrogels for enhanced cardiac regeneration [170]. The porcine ventricular-derived ECM hydrogel was used to culture rodent cardiac progenitor cells, which showed enhanced cardiac marker expression compared to when cultured on collagen hydrogel. The dECM was also used to culture human cardiac progenitor cells and extended for bioprinting applications when combined with GelMA hydrogels [171,172]. These studies prompted the researchers to develop a dECM-based medical gel, VentriGel, which was first tested on patients suffering from early and late myocardial infarction [173]. Despite all the advantages, the weaker mechanical properties of dECM have limited its ability to retain an ordered structural characteristic when used as an architectural scaffold. To overcome this limitation, hybrid composites of dECM with natural and synthetic derived polymers were fabricated. Jacqueline et al. present a highly reproducible composite material containing alginate and gelatin with dECM, which was found to be mechanically stable and bioprintable and have an elastic modulus similar to tumors induced by head and neck squamous cell carcinoma [174].

2.4.2. Matrigel

Despite the increased popularity of dECM in recent years, most of the in vitro 3D models in development and organotypic cultures currently use murine sarcoma-derived ECM hydrogel, which is commercially available under the trade name of Matrigel or EHS matrix [175]. Matrigel is an undefined mixture originating from the basement membrane of Engelbreth–Holm–Swarm mouse sarcoma and is composed of proteins, proteoglycans, and several growth factors [176]. Despite its well-known limitation, in the last decade, Matrigel has become a gold standard material for setting up 3D in vitro cultures and organoid development [177]. Embedding stem cells or primary cells in Matrigel under cell instructive conditions has created small but complex tissue architecture resembling the developmental stages of their organ of origin. Since the matrix is animal-derived, apart from compositional variation, it also contains the risk of carrying pathogens. ECM-based materials have indeed provided a very impressive and compelling means for cell growth and differentiation, which has enabled the scientific community to develop in vitro models that closely mimic the in vivo organ/tissue-like conditions. However, the compositional variation and complexity of the material has made it difficult to standardize protocols and to study the mechanistic biology of the targeted tissue. There is a need for a hydrogel-based biomaterial that has minimum variability, is non-immunogenic, and provides easy modification of cell instructive properties.

2.5. Engineered ECM (eECM) Hydrogels

In another recent approach, recombinant protein engineering was employed in the production of cell instructive hydrogels that have the advantages of both natural and synthetic polymers [178]. In this approach, the amino acid sequence of the target protein was encoded into a plasmid vector. The reengineered plasmid vector was transfected into a suitable host (e.g., *E. coli*), where the encoded genetic information was translated into the target protein. The technology allows the generation of tunable and user-defined modular protein that can integrate specific structural and functional domains of ECM. In another study, the ECM was engineered to contain an RGD motif sequence and elastin-like structural domain and used in the culture of mice intestine explant [179]. The result indicates that eECM has influence over tissue differentiation due to its adhesive and mechanical properties. The same engineered hydrogel was later used in a study to investigate the influence of the eECM's viscoelastic and degradation properties on the preservation of stemness and differentiation ability of adult murine neural stem cells [180].

2.6. Electroconductive Hydrogels

It is well established that the growth and function of many cells and tissues, including cardiac tissue, neural cells, and bone cells, can be affected by electrical stimuli [181]. However, the introduction of electrical signals into general biomaterials would be difficult, as the biomaterials are barely conductive [182]. In recent years, electrically conductive hydrogels have garnered attention due to their potential in a wide variety of biomedical applications, such as the development of biosensors [183], bio-actuators [184], and drug release devices [185]. Electroconductive hydrogels (EHs) are developed by doping conductive components into conventional hydrogels. Conductive polymers such as polypyrrole (PPy), polyaniline (PANI), and poly(3,4-ethylenedioxythiophene) (PEDOT) and conductive carbon materials such as carbon nanotubes and graphene are commonly used to synthesize EHs. There are numerous studies in which EHs are used as a cell culture matrix, which cannot only provide biochemical and structural support for cells, but can also introduce more functional monitoring and manipulation of cell activities. For example, Zhang et al. prepared an electroconductive scaffold using PPy and polycaprolactone [186]. They demonstrated that the application of a 200 uA direct current to human adipose-derived mesenchymal stem cells on this scaffold promoted the migration of these cells into the inner region of the scaffold and enhanced their osteogenic differentiation. Shin et al. used an EH as a scaffold to control the contraction function of formed cardiomyocyte tissue [184]. Cardiomyocytes were cultured on a multilayer hydrogel sheet impregnated with aligned carbon nanotube microelectrodes. Upon stimulation with electrical signals, the muscle tissue was able to mimic the contraction of the heart. The study demonstrates that multilayer hydrogels work as a 3D environment for cardiomyocyte growth and that their combination with muscle tissue remarkably emulates the bioactivity of an animal heart. By responding to various stimuli such as pH and chemical signaling, EHs are also an attractive candidate for the development of biosensors. Zhang et al. developed a 3D insulated cultured scaffold from poly(dimethylsiloxane) (PDMS) with electrochemical performance by uniformly coating it with PEDOT and platinum nanoparticles [187]. The setup demonstrated desirable biocompatibility for the 3D culture of cancer cells with an excellent catalytic ability for electrochemical sensing, allowing real-time monitoring of reactive oxygen species. This in vitro 3D platform shows promise for application in anticancer drug screening and cancer treatment. Despite several emerging studies on EHs, many technical and scientific challenges remain. Further enhancement of their biocompatibility, stability, and compatibility with microelectronics would bring EHs and EHs-based in vitro devices closer to being used in clinical therapy.

3. Preparation of Hydrogels

Hydrogels have been fabricated by structural modifications, physical interactions, and chemical crosslinking. The triggering of various environmental factors such as temperature, pH, ionic strength, and physicochemical interaction has also been used to form physical hydrogels. Additionally, reactions such as photo-polymerization, enzymatic reactions, and other chemical crosslinking methods have been used to fabricate chemical hydrogels where the main focus is to attain enhanced stability and superior mechanical strength. Polymer surface functionalization, chemical modification, and a few other promising approaches (e.g., click reactions) for the preparation of hybrid and composite hydrogels are briefly presented in this section.

3.1. Physical Interaction

The physical interactions used in the fabrication of hydrogels include hydrogen bonds, electrostatic interaction, coordination bonds, hydrophobic interactions (in many instances), as well as physicochemical interactions (stereo-complexation, supramolecular chemistry, etc.) [188]. The usual approach is to form a homogenous solution by modifying ionic strength, solvent composition, and the temperature and pH. These methods are reversible and enable the solution to change to the gelation form once it reaches its initial conditions [189]. Physical hydrogels are stimuli-responsive, but they possess poor mechanical strength and often show plastic flow [190]. Physical hydrogels are also developed by crystallization [191], protein interaction [192], and amphiphilic block and graft copolymers [193].

3.2. Crosslinking

The crosslinking of hydrogels can be approached in three ways. The first is physical crosslinking, which includes the multiplex process of coacervation (a frequent freeze-thawing cycle that leads to the formation of cryogels) and ionic interaction. The second approach is chemical crosslinking via polymerization, co-polymerization, and covalent chemical interaction using crosslinking agents such as glutaraldehyde, borate, and glyoxal. Emulsion techniques (reverse micro-emulsion, inverse mini-emulsion), radical polymerization, photolithographic crosslinking, Schiff base crosslinking (Nucleophile addition, Michaelis–Arbuzov reaction) [194] are additional techniques used to chemically crosslink hydrogels. The final approach is grafting using irradiation, which includes gamma radiation, UV radiation, and high-energy electron beam radiation, which depends on the time and intensity of the irradiation. Crosslinked hydrogels are highly stable, with a permanently fixed shape at rest, and are usually non-reversible. However, they exhibit low extendibility and low fracture toughness. In this regard, to contain the desired advantages of both approaches of crosslinking (physical and chemical), the development of double crosslinked hybrid hydrogels was proposed [195,196]. In a study, Kondo and the research group developed a homogenous network of double crosslinked gel with tetra arm star-shaped PEG and PDMS blocks that were built together by orthogonal cross-coupling [197]. Many other double-network (DN) hydrogels have been prepared using both physical and chemical crosslinking methods, but they have the major drawback of toxicity due to the crosslinking agents [198]. Further research is required in this area to create a new generation of DN gels.

3.3. Chemical Modification

Chemical modification includes the preparation of complex materials with specific functions or attributes. These can include modification to provide a variety of ligands for drug delivery in sustained or burst release and the inclusion of growth factors or other biologically active molecules for imparting additional features required for targeted therapy. Lim et al. presented a protein functionalized immobilized platform by conjugating the protein with the polymer backbone of the crosslinked hybrid network [199].

3.4. Functionalization

Functionalization is generally carried out on the surface of the polymers/hydrogels to impart certain features to the material. Hydrogels have been functionalized by conjugating certain small molecules to reduce toxicity or incorporating features such as electrical conductivity, and enhanced strength or elasticity [194]. Additionally, functionalization helps create hydrogels with various morphologies such as hollow, multilayered, and fiber microgels [200]. Furthermore, affixing certain bioactive molecules like peptides and ligands to hydrogel surfaces adds certain types of biological functionalization and regulates cell behavior for adhesion, proliferation, differentiation, and protein synthesis, in addition to promoting specific tissue regeneration.

3.5. New Approaches

Recent advances in polymer chemistry have led to the design of new materials with a wide range of applications. Among such chemical transformations, "click reactions" have proved to be a valuable tool in generating materials with tunable characteristics. Reactions such as thiol-maleimide Michael addition and thiol-norbornene click reactions have attracted the attention of scientists as they are orthogonal to many naturally occurring chemical functionalities, have fewer byproducts, and the formation of intermediate thioether succinimide linkage can be modulated to give them dynamic properties [201,202]. The chemical transformation has particular utility in fabricating biocompatible hydrogels with tunable viscoelastic properties, as the developed biomaterial can carry and transfer the encapsulated cargo in an instructive way to the surrounding tissues (triggered by a change in pH or glutathione in glutathione-mediated cleavage) [203]. For example, biocompatible HA hydrogels based on copper (I)-catalyzed azide-alkene cycloaddition (CuAAC) were utilized as a drug repository tissue construct [204]. Furthermore, with the assistance of a copper-free click chemistry reaction, it is reasonably easy to reduce the toxic catalysts used in copper-catalyzed reactions. Such facilitation have been observed in radical-mediated thiol-ene/yne chemistry [205], tetrazole alkene photo-click chemistry [206], and the oxime reaction [207].

4. Factors Considered in the Design of a Hydrogel-Based Substrate

Cells actively interact and sense the physical and biochemical signals from their surrounding matrix and accordingly change their properties and functions, including migration, growth, ECM production, and differentiation [208]. In vitro, the cell's functions and behavior are influenced by various substrate properties such as soluble molecules, mechanics, topography, stiffness, and degradation rate [209,210]. It should be noted that each tissue type and associated diseases differs in the types of cells present and their surrounding matrix. Therefore, the choice of biomaterial and the design of the model system are dependent on the tissue and disease type. Engineering techniques can be used to precisely control the substrate geometry, flow conditions, nutrient and oxygen supply, matrix stiffness, topography, and other local biochemical features.

4.1. Hydrogel Microenvironment

The in vitro development of specific 3D tissue constructs is majorly dependent on the biochemical properties of the surrounding material. Although the animal ECM-derived matrices such as dECM and Matrigel can guide cell proliferation and differentiation, their variability and poor characterization make it difficult to clearly define the cell–matrix interaction and the underlying mechanism. In this regard, chemically defined natural hydrogels are better suited for the generation of in vitro 3D tissue. For example, Broguiere et al. fabricated a fibrin-based hydrogel supplemented with laminin-111. The hydrogel was found to be conducive to the formation of both murine and human intestine epithelial 3D tissue and their long-term expansion [211]. It was thus reasoned that naturally occurring RGD motifs are crucial for maintaining stemness in intestinal cells. Additionally, defined synthetic hydrogels are used in engineering in vitro tissue models via precise control of their biochemical properties. For instance, PEG hydrogels are widely used in the formation of tissue engineering products due to their non-toxicity and ability to maintain hydration [212]. In one study, it was observed that compared to Matrigel, defined PEG-based hydrogels are better suited for inducing homogenous and polarized neuroepithelial colonies, which facilitate the formation of a dorsal–ventral patterned neural tubule-like structure [213]. The author also showed that when tuned to precise biochemical features, synthetic hydrogels can provide early neural morphogenesis.

4.2. Hydrogel Microarchitecture

With the help of the latest fabrication technologies such as wet-spinning, droplet microfluidics, and bioprinting, it is now possible to create a distinct internal structure of a hydrogel-based biomaterial that can provide a specific architectural environment resembling the targeted tissue. The specific topology and internal structure of hydrogels enable passive mechanical stimulation from the boundary surface of the biomaterial and allow regulation of tissue morphology, promote aggregation of cells, and promote multicellular interaction, which triggers the formation of 3D microtissues. Particular discrete structures such as microfibers and microgels fabricated by micro-technologies have been able to provide geometric confinements in the development of 3D tissues. For example, PEGDA-based microgels were formed to culture liver organoids with long-term culture and functional maturity when integrated with a perfusable chip [214]. Surface topographies have also been utilized to direct cell migration and alignment in 3D in vitro cell culture. For instance, PEG-based substrates that were formed into a specific pattern with plasma exposure and coated with Matrigel were used to generate human cardiac micro-chambers [215]. In a similar study, photopatterning was used in the development of PEG-based micro-wells, which were found to support the organization of stem cells derived from salivary glands. These micro-wells were coated with electrospun PCL nanofibers, which in turn were used in the generation of organoids with larger diameters [216]. In conclusion, hydrogel-based biomaterials can develop an in vitro cellular structure with distinct geometries and architecture by managing its topological structure, thereby guiding and improving the reproducibility of 3D tissues.

4.3. Hydrogel Mechanics

The mechanical properties of a biomaterial have a significant influence on the organization and development of 3D tissue. Matrices with specific mechanical properties are appealing biomaterial candidates for engineering 3D tissue formation and understanding its regulatory mechanism. Due to properties such as stiffness, elasticity, and durability, synthetic hydrogels are considered suitable analogs in the development of 3D culture. Studies have indicated that matrix stiffness is essential for regulating the integrin downstream pathway and other transcriptional activators in stem cells, which steer their organization and 3D tissue formation tendency [217,218]. In a study by Wu et al., the compressive modulus of hydrogels were prepared in lower (0.51 KPa) and upper (1.41 KPa) stiffness ranges of native brain tissue to encapsulate neural progenitor cells [219]. The result indicates increased neurite growth, proliferation, differentiation, and maturation of neural spheroids when cultured on the soft hydrogel as compared to a stiff one. In another study, PAm-based hydrogels of varying stiffness were prepared by controlling the ratio of acrylamide and cross-linkers (bis-acrylamide) to create cardiovascular organoid substrates with different mechanical conditioning [220]. The maximum differentiation efficiency and contractility percentage of cardiac organoid was obtained when the substrate had similar viscoelasticity to the native cardiac muscle (elastic modulus: 6 KPa). The phenomenon observed might be because the signal transmitted from the matrix rigidity leads to the further downstream signaling pathway in the cardiomyocytes' proliferation and differentiation. The mechanical properties of hydrogels can be further modulated using anisotropic modification and gelation, resulting in the generation of polarized tissues.

5. Types of Hydrogel Units, Platforms, and Fabrication Technologies

The diverse range of functionalities exhibited by in vivo native tissues and organs are characteristics of their numerous morphological features and topographical variations. Owing to their high degree of plasticity, hydrogels can imitate the geometrical and architectural features of targeted tissues, ranging from nanometers to micrometers. Such tunable attributes can be exploited in the reconstruction and reformation of hierarchical morphologies, including the internal structure and surface topologies of tissues and organs in in vitro 3D models. Figure 3 displays various hydrogel architectural units, including micro and nanogels, microfibers, patterned membranes, and other modular frameworks that have been developed with the help of various platforms and recent fabrication technologies, which provide precise control over the distributions of cells and molecules.

Figure 3. Schematic figure of existing fabrication technologies for designing various hydrogel units/platforms.

5.1. Cell-Laden Constructs

Encapsulations of cells in 3D bulk hydrogels are one of the most simplified, approachable, and by far the most successful methods in the development of in vitro tissue constructs and clinical transplantation [221]. A broad range of polymeric materials (both natural and synthetic) such as gelatin, PEG, and alginate are available for cellular encapsulation. It is expected that the materials be biocompatible; be easily tunable to incorporate molecules that can guide cell growth, proliferation, and differentiation; and create morphological and topographical features similar to those of the real tissue. In this approach, isolated cells are typically suspended with the bulk hydrogels and crosslinked under cell-favorable conditions to mimic the in vivo conditions of the tissue. Depending on the cross-linker and crosslinking mechanism, the hydrogel could be a long-term stable construct or rapidly degradable through hydrolysis or proteolysis [222–224]. While stable hydrogels often restrict cell movement, and tissue deposition, degradable hydrogels allow cell migration and growth, thus promoting tissue formation to replace the hydrogel [225]. However, native tissues in in vivo conditions are far more compositionally and architecturally complex than the in vitro cell-laden hydrogel constructs. Current research trends have been focused on microtechnologies, which enable the fabrication of complex architecture through the strategic placement of cells and tissue modules.

5.2. Microgels

Microgels are micrometer-sized hydrogel blocs that can be used to provide geometrical and spatial confinements during the development of 3D multicellular clusters. Although microgels are generally formed in spherical geometries, they can be developed in different sizes and morphologies, including hemispheres, capsules, disks, and spindles [226,227], which can affect cell growth, proliferation, and other metabolic functions. Microgels are developed using various precursors of hydrogels, including alginate, agarose, collagen, and PEGDA [228]. A droplet-based microfluidic platform is one of the common techniques used in the development of microgels [229]. The technique has the advantage of generating and engineering the formation of individual spheroids in the form of droplets in a highly controllable manner and can be used to encapsulate single cells and multiple-cell aggregates [230]. Most droplet-based microgel formation has been developed by encapsulating cell suspension within non-adherent hydrogels such as alginate, agarose, PEG, and their derivatives, which leads to the formation of multicellular spheroids [231]. Sabhachandani et al. developed cell-laden alginate microgels as 3D tumor spheroids and studied them as a more effective preclinical drug-resistant screening model [232]. Another technique is using multiple micrometer size compartments (microwells) to trap gel-laden cells to form microgels. Such wells are usually fabricated via soft lithography technique using a PDMS stamp or mold [233]. Microwell-based spheroid formation has also been applied to investigate the paracrine effect of hepatic stellate cells [234]. Electrostatic extrusion is yet another technique for developing microgels. At the extrusion point, the jet is subjected to an electric field that generates polymer-based microparticles. Jianjun et al. used electrostatic interaction to form a porous polyelectrolyte complex (PEC) from poly(L-glutamic acid) and chitosan. Upon culture with chondrocytes, the PEC microsphere produced a significant amount of cartilaginous matrix [235]. In another study, this technique was used with decellularized adipose tissue to fabricate microgels for dynamic culture and the expansion of human adipose-derived stromal/stem cells [236]. Presently, various other techniques are used in the development of microparticles/microgels, such as the coacervation method, vibrational jet method, air suspension method, and supercritical fluid precipitation method. The details of these techniques, the microgel process formation, and their 3D cell culture applications are not covered in this article, as they have been well described elsewhere [237,238].

5.3. Microfibers

Under physiological conditions, several tissues exhibit tubular configuration and a fibrous structure with tissue organization-specific functions. In the past few decades, methods such as wet spinning and microfluidic spinning have excelled at controlling the compositional and geometrical features when fabricating microfibers (solid, hollow, and core-shell fibers) [239,240]. In these techniques, rapidly cross-linked hydrogels (GelMA, alginate, collagen, etc.) are preferred, as the gelation of microfibers are obtained in an aqueous solution [241,242]. Microfluidic spinning employs a specially designed microchannel that is involved in the generation of 3D coaxial flow, which consists of sample and sheath flows. Through solidification of the coaxially flowing liquid using UV light, ionic and chemically crosslinking solidified fibers can be produced [243,244]. The potential of cell encapsulation using microfluidic spinning technology has enabled its applicability in the reconstruction of in vitro complex 3D tissues mimicking organs such as the liver and pancreas [245–247]. For example, Kang et al. cultured L929 fibroblasts and primary hepatocytes in the center and outer layers of coaxial alginate microfibers, respectively, to prepare a liver-mimicking tissue structure [244]. Electrospinning is another popular technique used to fabricate not only micro-sized but also nano-scale fibers that closely resemble the fibrous matrix of native ECM [248,249]. The method uses an electric force to draw a charged thread of polymeric solution up to a fiber diameter of some hundreds of nanometers. The basic materials are polymers or solvent, and most natural and synthetic polymers can be electrospun into fibers. The technique uses either a single type or multiple types of hydrogels to generate a randomly packed fibrous construct (either membranous or a 3D scaffold). In one study, Honkamaki et al. developed a layered 3D construct by embedding electrospun poly(L,D-lactide) fibers in collagen hydrogel. The layer-by-layer scaffold shows better resemblance to the native microenvironment, thereby guiding the growth and neurite development of iPSC-derived neuronal cells [250]. In another recent study, the authors used a 3D hydrogel structure as a base collector instead of a common metal collector to develop a unique hydrogel-assisted electrospinning process. With various nanofiber microstructures, the process permits the forging of various types of 3D structures. This new hydrogel-assisted electrospinning method is expected to be applied to drug delivery, tissue engineering and development, and the building of anisotropic in vitro 3D tissue models [251].

5.4. Transwell Platforms

Transwell devices consist of permeable membranes that form a physical barrier between two different cell types, only allowing the soluble molecules to pass through. They are considered important tools in developing a 3D in vitro model that can investigate the growth and interaction of two different cell types in two separate microenvironments. Transwell platforms have been assembled with other platforms such as microchannel or microfluidic devices [252]. Commercially available transwell culture plates consist of a porous membrane suspended between two chambers. This platform has been extensively used to study drug efficacy, cell migration, and cell invasion. The majority of in vitro brain–blood barrier models have used a transwell platform in which the upper chamber contains endothelial cells and the lower chamber contains astrocytes, which represent the neural tissue [253–255]. Kwak et al. used a transwell platform to model 3D skin tissue. In the setup, the authors used a GelMA hydrogel to encapsulate fibroblasts and added a transwell insert to mimic the dermal tissue. HaCaT cells were cultured on top of the GelMA hydrogel to simulate the keratinocytes layer over the dermal tissue [256]. Another good model for a transwell platform is sandwich cell configurations. The sandwich culture is considered a gold standard for developing liver models, as it allows long-term maintenance of differentiated hepatocytes. In hydrogel sandwich culture, cells are first seeded onto a hydrogel substrate. The cultured cells are given sufficient time to ensure full adherence to the substrate. Thereafter, another layer of hydrogel is placed on top of the adhered cells [257]. This system of sandwiching cells between the two layers of the hydrogel can be integrated with a transwell platform and thus can regulate different cell types spatiotemporally.

5.5. Microfluidic-Based Platforms/Organ-on-a-Chip (OOC)

In recent years, microfluidics have become an immensely popular and sought-after platform to develop 3D in vitro models. As the name suggests, this miniaturized platform is utilized with different types of cells that can be maintained in different micro-chambers and connected by arrays of various microchannels [258]. The major benefit of this platform is the continuous perfusive flow of media, which assists in the movement of nutrients, gases, and wastes. Several studies have proved the superiority of microfluidic systems over static culture in terms of cellular metabolism and functions [259].

In the past decade, microfluidic technology-based OOC have emerged as a frontrunner in the development of in vitro 3D model systems. The OOC systems comprise various living human primary cells lined with micrometer-sized compartments and channels that can imitate the structural, mechanical, biochemical, and functional features of in vivo tissue and its microenvironment [260]. Miniaturized versions of various tissues including liver [261,262], kidney [263], heart [107], lungs [264], and intestine [265] have been developed on this platform. The simplest OOC system consists of one type of cell with a single perfused channel that can perform the function of a single tissue. For example, demonstrations of functional activity by perfusion culture of 3D hepatocyte aggregates and their responses to shear stress through the exposure of vascular endothelial cells to medium flow in a microchannel encompass some of the initial-stage research on OOC systems [266,267]. Second, this simple model has designs with two or more compartments that are connected by microchannels or porous membranes and consist of two or more different types of cells. The challenge for OOC is to create a multi-OOC design to replace the animal model experiments in preclinical studies. In 2010, Van et al. for the first time combined liver and intestines models into a single microfluidic device. Through this model, they were able to demonstrate the applicability of organ–organ interaction study on a chip, including the regulation of bile acid synthesis. Since then, many organs have been concentrated onto individual chips [268]. A study by Maschmeyer et al. described the fabrication of a microphysiological system that maintains the functionality of four organs (liver, intestine, skin, and kidney) for over 28 days on a microsystem scale [269]. Benjamin et al. fabricated a chip formed with collagen I for the study of angiogenesis and thrombosis. In this microfluidic setup, they assembled a metabolically active liver, a free contracting cardiac tissue, and a metastatic solid tumor in a biodegradable scaffold. The system successfully demonstrated a complete cancer metastasis cascade across multiple organs [270]. Studies have shown that the OOC model can be utilized in the prediction of drug efficacy and toxicity, which can provide precise estimation compared to the animal models and therefore can bridge the gap between in vitro tests and clinical trials [271]. At present, polymeric materials such as PDMS and glass are used in the preparation of OOC devices. PDMS has the benefits of optical transparency and gas permeability, which enable the OOC platform to carry out real-time and high-resolution imaging of the in vitro setup. However, it was recently observed that PDMS-based material has a major drawback in that it absorbs hydrophobic molecules from the solution and can therefore provide a misleading outcome in cell-based experiments [272]. Many hydrophilic biomaterials and hydrogels with reduced ability to interact with the molecules of the solution have been used to provide a non-absorbent PDMS surface. For example, Kuddnnaya et al. addressed this issue by employing the (3-aminopropyl)triethoxysilane (APTES)-based crosslinking strategy to stabilize the ECM protein immobilization on PDMS. The surface modification supports long term viability and adhesion of neuronal and glial cells [273]. In comparison to conventional substrate materials, natural and synthetic polymers such as GelMA, alginate, and PEG are also considered a promising substitute in the fabrication of OOC systems, as they not only have the advantage of high permeability and tunable physicochemical properties, but are also highly biocompatible and have cell adhesive properties [274,275]. Aleman et al. fabricated a GelMA-based microfluidic device with an integrated endothelial microfluidic network [274]. In another study, Zhao et al. described the fabrication of a biofunctionalized microfluidic device based on a silk protein hydrogel elastomeric material. The device exhibited well-regulated mechanical properties, long term stability in various environmental conditions, and optical transparency [275]. For the achievement of a balanced and stable structure with intricate micro-sized channels in the development of hydrogel-based OOC, different microfabrication technologies have been applied, including micro-molding [276,277], photo-patterning [278,279], and bioprinting [272,280–282]. With the recent progress in composite hydrogels and micro-technologies and the use of an effective combination of them, the fabrication of functional miniaturized organs to be used as disease models or in drug screening applications can soon be realized.

5.6. 3D Bioprinting

Conceiving the complex hierarchical 3D structure demands a highly flexible method. In recent years, 3D printing technology has gained immense popularity among researchers, in fabricating structures with tunable size, geometry, and architecture with high spatial resolution. 3D bioprinters are capable of printing multiple materials simultaneously with enhanced control over the spatial placement of various material/hydrogel-laden cells in the same construct [283]. Based on the working method, 3D bioprinting can be classified into three types: the droplet method, the micro-extrusion based method, and the laser-assisted printing method [284,285]. The advantages and drawbacks of these methods are presented in Table 3. In bioprinting, the polymers (natural or synthetic) that encapsulate cells and are used to create 3D structures are called bio-inks [286]. The physicochemical properties of the polymeric hydrogels, such as rheological properties and crosslinking mechanism, define their suitability as a bio-ink [287]. The crosslinking of bio-inks can be triggered before (pre-crosslinking), after (post-crosslinking), or during the period of extrusion (in situ crosslinking) [288]. Among all the crosslinking methods used during bioprinting, photo polymerization (light-based radical polymerization) has become the most desired method due to its flexibility in stabilizing the shape and structure of the bioprinted construct [81]. Hydrogels such as gelatin, alginate, HA, and PEG-based polymers are presently used to develop bio-inks to help create tissue-like structures [289]. Some promising studies on self- and co-assembly of materials to form composite hydrogels as novel bio-inks are also emerging [116]. It is important to note that in cell-based bio-ink printing (cells with carrier hydrogels), apart from determining printability and structural fidelity, the priority is to ensure long-term cell viability in the 3D printed construct. Numerous factors can result in low viability in the bioprinted construct, such as increased shear force during extrusion of bio-ink or harmful crosslinking methods. In one study, Lee et al. presented a versatile and novel one-step fabrication method to develop an in vitro OOC platform using 3D bioprinting. The developed platform showed spatial heterogeneity, which was used to evaluate liver tissue functionality and drug testing efficacy [290]. Other 3D printing methods such as stereolithography (SLA), digital light processing (DLP), direct ink writing (DW), laser-induced forward transfer (LIFT), and multijet modeling are suitable to print 3D hydrogel-based constructs that can subsequently be populated with cells [291]. Although the progress in bioprinting technology has been well-paced, printing tissues and organs for clinical transplantation is still far from becoming a reality. However, the method has provided a platform for creating a more sophisticated in vitro 3D tissue model that will take us one step closer to the development of real tissue and organs in a dish.

Table 3. Types of bioprinting methods and their features.

Bioprinting Method	Working Method	Advantage	Limitation
Micro-extrusion method	Most common method. Physical force is used to distribute biomaterial and cells at a specific location through a nozzle.	Can print heterogeneous and complex structures	Low resolution printing
Droplet-based method	A controlled volume of cell suspension hydrogel is printed at the desired location. The print volume can be controlled via a magnetic field, an electric heating nozzle, and piezoelectric or acoustic actuators.	Much more accurate resolution than micro-extrusion printing	Difficult to print large-scale biological structures
Laser-assisted printing	Biological structures are printed by laser-guided front transfer. The solidification method uses a laser-induced photo-polymerization using UV, infrared, or visible light.	Prints at the highest resolution owing to the laser interference	Low cellular viability

5.7. Organoid Systems

Organoids are 3D multicellular micro-physiological systems that are formed by proliferation, differentiation, and self-organization of primary cells or stem cells that are placed close to each other [292]. The complex and organized structure that is formed recapitulates some of the structural and functional features of the real organs [293]. Initially, this strategy was used to form cancer cell spheroids to study tumorigenesis and cell metastasis and to scrutinize cancer drugs [294]. At present, this method is frequently used for developing 3D tissue or organ reconstruction. For example, intestinal organoids have been developed from the biopsies of intestinal tissues containing intestine stem cells [295]. In this method, after selection of a cell source, a homogenous medium (the following methods are more or less similar) is used for the culture of the cells. The cells need to be cultured in free form (either suspended or embedded in a 3D-conditioned surroundings), as it enables them to expand and remodel on their own [296]. Such free growth of cells can be achieved by culturing cells in a low-attachment substrate or in a 3D microenvironment by encapsulating them in a naturally derived hydrogel that can provide the necessary instructive signals (e.g., dECM or animal-based matrix) [297,298]. The cells under suspension form clusters, proliferating and differentiating to develop into organoids (Figure 4). In the matrix-supported method, ECM components such as fibronectin and laminin provide integrin receptors to cells that maintain cell integrity and functions during the formation of organoids. Matrigel has been used widely in the development of various organoids such as cerebral organoids [299], lung bud [300], liver bud [301], gastric organoids [302], kidney organoids [303], alveolar organoids [304], intestinal organoids [305], and pancreas organoids [306]. However, due to the uncontrollable microenvironment of Matrigel, in vitro spontaneous organoid morphogenesis is not easily controlled [307]. In this regard, engineering biomaterials to obtain precise control over nutrient supply and input and output flow conditions and tunable mechanical stimulation could provide the matrix support for the growth and development of organoids [92,308,309]. Defined natural hydrogels can act as a substitution of native ECM to promote specific organoid formation. For example, Broguiere et al. developed fibrin-based hydrogels supplemented with laminin-111, which was shown to support the epithelial organoid formation and expansion [211]. Moreover, the importance of laminin was well investigated as a major biological signaling factor in the ECM during organoid formation. In addition to natural hydrogels, well-defined synthetic hydrogels have been explored in the development of organoid culture due to their tunable biochemical properties. In a recent report, Lutof et al. developed a PEG-based synthetic hydrogel platform to assist the development of embryonic stem cell-derived 3D neural tube organoids that recapitulate the key features of neural morphogenesis [213]. In another study, a laminin I-functionalized PEG hydrogel system was shown to promote the formation and expansion of organoids from pancreatic progenitor cells [310]. To date, several organoid formations have been explored, including skin [311], pancreas [312], lungs [313], liver [314], kidney [315,316], and brain [299,317]. Even though these organoids lack complete features of real organs or tissue, their cellular assembly and heterogeneity have made them a suitable platform for screening drugs and developing disease models as an alternative to the present 2D cell-based assay and animal models [318].

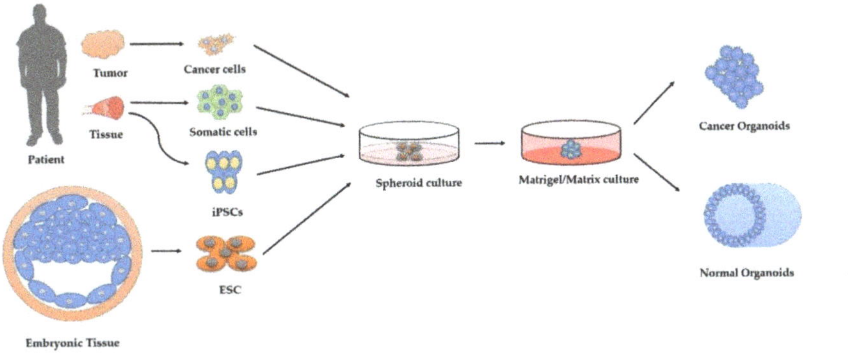

Figure 4. The development of organoids from primary cells, stem cells, and cancer cells. Embryonic stem cells (ESC) from embryonic tissue and iPSCs derived from adult primary tissue first experience direct differentiation into a floating spheroid culture. The culture is subsequently transferred into a matrix-based hydrogel with a specific medium to initiate the formation of organoids. Somatic stem cells can be obtained from patient tissue, which can be further cultured in a 3D medium to develop it into organoids. Cancer cells from tumor tissue can be developed into tumoroids in a specific 3D culture.

6. Engineered 3D In Vitro Models

In this section, we focus our discussion on various studies on the development of 3D models of specific tissues/organs in the past decade that have used hydrogels as biomaterials. A selected survey of recent in vitro models of various tissue types is provided in Table 4.

6.1. Skin

The skin is the largest organ of the human body, covering an area of ~2 m^2, and is the body's primary barrier against most environmental agents, including pathological microbes and chemicals. Culturing of human 3D skin ex vivo biopsies began early, in the middle of the twentieth century [319]. However, 2D culture later prevailed as standard lab practice, as it was much simpler and more reproducible than organ/ex vivo culture. Recent development in cell culture techniques and 3D fabrication technologies has seen an increased momentum toward research and development of in vitro 3D skin models, as indicated in numerous studies [320]. Skin equivalents containing the epidermis and dermis layers developed from keratinocytes and fibroblasts were the first type of 3D in vitro skin model [321,322]. Among several, collagen-blended hydrogels have emerged as the most worked on substrate when producing skin equivalents [323]. In one study, a collagen-based hydrogel containing dermal fibroblasts was deposited followed by another layer of keratinocytes and melanocytes to mimic skin tissue [324]. Stroebel et al. exhibited the development of scaffold-free spherical skin microtissues containing different layers of keratinocytes and a dermal fibroblast core [325]. Scaffold-free models are considered more advantageous due to their simplicity, low cost, easy reproducibility, and suitability for high-throughput biochemical analysis. However, they lack proper skin architecture and do not accurately recapitulate the liquid–air interface of physiological skin. In this regard, composite hydrogels are gaining popularity in skin tissue engineering, as the skin matrix is composed of several ECM components, such as collagen, fibronectin, elastin, vitronectin, and GAGs. In a recent report, a hydrogel produced from blended collagen and silk was presented as an ideal dermal material [326]. The silk helps stabilize the structure, whereas the combination of collagen presents the cell-binding domain to the construct. This blend proved more resistant to temporal degradation than one made from only collagen. In another study, using a mixture of gelatin, alginate, and fibrin, Pourchet and colleagues were able to bioprint a full-thickness skin model with a stratified epidermis [327]. Although extensive work has been done in 3D skin models, the key limiting factors in such skin equivalent models are vasculogenesis and functionally crucial appendages such as sebaceous glands [323,328]. Skin organoid models developed from iPSCs have been shown to develop sebaceous glands along with hair follicles in mice over 30 days [311]. Finally, issues such as the incorporation of immune systems and vasculature can be addressed with the current development and convergence of different fabrication techniques.

6.2. Bone

The musculoskeletal system determines the body's shape and is important for locomotion in vertebrates. Bone is one of the key components of the musculoskeletal system and plays a vital role in providing support, protecting organs, distributing force, and producing blood [329]. Bone is a natural composite and presents a unique hierarchical structural organization at multiple scales that grants it the required toughness [330]. To mimic the in vivo conditions of bone, researchers focused on the 3D spheroid construct. Gurumurthy et al. were able to show enhanced osteogenic functionality upon differentiation of MSC spheroids when compared to the conventional 2D culture [331]. To imitate the in vivo conditions more adequately, methodologies such as culturing of osteoblasts and endothelial cell types together to develop vascularized bone tissue were utilized. When HUVECs and osteoblast cells were co-cultured in a collagen matrix to form a spheroidal model, both osteogenic differentiation and vasculogenic tube-like structures were found [332]. It is well known that bone is a highly vascularized organ and engineering a vascularized and mineralized matrix requires synergistic interaction between osteogenic and endothelial precursors [333]. At present, vascularization strategies include using (i) angiogenic factors in combination with 3D scaffolds, (ii) pre-vascularization methods, and (iii) co-culture systems to engineer vascularized tissue constructs (Figure 5) [334,335]. For example, the Braghirolli group showed that the PCL scaffold constructed by electrospinning and loaded with VEGF molecules encouraged the penetration and growth of endothelial cells within the 3D matrix [336]. The co-culture method is comparatively more complex, as several parameters are considered for a successful vascularization outcome, including cell types, cell seeding methodology, 3D construct, media, and microenvironment. In recent work, Tsigkou and colleagues presented simultaneous osteogenic differentiation and vasculature development by combining bone marrow-derived MSCs and HUVECs in a 3D scaffold and hydrogel, respectively [337]. The result demonstrated capillary structure formation with three to seven days of culture when implanted in a mouse model. The hybrid scaffold has shown potential in the field of bone tissue engineering. Dhivya et al. constructed a hydrogel containing zinc-doped chitosan-HA-β-glycerophosphate and demonstrated the material bone formation ability both in vitro and in vivo without any toxic effect on cells [338]. In another instance, Zhai et al. constructed a bone cell-laden hydrogel consisting of PEGDA, HA, and nanoclay [339]. The nanocomposite showed enhanced osteoconductive properties in long-term culture, rationalized to be due to the presence of the bioactive ions of nanoclay. The use of 3D bioprinting using a composite hydrogel of alginate, gelatin, and hydroxyapatite as bio-ink has also been demonstrated to support mesenchymal stem cells and differentiation toward osteogenic lineage [340]. Among the various bioprinting techniques, extrusion-based bone printing is the most common, as it uses the hydrogel as bio-ink with varying viscosities and high cell densities. Levato and colleagues constructed a composite biomaterial consisting of PLA microcarriers with GelMA/gellan gum [341]. They used bio-ink combined with a high concentration of MSC microtissues to fabricate a bone construct via bioprinting. The in vitro studies demonstrated that the strategy allowed higher cell adhesion, proliferation, and differentiation with enhanced bone matrix deposition within the construct. However, due to the high mechanical properties of bone, hydrogels are not suitable for fabricating larger voids or hollow spaces within the construct, as they will collapse the structural features. In this regard, a sacrificial material, such as poloxamer F-127, can be introduced for printing constructs with voids for enhanced perfusion and subsequent vascularization [342,343].

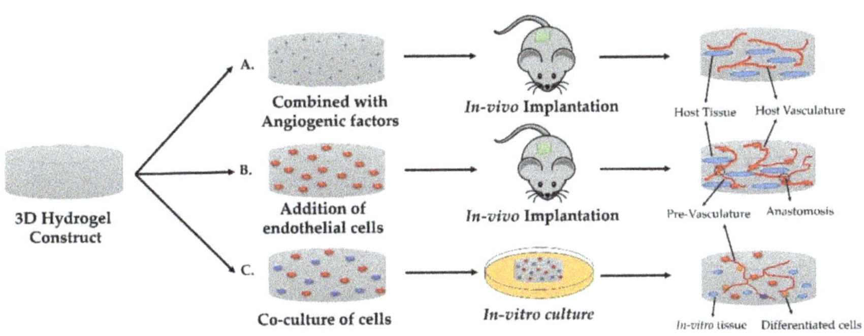

Figure 5. Vascularization strategies in 3D tissue construct. A. Hydrogel construct incorporated with angiogenic factors. Host vasculature developed in 3D construct upon in vivo implantation. B. Formation of pre-vascularization and anastomosis upon in vivo implantation of endothelial cell culture hydrogels. C. Co-culture study is more suitable in in vitro culture system.

6.3. Cartilage

Cartilage is smooth elastic tissue providing padding and protecting the end of the bone [344]. It consists of sparsely spread specialized cells called chondrocytes embedded in a large amount of ECM. Due to the avascular nature of cartilage, its repair and regeneration are slow and difficult [345]. With the increasing cases of arthritis and clinical demands, cartilage regeneration has gained much attention in the tissue engineering field. Several hydrogel-based methods have been used to develop chondrocyte niches and realistic microenvironments [342,346]. Within the last decade, lithography and 3D bioprinting have become a popular approach for this purpose. In a stereolithographic bioprinting approach, Lam et al. used methacrylated HA and methacrylated gelatin as bio-ink to create a cartilage model with varying chondrocyte densities. The model maintained its shape, cell distribution, and viability for over 14 days and demonstrated cartilage proteoglycans and type II collagen deposition. By increasing the densities of cells, the model showed a higher differentiation pattern that led to enhanced cartilage-typical zonal segmentation [347]. In another instance, HA/polyurethane was used as hydrogel material to print a 3D MSC-laden construct for cartilage formation [348]. The results showed that MSCs differentiate successfully into chondrocytes, whereas the matrix demonstrates mechanical properties similar to those of cartilage. Moreover, as cartilage is avascular and aneural, most of the stimuli received by chondrocytes are mechanical. In this regard, Paggi et al. developed a versatile microfluidic platform that supports evaluating the impact of various mechanical stimuli on chondrocytes up to single-cell resolution. This platform will be instrumental in studying the progression of various diseases and answering biological questions [349].

6.4. Liver

The liver is considered an important organ in detoxifying chemicals and metabolizing drugs in the body. The development of a 3D in vitro model of the liver can not only help to unravel the physiological phenomena in the liver, but also provide a platform to accurately predict drug effects and toxic responses [269]. Despite being capable of rapid regeneration, the liver remains in high demand for organ transplantation due to liver tissue damage from multiple factors [312]. From a material perspective, due to the low mechanical strength of the liver, hydrogel-based composites are an ideal candidate in engineering in vitro models. In one study, Ma et al. produced mechanically flexible composite hydrogels from GelMA to support iPSC-derived hepatic cells and glacial-methacrylate-HA to enhance endothelial cell growth and proliferation [350]. The research group used the above composite material as bio-ink to successfully bioprint a patient-specific hepatic model that closely mimics the architecture and cell functionality of native tissue. Mazzocchi et al. used a composite hydrogel of HA and collagen to print hepatic tissue that was shown to preserve the native tissue microenvironment [351]. The tissue construct used hepatocytes and stellate cells as cell sources and was found to be responsive against the effects of common liver toxicants. Although the 3D liver models can sustain in vivo-like conditions for several days or weeks, the static culture situations do not allow the movement of accumulated medium, which can have a toxic or self-inhibition effect on cell viability and functionality (urea or ammonia accumulation). In this regard, a microfluidic platform or liver-on-a-chip could be utilized to recapitulate the in vivo flow rate in removing the metabolites and other functional products. In a study by Rennert et al., a two-channel microfluidic platform was created to develop an in vitro 3D liver model [352]. In one of the channels, the vascular layer was developed composed of endothelial cells and macrophages, whereas the hepatic model was composed of HepaRG cells co-cultured with stellate cells. Both models were separated by a membrane mimicking the space of Disse. The results showed hepatocyte polarity and the allowance of hepatobiliary function. In a later follow-up study, an oxygen measurement system was incorporated for toxicological screening [353]. In another study conducted by Lee et al., porcine liver dECM was used as bio-ink to print a 3D liver-on-a-chip platform. Apart from hepatocytes, the study also included endothelial cells and cholangiocytes to mimic the vascular and biliary systems in the platform. The creation of a biliary system enhanced the liver-specific function and increased the drug sensitivity response many fold when compared to a chip without a biliary system [354]. In a later study, the same group developed a liver fibrosis-on-a-chip platform using dECM and gelatin bio-ink and activated stellate cells. The platform exhibited increased collagen accumulation, cell apoptosis, and reduced liver-specific functions, all of which are characteristic features of liver fibrosis [355]. In a multi-organ-level study, an OOC was developed to recreate the metabolic dynamics connecting gut epithelial cells to liver cells [356]. In such a multi-organ-level system, it is possible to study the gut–liver axis, which could also include immune cells, to study inflammatory responses, investigate diabetes, and develop fatty liver disease models [357].

6.5. Gastrointestinal Tract

In recent years, there has been a significantly growing interest in 3D in vitro model development of intestine tissue due to the increasing demand for food science and toxicology analysis and disease understanding. The intestinal epithelium is a multitasking tissue containing different cell types specialized in different functions: enteroendocrine cells, goblet cells, Paneth cells, and microfold cells [358]. Roughly 90% of the food digestion and absorption in the digestive tract happens in the small intestine [359]. The intestinal epithelium consists of highly polarized tissue with a defined 3D tissue architecture (Crypt-villus organization). Several materials and microfabrication techniques have been used in the generation of a 3D model for mimicking the topography of the gastrointestinal tract and to study its impact on cellular behavior. One of the first studies in this regard used a combination of molding techniques to generate a scaffold that represents intestinal villi in a collagen hydrogel [360]. Morphological similarities with human villi were found when Caco-2 cells were cultured on the construct for three weeks. The relevance of this system was further assessed by studying drug permeability and the role of mucin (MUC17) in antibacterial response [361,362]. Gjorevski et al. designed a composite hydrogel composed of PEG/laminin with RGD motifs [92]. The designed materials were initially optimized for stem cell proliferation, but were later observed to allow cell differentiation and organoid formation, presenting an alternative to Matrigel in intestinal organoid development. Kim et al. employed 3D bioprinting using collagen-based bio-ink to form a mesh structure with a crypt compartment and vertical protrusion to mimic the villi architecture [363]. The model was developed with external Caco-2 cells forming the epithelium and internal HUVECs to reproduce the capillary structure, and the results exhibited an increased proliferation rate and expression of differentiation markers. In another study, high-resolution stereolithography 3D printing was used in the formation of a PEG-DA-based hydrogel to form a 3D structure containing both crypts and villi [364]. The cultured Caco-2 cells exhibited increased polarization and expression of enterocyte differentiation markers, which also demonstrated a strong influence on the cell behavior of the hydrogel material and topology of the construct.

6.6. Cancer Model

Despite significant accomplishments in the field of biomedical treatments and therapy, cancer is still the leading cause of death worldwide. For the past four decades, the conventional 2D model was used to study tumor progression. However, it does not properly mimic the heterogeneous architecture of the native tumor microenvironment [365]. Therefore, there is a need to develop a 3D tumor model that can present all the necessary physiological characteristics [366]. The first 3D in vitro culture to be efficiently demonstrated was a cancer spheroid culture [367,368]. In recent years, with the help of 3D bioprinting techniques, in vitro cancer models have been developed that can mimic the 3D complexity of native tissue. Such models are used in the study of cancer pathophysiology as well as for the screening of anticancer drugs [369]. In this regard, numerous polymeric composite materials have been featured as bio-inks. In a study reported by Zhao et al., gelatin-alginate-fibrinogen composite bio-inks laden with cervical cancer (HeLa) cells were printed to form a 3D porous structure that stabilized by crosslinking with calcium chloride [370]. The sensitivity of the chemotherapeutic drug paclitaxel was evaluated, and it was found to be more effective on HeLa cells on 3D structures compared to 2D cultures. A study conducted by Beck and colleagues used a composite hydrogel of PEG/Matrigel to investigate cancer cell metastasis [371]. The matrix rigidity was controlled by increasing the degree of crosslinking of the PEG polymer, whereas the cellular adhesion signals were presented as peptide-conjugated cyclodextrin incorporated into the PEG networks. The study revealed that nominal values of cell adhesion and rigidity of the PEG matrix induce the migration of mammary malignant epithelial cells. Given the complexity and variability of tumor niches, researchers are striving to develop patient-specific tumor models that can be used in understanding tumor progression, diagnosis, and treatment [366,372]. The work performed to date has demonstrated the flexibility of hydrogels in investigating cancer pathophysiology. Further research is needed in the direction of the generation of matrices and their interaction with cancer cells to develop a better model mimicking in vitro tumors.

Table 4. Selected survey on hydrogel types and fabrication technology used in in vitro models of various tissue types.

Hydrogel Types	Fabrication Technology	Features	References
		Tissue Type: Skin	
Collagen, fibrinogen, and sodium hyaluronate	Bioprinting	A layer of thrombin was co-delivered to induce gelation of fibrinogen. The printed sheet could form in situ on murine and porcine wound skin with any topography.	[373]
Silk and collagen		Due to the presence of the cell-binding domain of collagen and stabilizing properties of silk, the scaffold exhibited more resistance to time-dependent degradation.	[326]
Gelatin, alginate, and fibrin	Bioprinting	A full-thickness skin model with a stratified epidermis was developed.	[327]
GelMA	Photocrosslinking	A skin model with an epidermal-like structure combined with an air–liquid interface was developed.	[374]
Silk and PCL	Electrospinning	The scaffold possessed a surface topography that promoted fibroblast-induced collagen deposition.	[375]
		Tissue Type: Bone	
Gelatin, chitosan, and hydroxyapatite	Freeze drying	The macroporous architecture allowed greater migration of MSC spheroids and led to a greater degree of mineralization of the construct.	[53]
PLA and collagen	3D printing	The porous disc-like construct was shown to support the growth and proliferation of osteoblasts, fibroblasts, and endothelial cells and induce neo-vessel formation.	[376]
Cellulose/BMP-2	Electrospinning	Bone marrow-derived MSCs showed oriented growth aligned with the underlying nanofiber morphology as well as increased alkaline phosphatase activity and calcium deposition with rapid rabbit calvaria wound repair.	[377]
Poly polystyrene sulfonate and collagen I	Ice templating	The interaction of human adipose-derived stem cells with electroactive 3D scaffolds was analyzed. The results highlighted the usefulness of porous conductive scaffolds as 3D in vitro platforms for bone tissue models.	[378]
PEG-DA	Stereolithography	An osteogenesis-on-a-chip device was developed that supports the proliferation, differentiation, and ECM production of human embryonic stem cell-derived mesenchymal progenitor cells for an extended period of 21 days.	[379]

Table 4. *Cont.*

Hydrogel Types	Fabrication Technology	Features	References
		Tissue Type: Cartilage	
HA	Electrospinning	A 3D nanofibrous scaffold was developed with crosslinked HA. The results showed a superabsorbent property, elastic behavior, and good cytocompatibility.	[380]
GelMA and methacrylated HA	Stereolithographic bioprinting	The 3D-printed model maintained chondrocyte distribution, differentiation, and ECM formation. Both materials showed cell viability and phenotype maintenance for a period of 21 days.	[347]
GelMA and tricalcium phosphate (TCP)	Co-axial extrusion bioprinting	The osteochondral defect was reconstructed by developing an in vitro 3D calcified cartilage tissue model. An investigation of a gene expression study confirmed the effects induced by ceramic nanoparticles in the differentiation of MSCs toward hypertrophic chondrocytes.	[381]
Agarose	Soft lithography	A versatile platform of articular cartilage-on-a-chip that can provide 3D multi-axial mechanical stimulation on a chondrocyte-loaded hydrogel was developed.	[349]
Methacrylated gelatin	Lithography using silicone mold	Microphysiological osteochondral tissue chips derived from human iPSCs were developed to model the pathologies of osteoarthritis (OA). Celecoxib, an OA drug, was shown to downregulate the proinflammatory cytokines of the OA model.	[382]
		Tissue Type: Liver	
Agarose	Self-aggregation of iPSCs	In vitro 3D liver tissue that exhibited a stable phenotype for over one year in culture was generated. The study presented an attractive resource for long-term liver in vitro studies	[383]
Agarose-chitosan	Liquid-cryo bath treatment of polymeric molds	At neutral pH, the negative charge of the scaffold surface ensured cell–cell interfacial interaction, followed by colonization of hepatocytes. The in vitro studies also indicated enhanced cellular metabolic activity.	[384]
Alginate and Pluronic F-127	3D bioprinting	iPSC-derived hepatocyte spheroids recapitulated liver epithelial parenchyma using 3D bioprinting.	[385]
Basement membrane extract	Soft lithography	A sinusoid-on-a-chip was established using four different types of liver cells (hepatocytes, endothelial cells, stellate cells, Kupffer cells). The study was the first to report the application of a liver chip in assessing the effect of hepatoprotective drugs.	[386]
Gelatin and porcine dECM	3D bioprinting	The study developed a liver fibrosis-on-a-chip platform using dECM and gelatin bio-ink and activated stellate cells. The platform exhibited increased collagen accumulation, cell apoptosis, and reduced liver-specific functions, which are characteristic features of liver fibrosis.	[355]
		Tissue Type: Gastrointestinal tract	
Silk	Freeze drying	Tissue characterization showed four differentiated epithelium cell types (enterocytes, goblet cells, Paneth cells, enteroendocrine cells) along with tight junction formation, microvilli polarization, low oxygen tension, and digestive enzyme secretion in the lumen.	[387]

Table 4. *Cont.*

Hydrogel Types	Fabrication Technology	Features	References
Thermo-responsive Novogel	3D bioprinting	The histological characterization of intestinal tissue demonstrated an injury response against the compound-induced toxicity and inflammation.	[388]
Collagen IV and Matrigel	Soft lithography	A human duodenum intestine-chip was developed. The in vitro tissue presented a polarized cell architecture with the presence of specialized cell subpopulations. It also demonstrated relevant expression and localization of major intestinal drug transporters.	[389]
Colon-derived dECM	3D bioprinting	The bioprinted intestinal tissue models showed spontaneous 3D morphogenesis of the human intestinal epithelium without any external stimuli.	[390]
Poly(3,4-ethylenedioxythiophene) doped with poly(styrene sulfonate) (PEDOT:PSS)	Freeze drying	A tubular electroactive scaffold served as a template for a 3D human intestine and enabled dynamic electrical monitoring of tissue formation over 1 month.	[391]
Tissue type: Cancer			
Poly-l-lactic acid	Thermally induced phase separation	The study generated scaffolds with different morphologies, porosities, and pore architectures and indicated that a pore size ranging from 40 to 50 μm induces tumor cell aggregation and the formation of the irregular tumor masses typically observed in vivo.	[392]
Matrigel	Organoid formation	A lung cancer organoid from patient tissue was established, the tissue architecture was recapitulated, and the genomic alterations of the original tumors were maintained during long-term expansion in vitro. The model responded to cancer drugs based on their genomic alterations and could be useful for predicting patient-specific drug responses.	[393]
Fibrinogen and Matrigel	Photolithography	A platform that imitates the mass transport near the arterial end of a tumor microenvironment was fabricated. An observation of the hallmark features of tumor progression was provided.	[394]
Matrigel	Photolithography	A colorectal tumor-on-a-chip model was developed. The platform can validate the efficacy of drug-loaded nanoparticles.	[395]
GelMA, alginate, and PEG-DA	Bioprinting	A tumor model that includes a hollow blood vessel and a lymphatic vessel was fabricated. The ability of imitating the transport mechanisms of drugs inside the tumor microenvironment was demonstrated.	[396]

7. Evaluation of In Vitro Models

The current biotech and pharmaceutical industry experiences high research and development costs and overall low success rates in the screening and development of new chemical and biological entities as therapeutic agents. Between 2005 and 2014, the primary reason for termination was poorly validated targets with little human relevance and lack of human efficacy at phase II/III (35%) [24]. There is a growing need for the development of human-based models that can be applied to healthy or disease states at the early stages of drug delivery. In recent years, numerous 3D models have been investigated and developed by a number of academic institutions and industries due to the emergence of innovative micro-technologies (spheroids, organoids, artificial scaffolds, lithography, and bioprinting). However, there is a significant gap between the development and qualification of such in vitro models, without which the end user cannot be confident enough about the physiological relevance of the system and subsequently the data produced from the model.

The standard rationale for any in vitro model is to streamline the experimental variables and fundamentally segregate different modules of organs or organ-like structures to study under well-organized and easily assessed conditions. How precisely these conditions recapitulate the in vivo conditions depends upon the study design and outcomes. A model should be qualified by deciding its domain of validity as defined by Scanell and Bosley based on rigorous scientific data and not on assumptions such as "primary cells are better than immortalized cell lines" or "a more complex model is more relevant" [397]. Different in vitro models represent different levels of cellular organization and behavior, and one classified as right must be fully qualified with regards to its qualities in contrast with the in vivo situation.

Many in vitro 3D models have been introduced into the literature with characterizations that have relied on techniques looking at a limited number of biological and physical descriptions, such as histological assessments, measurements of gene and protein expression at the subpopulation level by reverse transcriptase-polymerase chain reaction (RT-PCR), and enzyme-linked immunosorbent assay (ELISA)-based assays. More in-depth characterizations of models using the advances in omics technologies, including metabolomics, transcriptomics, and proteomics at the single-cell level, will enable a comprehensive understanding to score a model and assess whether it is fit for purpose. For example, a recent study by Kasendra et al. presented transcriptomic profiling, which showed there was an increased similarity in gene expression profiles when comparing a microfluidic intestinal organoid to adult duodenal tissue than to static duodenal organoids [389]. The study demonstrated a model that better replicates the in vivo tissue and thus increases the domain of validity for the model for possible drug transport metabolism and toxicity studies. The in vitro model also showed the presence of polarized cells, intestinal barrier function, the presence of specific cell populations, and in vivo–like expression, localization, and function of key intestinal drug transporters, which are necessary requirements for intestinal drug metabolism cellular models.

One of the important variables that determines the key aspect of developing, qualifying, and implementing in vitro models is cell type (cell lines, primary cells, stem cells). Immortalized cell lines have a distinct advantage, as they can be obtained in large numbers, which makes them an attractive candidate for in vitro model development for target validation work. With the aim of determining the relevance of the cells for use in research, the Open Targets project (in collaboration with Wellcome Trust Sanger Institute) has taken a number of cell lines used in research and performed a detailed investigation of gene expression with reference to primary cell data [398]. This study will provide a useful resource for the assessment of cell lines and physiological relevance with respect to different tissue types for use in in vitro models. Stem cells, on the other hand, have a number of unique advantages, as they retain the genetic information of the donor, but are scalable and amenable to gene editing and represent an infinite source of cells for target validation in in vitro models. Another benefit of stem cells from originator samples is the generation of organoid cultures, which is considered to mimic more closely the development and niche observed in vivo.

Development of a closely mimicking microenvironment in the model is another important criterion to evaluate. Many hundreds of studies to date have been published in which polymers from synthetic and natural sources and their blends have been used to create 3D constructs in many physical forms (e.g., gels, fibers, weaves, meshes, sponges, foams, channels), asserting their relevance based on cell adhesion, viability, proliferation, and superiority over 2D culture methods. Although comparisons of 3D models with traditional culture methods have merit, they alone cannot justify the significance of improved models. Similarly, many attempts have been made to seek a "one-size-fits-all" matrix, accommodating any cell type and many culture platforms, instead of developing organ-, tissue-, or pathology-specific models that can support clinically relevant pathways. Under normal physiological conditions, an ECM derived from different tissues exhibits its own unique architectures, mechanical properties (e.g., elastic energies, moduli), protein compositions, and molecular complexities [399]. These properties can be modulated as a function of normal processes, such as wound healing, or via pathological manifestations, such as cancer induction [400]. Instead of pursuing the development of a universal ECM mix, more focus should be placed on producing easily modified matrix components that can be mixed and matched to closely re-create native biochemistry and matrix rigidity. Another significant challenge in the development and implementation of in vitro models is the limitations imposed by static cellular models. Incorporating fluidics into models to mimic blood supply and interstitial flow is likely to increase the physiological relevance of the model [401]. The current microfluidics-based OOC models have solved this problem to a great extent [23]. Technological advances such as bioprinting have reproducibly demonstrated the ability to spatially control the deposition of multiple cell types and gels, resulting in the construction of tissues with architectures closer to those of organs [402]. Advances in big data and the capabilities of artificial intelligence (AI)-based largescale computing methods such as machine-learning (ML) and deep learning (DL) have the capacity to enable the multiparametric optimization of future in vitro models as well as to qualify and validate the clinical translatability of the models. Hepatotoxicity or drug-induced liver injury (DILI) is one of the most prominent areas in which such data-driven methods are used to support, optimize, and cross-validate the emerging complex in vitro models. Another good example is an ML-based platform, CANscript, developed by Mitra Biotech, where the data from in vitro models were used to train an ML-based algorithm to give a predictive translatability score representing the clinical efficacy of the drug response [403]. This shows the potential of computational and data-driven methods for building a more human-relevant in vitro model.

8. Challenges and Future Perspectives

Three-dimensional in vitro models are much more relevant compared to 2D culture in terms of imitating the complex physiological and pathological processes of native tissue and organs. These 3D in vitro platforms can potentially be used (i) to screen drugs and molecules for their safety and efficacy, (ii) to study the physiological processes in basic biology, and (iii) to facilitate clinical transplantation. As discussed in the previous section, several in vitro models targeting different tissues have been developed using various types and compositions of hydrogels. At this juncture, it is fair and appropriate to say that hydrogels are one of the basic requirements and logical choices in supporting 3D in vitro tissue model development. The simple reason behind this is they can inherently mimic the native tissue structure, which contains a complex polymeric network in a high-water-content environment. Hydrogels have been used in multiple 3D in vitro models to mimic the extracellular environment of the tissue. However, multiple factors need to be considered for composite hydrogels to recreate a 3D microenvironment of native tissue. In this section, we discuss the current challenges and future directions concerning hydrogel-based material design, recreation of the dynamic tissue microenvironment, and multi-tissue connectivity during the engineering of 3D in vitro models.

(i). The native ECM contains biochemical cues such as adhesion ligands, growth factors that are not evenly distributed throughout the matrix. Such anisotropic features in vivo are important in guiding cellular behavior and fate. Most of the hydrogel systems currently exhibit isotropic properties and completely lack anisotropy of the tissue microenvironment, which does not allow design flexibility in controlling hydrogel properties dynamically. In future studies, patterned systems could be developed containing gradient features over a hydrogel backbone for directing cell behavior. In this regard, stimuli-responsive smart hydrogels have attracted much attention, as they allow dynamic changes in their properties in the response under the defined stimulus. Novel hydrogel-based ECM mimetic formulations must be studied and investigated to contain features of anisotropy as well as flexibility (in stiffness, degradation, topography, etc.).

(ii). Tissue formation, disease development, and post-disease progression are all dynamic processes. Therefore, in vitro models must recapitulate such dynamic features of tissues and spatiotemporally control the features of the matrix, including presentation of biochemical features and topological and viscoelastic properties. In this regard, a reversible crosslinking strategy could be incorporated into the composite hydrogel system to control the degradation and mechanical properties of the synthetic matrix for supporting cell activity and long-term culture.

(iii). The major limitation of the present 3D in vitro model is its inability to completely mimic the complex features of tissue and disease microenvironments. The hierarchical design of organs, ranging from macroscale to micro and nanoscale, contains a complex structural arrangement of tissues that further contain a wide range of cell types with numerous cellular compositions and organizations. It is a challenge to design a tissue microenvironment in high resolution (concerning scale) that exhibits such variabilities and versatility of tissue arrangements. With the present technological advancement, it is quite difficult to exactly mimic a native tissue microenvironment. An ordered approach could be developed to determine the complexity needed for the 3D culture. In the long term, the development of an extensive database of various types of hydrogels, their interaction with cells, and other available knowledge of their features can be integrated with big data analytics and AI to predict definite factors in the development of a targeted model. Such an interdisciplinary approach would help us not only to understand the complexity of tissue niches, but also to develop and design composite hydrogels that would guide cell fate in the relevant microenvironment.

(iv). The last obstacle would be to integrate multiple tissue types or diseases in a single closed-loop platform. It is well understood that organs do not work in isolation, but are always working together and communicating with each other via biochemical cues, thus affecting and controlling each other directly or indirectly. For example, any toxic drug that affects the liver also influences the functionality of heart and lung tissue [281]. Similarly, in cancer, to metastasize the malignant cells, different tissue models are required, which can be connected via circulatory channels [404]. In this direction, multi-organ platforms have already been developed with some success [281]. However, the development of 3D systems with the above-mentioned features is highly desirable in establishing a functional and predictive model.

Although a 3D cell culture system provides a more physiologically relevant microenvironment as compared to 2D tissue, it can also be a source of shortcomings with respect to experimental analysis and detection. As most of the current methods of measurement are based on 2D cell culture, they might not be compatible with the 3D cell culture system. For instance, to determine cell number or quantify cell viability, most of the current methods rely on fluorescence-based analysis or manual counting of cells after trypsinization. These methods have a major limitation when applied in a 3D cell culture system. Therefore, novel methods and protocols that can provide accurate analysis in a 3D cell culture system must also be established.

9. Conclusions

Hydrogels are a class of biomaterial that has been developed to mimic the composition and structure of the native tissue matrix in a 3D in vitro tissue model. A range of biopolymers is used to fabricate hydrogels that can support and direct cell behavior and function in the desired way to generate 3D tissue-like culture. Due to the limitation of a single type of polymer, hybrid hydrogels are proposed with a combination of natural and synthetic polymers along with the amalgamation of a wide range of nanomaterials and biological factors that can impart specific tissue niche-like features. The development of dECM-derived hydrogels has presented a good alternative and opened up a new way to engineer the physiology of and simulate 3D in vitro models. This review article has summarized important advancements in the field of hydrogel biomaterial used in the generation of 3D tissues and found that hydrogels are becoming an attractive choice in emulating the 3D native matrices toward in vitro model development. Although research regarding hydrogels is still in its infancy, and many more technical challenges remain to be overcome, recognition of the reliability and reproducibility of hydrogels for specific in vitro models will be required for their acceptance in the therapeutic and pharmaceutical industries.

Author Contributions: Original draft preparation, S.M. and H.L.; Review and editing, S.M. and H.L.; Supervision, H.L. All authors have read and agreed to the published version of the manuscript.

Funding: This study was supported by a 2021 Research Grant from Kangwon National University. This work was also supported by National Research Foundation of Korea (NRF) grants funded by the Korean government (MSIT) (no. 2020R1C1C1011147 and no. 2020R1A4A1019475).

Data Availability Statement: The data presented in this study are available on request from the corresponding author.

Conflicts of Interest: The authors declare no conflict of interest.

References

1. Huang, G.; Li, F.; Zhao, X.; Ma, Y.; Li, Y.; Lin, M.; Jin, G.; Lu, T.J.; Genin, G.M.; Xu, F. Functional and biomimetic materials for engineering of the three-dimensional cell microenvironment. *Chem. Rev.* **2017**, *117*, 12764–12850. [CrossRef]
2. Ashammakhi, N.; Ahadian, S.; Darabi, M.A.; El Tahchi, M.; Lee, J.; Suthiwanich, K.; Sheikhi, A.; Dokmeci, M.R.; Oklu, R.; Khademhosseini, A. Minimally invasive and regenerative therapeutics. *Adv. Mater.* **2019**, *31*, 1804041. [CrossRef] [PubMed]
3. Edmondson, R.; Broglie, J.J.; Adcock, A.F.; Yang, L. Three-dimensional cell culture systems and their applications in drug discovery and cell-based biosensors. *Assay Drug Dev. Technol.* **2014**, *12*, 207–218. [CrossRef] [PubMed]
4. HogenEsch, H.; Nikitin, A.Y. Challenges in pre-clinical testing of anti-cancer drugs in cell culture and in animal models. *J. Control. Release* **2012**, *164*, 183–186. [CrossRef] [PubMed]
5. Alhaque, S.; Themis, M.; Rashidi, H. Three-dimensional cell culture: From evolution to revolution. *Philos. Trans. R. Soc. B Biol. Sci.* **2018**, *373*, 20170216. [CrossRef]
6. Hong, S.; Kim, J.S.; Jung, B.; Won, C.; Hwang, C. Coaxial bioprinting of cell-laden vascular constructs using a gelatin–tyramine bioink. *Biomater. Sci.* **2019**, *7*, 4578–4587. [CrossRef]
7. Kaushik, G.; Ponnusamy, M.P.; Batra, S.K. Concise review: Current status of three-dimensional Organoids as preclinical models. *Stem Cells* **2018**, *36*, 1329–1340. [CrossRef]
8. Burdick, J.A.; Vunjak-Novakovic, G. Engineered microenvironments for controlled stem cell differentiation. *Tissue Eng. Part A* **2009**, *15*, 205–219. [CrossRef]
9. Cukierman, E.; Pankov, R.; Yamada, K.M. Cell interactions with three-dimensional matrices. *Curr. Opin. Cell Biol.* **2002**, *14*, 633–640. [CrossRef]
10. Gelain, F. Novel opportunities and challenges offered by nanobiomaterials in tissue engineering. *Int. J. Nanomed.* **2008**, *3*, 415. [CrossRef]
11. Tibbitt, M.W.; Anseth, K.S. Hydrogels as extracellular matrix mimics for 3D cell culture. *Biotechnol. Bioeng.* **2009**, *103*, 655–663. [CrossRef]
12. Zahir, N.; Weaver, V.M. Death in the third dimension: Apoptosis regulation and tissue architecture. *Curr. Opin. Genet. Dev.* **2004**, *14*, 71–80. [CrossRef]
13. Hong, H.; McCullough, C.M.; Stegemann, J.P. The role of ERK signaling in protein hydrogel remodeling by vascular smooth muscle cells. *Biomaterials* **2007**, *28*, 3824–3833. [CrossRef]
14. McGuigan, A.P.; Bruzewicz, D.A.; Glavan, A.; Butte, M.; Whitesides, G.M. Cell encapsulation in sub-mm sized gel modules using replica molding. *PLoS ONE* **2008**, *3*, e2258. [CrossRef]
15. Saha, K.; Pollock, J.F.; Schaffer, D.V.; Healy, K.E. Designing synthetic materials to control stem cell phenotype. *Curr. Opin. Chem. Biol.* **2007**, *11*, 381–387. [CrossRef]
16. Song, Y.S.; Lin, R.L.; Montesano, G.; Durmus, N.G.; Lee, G.; Yoo, S.-S.; Kayaalp, E.; Hæggström, E.; Khademhosseini, A.; Demirci, U. Engineered 3D tissue models for cell-laden microfluidic channels. *Anal. Bioanal. Chem.* **2009**, *395*, 185–193. [CrossRef]
17. Jeong, M.K.; Choi, M.J.; Kwon, S.J.; Bae, M.H.; Paeng, D.G.; Zeqiri, B.; Wright, L.; Coleman, A. Ultrasonic characterization of thermal distribution in vicinity for a cylindrical thermal lesion in a biological tissue. *Key Eng. Mater.* **2006**, *321*, 1133–1138. [CrossRef]
18. Grossmann, J. Molecular mechanisms of "detachment-induced apoptosis—Anoikis". *Apoptosis* **2002**, *7*, 247–260. [CrossRef]
19. Bersini, S.; Gilardi, M.; Arrigoni, C.; Talò, G.; Zamai, M.; Zagra, L.; Caiolfa, V.; Moretti, M. Human in vitro 3D co-culture model to engineer vascularized bone-mimicking tissues combining computational tools and statistical experimental approach. *Biomaterials* **2016**, *76*, 157–172. [CrossRef]

20. Knight, E.; Przyborski, S. Advances in 3D cell culture technologies enabling tissue-like structures to be created in vitro. *J. Anat.* **2015**, *227*, 746–756. [CrossRef]
21. Sart, S.; Tsai, A.-C.; Li, Y.; Ma, T. Three-dimensional aggregates of mesenchymal stem cells: Cellular mechanisms, biological properties, and applications. *Tissue Eng. Part B Rev.* **2014**, *20*, 365–380. [CrossRef] [PubMed]
22. Houser, S.R.; Margulies, K.B.; Murphy, A.M.; Spinale, F.G.; Francis, G.S.; Prabhu, S.D.; Rockman, H.A.; Kass, D.A.; Molkentin, J.D.; Sussman, M.A. Animal models of heart failure: A scientific statement from the American Heart Association. *Circ. Res.* **2012**, *111*, 131–150. [CrossRef]
23. Bhatia, S.N.; Ingber, D.E. Microfluidic organs-on-chips. *Nat. Biotechnol.* **2014**, *32*, 760–772. [CrossRef] [PubMed]
24. Hay, M.; Thomas, D.W.; Craighead, J.L.; Economides, C.; Rosenthal, J. Clinical development success rates for investigational drugs. *Nat. Biotechnol.* **2014**, *32*, 40–51. [CrossRef]
25. Festing, S.; Wilkinson, R. The ethics of animal research: Talking point on the use of animals in scientific research. *EMBO Rep.* **2007**, *8*, 526–530. [CrossRef] [PubMed]
26. Takahashi, K.; Tanabe, K.; Ohnuki, M.; Narita, M.; Ichisaka, T.; Tomoda, K.; Yamanaka, S. Induction of pluripotent stem cells from adult human fibroblasts by defined factors. *Cell* **2007**, *131*, 861–872. [CrossRef]
27. Sung, J.H.; Esch, M.B.; Prot, J.-M.; Long, C.J.; Smith, A.; Hickman, J.J.; Shuler, M.L. Microfabricated mammalian organ systems and their integration into models of whole animals and humans. *Lab Chip* **2013**, *13*, 1201–1212. [CrossRef]
28. Tabata, Y. Biomaterial technology for tissue engineering applications. *J. R. Soc. Interface* **2009**, *6*, S311–S324. [CrossRef]
29. Kopeček, J. Hydrogel biomaterials: A smart future? *Biomaterials* **2007**, *28*, 5185–5192. [CrossRef]
30. Antman-Passig, M.; Shefi, O. Remote magnetic orientation of 3D collagen hydrogels for directed neuronal regeneration. *Nano Lett.* **2016**, *16*, 2567–2573. [CrossRef]
31. Kim, S.-H.; Lee, S.-H.; Lee, J.-E.; Park, S.J.; Kim, K.; Kim, I.S.; Lee, Y.-S.; Hwang, N.S.; Kim, B.-G. Tissue adhesive, rapid forming, and sprayable ECM hydrogel via recombinant tyrosinase crosslinking. *Biomaterials* **2018**, *178*, 401–412. [CrossRef]
32. Yu, Y.; Yuk, H.; Parada, G.A.; Wu, Y.; Liu, X.; Nabzdyk, C.S.; Youcef-Toumi, K.; Zang, J.; Zhao, X. Multifunctional "hydrogel skins" on diverse polymers with arbitrary shapes. *Adv. Mater.* **2019**, *31*, 1807101. [CrossRef]
33. Huang, J.; Liu, G.; Song, C.; Saiz, E.; Tomsia, A.P. Role of molecular chemistry of degradable pHEMA hydrogels in three-dimensional biomimetic mineralization. *Chem. Mater.* **2012**, *24*, 1331–1337. [CrossRef]
34. Liu, M.; Ishida, Y.; Ebina, Y.; Sasaki, T.; Aida, T. Photolatently modulable hydrogels using unilamellar titania nanosheets as photocatalytic crosslinkers. *Nat. Commun.* **2013**, *4*, 1–7. [CrossRef]
35. Fares, M.M.; Sani, E.S.; Lara, R.P.; Oliveira, R.B.; Khademhosseini, A.; Annabi, N. Interpenetrating network gelatin methacryloyl (GelMA) and pectin-g-PCL hydrogels with tunable properties for tissue engineering. *Biomater. Sci.* **2018**, *6*, 2938–2950. [CrossRef]
36. Karami, P.; Wyss, C.L.S.; Khoushabi, A.; Schmocker, A.; Broome, M.; Moser, C.; Bourban, P.-E.; Pioletti, D.P. Composite double-network hydrogels to improve adhesion on biological surfaces. *ACS Appl. Mater. Interfaces* **2018**, *10*, 38692–38699. [CrossRef]
37. Ahadian, S.; Sadeghian, R.B.; Salehi, S.; Ostrovidov, S.; Bae, H.; Ramalingam, M.; Khademhosseini, A. Bioconjugated hydrogels for tissue engineering and regenerative medicine. *Bioconjug. Chem.* **2015**, *26*, 1984–2001. [CrossRef]
38. Guvendiren, M.; Burdick, J.A. Engineering synthetic hydrogel microenvironments to instruct stem cells. *Curr. Opin. Biotechnol.* **2013**, *24*, 841–846. [CrossRef]
39. Yang, J.; Zhang, Y.S.; Yue, K.; Khademhosseini, A. Cell-laden hydrogels for osteochondral and cartilage tissue engineering. *Acta Biomater.* **2017**, *57*, 1–25. [CrossRef]
40. Nichol, J.W.; Koshy, S.T.; Bae, H.; Hwang, C.M.; Yamanlar, S.; Khademhosseini, A. Cell-laden microengineered gelatin methacrylate hydrogels. *Biomaterials* **2010**, *31*, 5536–5544. [CrossRef]
41. Farrukh, S.; Mustafa, K.; Hussain, A.; Ayoub, M. Synthesis and applications of carbohydrate-based hydrogels. In *Cellulose-Based Superabsorbent Hydrogels. Polymers and Polymeric Composites: A Reference Series*; Springer: New York, NY, USA, 2018. [CrossRef]
42. Griffith, L.G.; Swartz, M.A. Capturing complex 3D tissue physiology in vitro. *Nat. Rev. Mol. Cell Biol.* **2006**, *7*, 211–224. [CrossRef] [PubMed]
43. Shelke, N.B.; James, R.; Laurencin, C.T.; Kumbar, S.G. Polysaccharide biomaterials for drug delivery and regenerative engineering. *Polym. Adv. Technol.* **2014**, *25*, 448–460. [CrossRef]
44. Ivanovska, J.; Zehnder, T.; Lennert, P.; Sarker, B.; Boccaccini, A.R.; Hartmann, A.; Schneider-Stock, R.; Detsch, R. Biofabrication of 3D alginate-based hydrogel for cancer research: Comparison of cell spreading, viability, and adhesion characteristics of colorectal HCT116 tumor cells. *Tissue Eng. Part C Methods* **2016**, *22*, 708–715. [CrossRef] [PubMed]
45. Vanacker, J.; Amorim, C.A. Alginate: A versatile biomaterial to encapsulate isolated ovarian follicles. *Ann. Biomed. Eng.* **2017**, *45*, 1633–1649. [CrossRef] [PubMed]
46. Lee, K.Y.; Mooney, D.J. Alginate: Properties and biomedical applications. *Prog. Polym. Sci.* **2012**, *37*, 106–126. [CrossRef] [PubMed]
47. Webber, R.E.; Shull, K.R. Strain dependence of the viscoelastic properties of alginate hydrogels. *Macromolecules* **2004**, *37*, 6153–6160. [CrossRef]
48. Cavo, M.; Fato, M.; Peñuela, L.; Beltrame, F.; Raiteri, R.; Scaglione, S. Microenvironment complexity and matrix stiffness regulate breast cancer cell activity in a 3D in vitro model. *Sci. Rep.* **2016**, *6*, 1–13. [CrossRef]
49. Cavo, M.; Caria, M.; Pulsoni, I.; Beltrame, F.; Fato, M.; Scaglione, S. A new cell-laden 3D Alginate-Matrigel hydrogel resembles human breast cancer cell malignant morphology, spread and invasion capability observed "in vivo". *Sci. Rep.* **2018**, *8*, 1–12. [CrossRef]

50. Costa, R.; Urbani, A.; Salvalaio, M.; Bellesso, S.; Cieri, D.; Zancan, I.; Filocamo, M.; Bonaldo, P.; Szabò, I.; Tomanin, R. Perturbations in cell signaling elicit early cardiac defects in mucopolysaccharidosis type II. *Hum. Mol. Genet.* **2017**, *26*, 1643–1655. [CrossRef]
51. Xu, X.; Jha, A.K.; Harrington, D.A.; Farach-Carson, M.C.; Jia, X. Hyaluronic acid-based hydrogels: From a natural polysaccharide to complex networks. *Soft Matter* **2012**, *8*, 3280–3294. [CrossRef]
52. Zhu, X.; Gojgini, S.; Chen, T.-H.; Fei, P.; Dong, S.; Ho, C.-M.; Segura, T. Directing three-dimensional multicellular morphogenesis by self-organization of vascular mesenchymal cells in hyaluronic acid hydrogels. *J. Biol. Eng.* **2017**, *11*, 1–12. [CrossRef]
53. Maji, S.; Agarwal, T.; Das, J.; Maiti, T.K. Development of gelatin/carboxymethyl chitosan/nano-hydroxyapatite composite 3D macroporous scaffold for bone tissue engineering applications. *Carbohydr. Polym.* **2018**, *189*, 115–125. [CrossRef]
54. Skardal, A.; Devarasetty, M.; Rodman, C.; Atala, A.; Soker, S. Liver-tumor hybrid organoids for modeling tumor growth and drug response in vitro. *Ann. Biomed. Eng.* **2015**, *43*, 2361–2373. [CrossRef]
55. Wylie, R.G.; Shoichet, M.S. Three-dimensional spatial patterning of proteins in hydrogels. *Biomacromolecules* **2011**, *12*, 3789–3796. [CrossRef]
56. Kuo, C.K.; Marturano, J.E.; Tuan, R.S. Novel strategies in tendon and ligament tissue engineering: Advanced biomaterials and regeneration motifs. *BMC Sports Sci. Med. Rehabil.* **2010**, *2*, 20. [CrossRef]
57. Tabatabaei, F.; Moharamzadeh, K.; Tayebi, L. Fibroblast encapsulation in gelatin methacryloyl (GelMA) versus collagen hydrogel as substrates for oral mucosa tissue engineering. *J. Oral Biol. Craniofac. Res.* **2020**, *10*, 573–577. [CrossRef]
58. Li, S.; Lao, J.; Chen, B.P.; Li, Y.S.; Zhao, Y.; Chu, J.; Chen, K.D.; Tsou, T.C.; Peck, K.; Chien, S. Genomic analysis of smooth muscle cells in three-dimensional collagen matrix. *FASEB J.* **2003**, *17*, 97–99. [CrossRef]
59. Daya, S.; Loughlin, A.J.; MacQueen, H.A. Culture and differentiation of preadipocytes in two-dimensional and three-dimensional in vitro systems. *Differentiation* **2007**, *75*, 360–370. [CrossRef]
60. Qiao, J.; Sakurai, H.; Nigam, S.K. Branching morphogenesis independent of mesenchymal–epithelial contact in the developing kidney. *Proc. Natl. Acad. Sci. USA* **1999**, *96*, 7330–7335. [CrossRef]
61. Reichert, M.; Takano, S.; Heeg, S.; Bakir, B.; Botta, G.P.; Rustgi, A.K. Isolation, culture and genetic manipulation of mouse pancreatic ductal cells. *Nat. Protoc.* **2013**, *8*, 1354–1365. [CrossRef]
62. Miller, D.H.; Sokol, E.S.; Gupta, P.B. 3D primary culture model to study human mammary development. In *3D Cell Culture*; Humana Press: New York, NY, USA, 2017; pp. 139–147.
63. Thakuri, P.S.; Gupta, M.; Joshi, R.; Singh, S.; Tavana, H. Synergistic inhibition of kinase pathways overcomes resistance of colorectal cancer spheroids to cyclic targeted therapies. *ACS Pharmacol. Transl. Sci.* **2019**, *2*, 275–284. [CrossRef] [PubMed]
64. Lv, D.; Hu, Z.; Lu, L.; Lu, H.; Xu, X. Three-dimensional cell culture: A powerful tool in tumor research and drug discovery. *Oncol. Lett.* **2017**, *14*, 6999–7010. [CrossRef] [PubMed]
65. Devarasetty, M.; Skardal, A.; Cowdrick, K.; Marini, F.; Soker, S. Bioengineered submucosal organoids for in vitro modeling of colorectal cancer. *Tissue Eng. Part A* **2017**, *23*, 1026–1041. [CrossRef] [PubMed]
66. Davis, G.E.; Kim, D.J.; Meng, C.-X.; Norden, P.R.; Speichinger, K.R.; Davis, M.T.; Smith, A.O.; Bowers, S.L.; Stratman, A.N. Control of vascular tube morphogenesis and maturation in 3D extracellular matrices by endothelial cells and pericytes. In *Cell-Cell Interactions*; Humana Press: Totowa, NJ, USA, 2013; pp. 17–28.
67. Sheu, M.-T.; Huang, J.-C.; Yeh, G.-C.; Ho, H.-O. Characterization of collagen gel solutions and collagen matrices for cell culture. *Biomaterials* **2001**, *22*, 1713–1719. [CrossRef]
68. Brown, R.A.; Wiseman, M.; Chuo, C.B.; Cheema, U.; Nazhat, S.N. Ultrarapid engineering of biomimetic materials and tissues: Fabrication of nano-and microstructures by plastic compression. *Adv. Funct. Mater.* **2005**, *15*, 1762–1770. [CrossRef]
69. Besseau, L.; Coulomb, B.; Lebreton-Decoster, C.; Giraud-Guille, M.-M. Production of ordered collagen matrices for three-dimensional cell culture. *Biomaterials* **2002**, *23*, 27–36. [CrossRef]
70. Sapudom, J.; Kalbitzer, L.; Wu, X.; Martin, S.; Kroy, K.; Pompe, T. Fibril bending stiffness of 3D collagen matrices instructs spreading and clustering of invasive and non-invasive breast cancer cells. *Biomaterials* **2019**, *193*, 47–57. [CrossRef]
71. Mason, B.N.; Starchenko, A.; Williams, R.M.; Bonassar, L.J.; Reinhart-King, C.A. Tuning three-dimensional collagen matrix stiffness independently of collagen concentration modulates endothelial cell behavior. *Acta Biomater.* **2013**, *9*, 4635–4644. [CrossRef]
72. Jabaji, Z.; Brinkley, G.J.; Khalil, H.A.; Sears, C.M.; Lei, N.Y.; Lewis, M.; Stelzner, M.; Martín, M.G.; Dunn, J.C. Type I collagen as an extracellular matrix for the in vitro growth of human small intestinal epithelium. *PLoS ONE* **2014**, *9*, e107814.
73. Jee, J.H.; Lee, D.H.; Ko, J.; Hahn, S.; Jeong, S.Y.; Kim, H.K.; Park, E.; Choi, S.Y.; Jeong, S.; Lee, J.W. Development of collagen-based 3D matrix for gastrointestinal tract-derived organoid culture. *Stem Cells Int.* **2019**, *2019*, 8472712. [CrossRef]
74. Wimmer, R.A.; Leopoldi, A.; Aichinger, M.; Wick, N.; Hantusch, B.; Novatchkova, M.; Taubenschmid, J.; Hämmerle, M.; Esk, C.; Bagley, J.A. Human blood vessel organoids as a model of diabetic vasculopathy. *Nature* **2019**, *565*, 505–510. [CrossRef]
75. Francis, F.J. *Wiley Encyclopedia of Food Science and Technology*; John Wiley Sons Inc.: Hoboken, NJ, USA, 1999.
76. Amadori, S.; Torricelli, P.; Rubini, K.; Fini, M.; Panzavolta, S.; Bigi, A. Effect of sterilization and crosslinking on gelatin films. *J. Mater. Sci. Mater. Med.* **2015**, *26*, 69. [CrossRef]
77. Yue, K.; Trujillo-de Santiago, G.; Alvarez, M.M.; Tamayol, A.; Annabi, N.; Khademhosseini, A. Synthesis, properties, and biomedical applications of gelatin methacryloyl (GelMA) hydrogels. *Biomaterials* **2015**, *73*, 254–271. [CrossRef]
78. Lin, R.-Z.; Chen, Y.-C.; Moreno-Luna, R.; Khademhosseini, A.; Melero-Martin, J.M. Transdermal regulation of vascular network bioengineering using a photopolymerizable methacrylated gelatin hydrogel. *Biomaterials* **2013**, *34*, 6785–6796. [CrossRef]

79. Heo, D.N.; Castro, N.J.; Lee, S.-J.; Noh, H.; Zhu, W.; Zhang, L.G. Enhanced bone tissue regeneration using a 3D printed microstructure incorporated with a hybrid nano hydrogel. *Nanoscale* **2017**, *9*, 5055–5062. [CrossRef]
80. Kaemmerer, E.; Melchels, F.P.; Holzapfel, B.M.; Meckel, T.; Hutmacher, D.W.; Loessner, D. Gelatine methacrylamide-based hydrogels: An alternative three-dimensional cancer cell culture system. *Acta Biomater.* **2014**, *10*, 2551–2562. [CrossRef]
81. Zhao, X.; Lang, Q.; Yildirimer, L.; Lin, Z.Y.; Cui, W.; Annabi, N.; Ng, K.W.; Dokmeci, M.R.; Ghaemmaghami, A.M.; Khademhosseini, A. Photocrosslinkable gelatin hydrogel for epidermal tissue engineering. *Adv. Healthc. Mater.* **2016**, *5*, 108–118. [CrossRef]
82. Kolesky, D.B.; Truby, R.L.; Gladman, A.S.; Busbee, T.A.; Homan, K.A.; Lewis, J.A. 3D bioprinting of vascularized, heterogeneous cell-laden tissue constructs. *Adv. Mater.* **2014**, *26*, 3124–3130. [CrossRef]
83. Holland, C.; Numata, K.; Rnjak-Kovacina, J.; Seib, F.P. The biomedical use of silk: Past, present, future. *Adv. Healthc. Mater.* **2019**, *8*, 1800465. [CrossRef]
84. Hu, X.; Lu, Q.; Sun, L.; Cebe, P.; Wang, X.; Zhang, X.; Kaplan, D.L. Biomaterials from ultrasonication-induced silk fibroin-hyaluronic acid hydrogels. *Biomacromolecules* **2010**, *11*, 3178–3188. [CrossRef]
85. Hofmann, S.; Knecht, S.; Langer, R.; Kaplan, D.L.; Vunjak-Novakovic, G.; Merkle, H.P.; Meinel, L. Cartilage-like tissue engineering using silk scaffolds and mesenchymal stem cells. *Tissue Eng.* **2006**, *12*, 2729–2738. [CrossRef] [PubMed]
86. Tang-Schomer, M.D.; White, J.D.; Tien, L.W.; Schmitt, L.I.; Valentin, T.M.; Graziano, D.J.; Hopkins, A.M.; Omenetto, F.G.; Haydon, P.G.; Kaplan, D.L. Bioengineered functional brain-like cortical tissue. *Proc. Natl. Acad. Sci. USA* **2014**, *111*, 13811–13816. [CrossRef] [PubMed]
87. Reimers, K.; Liebsch, C.; Radtke, C.; Kuhbier, J.W.; Vogt, P.M. Silks as scaffolds for skin reconstruction. *Biotechnol. Bioeng.* **2015**, *112*, 2201–2205. [CrossRef] [PubMed]
88. Chen, Y.; Lin, Y.; Davis, K.M.; Wang, Q.; Rnjak-Kovacina, J.; Li, C.; Isberg, R.R.; Kumamoto, C.A.; Mecsas, J.; Kaplan, D.L. Robust bioengineered 3D functional human intestinal epithelium. *Sci. Rep.* **2015**, *5*, 1–11. [CrossRef] [PubMed]
89. Melke, J.; Midha, S.; Ghosh, S.; Ito, K.; Hofmann, S. Silk fibroin as biomaterial for bone tissue engineering. *Acta Biomater.* **2016**, *31*, 1–16. [CrossRef]
90. Jao, D.; Mou, X.; Hu, X. Tissue regeneration: A silk road. *J. Funct. Biomater.* **2016**, *7*, 22. [CrossRef]
91. Marklein, R.A.; Burdick, J.A. Controlling stem cell fate with material design. *Adv. Mater.* **2010**, *22*, 175–189. [CrossRef]
92. Gjorevski, N.; Sachs, N.; Manfrin, A.; Giger, S.; Bragina, M.E.; Ordóñez-Morán, P.; Clevers, H.; Lutolf, M.P. Designer matrices for intestinal stem cell and organoid culture. *Nature* **2016**, *539*, 560–564. [CrossRef]
93. Gjorevski, N.; Lutolf, M.P. Synthesis and characterization of well-defined hydrogel matrices and their application to intestinal stem cell and organoid culture. *Nat. Protoc.* **2017**, *12*, 2263–2274. [CrossRef]
94. Vedadghavami, A.; Minooei, F.; Mohammadi, M.H.; Khetani, S.; Kolahchi, A.R.; Mashayekhan, S.; Sanati-Nezhad, A. Manufacturing of hydrogel biomaterials with controlled mechanical properties for tissue engineering applications. *Acta Biomater.* **2017**, *62*, 42–63. [CrossRef]
95. Annabi, N.; Tamayol, A.; Uquillas, J.A.; Akbari, M.; Bertassoni, L.E.; Cha, C.; Camci-Unal, G.; Dokmeci, M.R.; Peppas, N.A.; Khademhosseini, A. 25th anniversary article: Rational design and applications of hydrogels in regenerative medicine. *Adv. Mater.* **2014**, *26*, 85–124. [CrossRef]
96. Thakuri, P.S.; Liu, C.; Luker, G.D.; Tavana, H. Biomaterials-Based Approaches to Tumor Spheroid and Organoid Modeling. *Adv. Healthc. Mater.* **2018**, *7*, 1700980. [CrossRef]
97. Wolf, M.T.; Dearth, C.L.; Sonnenberg, S.B.; Loboa, E.G.; Badylak, S.F. Naturally derived and synthetic scaffolds for skeletal muscle reconstruction. *Adv. Drug Deliv. Rev.* **2015**, *84*, 208–221. [CrossRef]
98. Wang, C.; Tong, X.; Jiang, X.; Yang, F. Effect of matrix metalloproteinase-mediated matrix degradation on glioblastoma cell behavior in 3D PEG-based hydrogels. *J. Biomed. Mater. Res. Part A* **2017**, *105*, 770–778. [CrossRef]
99. Grim, J.C.; Marozas, I.A.; Anseth, K.S. Thiol-ene and photo-cleavage chemistry for controlled presentation of biomolecules in hydrogels. *J. Control. Release* **2015**, *219*, 95–106. [CrossRef]
100. Zhu, J. Bioactive modification of poly (ethylene glycol) hydrogels for tissue engineering. *Biomaterials* **2010**, *31*, 4639–4656. [CrossRef]
101. Enemchukwu, N.O.; Cruz-Acuña, R.; Bongiorno, T.; Johnson, C.T.; García, J.R.; Sulchek, T.; García, A.J. Synthetic matrices reveal contributions of ECM biophysical and biochemical properties to epithelial morphogenesis. *J. Cell Biol.* **2016**, *212*, 113–124. [CrossRef]
102. Cruz-Acuña, R.; Quirós, M.; Huang, S.; Siuda, D.; Spence, J.R.; Nusrat, A.; García, A.J. PEG-4MAL hydrogels for human organoid generation, culture, and in vivo delivery. *Nat. Protoc.* **2018**, *13*, 2102–2119. [CrossRef]
103. Wang, Y.; Zhang, Z.; Xu, L.; Li, X.; Chen, H. Hydrogels of halogenated Fmoc-short peptides for potential application in tissue engineering. *Colloids Surf. B Biointerfaces* **2013**, *104*, 163–168. [CrossRef]
104. Wang, S.; Nagrath, D.; Chen, P.C.; Berthiaume, F.; Yarmush, M.L. Three-dimensional primary hepatocyte culture in synthetic self-assembling peptide hydrogel. *Tissue Eng. Part A* **2008**, *14*, 227–236. [CrossRef]
105. Chang, C.-W.; Hwang, Y.; Brafman, D.; Hagan, T.; Phung, C.; Varghese, S. Engineering cell–material interfaces for long-term expansion of human pluripotent stem cells. *Biomaterials* **2013**, *34*, 912–921. [CrossRef] [PubMed]
106. Sun, A.X.; Lin, H.; Fritch, M.R.; Shen, H.; Alexander, P.G.; DeHart, M.; Tuan, R.S. Chondrogenesis of human bone marrow mesenchymal stem cells in 3-dimensional, photocrosslinked hydrogel constructs: Effect of cell seeding density and material stiffness. *Acta Biomater.* **2017**, *58*, 302–311. [CrossRef] [PubMed]

107. Wang, Z.; Lee, S.J.; Cheng, H.-J.; Yoo, J.J.; Atala, A. 3D bioprinted functional and contractile cardiac tissue constructs. *Acta Biomater.* **2018**, *70*, 48–56. [CrossRef] [PubMed]
108. Matricardi, P.; Di Meo, C.; Coviello, T.; Hennink, W.E.; Alhaique, F. Interpenetrating polymer networks polysaccharide hydrogels for drug delivery and tissue engineering. *Adv. Drug Deliv. Rev.* **2013**, *65*, 1172–1187. [CrossRef] [PubMed]
109. Chaudhuri, O.; Gu, L.; Klumpers, D.; Darnell, M.; Bencherif, S.A.; Weaver, J.C.; Huebsch, N.; Lee, H.-P.; Lippens, E.; Duda, G.N. Hydrogels with tunable stress relaxation regulate stem cell fate and activity. *Nat. Mater.* **2016**, *15*, 326–334. [CrossRef] [PubMed]
110. Shin, S.; Ikram, M.; Subhan, F.; Kang, H.Y.; Lim, Y.; Lee, R.; Jin, S.; Jeong, Y.H.; Kwak, J.-Y.; Na, Y.-J. Alginate–marine collagen–agarose composite hydrogels as matrices for biomimetic 3D cell spheroid formation. *RSC Adv.* **2016**, *6*, 46952–46965. [CrossRef]
111. Chung, J.H.; Naficy, S.; Yue, Z.; Kapsa, R.; Quigley, A.; Moulton, S.E.; Wallace, G.G. Bio-ink properties and printability for extrusion printing living cells. *Biomater. Sci.* **2013**, *1*, 763–773. [CrossRef]
112. Ng, W.L.; Yeong, W.Y.; Naing, M.W. Polyelectrolyte gelatin-chitosan hydrogel optimized for 3D bioprinting in skin tissue engineering. *Int. J. Bioprint.* **2016**, *2*, 53–62. [CrossRef]
113. Das, S.; Pati, F.; Choi, Y.-J.; Rijal, G.; Shim, J.-H.; Kim, S.W.; Ray, A.R.; Cho, D.-W.; Ghosh, S. Bioprintable, cell-laden silk fibroin–gelatin hydrogel supporting multilineage differentiation of stem cells for fabrication of three-dimensional tissue constructs. *Acta Biomater.* **2015**, *11*, 233–246. [CrossRef]
114. Bartnikowski, M.; Wellard, R.M.; Woodruff, M.; Klein, T. Tailoring hydrogel viscoelasticity with physical and chemical crosslinking. *Polymers* **2015**, *7*, 2650–2669. [CrossRef]
115. Feng, Q.; Li, Q.; Wen, H.; Chen, J.; Liang, M.; Huang, H.; Lan, D.; Dong, H.; Cao, X. Injection and self-assembly of bioinspired stem cell-laden gelatin/hyaluronic acid hybrid microgels promote cartilage repair in vivo. *Adv. Funct. Mater.* **2019**, *29*, 1906690. [CrossRef]
116. Hedegaard, C.L.; Collin, E.C.; Redondo-Gómez, C.; Nguyen, L.T.; Ng, K.W.; Castrejón-Pita, A.A.; Castrejón-Pita, J.R.; Mata, A. Hydrodynamically Guided Hierarchical Self-Assembly of Peptide–Protein Bioinks. *Adv. Funct. Mater.* **2018**, *28*, 1703716. [CrossRef]
117. Yue, K.; Liu, Y.; Byambaa, B.; Singh, V.; Liu, W.; Li, X.; Sun, Y.; Zhang, Y.S.; Tamayol, A.; Zhang, P. Visible light crosslinkable human hair keratin hydrogels. *Bioeng. Transl. Med.* **2018**, *3*, 37–48. [CrossRef]
118. Shin, Y.M.; Kim, T.G.; Park, J.-S.; Gwon, H.-J.; Jeong, S.I.; Shin, H.; Kim, K.-S.; Kim, D.; Yoon, M.-H.; Lim, Y.-M. Engineered ECM-like microenvironment with fibrous particles for guiding 3D-encapsulated hMSC behaviours. *J. Mater. Chem. B* **2015**, *3*, 2732–2741. [CrossRef]
119. Li, X.; Cho, B.; Martin, R.; Seu, M.; Zhang, C.; Zhou, Z.; Choi, J.S.; Jiang, X.; Chen, L.; Walia, G. Nanofiber-hydrogel composite–mediated angiogenesis for soft tissue reconstruction. *Sci. Transl. Med.* **2019**, *11*, eaau6210. [CrossRef]
120. Bas, O.; De-Juan-Pardo, E.M.; Meinert, C.; D'Angella, D.; Baldwin, J.G.; Bray, L.J.; Wellard, R.M.; Kollmannsberger, S.; Rank, E.; Werner, C. Biofabricated soft network composites for cartilage tissue engineering. *Biofabrication* **2017**, *9*, 025014. [CrossRef]
121. Visser, J.; Melchels, F.P.; Jeon, J.E.; Van Bussel, E.M.; Kimpton, L.S.; Byrne, H.M.; Dhert, W.J.; Dalton, P.D.; Hutmacher, D.W.; Malda, J. Reinforcement of hydrogels using three-dimensionally printed microfibres. *Nat. Commun.* **2015**, *6*, 1–10. [CrossRef]
122. Lee, J.; Cha, M.-J.; Lim, K.S.; Kim, J.-K.; Lee, S.-K.; Kim, Y.-H.; Hwang, K.-C.; Lee, K.Y. Injectable microsphere/hydrogel hybrid system containing heat shock protein as therapy in a murine myocardial infarction model. *J. Drug Target.* **2013**, *21*, 822–829. [CrossRef]
123. Anjum, F.; Lienemann, P.S.; Metzger, S.; Biernaskie, J.; Kallos, M.S.; Ehrbar, M. Enzyme responsive GAG-based natural-synthetic hybrid hydrogel for tunable growth factor delivery and stem cell differentiation. *Biomaterials* **2016**, *87*, 104–117. [CrossRef]
124. Huang, J.; Ren, J.; Chen, G.; Li, Z.; Liu, Y.; Wang, G.; Wu, X. Tunable sequential drug delivery system based on chitosan/hyaluronic acid hydrogels and PLGA microspheres for management of non-healing infected wounds. *Mater. Sci. Eng. C* **2018**, *89*, 213–222. [CrossRef]
125. Lee, S.-H.; Moon, J.J.; West, J.L. Three-dimensional micropatterning of bioactive hydrogels via two-photon laser scanning photolithography for guided 3D cell migration. *Biomaterials* **2008**, *29*, 2962–2968. [CrossRef] [PubMed]
126. Freudenberg, U.; Zieris, A.; Chwalek, K.; Tsurkan, M.V.; Maitz, M.F.; Atallah, P.; Levental, K.R.; Eming, S.A.; Werner, C. Heparin desulfation modulates VEGF release and angiogenesis in diabetic wounds. *J. Control. Release* **2015**, *220*, 79–88. [CrossRef] [PubMed]
127. Byambaa, B.; Annabi, N.; Yue, K.; Trujillo-de Santiago, G.; Alvarez, M.M.; Jia, W.; Kazemzadeh-Narbat, M.; Shin, S.R.; Tamayol, A.; Khademhosseini, A. Bioprinted osteogenic and vasculogenic patterns for engineering 3D bone tissue. *Adv. Healthc. Mater.* **2017**, *6*, 1700015. [CrossRef] [PubMed]
128. Gurkan, U.A.; El Assal, R.; Yildiz, S.E.; Sung, Y.; Trachtenberg, A.J.; Kuo, W.P.; Demirci, U. Engineering anisotropic biomimetic fibrocartilage microenvironment by bioprinting mesenchymal stem cells in nanoliter gel droplets. *Mol. Pharm.* **2014**, *11*, 2151–2159. [CrossRef]
129. Lee, Y.-B.; Polio, S.; Lee, W.; Dai, G.; Menon, L.; Carroll, R.S.; Yoo, S.-S. Bio-printing of collagen and VEGF-releasing fibrin gel scaffolds for neural stem cell culture. *Exp. Neurol.* **2010**, *223*, 645–652. [CrossRef]
130. Poldervaart, M.T.; Gremmels, H.; van Deventer, K.; Fledderus, J.O.; Öner, F.C.; Verhaar, M.C.; Dhert, W.J.; Alblas, J. Prolonged presence of VEGF promotes vascularization in 3D bioprinted scaffolds with defined architecture. *J. Control. Release* **2014**, *184*, 58–66. [CrossRef]

131. Xiang, Y.; Mao, C.; Liu, X.; Cui, Z.; Jing, D.; Yang, X.; Liang, Y.; Li, Z.; Zhu, S.; Zheng, Y. Rapid and superior bacteria killing of carbon quantum dots/ZnO decorated injectable folic acid-conjugated PDA hydrogel through dual-light triggered ROS and membrane permeability. *Small* **2019**, *15*, 1900322. [CrossRef]
132. Si, Y.; Wang, L.; Wang, X.; Tang, N.; Yu, J.; Ding, B. Ultrahigh-water-content, superelastic, and shape-memory nanofiber-assembled hydrogels exhibiting pressure-responsive conductivity. *Adv. Mater.* **2017**, *29*, 1700339. [CrossRef]
133. Arslantunali, D.; Budak, G.; Hasirci, V. Multiwalled CNT-pHEMA composite conduit for peripheral nerve repair. *J. Biomed. Mater. Res. Part A* **2014**, *102*, 828–841. [CrossRef]
134. Liang, Y.; Zhao, X.; Hu, T.; Chen, B.; Yin, Z.; Ma, P.X.; Guo, B. Adhesive hemostatic conducting injectable composite hydrogels with sustained drug release and photothermal antibacterial activity to promote full-thickness skin regeneration during wound healing. *Small* **2019**, *15*, 1900046. [CrossRef]
135. Qu, J.; Zhao, X.; Liang, Y.; Zhang, T.; Ma, P.X.; Guo, B. Antibacterial adhesive injectable hydrogels with rapid self-healing, extensibility and compressibility as wound dressing for joints skin wound healing. *Biomaterials* **2018**, *183*, 185–199. [CrossRef]
136. Narayanan, L.K.; Huebner, P.; Fisher, M.B.; Spang, J.T.; Starly, B.; Shirwaiker, R.A. 3D-bioprinting of polylactic acid (PLA) nanofiber–alginate hydrogel bioink containing human adipose-derived stem cells. *ACS Biomater. Sci. Eng.* **2016**, *2*, 1732–1742. [CrossRef]
137. Markstedt, K.; Mantas, A.; Tournier, I.; Martínez Ávila, H.; Hagg, D.; Gatenholm, P. 3D bioprinting human chondrocytes with nanocellulose–alginate bioink for cartilage tissue engineering applications. *Biomacromolecules* **2015**, *16*, 1489–1496. [CrossRef]
138. Afewerki, S.; Sheikhi, A.; Kannan, S.; Ahadian, S.; Khademhosseini, A. Gelatin-polysaccharide composite scaffolds for 3D cell culture and tissue engineering: Towards natural therapeutics. *Bioeng. Transl. Med.* **2019**, *4*, 96–115. [CrossRef]
139. Zhang, H.J.; Sun, T.L.; Zhang, A.K.; Ikura, Y.; Nakajima, T.; Nonoyama, T.; Kurokawa, T.; Ito, O.; Ishitobi, H.; Gong, J.P. Tough physical double-network hydrogels based on amphiphilic triblock copolymers. *Adv. Mater.* **2016**, *28*, 4884–4890. [CrossRef]
140. Jang, K.-I.; Chung, H.U.; Xu, S.; Lee, C.H.; Luan, H.; Jeong, J.; Cheng, H.; Kim, G.-T.; Han, S.Y.; Lee, J.W. Soft network composite materials with deterministic and bio-inspired designs. *Nat. Commun.* **2015**, *6*, 1–11. [CrossRef]
141. Dalby, M.J.; Gadegaard, N.; Oreffo, R.O. Harnessing nanotopography and integrin–matrix interactions to influence stem cell fate. *Nat. Mater.* **2014**, *13*, 558–569. [CrossRef]
142. Chimene, D.; Alge, D.L.; Gaharwar, A.K. Two-dimensional nanomaterials for biomedical applications: Emerging trends and future prospects. *Adv. Mater.* **2015**, *27*, 7261–7284. [CrossRef]
143. Mehrali, M.; Thakur, A.; Pennisi, C.P.; Talebian, S.; Arpanaei, A.; Nikkhah, M.; Dolatshahi-Pirouz, A. Nanoreinforced hydrogels for tissue engineering: Biomaterials that are compatible with load-bearing and electroactive tissues. *Adv. Mater.* **2017**, *29*, 1603612. [CrossRef]
144. Thoniyot, P.; Tan, M.J.; Karim, A.A.; Young, D.J.; Loh, X.J. Nanoparticle–hydrogel composites: Concept, design, and applications of these promising, multi-functional materials. *Adv. Sci.* **2015**, *2*, 1400010. [CrossRef]
145. Murthy, P.K.; Mohan, Y.M.; Varaprasad, K.; Sreedhar, B.; Raju, K.M. First successful design of semi-IPN hydrogel–silver nanocomposites: A facile approach for antibacterial application. *J. Colloid Interface Sci.* **2008**, *318*, 217–224. [CrossRef]
146. Li, Y.; Huang, G.; Zhang, X.; Li, B.; Chen, Y.; Lu, T.; Lu, T.J.; Xu, F. Magnetic hydrogels and their potential biomedical applications. *Adv. Funct. Mater.* **2013**, *23*, 660–672. [CrossRef]
147. Xin, Q.; Shah, H.; Nawaz, A.; Xie, W.; Akram, M.Z.; Batool, A.; Tian, L.; Jan, S.U.; Boddula, R.; Guo, B. Antibacterial carbon-based nanomaterials. *Adv. Mater.* **2019**, *31*, 1804838. [CrossRef]
148. Zhao, X.; Guo, B.; Wu, H.; Liang, Y.; Ma, P.X. Injectable antibacterial conductive nanocomposite cryogels with rapid shape recovery for noncompressible hemorrhage and wound healing. *Nat. Commun.* **2018**, *9*, 1–17. [CrossRef]
149. Tai, Z.; Yang, J.; Qi, Y.; Yan, X.; Xue, Q. Synthesis of a graphene oxide–polyacrylic acid nanocomposite hydrogel and its swelling and electroresponsive properties. *Rsc Adv.* **2013**, *3*, 12751–12757. [CrossRef]
150. Liu, R.; Liang, S.; Tang, X.-Z.; Yan, D.; Li, X.; Yu, Z.-Z. Tough and highly stretchable graphene oxide/polyacrylamide nanocomposite hydrogels. *J. Mater. Chem.* **2012**, *22*, 14160–14167. [CrossRef]
151. Maji, S.; Agarwal, T.; Maiti, T.K. PAMAM (generation 4) incorporated gelatin 3D matrix as an improved dermal substitute for skin tissue engineering. *Colloids Surf. B Biointerfaces* **2017**, *155*, 128–134. [CrossRef]
152. Ding, Q.; Xu, X.; Yue, Y.; Mei, C.; Huang, C.; Jiang, S.; Wu, Q.; Han, J. Nanocellulose-mediated electroconductive self-healing hydrogels with high strength, plasticity, viscoelasticity, stretchability, and biocompatibility toward multifunctional applications. *ACS Appl. Mater. Interfaces* **2018**, *10*, 27987–28002. [CrossRef] [PubMed]
153. Kontturi, E.; Laaksonen, P.; Linder, M.B.; Gröschel, A.H.; Rojas, O.J.; Ikkala, O. Advanced materials through assembly of nanocelluloses. *Adv. Mater.* **2018**, *30*, 1703779. [CrossRef] [PubMed]
154. Chang, C.; Peng, J.; Zhang, L.; Pang, D.-W. Strongly fluorescent hydrogels with quantum dots embedded in cellulose matrices. *J. Mater. Chem.* **2009**, *19*, 7771–7776. [CrossRef]
155. Han, L.; Lu, X.; Liu, K.; Wang, K.; Fang, L.; Weng, L.-T.; Zhang, H.; Tang, Y.; Ren, F.; Zhao, C. Mussel-inspired adhesive and tough hydrogel based on nanoclay confined dopamine polymerization. *ACS Nano* **2017**, *11*, 2561–2574. [CrossRef] [PubMed]
156. Kim, H.H.; Park, J.B.; Kang, M.J.; Park, Y.H. Surface-modified silk hydrogel containing hydroxyapatite nanoparticle with hyaluronic acid–dopamine conjugate. *Int. J. Biol. Macromol.* **2014**, *70*, 516–522. [CrossRef] [PubMed]
157. Kuśtrowski, P.; Natkański, P.; Rokicińska, A.; Witek, E. Polymer hydrogel-clay (Nano) composites. In *Polymer Gels*; Springer: Singapore, 2018; pp. 1–62.

158. Timofejeva, A.; D'Este, M.; Loca, D. Calcium phosphate/polyvinyl alcohol composite hydrogels: A review on the freeze-thawing synthesis approach and applications in regenerative medicine. *Eur. Polym. J.* **2017**, *95*, 547–565. [CrossRef]
159. Grillo, H.C.; McKhann, C.F. The acceptance and evolution of dermal homografts freed of viable cells. *Transplantation* **1964**, *2*, 48–59. [CrossRef]
160. Poel, W.E. Preparation of acellular homogenates from muscle samples. *Science* **1948**, *108*, 390–391. [CrossRef]
161. Mazza, G.; Rombouts, K.; Hall, A.R.; Urbani, L.; Luong, T.V.; Al-Akkad, W.; Longato, L.; Brown, D.; Maghsoudlou, P.; Dhillon, A.P. Decellularized human liver as a natural 3D-scaffold for liver bioengineering and transplantation. *Sci. Rep.* **2015**, *5*, 1–15. [CrossRef]
162. Sánchez, P.L.; Fernández-Santos, M.E.; Espinosa, M.A.; González-Nicolas, M.A.; Acebes, J.R.; Costanza, S.; Moscoso, I.; Rodríguez, H.; García, J.; Romero, J. Data from acellular human heart matrix. *Data Brief* **2016**, *8*, 211–219. [CrossRef]
163. Dorrello, N.V.; Guenthart, B.A.; O'Neill, J.D.; Kim, J.; Cunningham, K.; Chen, Y.-W.; Biscotti, M.; Swayne, T.; Wobma, H.M.; Huang, S.X. Functional vascularized lung grafts for lung bioengineering. *Sci. Adv.* **2017**, *3*, e1700521. [CrossRef]
164. Song, J.J.; Guyette, J.P.; Gilpin, S.E.; Gonzalez, G.; Vacanti, J.P.; Ott, H.C. Regeneration and experimental orthotopic transplantation of a bioengineered kidney. *Nat. Med.* **2013**, *19*, 646–651. [CrossRef]
165. Orlando, G.; Booth, C.; Wang, Z.; Totonelli, G.; Ross, C.L.; Moran, E.; Salvatori, M.; Maghsoudlou, P.; Turmaine, M.; Delario, G. Discarded human kidneys as a source of ECM scaffold for kidney regeneration technologies. *Biomaterials* **2013**, *34*, 5915–5925. [CrossRef]
166. Lee, H.; Han, W.; Kim, H.; Ha, D.-H.; Jang, J.; Kim, B.S.; Cho, D.-W. Development of liver decellularized extracellular matrix bioink for three-dimensional cell printing-based liver tissue engineering. *Biomacromolecules* **2017**, *18*, 1229–1237. [CrossRef]
167. Harris, G.M.; Raitman, I.; Schwarzbauer, J.E. Cell-derived decellularized extracellular matrices. *Methods Cell Biol.* **2018**, *143*, 97–114.
168. Sundarakrishnan, A.; Chen, Y.; Black, L.D.; Aldridge, B.B.; Kaplan, D.L. Engineered cell and tissue models of pulmonary fibrosis. *Adv. Drug Deliv. Rev.* **2018**, *129*, 78–94. [CrossRef]
169. Jiang, K.; Chaimov, D.; Patel, S.; Liang, J.-P.; Wiggins, S.; Samojlik, M.; Rubiano, A.; Simmons, C.; Stabler, C. 3-D physiomimetic extracellular matrix hydrogels provide a supportive microenvironment for rodent and human islet culture. *Biomaterials* **2019**, *198*, 37–48. [CrossRef]
170. French, K.M.; Boopathy, A.V.; DeQuach, J.A.; Chingozha, L.; Lu, H.; Christman, K.L.; Davis, M.E. A naturally derived cardiac extracellular matrix enhances cardiac progenitor cell behavior in vitro. *Acta Biomater.* **2012**, *8*, 4357–4364. [CrossRef]
171. Bejleri, D.; Streeter, B.W.; Nachlas, A.L.; Brown, M.E.; Gaetani, R.; Christman, K.L.; Davis, M.E. A bioprinted cardiac patch composed of cardiac-specific extracellular matrix and progenitor cells for heart repair. *Adv. Healthc. Mater.* **2018**, *7*, 1800672. [CrossRef]
172. Gaetani, R.; Yin, C.; Srikumar, N.; Braden, R.; Doevendans, P.A.; Sluijter, J.P.; Christman, K.L. Cardiac-derived extracellular matrix enhances cardiogenic properties of human cardiac progenitor cells. *Cell Transplant.* **2016**, *25*, 1653–1663. [CrossRef]
173. Traverse, J.H.; Henry, T.D.; Dib, N.; Patel, A.N.; Pepine, C.; Schaer, G.L.; DeQuach, J.A.; Kinsey, A.M.; Chamberlin, P.; Christman, K.L. First-in-man study of a cardiac extracellular matrix hydrogel in early and late myocardial infarction patients. *JACC Basic Transl. Sci.* **2019**, *4*, 659–669. [CrossRef]
174. Kort-Mascort, J.; Bao, G.; Elkashty, O.; Flores-Torres, S.; Munguia-Lopez, J.G.; Jiang, T.; Ehrlicher, A.J.; Mongeau, L.; Tran, S.D.; Kinsella, J.M. Decellularized Extracellular Matrix Composite Hydrogel Bioinks for the Development of 3D Bioprinted Head and Neck in Vitro Tumor Models. *ACS Biomater. Sci. Eng.* **2021**, *7*, 5288–5300. [CrossRef]
175. Benton, G.; Arnaoutova, I.; George, J.; Kleinman, H.K.; Koblinski, J. Matrigel: From discovery and ECM mimicry to assays and models for cancer research. *Adv. Drug Deliv. Rev.* **2014**, *79*, 3–18. [CrossRef]
176. Hughes, C.S.; Postovit, L.M.; Lajoie, G.A. Matrigel: A complex protein mixture required for optimal growth of cell culture. *Proteomics* **2010**, *10*, 1886–1890. [CrossRef] [PubMed]
177. Blondel, D.; Lutolf, M.P. Bioinspired hydrogels for 3D organoid culture. *CHIMIA Int. J. Chem.* **2019**, *73*, 81–85. [CrossRef]
178. DiMarco, R.L.; Heilshorn, S.C. Multifunctional materials through modular protein engineering. *Adv. Mater.* **2012**, *24*, 3923–3940. [CrossRef]
179. DiMarco, R.L.; Dewi, R.E.; Bernal, G.; Kuo, C.; Heilshorn, S.C. Protein-engineered scaffolds for in vitro 3D culture of primary adult intestinal organoids. *Biomater. Sci.* **2015**, *3*, 1376–1385. [CrossRef]
180. Madl, C.M.; LeSavage, B.L.; Dewi, R.E.; Dinh, C.B.; Stowers, R.S.; Khariton, M.; Lampe, K.J.; Nguyen, D.; Chaudhuri, O.; Enejder, A. Maintenance of neural progenitor cell stemness in 3D hydrogels requires matrix remodelling. *Nat. Mater.* **2017**, *16*, 1233–1242. [CrossRef] [PubMed]
181. Tandon, N.; Cannizzaro, C.; Chao, P.-H.G.; Maidhof, R.; Marsano, A.; Au, H.T.H.; Radisic, M.; Vunjak-Novakovic, G. Electrical stimulation systems for cardiac tissue engineering. *Nat. Protoc.* **2009**, *4*, 155–173. [CrossRef] [PubMed]
182. Guiseppi-Elie, A. Electroconductive hydrogels: Synthesis, characterization and biomedical applications. *Biomaterials* **2010**, *31*, 2701–2716. [CrossRef]
183. Li, L.; Wang, Y.; Pan, L.; Shi, Y.; Cheng, W.; Shi, Y.; Yu, G. A nanostructured conductive hydrogels-based biosensor platform for human metabolite detection. *Nano Lett.* **2015**, *15*, 1146–1151. [CrossRef]

184. Shin, S.R.; Shin, C.; Memic, A.; Shadmehr, S.; Miscuglio, M.; Jung, H.Y.; Jung, S.M.; Bae, H.; Khademhosseini, A.; Tang, X. Aligned carbon nanotube–based flexible gel substrates for engineering biohybrid tissue actuators. *Adv. Funct. Mater.* **2015**, *25*, 4486–4495. [CrossRef]
185. Dong, R.; Zhao, X.; Guo, B.; Ma, P.X. Self-healing conductive injectable hydrogels with antibacterial activity as cell delivery carrier for cardiac cell therapy. *ACS Appl. Mater. Interfaces* **2016**, *8*, 17138–17150. [CrossRef]
186. Zhang, J.; Li, M.; Kang, E.-T.; Neoh, K.G. Electrical stimulation of adipose-derived mesenchymal stem cells in conductive scaffolds and the roles of voltage-gated ion channels. *Acta Biomater.* **2016**, *32*, 46–56. [CrossRef] [PubMed]
187. Zhang, H.-W.; Hu, X.-B.; Qin, Y.; Jin, Z.-H.; Zhang, X.-W.; Liu, Y.-L.; Huang, W.-H. Conductive polymer coated scaffold to integrate 3D cell culture with electrochemical sensing. *Anal. Chem.* **2019**, *91*, 4838–4844. [CrossRef] [PubMed]
188. Hoare, T.R.; Kohane, D.S. Hydrogels in drug delivery: Progress and challenges. *Polymer* **2008**, *49*, 1993–2007. [CrossRef]
189. Ebara, M.; Kotsuchibashi, Y.; Narain, R.; Idota, N.; Kim, Y.-J.; Hoffman, J.M.; Uto, K.; Aoyagi, T. *Smart Biomaterials*; Springer: Tokyo, Japan, 2014.
190. Czarnecki, S.; Rossow, T.; Seiffert, S. Hybrid polymer-network hydrogels with tunable mechanical response. *Polymers* **2016**, *8*, 82. [CrossRef]
191. Amini, A.A.; Nair, L.S. Injectable hydrogels for bone and cartilage repair. *Biomed. Mater.* **2012**, *7*, 024105. [CrossRef]
192. Augst, A.D.; Kong, H.J.; Mooney, D.J. Alginate hydrogels as biomaterials. *Macromol. Biosci.* **2006**, *6*, 623–633. [CrossRef]
193. Jin, R. In-situ forming biomimetic hydrogels for tissue regeneration. *Biomedicine* **2012**, *2*, 35–58.
194. Gulrez, S.K.; Al-Assaf, S.; Phillips, G.O. Hydrogels: Methods of preparation, characterisation and applications. In *Progress in Molecular and Environmental Bioengineering*; IntechOpen: London, UK, 2011; pp. 117–150.
195. Lin, P.; Ma, S.; Wang, X.; Zhou, F. Molecularly engineered dual-crosslinked hydrogel with ultrahigh mechanical strength, toughness, and good self-recovery. *Adv. Mater.* **2015**, *27*, 2054–2059. [CrossRef]
196. Narita, T.; Mayumi, K.; Ducouret, G.; Hebraud, P. Viscoelastic properties of poly (vinyl alcohol) hydrogels having permanent and transient cross-links studied by microrheology, classical rheometry, and dynamic light scattering. *Macromolecules* **2013**, *46*, 4174–4183. [CrossRef]
197. Kondo, S.; Hiroi, T.; Han, Y.S.; Kim, T.H.; Shibayama, M.; Chung, U.I.; Sakai, T. Reliable hydrogel with mechanical "fuse link" in an aqueous environment. *Adv. Mater.* **2015**, *27*, 7407–7411. [CrossRef]
198. Vasile, C.; Pamfil, D.; Stoleru, E.; Baican, M. New developments in medical applications of hybrid hydrogels containing natural polymers. *Molecules* **2020**, *25*, 1539. [CrossRef]
199. Lim, S.; Jung, G.A.; Muckom, R.J.; Glover, D.J.; Clark, D.S. Engineering bioorthogonal protein–polymer hybrid hydrogel as a functional protein immobilization platform. *Chem. Commun.* **2019**, *55*, 806–809. [CrossRef]
200. Sanson, N.; Rieger, J. Synthesis of nanogels/microgels by conventional and controlled radical crosslinking copolymerization. *Polym. Chem.* **2010**, *1*, 965–977. [CrossRef]
201. DeForest, C.A.; Anseth, K.S. Cytocompatible click-based hydrogels with dynamically tunable properties through orthogonal photoconjugation and photocleavage reactions. *Nat. Chem.* **2011**, *3*, 925–931. [CrossRef]
202. Koshy, S.T.; Desai, R.M.; Joly, P.; Li, J.; Bagrodia, R.K.; Lewin, S.A.; Joshi, N.S.; Mooney, D.J. Click-Crosslinked Injectable Gelatin Hydrogels. *Adv. Healthc. Mater.* **2016**, *5*, 541–547. [CrossRef]
203. Baldwin, A.D.; Kiick, K.L. Tunable degradation of maleimide–thiol adducts in reducing environments. *Bioconjug. Chem.* **2011**, *22*, 1946–1953. [CrossRef]
204. Crescenzi, V.; Cornelio, L.; Di Meo, C.; Nardecchia, S.; Lamanna, R. Novel hydrogels via click chemistry: Synthesis and potential biomedical applications. *Biomacromolecules* **2007**, *8*, 1844–1850. [CrossRef]
205. Zhu, W.; Xiong, L.; Wang, H.; Zha, G.; Du, H.; Li, X.; Shen, Z. Sustained drug release from an ultrathin hydrogel film. *Polym. Chem.* **2015**, *6*, 7097–7099. [CrossRef]
206. Fan, Y.; Deng, C.; Cheng, R.; Meng, F.; Zhong, Z. In situ forming hydrogels via catalyst-free and bioorthogonal "tetrazole–alkene" photo-click chemistry. *Biomacromolecules* **2013**, *14*, 2814–2821. [CrossRef]
207. Mukherjee, S.; Hill, M.R.; Sumerlin, B.S. Self-healing hydrogels containing reversible oxime crosslinks. *Soft Matter* **2015**, *11*, 6152–6161. [CrossRef]
208. Seib, F.P.; Prewitz, M.; Werner, C.; Bornhäuser, M. Matrix elasticity regulates the secretory profile of human bone marrow-derived multipotent mesenchymal stromal cells (MSCs). *Biochem. Biophys. Res. Commun.* **2009**, *389*, 663–667. [CrossRef] [PubMed]
209. Rozario, T.; DeSimone, D.W. The extracellular matrix in development and morphogenesis: A dynamic view. *Dev. Biol.* **2010**, *341*, 126–140. [CrossRef]
210. Xu, Y.; Zhu, X.; Hahm, H.S.; Wei, W.; Hao, E.; Hayek, A.; Ding, S. Revealing a core signaling regulatory mechanism for pluripotent stem cell survival and self-renewal by small molecules. *Proc. Natl. Acad. Sci. USA* **2010**, *107*, 8129–8134. [CrossRef] [PubMed]
211. Broguiere, N.; Isenmann, L.; Hirt, C.; Ringel, T.; Placzek, S.; Cavalli, E.; Ringnalda, F.; Villiger, L.; Züllig, R.; Lehmann, R. Growth of epithelial organoids in a defined hydrogel. *Adv. Mater.* **2018**, *30*, 1801621. [CrossRef] [PubMed]
212. Hutson, C.B.; Nichol, J.W.; Aubin, H.; Bae, H.; Yamanlar, S.; Al-Haque, S.; Koshy, S.T.; Khademhosseini, A. Synthesis and characterization of tunable poly (ethylene glycol): Gelatin methacrylate composite hydrogels. *Tissue Eng. Part A* **2011**, *17*, 1713–1723. [CrossRef] [PubMed]
213. Ranga, A.; Girgin, M.; Meinhardt, A.; Eberle, D.; Caiazzo, M.; Tanaka, E.M.; Lutolf, M.P. Neural tube morphogenesis in synthetic 3D microenvironments. *Proc. Natl. Acad. Sci. USA* **2016**, *113*, E6831–E6839. [CrossRef] [PubMed]

214. Schepers, A.; Li, C.; Chhabra, A.; Seney, B.T.; Bhatia, S. Engineering a perfusable 3D human liver platform from iPS cells. *Lab Chip* **2016**, *16*, 2644–2653. [CrossRef]
215. Hoang, P.; Wang, J.; Conklin, B.R.; Healy, K.E.; Ma, Z. Generation of spatial-patterned early-developing cardiac organoids using human pluripotent stem cells. *Nat. Protoc.* **2018**, *13*, 723–737. [CrossRef]
216. Shin, H.-S.; Hong, H.J.; Koh, W.-G.; Lim, J.-Y. Organotypic 3D culture in nanoscaffold microwells supports salivary gland stem-cell-based organization. *ACS Biomater. Sci. Eng.* **2018**, *4*, 4311–4320. [CrossRef]
217. Brusatin, G.; Panciera, T.; Gandin, A.; Citron, A.; Piccolo, S. Biomaterials and engineered microenvironments to control YAP/TAZ-dependent cell behaviour. *Nat. Mater.* **2018**, *17*, 1063–1075. [CrossRef]
218. Lv, H.; Li, L.; Sun, M.; Zhang, Y.; Chen, L.; Rong, Y.; Li, Y. Mechanism of regulation of stem cell differentiation by matrix stiffness. *Stem Cell Res. Ther.* **2015**, *6*, 1–11. [CrossRef]
219. Wu, S.; Xu, R.; Duan, B.; Jiang, P. Three-dimensional hyaluronic acid hydrogel-based models for in vitro human iPSC-derived NPC culture and differentiation. *J. Mater. Chem. B* **2017**, *5*, 3870–3878. [CrossRef]
220. Shkumatov, A.; Baek, K.; Kong, H. Matrix rigidity-modulated cardiovascular organoid formation from embryoid bodies. *PLoS ONE* **2014**, *9*, e94764. [CrossRef]
221. Khetan, S.; Burdick, J. Cellular encapsulation in 3D hydrogels for tissue engineering. *J. Vis. Exp.* **2009**, *32*, e1590. [CrossRef] [PubMed]
222. Huebsch, N.; Arany, P.R.; Mao, A.S.; Shvartsman, D.; Ali, O.A.; Bencherif, S.A.; Rivera-Feliciano, J.; Mooney, D.J. Harnessing traction-mediated manipulation of the cell/matrix interface to control stem-cell fate. *Nat. Mater.* **2010**, *9*, 518–526. [CrossRef]
223. Khetan, S.; Guvendiren, M.; Legant, W.R.; Cohen, D.M.; Chen, C.S.; Burdick, J.A. Degradation-mediated cellular traction directs stem cell fate in covalently crosslinked three-dimensional hydrogels. *Nat. Mater.* **2013**, *12*, 458–465. [CrossRef]
224. Sahoo, S.; Chung, C.; Khetan, S.; Burdick, J.A. Hydrolytically degradable hyaluronic acid hydrogels with controlled temporal structures. *Biomacromolecules* **2008**, *9*, 1088–1092. [CrossRef]
225. Huebsch, N.; Lippens, E.; Lee, K.; Mehta, M.; Koshy, S.T.; Darnell, M.C.; Desai, R.M.; Madl, C.M.; Xu, M.; Zhao, X. Matrix elasticity of void-forming hydrogels controls transplanted-stem-cell-mediated bone formation. *Nat. Mater.* **2015**, *14*, 1269–1277. [CrossRef]
226. Choi, A.; Seo, K.D.; Kim, B.C.; Kim, D.S. Recent advances in engineering microparticles and their nascent utilization in biomedical delivery and diagnostic applications. *Lab Chip* **2017**, *17*, 591–613. [CrossRef]
227. Li, W.; Zhang, L.; Ge, X.; Xu, B.; Zhang, W.; Qu, L.; Choi, C.-H.; Xu, J.; Zhang, A.; Lee, H. Microfluidic fabrication of microparticles for biomedical applications. *Chem. Soc. Rev.* **2018**, *47*, 5646–5683. [CrossRef]
228. Jiang, W.; Li, M.; Chen, Z.; Leong, K.W. Cell-laden microfluidic microgels for tissue regeneration. *Lab Chip* **2016**, *16*, 4482–4506. [CrossRef]
229. Theberge, A.B.; Courtois, F.; Schaerli, Y.; Fischlechner, M.; Abell, C.; Hollfelder, F.; Huck, W.T. Microdroplets in microfluidics: An evolving platform for discoveries in chemistry and biology. *Angew. Chem. Int. Ed.* **2010**, *49*, 5846–5868. [CrossRef]
230. Shang, L.; Cheng, Y.; Zhao, Y. Emerging droplet microfluidics. *Chem. Rev.* **2017**, *117*, 7964–8040. [CrossRef]
231. Jang, M.; Yang, S.; Kim, P. Microdroplet-based cell culture models and their application. *BioChip J.* **2016**, *10*, 310–317. [CrossRef]
232. Sabhachandani, P.; Motwani, V.; Cohen, N.; Sarkar, S.; Torchilin, V.; Konry, T. Generation and functional assessment of 3D multicellular spheroids in droplet based microfluidics platform. *Lab Chip* **2016**, *16*, 497–505. [CrossRef]
233. Williams, C.M.; Mehta, G.; Peyton, S.R.; Zeiger, A.S.; Van Vliet, K.J.; Griffith, L.G. Autocrine-controlled formation and function of tissue-like aggregates by primary hepatocytes in micropatterned hydrogel arrays. *Tissue Eng. Part A* **2011**, *17*, 1055–1068. [CrossRef]
234. Lee, S.-A.; Kang, E.; Ju, J.; Kim, D.-S.; Lee, S.-H. Spheroid-based three-dimensional liver-on-a-chip to investigate hepatocyte–hepatic stellate cell interactions and flow effects. *Lab Chip* **2013**, *13*, 3529–3537. [CrossRef]
235. Fang, J.; Zhang, Y.; Yan, S.; Liu, Z.; He, S.; Cui, L.; Yin, J. Poly (L-glutamic acid)/chitosan polyelectrolyte complex porous microspheres as cell microcarriers for cartilage regeneration. *Acta Biomater.* **2014**, *10*, 276–288. [CrossRef]
236. Yu, C.; Kornmuller, A.; Brown, C.; Hoare, T.; Flynn, L.E. Decellularized adipose tissue microcarriers as a dynamic culture platform for human adipose-derived stem/stromal cell expansion. *Biomaterials* **2017**, *120*, 66–80. [CrossRef]
237. Huang, L.; Abdalla, A.M.; Xiao, L.; Yang, G. Biopolymer-based microcarriers for three-dimensional cell culture and engineered tissue formation. *Int. J. Mol. Sci.* **2020**, *21*, 1895. [CrossRef]
238. Lengyel, M.; Kállai-Szabó, N.; Antal, V.; Laki, A.J.; Antal, I. Microparticles, microspheres, and microcapsules for advanced drug delivery. *Sci. Pharm.* **2019**, *87*, 20. [CrossRef]
239. Cheng, J.; Jun, Y.; Qin, J.; Lee, S.-H. Electrospinning versus microfluidic spinning of functional fibers for biomedical applications. *Biomaterials* **2017**, *114*, 121–143. [CrossRef]
240. Pedde, R.D.; Mirani, B.; Navaei, A.; Styan, T.; Wong, S.; Mehrali, M.; Thakur, A.; Mohtaram, N.K.; Bayati, A.; Dolatshahi-Pirouz, A. Emerging biofabrication strategies for engineering complex tissue constructs. *Adv. Mater.* **2017**, *29*, 1606061. [CrossRef]
241. Costantini, M.; Colosi, C.; Święszkowski, W.; Barbetta, A. Co-axial wet-spinning in 3D bioprinting: State of the art and future perspective of microfluidic integration. *Biofabrication* **2018**, *11*, 012001. [CrossRef]
242. Daniele, M.A.; Boyd, D.A.; Adams, A.A.; Ligler, F.S. Microfluidic strategies for design and assembly of microfibers and nanofibers with tissue engineering and regenerative medicine applications. *Adv. Healthc. Mater.* **2015**, *4*, 11–28. [CrossRef]
243. Jun, Y.; Kang, E.; Chae, S.; Lee, S.-H. Microfluidic spinning of micro-and nano-scale fibers for tissue engineering. *Lab Chip* **2014**, *14*, 2145–2160. [CrossRef]

244. Kang, E.; Jeong, G.S.; Choi, Y.Y.; Lee, K.H.; Khademhosseini, A.; Lee, S.-H. Digitally tunable physicochemical coding of material composition and topography in continuous microfibres. *Nat. Mater.* **2011**, *10*, 877–883. [CrossRef]
245. Jun, Y.; Kang, A.R.; Lee, J.S.; Park, S.-J.; Lee, D.Y.; Moon, S.-H.; Lee, S.-H. Microchip-based engineering of super-pancreatic islets supported by adipose-derived stem cells. *Biomaterials* **2014**, *35*, 4815–4826. [CrossRef]
246. Kobayashi, A.; Yamakoshi, K.; Yajima, Y.; Utoh, R.; Yamada, M.; Seki, M. Preparation of stripe-patterned heterogeneous hydrogel sheets using microfluidic devices for high-density coculture of hepatocytes and fibroblasts. *J. Biosci. Bioeng.* **2013**, *116*, 761–767. [CrossRef]
247. Yamada, M.; Utoh, R.; Ohashi, K.; Tatsumi, K.; Yamato, M.; Okano, T.; Seki, M. Controlled formation of heterotypic hepatic micro-organoids in anisotropic hydrogel microfibers for long-term preservation of liver-specific functions. *Biomaterials* **2012**, *33*, 8304–8315. [CrossRef]
248. Bosworth, L.A.; Turner, L.-A.; Cartmell, S.H. State of the art composites comprising electrospun fibres coupled with hydrogels: A review. *Nanomed. Nanotechnol. Biol. Med.* **2013**, *9*, 322–335. [CrossRef]
249. Wade, R.J.; Burdick, J.A. Advances in nanofibrous scaffolds for biomedical applications: From electrospinning to self-assembly. *Nano Today* **2014**, *9*, 722–742. [CrossRef]
250. Honkamäki, L.; Joki, T.; Grigoryev, N.A.; Levon, K.; Ylä-Outinen, L.; Narkilahti, S. Novel method to produce a layered 3D scaffold for human pluripotent stem cell-derived neuronal cells. *J. Neurosci. Methods* **2021**, *350*, 109043. [CrossRef]
251. Eom, S.; Park, S.M.; Hong, H.; Kwon, J.; Oh, S.-R.; Kim, J.; Kim, D.S. Hydrogel-assisted electrospinning for fabrication of a 3D complex tailored nanofiber macrostructure. *ACS Appl. Mater. Interfaces* **2020**, *12*, 51212–51224. [CrossRef]
252. Lee, P.J.; Hung, P.J.; Lee, L.P. An artificial liver sinusoid with a microfluidic endothelial-like barrier for primary hepatocyte culture. *Biotechnol. Bioeng.* **2007**, *97*, 1340–1346. [CrossRef]
253. Deli, M.A.; Ábrahám, C.S.; Kataoka, Y.; Niwa, M. Permeability studies on in vitro blood–brain barrier models: Physiology, pathology, and pharmacology. *Cell. Mol. Neurobiol.* **2005**, *25*, 59–127. [CrossRef]
254. Naik, P.; Cucullo, L. In vitro blood–brain barrier models: Current and perspective technologies. *J. Pharm. Sci.* **2012**, *101*, 1337–1354. [CrossRef]
255. Nakagawa, S.; Deli, M.A.; Kawaguchi, H.; Shimizudani, T.; Shimono, T.; Kittel, A.; Tanaka, K.; Niwa, M. A new blood–brain barrier model using primary rat brain endothelial cells, pericytes and astrocytes. *Neurochem. Int.* **2009**, *54*, 253–263. [CrossRef]
256. Kwak, B.S.; Choi, W.; Jeon, J.-W.; Won, J.-I.; Sung, G.Y.; Kim, B.; Sung, J.H. In vitro 3D skin model using gelatin methacrylate hydrogel. *J. Ind. Eng. Chem.* **2018**, *66*, 254–261. [CrossRef]
257. Kang, Y.B.A.; Rawat, S.; Cirillo, J.; Bouchard, M.; Noh, H.M. Layered long-term co-culture of hepatocytes and endothelial cells on a transwell membrane: Toward engineering the liver sinusoid. *Biofabrication* **2013**, *5*, 045008. [CrossRef]
258. Shin, Y.; Han, S.; Jeon, J.S.; Yamamoto, K.; Zervantonakis, I.K.; Sudo, R.; Kamm, R.D.; Chung, S. Microfluidic assay for simultaneous culture of multiple cell types on surfaces or within hydrogels. *Nat. Protoc.* **2012**, *7*, 1247–1259. [CrossRef]
259. Wan, X.; Ball, S.; Willenbrock, F.; Yeh, S.; Vlahov, N.; Koennig, D.; Green, M.; Brown, G.; Jeyaretna, S.; Li, Z. Perfused three-dimensional organotypic culture of human cancer cells for therapeutic evaluation. *Sci. Rep.* **2017**, *7*, 1–13. [CrossRef]
260. Esch, E.W.; Bahinski, A.; Huh, D. Organs-on-chips at the frontiers of drug discovery. *Nat. Rev. Drug Discov.* **2015**, *14*, 248–260. [CrossRef]
261. Ai, Y.; Zhang, F.; Wang, C.; Xie, R.; Liang, Q. Recent progress in lab-on-a-chip for pharmaceutical analysis and pharmacological/toxicological test. *TrAC Trends Anal. Chem.* **2019**, *117*, 215–230. [CrossRef]
262. Lee, J.; Kim, S.H.; Kim, Y.-C.; Choi, I.; Sung, J.H. Fabrication and characterization of microfluidic liver-on-a-chip using microsomal enzymes. *Enzym. Microb. Technol.* **2013**, *53*, 159–164. [CrossRef]
263. Wang, L.; Tao, T.; Su, W.; Yu, H.; Yu, Y.; Qin, J. A disease model of diabetic nephropathy in a glomerulus-on-a-chip microdevice. *Lab Chip* **2017**, *17*, 1749–1760. [CrossRef]
264. Zhang, M.; Xu, C.; Jiang, L.; Qin, J. A 3D human lung-on-a-chip model for nanotoxicity testing. *Toxicol. Res.* **2018**, *7*, 1048–1060. [CrossRef]
265. Wang, C.; Tanataweethum, N.; Karnik, S.; Bhushan, A. Novel microfluidic colon with an extracellular matrix membrane. *ACS Biomater. Sci. Eng.* **2018**, *4*, 1377–1385. [CrossRef]
266. Borenstein, J.T.; Terai, H.; King, K.R.; Weinberg, E.; Kaazempur-Mofrad, M.; Vacanti, J. Microfabrication technology for vascularized tissue engineering. *Biomed. Microdevices* **2002**, *4*, 167–175. [CrossRef]
267. Griffith, L.G.; Naughton, G. Tissue engineering—Current challenges and expanding opportunities. *Science* **2002**, *295*, 1009–1014. [CrossRef]
268. Van Midwoud, P.M.; Merema, M.T.; Verpoorte, E.; Groothuis, G.M. A microfluidic approach for in vitro assessment of interorgan interactions in drug metabolism using intestinal and liver slices. *Lab Chip* **2010**, *10*, 2778–2786. [CrossRef]
269. Maschmeyer, I.; Lorenz, A.K.; Schimek, K.; Hasenberg, T.; Ramme, A.P.; Hübner, J.; Lindner, M.; Drewell, C.; Bauer, S.; Thomas, A. A four-organ-chip for interconnected long-term co-culture of human intestine, liver, skin and kidney equivalents. *Lab Chip* **2015**, *15*, 2688–2699. [CrossRef]
270. Lai, B.F.L.; Huyer, L.D.; Lu, R.X.Z.; Drecun, S.; Radisic, M.; Zhang, B. InVADE: Integrated vasculature for assessing dynamic events. *Adv. Funct. Mater.* **2017**, *27*, 1703524. [CrossRef]
271. Ma, C.; Peng, Y.; Li, H.; Chen, W. Organ-on-a-chip: A new paradigm for drug development. *Trends Pharmacol. Sci.* **2021**, *42*, 119–133. [CrossRef]

272. Zhang, B.; Korolj, A.; Lai, B.F.L.; Radisic, M. Advances in organ-on-a-chip engineering. *Nat. Rev. Mater.* **2018**, *3*, 257–278. [CrossRef]
273. Kuddannaya, S.; Bao, J.; Zhang, Y. Enhanced in vitro biocompatibility of chemically modified poly (dimethylsiloxane) surfaces for stable adhesion and long-term investigation of brain cerebral cortex cells. *ACS Appl. Mater. Interfaces* **2015**, *7*, 25529–25538. [CrossRef]
274. Aleman, J.; Zhang, Y.; Skardal, A.; Khademhosseini, A. Lego-inspired organ-on-a-chip gelatin methacryloyl microfluidic system. In Proceedings of the 10th World Biomaterials Congress, Montreal, QB, Canada, 17–22 May 2016. [CrossRef]
275. Zhao, S.; Chen, Y.; Partlow, B.P.; Golding, A.S.; Tseng, P.; Coburn, J.; Applegate, M.B.; Moreau, J.E.; Omenetto, F.G.; Kaplan, D.L. Bio-functionalized silk hydrogel microfluidic systems. *Biomaterials* **2016**, *93*, 60–70. [CrossRef]
276. Agarwal, A.; Farouz, Y.; Nesmith, A.P.; Deravi, L.F.; McCain, M.L.; Parker, K.K. Micropatterning alginate substrates for in vitro cardiovascular muscle on a chip. *Adv. Funct. Mater.* **2013**, *23*, 3738–3746. [CrossRef]
277. Wang, Y.; Gunasekara, D.B.; Reed, M.I.; DiSalvo, M.; Bultman, S.J.; Sims, C.E.; Magness, S.T.; Allbritton, N.L. A microengineered collagen scaffold for generating a polarized crypt-villus architecture of human small intestinal epithelium. *Biomaterials* **2017**, *128*, 44–55. [CrossRef]
278. Fan, Y.; Nguyen, D.T.; Akay, Y.; Xu, F.; Akay, M. Engineering a brain cancer chip for high-throughput drug screening. *Sci. Rep.* **2016**, *6*, 1–12. [CrossRef]
279. Huval, R.M.; Miller, O.H.; Curley, J.L.; Fan, Y.; Hall, B.J.; Moore, M.J. Microengineered peripheral nerve-on-a-chip for preclinical physiological testing. *Lab Chip* **2015**, *15*, 2221–2232. [CrossRef]
280. Johnson, B.N.; Lancaster, K.Z.; Hogue, I.B.; Meng, F.; Kong, Y.L.; Enquist, L.W.; McAlpine, M.C. Correction: 3D printed nervous system on a chip. *Lab Chip* **2016**, *16*, 1946. [CrossRef]
281. Skardal, A.; Murphy, S.V.; Devarasetty, M.; Mead, I.; Kang, H.-W.; Seol, Y.-J.; Zhang, Y.S.; Shin, S.-R.; Zhao, L.; Aleman, J. Multi-tissue interactions in an integrated three-tissue organ-on-a-chip platform. *Sci. Rep.* **2017**, *7*, 1–16. [CrossRef]
282. Zhang, Y.S.; Arneri, A.; Bersini, S.; Shin, S.-R.; Zhu, K.; Goli-Malekabadi, Z.; Aleman, J.; Colosi, C.; Busignani, F.; Dell'Erba, V. Bioprinting 3D microfibrous scaffolds for engineering endothelialized myocardium and heart-on-a-chip. *Biomaterials* **2016**, *110*, 45–59. [CrossRef]
283. Guvendiren, M.; Molde, J.; Soares, R.M.; Kohn, J. Designing biomaterials for 3D printing. *ACS Biomater. Sci. Eng.* **2016**, *2*, 1679–1693. [CrossRef]
284. Choi, Y.-J.; Park, H.; Ha, D.-H.; Yun, H.-S.; Yi, H.-G.; Lee, H. 3D bioprinting of in vitro models using hydrogel-based bioinks. *Polymers* **2021**, *13*, 366. [CrossRef]
285. Yi, H.-G.; Lee, H.; Cho, D.-W. 3D printing of organs-on-chips. *Bioengineering* **2017**, *4*, 10. [CrossRef]
286. Gungor-Ozkerim, P.S.; Inci, I.; Zhang, Y.S.; Khademhosseini, A.; Dokmeci, M.R. Bioinks for 3D bioprinting: An overview. *Biomater. Sci.* **2018**, *6*, 915–946. [CrossRef]
287. Foyt, D.A.; Norman, M.D.; Yu, T.T.; Gentleman, E. Exploiting advanced hydrogel technologies to address key challenges in regenerative medicine. *Adv. Healthc. Mater.* **2018**, *7*, 1700939. [CrossRef]
288. Moroni, L.; Burdick, J.A.; Highley, C.; Lee, S.J.; Morimoto, Y.; Takeuchi, S.; Yoo, J.J. Biofabrication strategies for 3D in vitro models and regenerative medicine. *Nat. Rev. Mater.* **2018**, *3*, 21–37. [CrossRef]
289. Jang, J.; Park, J.Y.; Gao, G.; Cho, D.-W. Biomaterials-based 3D cell printing for next-generation therapeutics and diagnostics. *Biomaterials* **2018**, *156*, 88–106. [CrossRef]
290. Lee, H.; Cho, D.-W. One-step fabrication of an organ-on-a-chip with spatial heterogeneity using a 3D bioprinting technology. *Lab Chip* **2016**, *16*, 2618–2625. [CrossRef]
291. Ji, S.; Guvendiren, M. Recent advances in bioink design for 3D bioprinting of tissues and organs. *Front. Bioeng. Biotechnol.* **2017**, *5*, 23. [CrossRef]
292. Hu, J.L.; Todhunter, M.E.; LaBarge, M.A.; Gartner, Z.J. Opportunities for organoids as new models of aging. *J. Cell Biol.* **2018**, *217*, 39–50. [CrossRef]
293. Clevers, H. Modeling development and disease with organoids. *Cell* **2016**, *165*, 1586–1597. [CrossRef]
294. Leong, D.T.; Ng, K.W. Probing the relevance of 3D cancer models in nanomedicine research. *Adv. Drug Deliv. Rev.* **2014**, *79*, 95–106. [CrossRef]
295. Dekkers, J.F.; Wiegerinck, C.L.; De Jonge, H.R.; Bronsveld, I.; Janssens, H.M.; De Winter-de Groot, K.M.; Brandsma, A.M.; De Jong, N.W.; Bijvelds, M.J.; Scholte, B.J. A functional CFTR assay using primary cystic fibrosis intestinal organoids. *Nat. Med.* **2013**, *19*, 939–945. [CrossRef]
296. Yin, X.; Mead, B.E.; Safaee, H.; Langer, R.; Karp, J.M.; Levy, O. Engineering stem cell organoids. *Cell Stem Cell* **2016**, *18*, 25–38. [CrossRef]
297. Turner, D.A.; Baillie-Johnson, P.; Martinez Arias, A. Organoids and the genetically encoded self-assembly of embryonic stem cells. *BioEssays* **2016**, *38*, 181–191. [CrossRef]
298. Woei Ng, K.; Speicher, T.; Dombrowski, C.; Helledie, T.; Haupt, L.M.; Nurcombe, V.; Cool, S.M. Osteogenic differentiation of murine embryonic stem cells is mediated by fibroblast growth factor receptors. *Stem Cells Dev.* **2007**, *16*, 305–318. [CrossRef]
299. Qian, X.; Nguyen, H.N.; Song, M.M.; Hadiono, C.; Ogden, S.C.; Hammack, C.; Yao, B.; Hamersky, G.R.; Jacob, F.; Zhong, C. Brain-region-specific organoids using mini-bioreactors for modeling ZIKV exposure. *Cell* **2016**, *165*, 1238–1254. [CrossRef]

300. Chen, Y.-W.; Huang, S.X.; de Carvalho, A.L.R.T.; Ho, S.-H.; Islam, M.N.; Volpi, S.; Notarangelo, L.D.; Ciancanelli, M.; Casanova, J.-L.; Bhattacharya, J. A three-dimensional model of human lung development and disease from pluripotent stem cells. *Nat. Cell Biol.* **2017**, *19*, 542–549. [CrossRef]
301. Takebe, T.; Sekine, K.; Enomura, M.; Koike, H.; Kimura, M.; Ogaeri, T.; Zhang, R.-R.; Ueno, Y.; Zheng, Y.-W.; Koike, N. Vascularized and functional human liver from an iPSC-derived organ bud transplant. *Nature* **2013**, *499*, 481–484. [CrossRef]
302. McCracken, K.W.; Catá, E.M.; Crawford, C.M.; Sinagoga, K.L.; Schumacher, M.; Rockich, B.E.; Tsai, Y.-H.; Mayhew, C.N.; Spence, J.R.; Zavros, Y. Modelling human development and disease in pluripotent stem-cell-derived gastric organoids. *Nature* **2014**, *516*, 400–404. [CrossRef]
303. Forbes, T.A.; Howden, S.E.; Lawlor, K.; Phipson, B.; Maksimovic, J.; Hale, L.; Wilson, S.; Quinlan, C.; Ho, G.; Holman, K. Patient-iPSC-derived kidney organoids show functional validation of a ciliopathic renal phenotype and reveal underlying pathogenetic mechanisms. *Am. J. Hum. Genet.* **2018**, *102*, 816–831. [CrossRef]
304. Yamamoto, Y.; Gotoh, S.; Korogi, Y.; Seki, M.; Konishi, S.; Ikeo, S.; Sone, N.; Nagasaki, T.; Matsumoto, H.; Muro, S. Long-term expansion of alveolar stem cells derived from human iPS cells in organoids. *Nat. Methods* **2017**, *14*, 1097–1106. [CrossRef]
305. Sato, T.; Vries, R.G.; Snippert, H.J.; Van De Wetering, M.; Barker, N.; Stange, D.E.; Van Es, J.H.; Abo, A.; Kujala, P.; Peters, P.J. Single Lgr5 stem cells build crypt-villus structures in vitro without a mesenchymal niche. *Nature* **2009**, *459*, 262–265. [CrossRef]
306. Broutier, L.; Andersson-Rolf, A.; Hindley, C.J.; Boj, S.F.; Clevers, H.; Koo, B.-K.; Huch, M. Culture and establishment of self-renewing human and mouse adult liver and pancreas 3D organoids and their genetic manipulation. *Nat. Protoc.* **2016**, *11*, 1724–1743. [CrossRef]
307. Czerwinski, M.; Spence, J.R. Hacking the matrix. *Cell Stem Cell* **2017**, *20*, 9–10. [CrossRef]
308. Cruz-Acuña, R.; Quirós, M.; Farkas, A.E.; Dedhia, P.H.; Huang, S.; Siuda, D.; García-Hernández, V.; Miller, A.J.; Spence, J.R.; Nusrat, A. Synthetic hydrogels for human intestinal organoid generation and colonic wound repair. *Nat. Cell Biol.* **2017**, *19*, 1326–1335. [CrossRef]
309. Kratochvil, M.J.; Seymour, A.J.; Li, T.L.; Paşca, S.P.; Kuo, C.J.; Heilshorn, S.C. Engineered materials for organoid systems. *Nat. Rev. Mater.* **2019**, *4*, 606–622. [CrossRef]
310. Greggio, C.; De Franceschi, F.; Figueiredo-Larsen, M.; Gobaa, S.; Ranga, A.; Semb, H.; Lutolf, M.; Grapin-Botton, A. Artificial three-dimensional niches deconstruct pancreas development in vitro. *Development* **2013**, *140*, 4452–4462. [CrossRef]
311. Lee, J.; Böscke, R.; Tang, P.-C.; Hartman, B.H.; Heller, S.; Koehler, K.R. Hair follicle development in mouse pluripotent stem cell-derived skin organoids. *Cell Rep.* **2018**, *22*, 242–254. [CrossRef]
312. Dorrell, C.; Tarlow, B.; Wang, Y.; Canaday, P.S.; Haft, A.; Schug, J.; Streeter, P.R.; Finegold, M.J.; Shenje, L.T.; Kaestner, K.H. The organoid-initiating cells in mouse pancreas and liver are phenotypically and functionally similar. *Stem Cell Res.* **2014**, *13*, 275–283. [CrossRef]
313. Dye, B.R.; Hill, D.R.; Ferguson, M.A.; Tsai, Y.-H.; Nagy, M.S.; Dyal, R.; Wells, J.M.; Mayhew, C.N.; Nattiv, R.; Klein, O.D. In vitro generation of human pluripotent stem cell derived lung organoids. *eLife* **2015**, *4*, e05098. [CrossRef]
314. Zhang, R.-R.; Koido, M.; Tadokoro, T.; Ouchi, R.; Matsuno, T.; Ueno, Y.; Sekine, K.; Takebe, T.; Taniguchi, H. Human iPSC-derived posterior gut progenitors are expandable and capable of forming gut and liver organoids. *Stem Cell Rep.* **2018**, *10*, 780–793. [CrossRef]
315. Taguchi, A.; Nishinakamura, R. Higher-order kidney organogenesis from pluripotent stem cells. *Cell Stem Cell* **2017**, *21*, 730–746.e6. [CrossRef]
316. Takasato, M.; Pei, X.E.; Chiu, H.S.; Maier, B.; Baillie, G.J.; Ferguson, C.; Parton, R.G.; Wolvetang, E.J.; Roost, M.S.; de Sousa Lopes, S.M.C. Kidney organoids from human iPS cells contain multiple lineages and model human nephrogenesis. *Nature* **2015**, *526*, 564–568. [CrossRef]
317. Paşca, A.M.; Sloan, S.A.; Clarke, L.E.; Tian, Y.; Makinson, C.D.; Huber, N.; Kim, C.H.; Park, J.-Y.; O'rourke, N.A.; Nguyen, K.D. Functional cortical neurons and astrocytes from human pluripotent stem cells in 3D culture. *Nat. Methods* **2015**, *12*, 671–678. [CrossRef]
318. Nguyen, L.T.; Muktabar, A.; Tang, J.; Wong, Y.S.; Thaxton, C.S.; Venkatraman, S.S.; Ng, K.W. The potential of fluocinolone acetonide to mitigate inflammation and lipid accumulation in 2D and 3D foam cell cultures. *BioMed Res. Int.* **2018**, *2018*, 3739251. [CrossRef]
319. Medawar, P. The cultivation of adult mammalian skin epithelium in vitro. *J. Cell Sci.* **1948**, *3*, 187–196. [CrossRef]
320. Randall, M.J.; Jüngel, A.; Rimann, M.; Wuertz-Kozak, K. Advances in the Biofabrication of 3D Skin in vitro: Healthy and Pathological Models. *Front. Bioeng. Biotechnol.* **2018**, *6*, 154. [CrossRef]
321. Ng, K.W.; Hutmacher, D.W. Reduced contraction of skin equivalent engineered using cell sheets cultured in 3D matrices. *Biomaterials* **2006**, *27*, 4591–4598. [CrossRef]
322. Sun, B.K.; Siprashvili, Z.; Khavari, P.A. Advances in skin grafting and treatment of cutaneous wounds. *Science* **2014**, *346*, 941–945. [CrossRef]
323. Kim, B.S.; Kwon, Y.W.; Kong, J.-S.; Park, G.T.; Gao, G.; Han, W.; Kim, M.-B.; Lee, H.; Kim, J.H.; Cho, D.-W. 3D cell printing of in vitro stabilized skin model and in vivo pre-vascularized skin patch using tissue-specific extracellular matrix bioink: A step towards advanced skin tissue engineering. *Biomaterials* **2018**, *168*, 38–53. [CrossRef]
324. Min, D.; Lee, W.; Bae, I.H.; Lee, T.R.; Croce, P.; Yoo, S.S. Bioprinting of biomimetic skin containing melanocytes. *Exp. Dermatol.* **2018**, *27*, 453–459. [CrossRef]

325. Ströbel, S.; Buschmann, N.; Neeladkandhan, A.; Messner, S.; Kelm, J. Characterization of a novel in vitro 3D skin microtissue model for efficacy and toxicity testing. *Toxicol. Lett.* **2016**, S156–S157. [CrossRef]
326. Vidal, S.E.L.; Tamamoto, K.A.; Nguyen, H.; Abbott, R.D.; Cairns, D.M.; Kaplan, D.L. 3D biomaterial matrix to support long term, full thickness, immuno-competent human skin equivalents with nervous system components. *Biomaterials* **2019**, *198*, 194–203. [CrossRef]
327. Pourchet, L.J.; Thepot, A.; Albouy, M.; Courtial, E.J.; Boher, A.; Blum, L.J.; Marquette, C.A. Human skin 3D bioprinting using scaffold-free approach. *Adv. Healthc. Mater.* **2017**, *6*, 1601101. [CrossRef] [PubMed]
328. Gurtner, G.C.; Werner, S.; Barrandon, Y.; Longaker, M.T. Wound repair and regeneration. *Nature* **2008**, *453*, 314–321. [CrossRef]
329. Zimmermann, E.A.; Ritchie, R.O. Bone as a structural material. *Adv. Healthc. Mater.* **2015**, *4*, 1287–1304. [CrossRef] [PubMed]
330. Reznikov, N.; Bilton, M.; Lari, L.; Stevens, M.M.; Kröger, R. Fractal-like hierarchical organization of bone begins at the nanoscale. *Science* **2018**, *360*, eaao2189. [CrossRef] [PubMed]
331. Gurumurthy, B.; Bierdeman, P.C.; Janorkar, A.V. Spheroid model for functional osteogenic evaluation of human adipose derived stem cells. *J. Biomed. Mater. Res. Part A* **2017**, *105*, 1230–1236. [CrossRef] [PubMed]
332. Wenger, A.; Stahl, A.; Weber, H.; Finkenzeller, G.; Augustin, H.; Stark, G.; Kneser, U. Modulation of in vitro angiogenesis in a three-dimensional spheroidal coculture model for bone tissue engineering. *Tissue Eng.* **2004**, *10*, 1536–1547. [CrossRef] [PubMed]
333. Correia, C.; Grayson, W.L.; Park, M.; Hutton, D.; Zhou, B.; Guo, X.E.; Niklason, L.; Sousa, R.A.; Reis, R.L.; Vunjak-Novakovic, G. In vitro model of vascularized bone: Synergizing vascular development and osteogenesis. *PLoS ONE* **2011**, *6*, e28352. [CrossRef]
334. Bayer, E.; Gottardi, R.; Fedorchak, M.; Little, S. The scope and sequence of growth factor delivery for vascularized bone tissue regeneration. *J. Control. Release* **2015**, *219*, 129–140. [CrossRef]
335. Liu, Y.; Chan, J.K.; Teoh, S.H. Review of vascularised bone tissue-engineering strategies with a focus on co-culture systems. *J. Tissue Eng. Regen. Med.* **2015**, *9*, 85–105. [CrossRef]
336. Braghirolli, D.; Helfer, V.; Chagastelles, P.; Dalberto, T.; Gamba, D.; Pranke, P. Electrospun scaffolds functionalized with heparin and vascular endothelial growth factor increase the proliferation of endothelial progenitor cells. *Biomed. Mater.* **2017**, *12*, 025003. [CrossRef]
337. Tsigkou, O.; Pomerantseva, I.; Spencer, J.A.; Redondo, P.A.; Hart, A.R.; O'Doherty, E.; Lin, Y.; Friedrich, C.C.; Daheron, L.; Lin, C.P. Engineered vascularized bone grafts. *Proc. Natl. Acad. Sci. USA* **2010**, *107*, 3311–3316. [CrossRef]
338. Dhivya, S.; Saravanan, S.; Sastry, T.; Selvamurugan, N. Nanohydroxyapatite-reinforced chitosan composite hydrogel for bone tissue repair in vitro and in vivo. *J. Nanobiotechnol.* **2015**, *13*, 1–13. [CrossRef]
339. Zhai, X.; Ruan, C.; Ma, Y.; Cheng, D.; Wu, M.; Liu, W.; Zhao, X.; Pan, H.; Lu, W.W. 3D-bioprinted osteoblast-laden nanocomposite hydrogel constructs with induced microenvironments promote cell viability, differentiation, and osteogenesis both in vitro and in vivo. *Adv. Sci.* **2018**, *5*, 1700550. [CrossRef]
340. Wüst, S.; Godla, M.E.; Müller, R.; Hofmann, S. Tunable hydrogel composite with two-step processing in combination with innovative hardware upgrade for cell-based three-dimensional bioprinting. *Acta Biomater.* **2014**, *10*, 630–640. [CrossRef]
341. Levato, R.; Visser, J.; Planell, J.A.; Engel, E.; Malda, J.; Mateos-Timoneda, M.A. Biofabrication of tissue constructs by 3D bioprinting of cell-laden microcarriers. *Biofabrication* **2014**, *6*, 035020. [CrossRef]
342. Kang, H.-W.; Lee, S.J.; Ko, I.K.; Kengla, C.; Yoo, J.J.; Atala, A. A 3D bioprinting system to produce human-scale tissue constructs with structural integrity. *Nat. Biotechnol.* **2016**, *34*, 312–319. [CrossRef]
343. Kolesky, D.B.; Homan, K.A.; Skylar-Scott, M.A.; Lewis, J.A. Three-dimensional bioprinting of thick vascularized tissues. *Proc. Natl. Acad. Sci. USA* **2016**, *113*, 3179–3184. [CrossRef]
344. Benders, K.E.; van Weeren, P.R.; Badylak, S.F.; Saris, D.B.; Dhert, W.J.; Malda, J. Extracellular matrix scaffolds for cartilage and bone regeneration. *Trends Biotechnol.* **2013**, *31*, 169–176. [CrossRef]
345. Levato, R.; Webb, W.R.; Otto, I.A.; Mensinga, A.; Zhang, Y.; van Rijen, M.; van Weeren, R.; Khan, I.M.; Malda, J. The bio in the ink: Cartilage regeneration with bioprintable hydrogels and articular cartilage-derived progenitor cells. *Acta Biomater.* **2017**, *61*, 41–53. [CrossRef]
346. Hutmacher, D.W.; Ng, K.W.; Kaps, C.; Sittinger, M.; Kläring, S. Elastic cartilage engineering using novel scaffold architectures in combination with a biomimetic cell carrier. *Biomaterials* **2003**, *24*, 4445–4458. [CrossRef]
347. Lam, T.; Dehne, T.; Krüger, J.P.; Hondke, S.; Endres, M.; Thomas, A.; Lauster, R.; Sittinger, M.; Kloke, L. Photopolymerizable gelatin and hyaluronic acid for stereolithographic 3D bioprinting of tissue-engineered cartilage. *J. Biomed. Mater. Res. Part B Appl. Biomater.* **2019**, *107*, 2649–2657. [CrossRef]
348. Shie, M.-Y.; Chang, W.-C.; Wei, L.-J.; Huang, Y.-H.; Chen, C.-H.; Shih, C.-T.; Chen, Y.-W.; Shen, Y.-F. 3D printing of cytocompatible water-based light-cured polyurethane with hyaluronic acid for cartilage tissue engineering applications. *Materials* **2017**, *10*, 136. [CrossRef]
349. Paggi, C.; Venzac, B.; Leijten, J.; Leijten, L.M.T.; Karperien, M. Cartilage-on-chip: A multi-modal platform to study human chondrocyte's response to mechanical stimuli. *Osteoarthr. Cartil.* **2020**, *28*, S176–S177. [CrossRef]
350. Ma, X.; Qu, X.; Zhu, W.; Li, Y.-S.; Yuan, S.; Zhang, H.; Liu, J.; Wang, P.; Lai, C.S.E.; Zanella, F. Deterministically patterned biomimetic human iPSC-derived hepatic model via rapid 3D bioprinting. *Proc. Natl. Acad. Sci. USA* **2016**, *113*, 2206–2211. [CrossRef] [PubMed]
351. Mazzocchi, A.; Devarasetty, M.; Huntwork, R.; Soker, S.; Skardal, A. Optimization of collagen type I-hyaluronan hybrid bioink for 3D bioprinted liver microenvironments. *Biofabrication* **2018**, *11*, 015003. [CrossRef] [PubMed]

352. Rennert, K.; Steinborn, S.; Gröger, M.; Ungerböck, B.; Jank, A.-M.; Ehgartner, J.; Nietzsche, S.; Dinger, J.; Kiehntopf, M.; Funke, H. A microfluidically perfused three dimensional human liver model. *Biomaterials* **2015**, *71*, 119–131. [CrossRef] [PubMed]
353. Danoy, M.; Poulain, S.; Lereau-Bernier, M.; Kato, S.; Scheidecker, B.; Kido, T.; Miyajima, A.; Sakai, Y.; Plessy, C.; Leclerc, E. Characterization of liver zonation-like transcriptomic patterns in HLCs derived from hiPSCs in a microfluidic biochip environment. *Biotechnol. Prog.* **2020**, *36*, e3013. [CrossRef]
354. Lee, H.; Chae, S.; Kim, J.Y.; Han, W.; Kim, J.; Choi, Y.; Cho, D.-W. Cell-printed 3D liver-on-a-chip possessing a liver microenvironment and biliary system. *Biofabrication* **2019**, *11*, 025001. [CrossRef]
355. Lee, H.; Kim, J.; Choi, Y.; Cho, D.-W. Application of gelatin bioinks and cell-printing technology to enhance cell delivery capability for 3D liver fibrosis-on-a-chip development. *ACS Biomater. Sci. Eng.* **2020**, *6*, 2469–2477. [CrossRef]
356. Choe, A.; Ha, S.K.; Choi, I.; Choi, N.; Sung, J.H. Microfluidic Gut-liver chip for reproducing the first pass metabolism. *Biomed. Microdevices* **2017**, *19*, 4. [CrossRef]
357. Jeon, J.-w.; Choi, N.; Lee, S.H.; Sung, J.H. Three-tissue microphysiological system for studying inflammatory responses in gut-liver Axis. *Biomed. Microdevices* **2020**, *22*, 1–11. [CrossRef]
358. Crosnier, C.; Stamataki, D.; Lewis, J. Organizing cell renewal in the intestine: Stem cells, signals and combinatorial control. *Nat. Rev. Genet.* **2006**, *7*, 349–359. [CrossRef]
359. Balimane, P.V.; Chong, S. Cell culture-based models for intestinal permeability: A critique. *Drug Discov. Today* **2005**, *10*, 335–343. [CrossRef]
360. Sung, J.H.; Yu, J.; Luo, D.; Shuler, M.L.; March, J.C. Microscale 3-D hydrogel scaffold for biomimetic gastrointestinal (GI) tract model. *Lab Chip* **2011**, *11*, 389–392. [CrossRef]
361. Kim, S.H.; Chi, M.; Yi, B.; Kim, S.H.; Oh, S.; Kim, Y.; Park, S.; Sung, J.H. Three-dimensional intestinal villi epithelium enhances protection of human intestinal cells from bacterial infection by inducing mucin expression. *Integr. Biol.* **2014**, *6*, 1122–1131. [CrossRef]
362. Yu, J.; Peng, S.; Luo, D.; March, J.C. In vitro 3D human small intestinal villous model for drug permeability determination. *Biotechnol. Bioeng.* **2012**, *109*, 2173–2178. [CrossRef]
363. Kim, W.; Kim, G.H. An innovative cell-printed microscale collagen model for mimicking intestinal villus epithelium. *Chem. Eng. J.* **2018**, *334*, 2308–2318. [CrossRef]
364. Creff, J.; Courson, R.; Mangeat, T.; Foncy, J.; Souleille, S.; Thibault, C.; Besson, A.; Malaquin, L. Fabrication of 3D scaffolds reproducing intestinal epithelium topography by high-resolution 3D stereolithography. *Biomaterials* **2019**, *221*, 119404. [CrossRef]
365. Benien, P.; Swami, A. 3D tumor models: History, advances and future perspectives. *Future Oncol.* **2014**, *10*, 1311–1327. [CrossRef]
366. Li, Y.; Kumacheva, E. Hydrogel microenvironments for cancer spheroid growth and drug screening. *Sci. Adv.* **2018**, *4*, eaas8998. [CrossRef]
367. Hutmacher, D.W. Biomaterials offer cancer research the third dimension. *Nat. Mater.* **2010**, *9*, 90–93. [CrossRef]
368. Weaver, V.M.; Petersen, O.; Wang, F.; Larabell, C.; Briand, P.; Damsky, C.; Bissell, M. Reversion of the malignant phenotype of human breast cells in three-dimensional culture and in vivo by integrin blocking antibodies. *J. Cell Biol.* **1997**, *137*, 231–245. [CrossRef]
369. Shafiee, A.; Atala, A. Printing technologies for medical applications. *Trends Mol. Med.* **2016**, *22*, 254–265. [CrossRef]
370. Zhao, Y.; Yao, R.; Ouyang, L.; Ding, H.; Zhang, T.; Zhang, K.; Cheng, S.; Sun, W. Three-dimensional printing of Hela cells for cervical tumor model in vitro. *Biofabrication* **2014**, *6*, 035001. [CrossRef]
371. Beck, J.N.; Singh, A.; Rothenberg, A.R.; Elisseeff, J.H.; Ewald, A.J. The independent roles of mechanical, structural and adhesion characteristics of 3D hydrogels on the regulation of cancer invasion and dissemination. *Biomaterials* **2013**, *34*, 9486–9495. [CrossRef]
372. Drost, J.; Clevers, H. Organoids in cancer research. *Nat. Rev. Cancer* **2018**, *18*, 407–418. [CrossRef]
373. Hakimi, N.; Cheng, R.; Leng, L.; Sotoudehfar, M.; Ba, P.Q.; Bakhtyar, N.; Amini-Nik, S.; Jeschke, M.G.; Günther, A. Handheld skin printer: In situ formation of planar biomaterials and tissues. *Lab Chip* **2018**, *18*, 1440–1451. [CrossRef]
374. Rimann, M.; Bono, E.; Annaheim, H.; Bleisch, M.; Graf-Hausner, U. Standardized 3D bioprinting of soft tissue models with human primary cells. *J. Lab. Autom.* **2016**, *21*, 496–509. [CrossRef]
375. Lee, J.M.; Chae, T.; Sheikh, F.A.; Ju, H.W.; Moon, B.M.; Park, H.J.; Park, Y.R.; Park, C.H. Three dimensional poly (ε-caprolactone) and silk fibroin nanocomposite fibrous matrix for artificial dermis. *Mater. Sci. Eng. C* **2016**, *68*, 758–767. [CrossRef]
376. Ritz, U.; Gerke, R.; Götz, H.; Stein, S.; Rommens, P.M. A new bone substitute developed from 3D-prints of polylactide (PLA) loaded with collagen I: An in vitro study. *Int. J. Mol. Sci.* **2017**, *18*, 2569. [CrossRef]
377. Zhang, X.; Wang, C.; Liao, M.; Dai, L.; Tang, Y.; Zhang, H.; Coates, P.; Sefat, F.; Zheng, L.; Song, J. Aligned electrospun cellulose scaffolds coated with rhBMP-2 for both in vitro and in vivo bone tissue engineering. *Carbohydr. Polym.* **2019**, *213*, 27–38. [CrossRef] [PubMed]
378. Iandolo, D.; Ravichandran, A.; Liu, X.; Wen, F.; Chan, J.K.; Berggren, M.; Teoh, S.H.; Simon, D.T. Development and characterization of organic electronic scaffolds for bone tissue engineering. *Adv. Healthc. Mater.* **2016**, *5*, 1505–1512. [CrossRef] [PubMed]
379. Bahmaee, H.; Owen, R.; Boyle, L.; Perrault, C.M.; Garcia-Granada, A.A.; Reilly, G.C.; Claeyssens, F. Design and evaluation of an osteogenesis-on-a-chip microfluidic device incorporating 3D cell culture. *Front. Bioeng. Biotechnol.* **2020**, *8*, 1042. [CrossRef] [PubMed]
380. Chen, W.; Chen, S.; Morsi, Y.; El-Hamshary, H.; El-Newhy, M.; Fan, C.; Mo, X. Superabsorbent 3D scaffold based on electrospun nanofibers for cartilage tissue engineering. *ACS Appl. Mater. Interfaces* **2016**, *8*, 24415–24425. [CrossRef] [PubMed]

381. Kosik-Kozioł, A.; Costantini, M.; Mróz, A.; Idaszek, J.; Heljak, M.; Jaroszewicz, J.; Kijeńska, E.; Szöke, K.; Frerker, N.; Barbetta, A. 3D bioprinted hydrogel model incorporating β-tricalcium phosphate for calcified cartilage tissue engineering. *Biofabrication* **2019**, *11*, 035016. [CrossRef] [PubMed]
382. Lin, Z.; Li, Z.; Li, E.N.; Li, X.; Del Duke, C.J.; Shen, H.; Hao, T.; O'Donnell, B.; Bunnell, B.A.; Goodman, S.B. Osteochondral tissue chip derived from iPSCs: Modeling OA pathologies and testing drugs. *Front. Bioeng. Biotechnol.* **2019**, *7*, 411. [CrossRef] [PubMed]
383. Rashidi, H.; Luu, N.-T.; Alwahsh, S.M.; Ginai, M.; Alhaque, S.; Dong, H.; Tomaz, R.A.; Vernay, B.; Vigneswara, V.; Hallett, J.M. 3D human liver tissue from pluripotent stem cells displays stable phenotype in vitro and supports compromised liver function in vivo. *Arch. Toxicol.* **2018**, *92*, 3117–3129. [CrossRef]
384. Tripathi, A.; Melo, J.S. Preparation of a sponge-like biocomposite agarose–chitosan scaffold with primary hepatocytes for establishing an in vitro 3D liver tissue model. *RSC Adv.* **2015**, *5*, 30701–30710. [CrossRef]
385. Goulart, E.; de Caires-Junior, L.C.; Telles-Silva, K.A.; Araujo, B.H.S.; Rocco, S.A.; Sforca, M.; de Sousa, I.L.; Kobayashi, G.S.; Musso, C.M.; Assoni, A.F. 3D bioprinting of liver spheroids derived from human induced pluripotent stem cells sustain liver function and viability in vitro. *Biofabrication* **2019**, *12*, 015010. [CrossRef]
386. Deng, J.; Cong, Y.; Han, X.; Wei, W.; Lu, Y.; Liu, T.; Zhao, W.; Lin, B.; Luo, Y.; Zhang, X. A liver-on-a-chip for hepatoprotective activity assessment. *Biomicrofluidics* **2020**, *14*, 064107. [CrossRef]
387. Chen, Y.; Zhou, W.; Roh, T.; Estes, M.K.; Kaplan, D.L. In vitro enteroid-derived three-dimensional tissue model of human small intestinal epithelium with innate immune responses. *PLoS ONE* **2017**, *12*, e0187880. [CrossRef]
388. Madden, L.R.; Nguyen, T.V.; Garcia-Mojica, S.; Shah, V.; Le, A.V.; Peier, A.; Visconti, R.; Parker, E.M.; Presnell, S.C.; Nguyen, D.G. Bioprinted 3D primary human intestinal tissues model aspects of native physiology and ADME/Tox functions. *IScience* **2018**, *2*, 156–167. [CrossRef]
389. Kasendra, M.; Luc, R.; Yin, J.; Manatakis, D.V.; Kulkarni, G.; Lucchesi, C.; Sliz, J.; Apostolou, A.; Sunuwar, L.; Obrigewitch, J. Duodenum intestine-chip for preclinical drug assessment in a human relevant model. *Elife* **2020**, *9*, e50135. [CrossRef]
390. Han, H.; Park, Y.; Choi, Y.m.; Yong, U.; Kang, B.; Shin, W.; Min, S.; Kim, H.J.; Jang, J. A Bioprinted Tubular Intestine Model using a Colon-Specific Extracellular Matrix Bioink. *Adv. Healthc. Mater.* **2021**, *11*, 2101768. [CrossRef]
391. Moysidou, C.M.; Pitsalidis, C.; Al-Sharabi, M.; Withers, A.M.; Zeitler, J.A.; Owens, R.M. 3D Bioelectronic Model of the Human Intestine. *Adv. Biol.* **2021**, *5*, 2000306. [CrossRef]
392. Lombardo, M.E.; Pavia, F.C.; Vitrano, I.; Ghersi, G.; Brucato, V.; Rosei, F.; La Carrubba, V. PLLA scaffolds with controlled architecture as potential microenvironment for in vitro tumor model. *Tissue Cell* **2019**, *58*, 33–41. [CrossRef]
393. Kim, M.; Mun, H.; Sung, C.O.; Cho, E.J.; Jeon, H.-J.; Chun, S.-M.; Shin, T.H.; Jeong, G.S.; Kim, D.K.; Choi, E.K. Patient-derived lung cancer organoids as in vitro cancer models for therapeutic screening. *Nat. Commun.* **2019**, *10*, 1–15. [CrossRef]
394. Shirure, V.S.; Bi, Y.; Curtis, M.B.; Lezia, A.; Goedegebuure, M.M.; Goedegebuure, S.P.; Aft, R.; Fields, R.C.; George, S.C. Tumor-on-a-chip platform to investigate progression and drug sensitivity in cell lines and patient-derived organoids. *Lab Chip* **2018**, *18*, 3687–3702. [CrossRef]
395. Carvalho, M.; Barata, D.; Teixeira, L.; Giselbrecht, S.; Reis, R.; Oliveira, J.; Truckenmüller, R.; Habibovic, P. Colorectal tumor-on-a-chip system: A 3D tool for precision onco-nanomedicine. *Sci. Adv.* **2019**, *5*, eaaw1317. [CrossRef]
396. Cao, X.; Ashfaq, R.; Cheng, F.; Maharjan, S.; Li, J.; Ying, G.; Hassan, S.; Xiao, H.; Yue, K.; Zhang, Y.S. A tumor-on-a-chip system with bioprinted blood and lymphatic vessel pair. *Adv. Funct. Mater.* **2019**, *29*, 1807173. [CrossRef]
397. Scannell, J.W.; Bosley, J. When quality beats quantity: Decision theory, drug discovery, and the reproducibility crisis. *PLoS ONE* **2016**, *11*, e0147215. [CrossRef]
398. Najgebauer, H.; Yang, M.; Francies, H.E.; Pacini, C.; Stronach, E.A.; Garnett, M.J.; Saez-Rodriguez, J.; Iorio, F. CELLector: Genomics-guided selection of cancer in vitro models. *Cell Syst.* **2020**, *10*, 424–432.e6. [CrossRef]
399. Geiger, B. Encounters in space. *Science* **2001**, *294*, 1661–1663. [CrossRef] [PubMed]
400. Amatangelo, M.D.; Bassi, D.E.; Klein-Szanto, A.J.; Cukierman, E. Stroma-derived three-dimensional matrices are necessary and sufficient to promote desmoplastic differentiation of normal fibroblasts. *Am. J. Pathol.* **2005**, *167*, 475–488. [CrossRef]
401. Blume, C.; Swindle, E.J.; Dennison, P.; Jayasekera, N.P.; Dudley, S.; Monk, P.; Behrendt, H.; Schmidt-Weber, C.B.; Holgate, S.T.; Howarth, P.H. Barrier responses of human bronchial epithelial cells to grass pollen exposure. *Eur. Respir. J.* **2013**, *42*, 87–97. [CrossRef] [PubMed]
402. Vijayavenkataraman, S.; Yan, W.-C.; Lu, W.F.; Wang, C.-H.; Fuh, J.Y.H. 3D bioprinting of tissues and organs for regenerative medicine. *Adv. Drug Deliv. Rev.* **2018**, *132*, 296–332. [CrossRef]
403. Majumder, B.; Baraneedharan, U.; Thiyagarajan, S.; Radhakrishnan, P.; Narasimhan, H.; Dhandapani, M.; Brijwani, N.; Pinto, D.D.; Prasath, A.; Shanthappa, B.U. Predicting clinical response to anticancer drugs using an ex vivo platform that captures tumour heterogeneity. *Nat. Commun.* **2015**, *6*, 1–14. [CrossRef]
404. Skardal, A.; Devarasetty, M.; Forsythe, S.; Atala, A.; Soker, S. A reductionist metastasis-on-a-chip platform for in vitro tumor progression modeling and drug screening. *Biotechnol. Bioeng.* **2016**, *113*, 2020–2032. [CrossRef]

International Journal of *Molecular Sciences*

MDPI

Article

Injectable Human Hair Keratin–Fibrinogen Hydrogels for Engineering 3D Microenvironments to Accelerate Oral Tissue Regeneration

Hyeon Jeong Kang [1,†], Nare Ko [2,3,†], Seung Jun Oh [3], Seong Yeong An [1], Yu-Shik Hwang [1,*] and So Yeon Kim [4,*]

1 Department of Maxillofacial Biomedical Engineering and Institute of Oral Biology, School of Dentistry, Kyung Hee University, Seoul 02447, Korea; khj001@khu.ac.kr (H.J.K.); syan0426@gmail.com (S.Y.A.)
2 Biomedical Research Center, Asan Institute for Life Sciences, 88 Olympic-ro 43-gil, Songpa-gu, Seoul 05505, Korea; nare.ko@amc.seoul.kr
3 Department of Nuclear Medicine, Asan Medical Center, University of Ulsan College of Medicine, 88 Olympic-ro 43-gil, Songpa-gu, Seoul 05505, Korea; sjoh@amc.seoul.kr
4 Department of Dental Hygiene, College of Health & Medical Sciences, Cheongju University, Cheongju 28503, Korea
* Correspondence: yshwang@khu.ac.kr (Y.-S.H.); ksy0615@cju.ac.kr (S.Y.K.)
† These authors contributed equally to this study.

Citation: Kang, H.J.; Ko, N.; Oh, S.J.; An, S.Y.; Hwang, Y.-S.; Kim, S.Y. Injectable Human Hair Keratin–Fibrinogen Hydrogels for Engineering 3D Microenvironments to Accelerate Oral Tissue Regeneration. *Int. J. Mol. Sci.* **2021**, *22*, 13269. https://doi.org/10.3390/ijms222413269

Academic Editor: Peter John Jervis

Received: 6 October 2021
Accepted: 2 December 2021
Published: 9 December 2021

Abstract: Traumatic injury of the oral cavity is atypical and often accompanied by uncontrolled bleeding and inflammation. Injectable hydrogels have been considered to be promising candidates for the treatment of oral injuries because of their simple formulation, minimally invasive application technique, and site-specific delivery. Fibrinogen-based hydrogels have been widely explored as effective materials for wound healing in tissue engineering due to their uniqueness. Recently, an injectable foam has taken the spotlight. However, the fibrin component of this biomaterial is relatively stiff. To address these challenges, we created keratin-conjugated fibrinogen (KRT-FIB). This study aimed to develop a novel keratin biomaterial and assess cell–biomaterial interactions. Consequently, a novel injectable KRT-FIB hydrogel was optimized through rheological measurements, and its injection performance, swelling behavior, and surface morphology were investigated. We observed an excellent cell viability, proliferation, and migration/cell–cell interaction, indicating that the novel KRT-FIB-injectable hydrogel is a promising platform for oral tissue regeneration with a high clinical applicability.

Keywords: human hair keratin; fibrinogen; injectable hydrogel; human gingival fibroblast; biomaterials; biocompatibility

1. Introduction

Keratin (KRT) is an insoluble protein that forms a part of intermediate filaments in epidermal appendageal structures, hair, nails, horn, hoofs, wool, and feathers [1]. KRT serves important structural and protective functions, particularly in the epithelium [2,3]. In particular, human hair-derived KRTs contain excellent cell adhesion motif Leu-Asp-Val (LDV) that enables the extracellular substrate for cell attachment and support [4,5]. We have recently reported that human hair-derived KRT is highly attractive for wound healing therapy because of its inherent bioactivity, biocompatibility, and physical properties [6]. Despite these advantages, previous KRT-based hydrogel studies are only limited to wool- or feather-based KRT proteins [7]. In addition, hydrogels based on human hair-derived KRTs, especially injectable platforms, have never been studied for oral tissue regeneration.

Fibrinogen (FIB) is a promising candidate due to its innate wound healing process that stimulates reparative cell activity, as well as exhibits both angiogenic and anti-inflammatory properties that are essential for wound healing and tissue repair [8–10]. In addition, FIB is known as a key protein for the regulation of angiogenesis and bone regeneration. The first clinically approved FIB hydrogel for tissue repair is produced by the formation of FIB polymers between transglutaminase factor XIII and calcium ions in the presence of thrombin (THR) [11]. However, current FIB hydrogels exhibit weak mechanical strength with a lack of porosity and adhesive properties due to their limited design ability of the protein backbone [12]. Hydrogels comprising FIB alone are known to increase scaffold contraction and do not support continuous cell growth [13]. Thus, such FIB hydrogels are relatively stiff and difficult to handle when implanted.

To resolve these problems, we developed a novel injectable KRT-FIB hydrogel (KFH) for oral tissue regeneration (Figure 1). To achieve the homogeneous formation of the KFH, human hair-driven KRT proteins were covalently conjugated with the FIB proteins via a facile coupling reaction. Importantly, the resulting KRT-FIB precursor is the first example of KRT-conjugated FIB material that enables the formation of well-established KFH in the presence of THR. With the optimal ratio of KRT: FIB, KFH demonstrated great injectable performance, swelling behavior, and high porosity. Human gingival fibroblasts (HGFs) encapsulated in the KFH showed excellent cell viability, proliferation, and migration with cell-cell interaction. These results revealed that the KFH has a great potential to promote oral tissue regeneration by the formation of the HGFs network involved in the basal layer of the oral gingival dermis. Moreover, the injectable KFH provides high clinical applicability with reducing reduce patient pain, treatment cost, and defect recovery time.

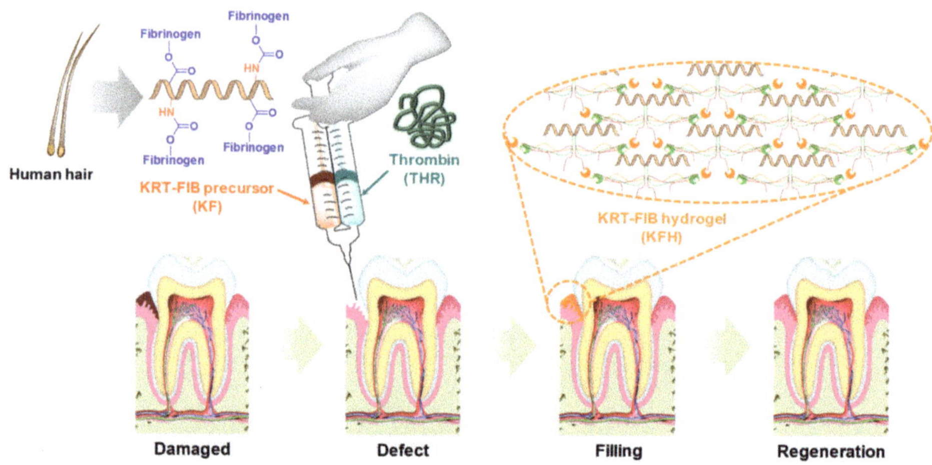

Figure 1. Schematic illustration of keratin-conjugated fibrinogen (KRT-FIB) hydrogels (KFHs) for oral tissue regeneration.

2. Results and Discussion

2.1. Synthesis of KRT-FIB Precursors

Although both KRT and FIB are hydrophilic proteins, they cannot be integrated into homogeneous hydrogels in the presence of THR. To address this challenge, covalently linked KRT-FIB precursors were synthesized via a facile coupling reaction (Figure 2a). The human hair-driven KRT protein was first succinylated to introduce carboxylic acids as terminal functional groups. The carboxyl groups on the resulting succinyl KRT protein (KRT-COOH) enable conjugation to the amine groups on FIB via a carbodiimide/N-hydroxysuccinimide (EDC/NHS) reaction. With varying molar ratios of KRT and FIB, a series of precursors were prepared, abbreviated as KF-1, KF-3, and KF-6, which stand for the 1:1, 3:1, and 6:1 mol ratios of KRT: FIB. Such direct conjugation of KRT and FIB facilitates the formation of well-defined KFH without phase separation. Sodium dodecyl sulfate-polyacrylamide gel electrophoresis (SDS-PAGE) was performed to study the change in molecular weight of the KRT-FIB precursors during the reaction (Figure 2b). The KRT protein weighs less than 40 kDa and is not shown on the gel, and the band of the FIB protein appears at 350 kDa. After conjugation, the molecular weight of KRT significantly increased and thick bands corresponding to the KRT-FIB precursors were positioned higher than 350 kDa. Note that precursors larger than 500 kDa were not capable of passing through the polyacrylamide gel and were positioned at the top of the stacking gel, indicating the formation of precursors with a molecular weight larger than 500 kDa.

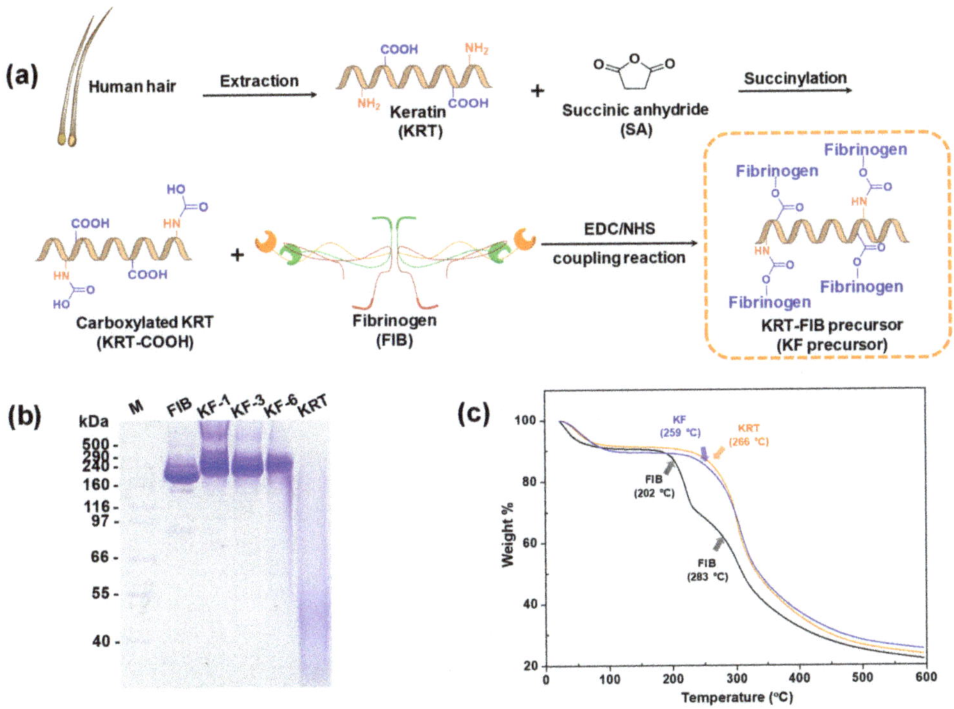

Figure 2. (**a**) Synthetic scheme of KRT-FIB precursor. (**b**) Thermogravimetric analysis diagrams of KRT, FIB, and KRT-FIB precursors. Each arrow indicates a temperature where the major weight loss starts. (**c**) Sodium dodecyl sulfate-polyacrylamide gel electrophoresis results obtained for the KRT; FIB; and KF-1, 3, 6 precursors.

Furthermore, the effect of KRT conjugation was evaluated by analyzing its thermal properties using thermogravimetric analysis (TGA). Figure 2c shows the TGA data of the weight loss of the KRT-FIB precursor upon heating, compared with the KRT and FIB proteins as controls. For the precursor, major weight loss started at 259 °C. This temperature is lower than that for the KRT protein (266 °C), but higher than that for the FIB protein (202 °C, the first curve), indicating that the introduction of KRT enhances the thermal stability of FIB for the formation of hydrogels.

2.2. Preparation and Rheological Studies of KFHs

The gelation rate is a critical parameter for hydrogels for injectable applications. Slow gelation causes the lateral spreading of precursors to non-target sites, whereas fast gelation clogs needles due to the formation of entanglements in the syringe [14,15] To demonstrate rapid gelation with a high efficiency, KFHs were prepared via THR-induced crosslinking (Figure 3a). In the presence of THR, FIB proteins are degraded into fibrin monomers by the cleavage of fibrinopeptide A/B. N-terminal fragments of α chains (known as "knobs") on the fibrin monomers bind to the complementary C-terminal of the β/γ chains (known as "holes") and the fibrin monomers are polymerized via intermolecular interactions. Such THR-mediated knob-hole interactions formulate the 3D network structure of the KFH, crosslinked by fibrin polymers (Figure 3b). A series of KFHs with varying KRT: FIB mole ratios were prepared (Figure 3c). In the presence of 5 unit/mL of THR, the gelation of KRT-FIB precursors occurred spontaneously and was completed within 1 min.

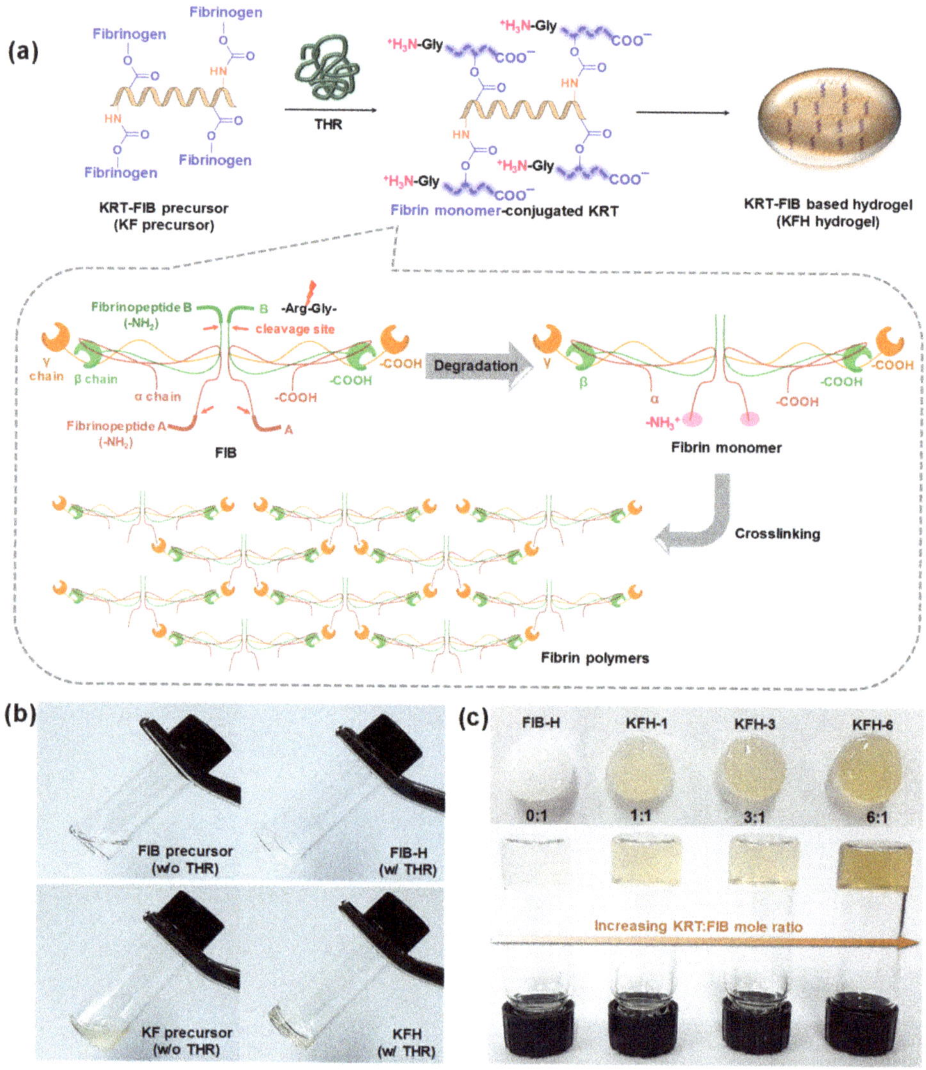

Figure 3. (**a**) Schematic illustration of KFH preparation and the mechanism of thrombin (THR)-induced fibrin cross-linking. (**b**) Gelation images of FIB and KRT-FIB precursors with or without THR. (**c**) Digital pictures of KFH samples with varying KRT:FIB mole ratios. White-colored turbid FIB-based hydrogel (FIB-H) gradually becomes brownish and transparent as the KRT content in the hydrogels increases.

To achieve clinically acceptable injectability, hydrogels must be designed considering several critical physical properties: (1) shear-thinning behavior [16], (2) sufficient strength to resist deformations [17] and (3) a higher frequency-dependent storage modulus (G′) than the corresponding loss modulus (G″) [18]. To ascertain mechanical properties for injectable applications, rheological measurements of KFHs (1, 3, and 6) were carried out. First, the change in the complex viscosity of the KFH was measured with an oscillatory frequency sweep at 37 °C. Complex viscosity is an important parameter to consider in the design of injectable platforms, as it measures the ability of the hydrogels not only to respond to changes in shear stress during the injection but also to resist deformation within a tissue after injection. As depicted in Figure 4a, the hydrogels show a continuous decrease in complex viscosity with an increasing frequency sweep, which indicates shear-thinning viscoelastic behavior. In addition, the complex viscosity of the KFH gradually increased with the increasing molar ratio of KRT to FIB. This result reveals that the KFHs are suitable for injection through needle extrusion and that the viscosity can be regulated by the KRT content [18].

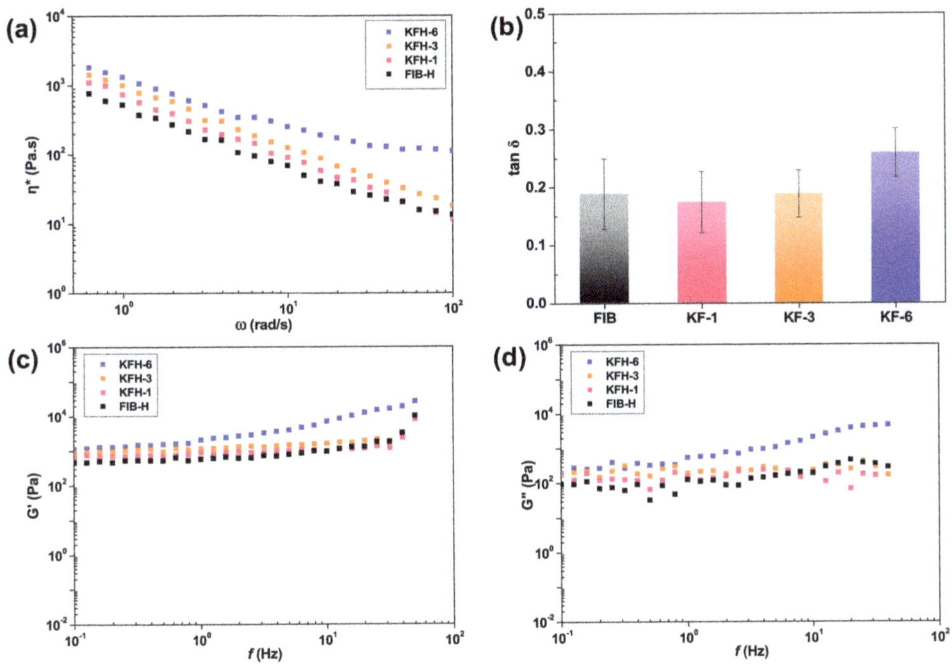

Figure 4. Characterization of THR-induced hydrogels: (**a**) Gelation images of hydrogels with different KRT compositions. Changes in (**b**) complex viscosity following the increase in frequency sweep, (**c**) storage modulus, and (**d**) loss modulus following the frequency. All measurements are performed with FIB-Hs as controls.

To further study the elastic and viscous behaviors, the G′ and G″ moduli of the KFHs were examined (Figure 4b–d). Constant G′ and G″ values and low tan δ (the ratio of G″/G′; tan δ <1) revealed the gel-like behavior of the KFH. As the G′ value was proportional to the KRT content with increasing stiffness, the KFH-3 hydrogel fabricated with the KF-3 precursor (mole ratio of KRT: FIB = 3:1) was selected and used for further studies. These rheological results suggest that KFHs have great potential as injectable platforms.

2.3. *Injectable Performance, Swelling, and Degradation Behavior of KFH-3*

Excellent injectable performance and swelling behavior of a hydrogel are important characteristics to facilitate the filling of irregularly shaped defects, simple formulation, non-invasive technique, and site-specific action [19,20]. Figure 5a indicates that the KFH-3 hydrogel passed smoothly through an 18 G needle and immediately acquired a gel-like structure. Such an excellent injectable performance indicates that the KFHs facilitate the filling of irregular defects. In addition, the swelling ratio of the KFH-3 hydrogel was almost six times higher than that of the FIB-H hydrogel (Figure 5b). As well-defined hydrogels do not dissolve but swell in water, this result revealed that the swelling rate of the hydrogels was significantly improved by the introduction of KRT. Such water absorption ability and swelling kinetics of hydrogels can be modulated by porosity [21]. Scanning electron microscopy (SEM) images show that KFH-3 has a highly porous structure with a 10–100 μm diameter compared to the non-porous FIB-H (Figure 5c). These results indicate that the swelling ratio of the KFH-3 strongly depends on the porous morphology of the hydrogel.

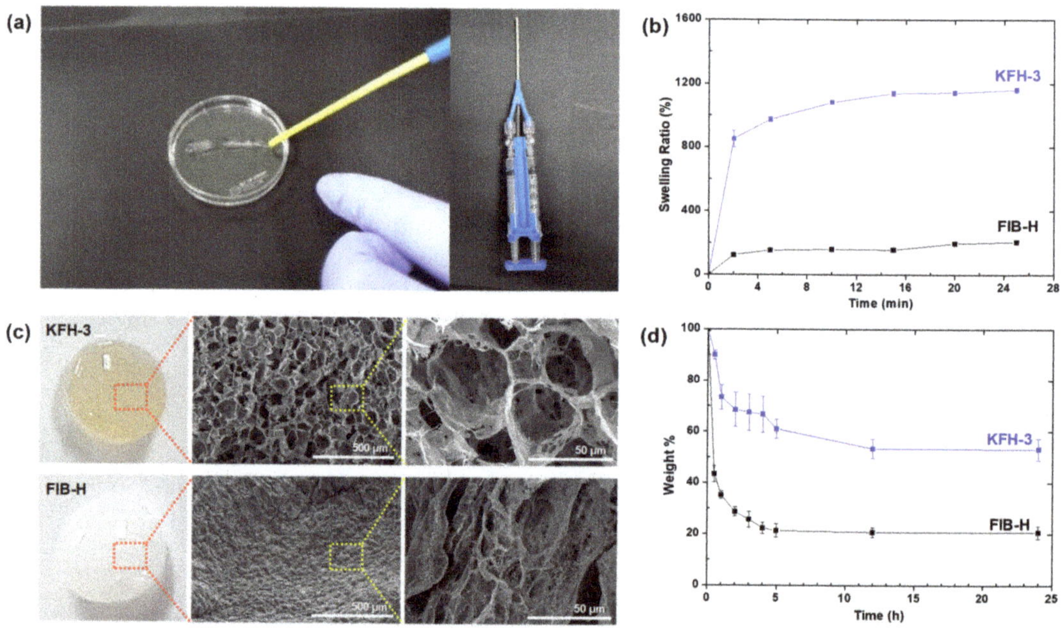

Figure 5. (**a**) The KFH-3 hydrogel was able to pass through a needle without clogging, indicating its high injectability (**left**); the precursor solution was loaded onto one side of a dual-chambered syringe with an 18 G needle (diameter = 1.3 mm), and the THR solution was loaded onto the other side (**right**). (**b**) Swelling kinetics of KFH-3 and FIB-H in deionized water at 37 °C. (**c**) Digital pictures and SEM images of KFH-3 and FIB-H at different magnifications (scale bar = 500 μm and 50 μm, respectively). (**d**) Temperature- and pH-responsive degradation study of hydrogels was evaluated by the dry weight ratio of the hydrogel samples.

The in vitro degradation behavior of KFH-3 and FIB-H hydrogels in warm cell media (50 °C) at pH 5.5 was monitored by a comparison of the weight change. As shown in Figure 5d, the weight of both hydrogel samples steadily decreased. In addition, a significant difference in weight between KFH-3 and FIB-H hydrogels was observed. After 12 h of incubation, the weight of KFH-3 hydrogel decreased to 20% and that of FIB-H decreased to 53.3%. This result indicated that KRT-conjugated fibrin-based hydrogel was more resistant to high-temperature and acidic conditions than pristine fibrin-based hydrogel, protecting itself from the integrity loss.

2.4. HGF Viability in KRT-Based Fibrin Hydrogels: Cytotoxicity Studies and 3D Cell Encapsulation in KRT-Based Fibrin Hydrogels

Porosity is a property of hydrogels that is important for tissue regeneration [22,23]. Porous hydrogels provide the necessary space for cell growth and vascularization [24–26]. It is well-known that such a three-dimensional (3D) structure of hydrogels is necessary to achieve successful oral tissue regeneration because it can be used as a delivery vehicle for bioactive substances in cells [22]. The efficient transportation of nutrients and oxygen through interconnected pathways promotes cell proliferation, migration, and cell–cell contact [27,28].

The cytotoxicity analysis of cell adhesion and proliferation was performed using HGFs for the LIVE/DEAD (Figure 6a,b) and CCK-8 assay (Figure 6f). The cell proliferation of all groups increased with the culture time. Furthermore, except for the samples from day 1, much higher cell viabilities in the KFH-3 were detected than those of the FIB-H at the other time intervals, indicating better cytocompatibility to support cell proliferation. In addition, we observed that HGFs proliferated and exhibited spreading morphology in encapsulated KFH. Phalloidin staining (Figure 6c,d), used to perform cellular interaction studies using HGF cells, demonstrated the significant free flow of biological fluids and cell migration and growth inside the KFH.

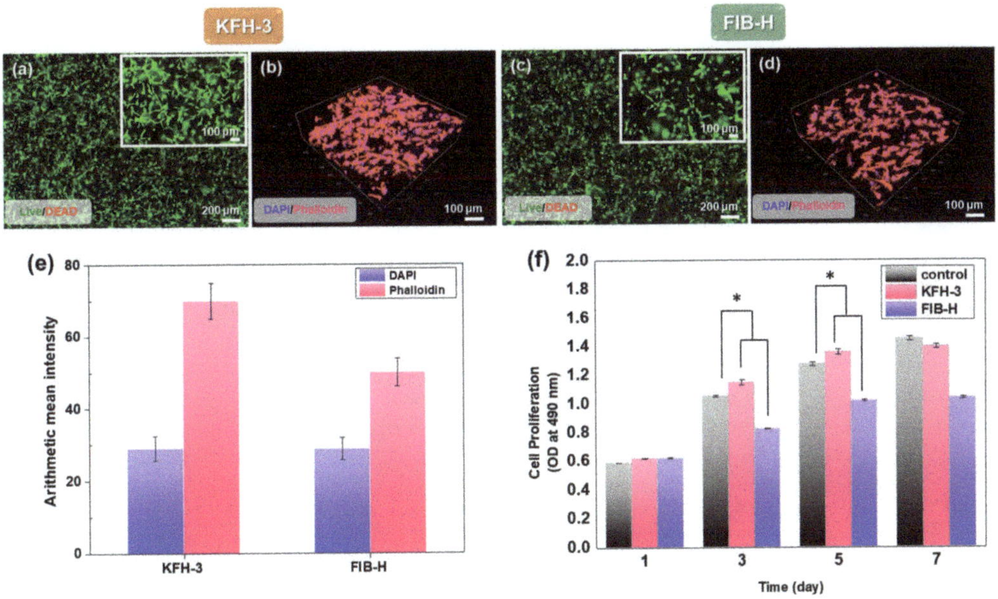

Figure 6. Characterization of cell adhesion and proliferation in three-dimensional cell encapsulation in FIB-H and KFH-3. For the cell viability assay, human gingiva fibroblasts (HGFs) embedded in FIB-H and KFH-3 were stained with calce-in_Am (green)/ethidium homodimer (red). LIVE/DEAD assay 24 h after encapsulation (**a**,**c**) shown at low (scale bar = 200 μm) and high (scale bar = 100 μm) magnification. (**b**,**d**) Confocal pictures of encapsulated HGF cells labeled with 2-(4-aminophenyl)-1H-indole-6-carboxamidine (DAPI) and Phalloidin (F-actin) (scale bar = 100 μm). (**e**) Expression of DAPI and phalloidin was quantified by immunofluorescence after 3 days in culture. (**f**) The proliferation of HGF cells encapsulated in KFH-3 and FIB-H for 7 days, as measured in a CCK-8 assay. Both data were evaluated for cyto-compatibility at a concentration of 20 mg/mL. Data are presented as the mean ± SD of triplicate experiments: * $p < 0.05$.

KFH-3, possessing its inherent interconnected porous structure, showed a high cell proliferation and cell viability, while the non-porous FIB-H showed a lower cytocompatibility than KFH-3. HGFs elongated and spread out to form interconnected networks in the KFH-3 hydrogel, compared with FIB-H. Notably, hydrogels with small pores of less than 10 μm are undesirable for use in tissue engineering applications, as they limit cell migration and nutrient diffusion [29–32]. FIB-H has small pores and large, non-uniform, and different surface areas. After 3 days of culture, the live cell density in the hydrogel with KRT was also higher than that in the hydrogel without KRT. Thus, KFH-3 induced a higher cell proliferation and exhibited a larger adhesion area than the FIB-H hydrogel. The hydrogel developed in this study has the porosity and microarchitecture required for use in an in vitro 3D environment, allowing the free flow of biological fluids and cell migration and growth inside the material.

3. Materials and Methods

3.1. Materials

The KRT protein was extracted from human hair and the detailed procedure has been described in our previous study [33]. Human plasma FIB (Merck Millipore, MA, USA) and human plasma THR (Calbiochem, CA, USA) were used to prepare the fibrin gel. Dulbecco's phosphate-buffered saline (PBS) was obtained from Thermo Fisher Scientific (Waltham, MA, USA). EDC/NHS, Tris-buffered saline, and sodium hydroxide (NaOH) were obtained from Sigma-Aldrich.

HGFs were purchased from ScienCell Research Laboratories (Carlsbad, CA, USA). All chemicals were used without further purification.

3.2. Succinylation of KRT Protein (KRT-COOH)

Succinic anhydride (25 mg) dissolved in PBS was added dropwise to a solution of KRT (200 mg) in 100 mL PBS (pH 6–7). The resulting mixture was stirred at room temperature for 1 h and distilled water (DW) for 4 h. The purified KRT-COOH was lyophilized and stored at −20 °C until further use.

3.3. Synthesis of KRT-FIB Precursors via An EDC/NHS Coupling Reaction

To synthesize a series of KRT-FIB precursors, different amounts of KRT-COOH (0, 5, 10, 30, and 60 mg) were activated in the presence of EDC/NHS. First, 19.25 mg of EDC was added to a solution of KRT-COOH in 100 mL of PBS. Then, 54.25 mg of sulfur-NHS was added to the solution and stirred at room temperature for 15 min. Then, 62.5 μL of 2-mercaptoethanol was added drop-wise. Subsequently, the pH of the resulting mixture was adjusted to 7.0 via the addition of 10X PBS (10 mL), and 100 mg of FIB was added to the mixture while stirring at room temperature. The reaction was allowed to proceed for 2 h and was then quenched by the addition of a Tris buffer. The pH of the quenched solution was increased to 8.0 by adding 1 N NaOH. The resulting precursor solutions were placed in a dialysis membrane with a molecular weight cut-off of 25,000 g/mol (Spectrum Laboratories Inc., Rancho Dominguez, CA, USA) and dialyzed over DW for three days. The purified precursors were then lyophilized and stored at −20 °C until further use.

3.4. SDS-PAGE

SDS-PAGE was performed to confirm the formation of the precursors. All samples (KF-1, 3, and 6) were run on a 7% Tris-acrylamide running gel with 4% stacking, for 90 min at 150 V, and the electrophoretic bands were stained with Coomassie Brilliant Blue R-250.

3.5. Preparation of KFHs and FIB-H Hydrogels

A series of KFHs (1, 3, and 6) were prepared using the corresponding precursors with different KRT contents (KF-1, 3, 6). First, solutions including each precursor (20 mg/mL), THR (5 IU/mL), and $CaCl_2$ (2.5 mg/mL) were prepared. Then, 2 mL of the precursor solution was loaded onto one side of a dual-chambered syringe with an 18 G needle (diameter = 1.3 mm), and 0.1 mL of THR solution was loaded onto the other side. The pre-filled syringe was slowly extruded on a glass plate at room temperature. As a control, FIB-H was prepared following the same procedure as the FIB solution (20 mg/mL). To evaluate the injection performance, the solutions of the precursor and THR prefilled in the dual-chambered syringe were injected into DW and on a glass plate.

3.6. Characterization of Injectable Hydrogels

The thermal properties of a series of KRT-FIB precursors and the corresponding hydrogels (KFHs) were determined by TGA (SDT Q600, TA Instruments, New Castle, PA, USA) under nitrogen gas flow, and the temperature ranged from 20 to 600 °C at a heating rate of 10 °C/min. The morphology of the hydrogels was observed by SEM (S-4700, Hitachi, Japan) with a gold coating. The rheological properties of the hydrogels were investigated using an ARES-G2 rheometer (TA Instruments, New Castle, PA, USA) at 37 °C with a parallel-plate geometry of 8 mm in diameter

To examine the injectable performance of the KFH, an extrusion experiment was performed. A dual-chambered syringe was used to make injectable hydrogels. In all cases, the precursor solutions/THR in each dual-chambered syringe were liquids with very low viscosity and were easily extruded, first through a static mixer placed at the outlet of the dual-chambered syringe and then through an 18 G needle. A dual-chambered syringe loaded with a FIB and KF-3 precursor, and THR solution was slowly extruded by hand into DW and on a glass plate. The extrusion was carried out by pressing down on the syringe plunger, by hand. The rheology of the injectable hydrogels was tested using a rheometer (HAAKE, Model MARS, Karlsruhe, Germany). Rheology Advantage Data Analysis (TA Instruments, Karlsruhe, Germany) software was then used to plot the frequency sweeps. All hydrogels were cylindrical with a diameter of 20 mm and a thickness of 1 mm. A dynamic frequency sweep test from 0.1 to 100 rad/s was performed to determine the dynamic storage modulus (G′) of each hydrogel, at a strain rate confirmed to be in the linear viscoelastic range for each type of hydrogel prepared. The temperature was maintained at 37 °C during all measurements. At least five different hydrogels were tested with the same experimental settings; average values are presented.

3.7. Swelling and Degradation Studies

The swelling ratios of KFH-3 and FIB-H were evaluated based on their weight change. First, cylindrical hydrogel samples (diameter × height = 5 × 2 mm) were prepared and air-dried to measure the dry weight ($\boldsymbol{W}_{\mathrm{dry}}$) of the samples. The samples were then immersed in DW at room temperature, and the weights of wet samples ($\boldsymbol{W}_{\mathrm{wet}}$) were measured at certain intervals. All measurements were performed in triplicate and the swelling ratio (%) was calculated using the following formula:

$$\textbf{Swelling ratio}\ (\%) = \frac{W_{wet} - W_{dry}}{W_{dry}} \times 100\% \quad (1)$$

Temperature and pH-responsive degradation studies of hydrogels were evaluated by the dry weight ratio of hydrogel samples. Well-established cylindrical KFH-3 and FIB-H hydrogels (diameter × height = 5 × 2 mm) were soaked in cell media containing plasmin and non-plasmin proteases, prepared at 5.5 pH and incubated at 50 °C, under mild shaking conditions. The hydrogels were taken at different time points (0.5, 1, 2, 3, 4, 5, 12, 24 h) and dried. The dried hydrogels were weighted ($\boldsymbol{W}_{\mathrm{t}}$) and compared with the initial weights ($\boldsymbol{W}_{0}$) of samples. The weight change of hydrogels was calculated as follows:

$$\textbf{Weight}\ (\%) = \frac{(w_o - w_t)}{w_o} \times 100 \quad (2)$$

3.8. Cell Culture

Primary HGF cells were purchased from ScienCell Research Laboratories. Cells were propagated in fibroblast medium, supplemented with 2% (*v*/*v*) fetal bovine serum, fibroblast growth supplement, and penicillin (100 U/mL)-streptomycin (100 µg/mL), on poly-l-lysine–coated flasks (ScienCell Research Laboratories, Carlsbad, CA, USA). The cells were cultured in a humidified incubator at 37 °C and 5% CO_2. The culture medium was replaced every two days. HGF passages 4–7 were used in the experiments.

3.9. Cell Encapsulation

The KFH-3 was evaluated by the encapsulation test of HGF cells using the FIB hydrogel as a control. Briefly, HGF cells were suspended in media at a density of 5×10^5 cells/mL. The cell suspension (200 µL) was then mixed with the same volume of FIB solution, or

KRT-FIB solution with THR. After gel formation, 2% (*v*/*v*) fetal bovine serum, fibroblast growth supplement and penicillin (100 U/mL)-streptomycin (100 μg/mL) was added. Then, HGF-embedded hydrogels were maintained at 37 °C and 5% CO_2 in a humidified incubator until processing.

3.10. Cell Viability

A LIVE/DEAD assay was used to investigate the viability of encapsulated HGF cells. After encapsulation, the samples were incubated in a solution of calcein-AM/ethidium homodimer for 30 min at 37 °C, immediately and after 3 days of culture. A calcein-AM/ethidium homodimer LIVE/DEAD assay was used to quantify cell viability according to the manufacturer's instructions. Cell morphologies were observed under a fluorescence microscope (IX71; Olympus Life Science; Tokyo, Japan). The cell proliferation was evaluated using Cell Counting Kit-8 (CCK-8, DOJINDO, Kumamot, Japan), after incubation of 1, 3, 5, and 7 days. Briefly, the cultured cells were incubated with 10% CCK-8 working solution in a cell culture medium for 3 h at 37 °C, in the dark. Then, 100 μL of the supernatant was extracted to a new 96-well plate and the absorbance at 450 nm was measured using Plate Readers (PerkinElmer, MA, USA).

3.11. Confocal Laser Scanning Microscopy

HGF cells were seeded into the hydrogels and placed in 6-well plates. After 5 days of culture, the cells were encapsulated in a fixed hydrogel and stained with rhodamine-phalloidin (Invitrogen) and 2-(4-aminophenyl)-1H-indole-6-carboxamidine to visualize F-actin filaments and cell nuclei, respectively. Fluorescence images were obtained using an LSM 980 with Airyscan 2 (Carl Zeiss, Jena, Germany). The imaging setup comprised a water immersion lens with a 40× objective, a slice thickness of 5 μm, and a total thickness of 205 μm. The hydrogel images of the entire volume were obtained using fluorescence emission intensity, processed, and combined into a 3D volume using image processing software (ZEN black software, Oberkochen, Germany).

3.12. Statistical Analysis

Statistical analyses were carried out by one-way analysis of variance with Tukey's post hoc analysis for significance. Statistical significance was set at $p < 0.05$. For rheology and image analysis, statistical significance was analyzed using the Origin software (Origin Software, CA, USA).

4. Conclusions

In this study, the injectable KFH was developed for improving the porosity and viscosity by controlling the molar ratio of KRT and FIB. The resulting KFH showed the synergistically improved proliferation of encapsulated HGFs, and it was also observed that HGFs exhibited a diffuse morphology when they were encapsulated in KFH. These results suggest that the novel KRT-based hydrogel has great potential for use as a scaffold for tissue regeneration for biomedical applications.

Author Contributions: Conceptualization, H.J.K., N.K., Y.-S.H. and S.Y.K.; Data curation, H.J.K., N.K., S.J.O., Y.-S.H. and S.Y.K.; Formal Analysis, H.J.K., N.K., S.J.O., S.Y.A., Y.-S.H. and S.Y.K.; Funding Acquisition, H.J.K., N.K., Y.-S.H. and S.Y.K.; Methodology, N.K., S.J.O., Y.-S.H. and S.Y.K.; Writing—original draft, H.J.K., N.K., Y.-S.H. and S.Y.K.; Revision of manuscript, S.Y.A., Y.-S.H. and S.Y.K.; H.J.K. and N.K. contributed equally as first authors; Y.-S.H. and S.Y.K. contributed equally as a corresponding author. All authors have read and agreed to the published version of the manuscript.

Funding: This research was supported by a grant of the Korea Health Technology R&D Project through the Korea Health Industry Development Institute (KHIDI), funded by the Ministry of Health & Welfare, Republic of Korea (grant number: HI18C2383). This work was supported by the National Research Foundation of Korea (NRF) grant funded by the Korea government (MSIT) (NRF-2019R1F1A1061704). This work was supported by the Technology Innovation Program (20008650,

Development of Recombinant Keratin Protein-based Biomaterials for Soft Tissue Regeneration) funded by the Ministry of Trade, Industry & Energy (MOTIE, Korea).

Conflicts of Interest: The authors declare no conflict of interest.

References

1. Moll, R.; Divo, M.; Langbein, L. The human keratins: Biology and pathology. *Histochem. Cell Biol.* **2008**, *129*, 705. [CrossRef] [PubMed]
2. Marceau, N.; Loranger, A.; Gilbert, S.; Daigle, N.; Champetier, S. Keratin-mediated resistance to stress and apoptosis in simple epithelial cells in relation to health and disease. *Biochem. Cell Biol.* **2001**, *79*, 543–555. [CrossRef] [PubMed]
3. Sierpinski, P.; Garrett, J.; Ma, J.; Apel, P.; Klorig, D.; Smith, T.; Koman, L.A.; Atala, A.; Van Dyke, M. The use of keratin biomaterials derived from human hair for the promotion of rapid regeneration of peripheral nerves. *Biomaterials* **2008**, *29*, 118–128. [CrossRef] [PubMed]
4. Richter, J.R.; De Guzman, R.C.; Greengauz-Roberts, O.K.; Van Dyke, M. Structure-property relationships of meta-kerateine biomaterials derived from human hair. *Acta Biomater.* **2012**, *8*, 274–281. [CrossRef]
5. Rahmany, M.B.; Hantgan, R.R.; Van Dyke, M. A mechanistic investigation of the effect of keratin-based hemostatic agents on coagulation. *Biomaterials* **2013**, *34*, 2492–2500. [CrossRef]
6. Rouse, J.G.; Van Dyke, M.E. A review of keratin-based biomaterials for biomedical applications. *Materials* **2010**, *3*, 999–1014. [CrossRef]
7. Verma, V.; Verma, P.; Ray, P.; Ray, A.R. Preparation of scaffolds from human hair proteins for tissue-engineering applications. *Biomed. Mater.* **2008**, *3*, 025007. [CrossRef]
8. Hsieh, J.Y.; Smith, T.D.; Meli, V.S.; Tran, T.N.; Botvinick, E.L.; Liu, W.F. Differential regulation of macrophage inflammatory activation by fibrin and fibrinogen. *Acta Biomater.* **2017**, *47*, 14–24. [CrossRef]
9. Horbett, T.A. Fibrinogen adsorption to biomaterials. *J. Biomed. Mater. Res. A* **2018**, *106*, 2777–2788. [CrossRef]
10. Solovieva, E.V.; Fedotov, A.Y.; Mamonov, V.E.; Komlev, V.S.; Panteleyev, A.A. Fibrinogen-modified sodium alginate as a scaffold material for skin tissue engineering. *Biomed. Mater.* **2018**, *13*, 025007. [CrossRef]
11. Litvinov, R.I.; Weisel, J.W. Fibrin mechanical properties and their structural origins. *Matrix Biology* **2017**, *60*, 110–123. [CrossRef]
12. Yu, Z.; Li, H.; Xia, P.; Kong, W.; Chang, Y.; Fu, C.; Wang, K.; Yang, X.; Qi, Z. Application of fibrin-based hydrogels for nerve protection and regeneration after spinal cord injury. *J. Biol. Eng.* **2020**, *14*, 1–15. [CrossRef]
13. Cao, Y.; Yao, Y.; Li, Y.; Yang, X.; Cao, Z.; Yang, G. Tunable keratin hydrogel based on disulfide shuffling strategy for drug delivery and tissue engineering. *J. Colloid Interface Sci.* **2019**, *544*, 121–129. [CrossRef]
14. Lou, J.; Liu, F.; Lindsay, C.D.; Chaudhuri, O.; Heilshorn, S.C.; Xia, Y. Dynamic hyaluronan hydrogels with temporally modulated high injectability and stability using a biocompatible catalyst. *Adv. Mater.* **2018**, *30*, 1705215. [CrossRef]
15. Han, C.; Zhang, H.; Wu, Y.; He, X.; Chen, X. Dual-crosslinked hyaluronan hydrogels with rapid gelation and high injectability for stem cell protection. *Sci. Rep.* **2020**, *10*, 1–7.
16. Zhu, Y.; Luo, Q.; Zhang, H.; Cai, Q.; Li, X.; Shen, Z.; Zhu, W. A shear-thinning electrostatic hydrogel with antibacterial activity by nanoengineering of polyelectrolytes. *Biomater. Sci.* **2020**, *8*, 1394–1404. [CrossRef]
17. Slaughter, B.V.; Khurshid, S.S.; Fisher, O.Z.; Khademhosseini, A.; Peppas, N.A. Hydrogels in regenerative medicine. *Adv. Mater.* **2009**, *21*, 3307–3329. [CrossRef]
18. Yan, C.; Pochan, D.J. Rheological properties of peptide-based hydrogels for biomedical and other applications. *Chem. Soc. Rev.* **2010**, *39*, 3528–3540. [CrossRef]
19. Mendez, J.J.; Ghaedi, M.; Sivarapatna, A.; Dimitrievska, S.; Shao, Z.; Osuji, C.O.; Steinbacher, D.M.; Leffell, D.J.; Niklason, L.E. Mesenchymal stromal cells form vascular tubes when placed in fibrin sealant and accelerate wound healing in vivo. *Biomaterials* **2015**, *40*, 61–71. [CrossRef]
20. Dai, Y.; Liu, G.; Ma, L.; Wang, D.; Gao, C. Cell-free macro-porous fibrin scaffolds for in situ inductive regeneration of full-thickness cartilage defects. *J. Mater. Chem. B* **2016**, *4*, 4410–4419. [CrossRef]
21. Sauerwein, M.; Steeb, H. Modeling of dynamic hydrogel swelling within the pore space of a porous medium. *Int. J. Eng. Sci.* **2020**, *155*, 103353. [CrossRef]
22. Reinertsen, E.; Skinner, M.; Wu, B.; Tawil, B. Concentration of fibrin and presence of plasminogen affect proliferation, fibrinolytic activity, and morphology of human fibroblasts and keratinocytes in 3D fibrin constructs. *Tissue Eng. Part A* **2014**, *20*, 2860–2869. [CrossRef]
23. Tsai, T.-Y.; Shen, K.-H.; Chang, C.-W.; Jovanska, L.; Wang, R.; Yeh, Y.-C. In situ formation of nanocomposite double-network hydrogels with shear-thinning and self-healing properties. *Biomater. Sci.* **2021**, *9*, 985–999. [CrossRef]
24. Rahman, M.S.; Islam, M.M.; Islam, M.S.; Zaman, A.; Ahmed, T.; Biswas, S.; Sharmeen, S.; Rashid, T.U.; Rahman, M.M. Morphological Characterization of Hydrogels. In *Cellulose-Based Superabsorbent Hydrogels*; Mondal, M.I.H., Ed.; Springer International Publishing: Cham, Switzerland, 2019; pp. 819–863.
25. Gao, L.; Chen, J.; Feng, W.; Song, Q.; Huo, J.; Yu, L.; Liu, N.; Wang, T.; Li, P.; Huang, W. A multifunctional shape-adaptive and biodegradable hydrogel with hemorrhage control and broad-spectrum antimicrobial activity for wound healing. *Biomater. Sci.* **2020**, *8*, 6930–6945. [CrossRef]

26. Akther, F.; Little, P.; Li, Z.; Nguyen, N.-T.; Ta, H.T. Hydrogels as artificial matrices for cell seeding in microfluidic devices. *RSC Adv.* **2020**, *10*, 43682–43703. [CrossRef]
27. Annabi, N.; Mithieux, S.M.; Weiss, A.S.; Dehghani, F. Cross-linked open-pore elastic hydrogels based on tropoelastin, elastin and high-pressure CO2. *Biomaterials* **2010**, *31*, 1655–1665. [CrossRef]
28. Rosser, J.; Calvo, I.; Schlager, M.; Purtscher, M.; Jenner, F.; Ertl, P. Recent advances of biologically inspired 3D microfluidic hydrogel cell culture systems. *J. Cell Biol. Cell Metab.* **2015**, *2*, 1–14.
29. Lee, J.-Y.; Tan, B.; Cooper, A.I. CO2-in-water emulsion-templated poly (vinyl alcohol) hydrogels using poly (vinyl acetate)-based surfactants. *Macromolecules* **2007**, *40*, 1955–1961. [CrossRef]
30. Murphy, C.M.; O'Brien, F.J. Understanding the effect of mean pore size on cell activity in collagen-glycosaminoglycan scaffolds. *Cell Adhes. Migr.* **2010**, *4*, 377–381. [CrossRef]
31. Gun'Ko, V.M.; Mikhalovska, L.I.; Savina, I.N.; Shevchenko, R.V.; James, S.L.; Tomlins, P.E.; Mikhalovsky, S.V. Characterisation and performance of hydrogel tissue scaffolds. *Soft Matter* **2010**, *6*, 5351–5358. [CrossRef]
32. Rivero, R.E.; Capella, V.; Liaudat, A.C.; Bosch, P.; Barbero, C.A.; Rodríguez, N.; Rivarola, C.R. Mechanical and physicochemical behavior of a 3D hydrogel scaffold during cell growth and proliferation. *RSC Adv.* **2020**, *10*, 5827–5837. [CrossRef]
33. Kim, S.Y.; Park, B.J.; Lee, Y.; Park, N.J.; Park, K.M.; Hwang, Y.-S.; Park, K.D. Human hair keratin-based hydrogels as dynamic matrices for facilitating wound healing. *J. Ind. Eng. Chem.* **2019**, *73*, 142–151. [CrossRef]

International Journal of
Molecular Sciences

MDPI

Article

Development of Polyvinyl Alcohol/Kaolin Sponges Stimulated by Marjoram as Hemostatic, Antibacterial, and Antioxidant Dressings for Wound Healing Promotion

Tamer M. Tamer [1,*,†], Mosa H. Alsehli [2], Ahmed M. Omer [1], Tarek H. Afifi [2], Maysa M. Sabet [1], Mohamed S. Mohy-Eldin [1] and Mohamed A. Hassan [3,4,*,†]

1 Polymer Materials Research Department, Advanced Technology and New Materials Research Institute (ATNMRI), City of Scientific Research and Technological Applications (SRTA-City), New Borg El-Arab City, Alexandria 21934, Egypt; Ahmedomer_81@yahoo.com (A.M.O.); Maysamohamed19@yahoo.com (M.M.S.); mohyeldinmohamed@gmail.com (M.S.M.-E.)
2 Department of Chemistry, Taibah University, Madinah 30002, Saudi Arabia; mosa_alsehli@hotmail.com (M.H.A.); afifith@yahoo.com (T.H.A.)
3 Protein Research Department, Genetic Engineering and Biotechnology Research Institute (GEBRI), City of Scientific Research and Technological Applications (SRTA-City), New Borg El-Arab City, Alexandria 21934, Egypt
4 University Medical Center Göttingen, Georg-August-University, 37073 Göttingen, Germany
* Correspondence: ttamer85@gmail.com (T.M.T.); madel@srtacity.sci.eg (M.A.H.); Tel.: +20-34593414 (T.M.T.); +20-34593422 (M.A.H.)
† These authors equally contributed to this work.

Citation: Tamer, T.M.; Alsehli, M.H.; Omer, A.M.; Afifi, T.H.; Sabet, M.M.; Mohy-Eldin, M.S.; Hassan, M.A. Development of Polyvinyl Alcohol/Kaolin Sponges Stimulated by Marjoram as Hemostatic, Antibacterial, and Antioxidant Dressings for Wound Healing Promotion. *Int. J. Mol. Sci.* **2021**, 22, 13050. https://doi.org/10.3390/ijms222313050

Academic Editor: Peter John Jervis

Received: 21 October 2021
Accepted: 24 November 2021
Published: 2 December 2021

Abstract: The predominant impediments to cutaneous wound regeneration are hemorrhage and bacterial infections that lead to extensive inflammation with lethal impact. We thus developed a series of composite sponges based on polyvinyl alcohol (PVA) inspired by marjoram essential oil and kaolin (PVA/marjoram/kaolin), adopting a freeze–thaw method to treat irregular wounds by thwarting lethal bleeding and microbial infections. Microstructure analyses manifested three-dimensional interconnected porous structures for PVA/marjoram/kaolin. Additionally, upon increasing marjoram and kaolin concentrations, the pore diameters of the sponges significantly increased, recording a maximum of 34 ± 5.8 μm for PVA-M0.5-K0.1. Moreover, the porosity and degradation properties of PVA/marjoram/kaolin sponges were markedly enhanced compared with the PVA sponge with high swelling capacity. Furthermore, the PVA/marjoram/kaolin sponges exerted exceptional antibacterial performance against *Escherichia coli* and *Bacillus cereus*, along with remarkable antioxidant properties. Moreover, PVA/marjoram/kaolin sponges demonstrated significant thrombogenicity, developing high thrombus mass and hemocompatibility, in addition to their remarkable safety toward fibroblast cells. Notably, this is the first study to our knowledge investigating the effectiveness of marjoram in a polymeric carrier for prospective functioning as a wound dressing. Collectively, the findings suggest the prospective usage of the PVA-M0.5-K0.1 sponge in wound healing for hemorrhage and bacterial infection control.

Keywords: PVA; marjoram oil; kaolin; hemostatic dressing; antibacterial and antioxidant wound dressing

1. Introduction

Wound healing is an inherent rejuvenating reaction to skin damage. Essentially, the cutaneous healing process is orchestrated into four overlapping phases: hemostasis, inflammation, proliferation, and tissue remodeling [1,2].

Instantly following a skin injury, avascular constriction and blood clotting cascade preclude the hemorrhage and infiltration of microorganisms. Furthermore, blood clots also serve as a reservoir for growth factors and cytokines as well as a scaffold for immigrant cells recruited for wound recovery and tissue regeneration [3,4]. On the other hand, bleeding has been the predominant cause of mortality in civilian and military communities over past decades. In trauma patients, instant hemorrhage management through blood clotting intervention is imperative since most deaths emerge within the first hour following severe trauma [5,6]. Moreover, uncontrolled bleeding impedes wound healing and provokes significant consequences, including severe inflammation and microbial infections of the wound [7].

The blood clotting pathway is coordinated in three sequential phases: (I) initiation, which involves the production of thrombin, (II) amplification, which is characterized by activation and aggregation of platelets; and (II) proliferation, which is characterized by the development of fibrin alongside stability of platelet clot. Most applied hemostatic agents play a crucial role by enhancing platelet aggregation and coagulation within the amplification and proliferation phases [8].

A hemostatic agent should encounter specific vital characteristics, including ease of use, cost-effectiveness, hemocompatibility, cytocompatibility, and intrinsic biodegradability [9]. Injectable hydrogels have revealed good healing with hemostatic properties [10]; nevertheless, they typically have insufficient mechanical features, curtailing their implementation as hemostatic materials [11]. Furthermore, such injectable hydrogels might incite adverse complications for patients, aggravating the treated wound since it is painful either to change or remove them, particularly from profound injuries [12]. Thus, developing porous sponges with hemostatic, antibacterial, and antioxidant performances is advantageous because of their distinct three-dimensional structures, which enable them to have mechanical stability. Moreover, it is worth pointing out that the three-dimensional structure fortifies the adhesion and proliferation of cells recruited for wound healing [9].

Polyvinyl alcohol (PVA)-based porous sponges have demonstrated outstanding mechanical properties and remarkable biocompatibility [13]. Physical cross-linking could be adopted to formulate PVA sponges to evade the utilization of chemical cross-linking solvents, which might be hazardous for biomedical applications. Therefore, sequential freeze–thawing could cross-link PVA-based sponges, which could develop crystalline clusters as a cross-linking point [14,15]. Given that the PVA sponge lacks hemostatic and antibacterial capacities, it is thus indispensable to adapt the PVA sponge by incorporating a hemostatic agent, such as kaolin to stimulate blood cells and platelet accumulation, and another bioactive compound (marjoram essential oil) to endow the sponge with antibacterial and antioxidant capacities to frustrate the inflammation in the wound bed.

Kaolin has thus been identified as one of the significant hemostatic agents, which could substantially promote blood coagulation [16]. Kaolin, often known as China clay, principally comprises kaolinite and aluminium silicate [16]. Given that negative charges on the kaolin surface could substantially promote blood coagulation, kaolin was successfully implemented as active material in surgical hemostasis. Furthermore, kaolin could induce factor XII and platelets, further initiating the blood clotting cascade [17].

Marjoram (*Origanum majorana* L.) is one of the promising essential oil candidates for several applications. It is an herbaceous and perennial plant broadly cultivated in various areas of the globe and could be used for traditional medicine [18]. In addition, previous studies have evidenced marjoram's antibacterial, antioxidant, anti-inflammatory, and anticancer properties [19]. The major constituents of marjoram essential oil are terpinen-4-ol (which accounts for more than 20% of the total), (+)-cis-sabinene hydrate (3–18%), γ-terpinene and terpinolene, thymol, and carvacrol [20]. In addition, the high antioxidant activity of oil derived from marjoram accounted for the predominance of phenolic acids and terpenoids [21].

We presumed that the synergistic effect of the sponge composite based on PVA, marjoram essential oil, and kaolin could effectively frustrate the hemorrhage and the microbial infections of injury alongside the competency of PVA to seal the wound, which could significantly accelerate the action of the devised sponge. We thus fabricated a series of PVA/marjoram/kaolin porous sponge composites as hemostatic and antibacterial wound dressings. The designed sponges were characterized to determine their physical and chemical attributes in order to explore the extent to which these sponges correspond with the paramount traits of the ideal wound dressing. Additionally, the antibacterial and antioxidant activities and the hemostatic capability of the developed sponges were assessed. Moreover, the cytotoxicity of the sponges was studied in relation to the human dermal fibroblast. The findings of these in vitro investigations will ascertain the favorable sponge, which will require further in vitro and in vivo investigations to fully validate its application as a promising hemostatic and antibacterial dressing.

2. Results and Discussion

2.1. Development of PVA/Marjoram/Kaolin Sponges

In our previous investigations, we manifested the potency of hemostatic and antibacterial performances of PVA/kaolin, which were reinforced by penicillin–streptomycin [22]. However, the antibacterial activities of these sponges substantially depend on the presence of antibiotics that might be frustrated by the emergence of resistant bacteria. Furthermore, the antioxidant activity of these sponges has not been reported.

Due to these shortcomings, we aspired in this study to devise non-antibiotic sponges for further implementation in the wound healing process. Given the particular structure and tunable traits of PVA, it is predominantly recruited for hydrogel development by cyclical freezing–thawing to develop crystalline clusters alongside its competency acting as a carrier of drugs [23]. As portrayed in Figure 1, we formulated new crosslinked sponges using PVA inspired by marjoram extract and kaolin employing a freezing-thawing approach. Moreover, due to the antibacterial, antioxidant, and hemostatic deficiencies of PVA, marjoram extract and kaolin were incorporated into the designed sponges. After freezing-thawing, PVA, marjoram extract, and kaolin, in addition to high water molecules, were entrapped and entangled into the physically crosslinked three-dimensional network of sponges. The developed sponges with various concentrations of marjoram extract were labelled as PVA-M0.1, PVA-M0.25, and PVA-M0.5, while those boosted by marjoram extract and kaolin were denominated as PVA-M0.5-K0.1, PVA-M0.5-K0.25, and PVA-M0.5-K0.5.

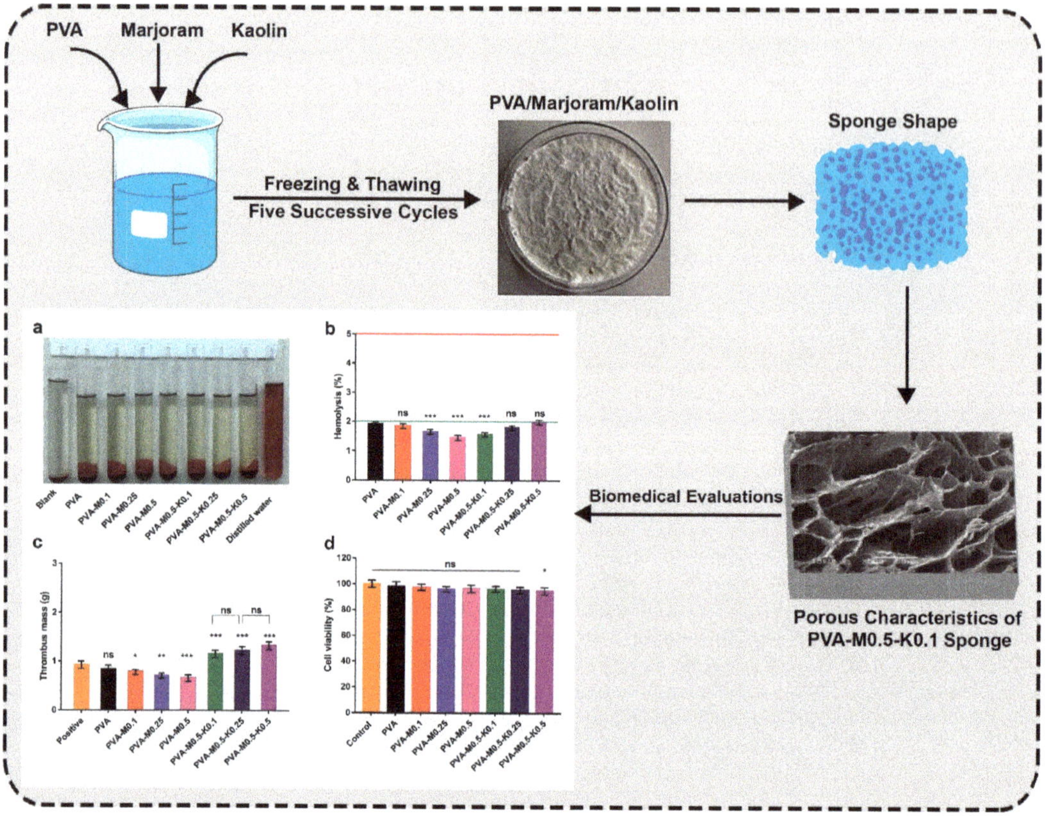

Figure 1. Schematic illustration of the fabrication and biomedical evaluations of hemostatic and antibacterial PVA/marjoram/kaolin sponges. ((**a–d**) in the scheme depict some of the biomedical evaluations performed for PVA/marjoram/kaolin composite sponges, including biocompatibility and thrombogenicity with statistical analysis (*** $p < 0.001$, ** $p < 0.01$, * $p < 0.05$, and (ns) points to a non-significant difference).

2.2. Characterization of PVA/Marjoram/Kaolin Sponges

2.2.1. FT-IR Analysis

As illustrated in Figure 2, the structures of the PVA sponge and the alterations of PVA composite sponges, including marjoram oil and kaolin with different contents, were studied by means of FT-IR. The appearance of characteristic bands at 3387 cm^{-1} is assigned to –OH on hydrogen bonds between –OH groups among PVA chains, bestowing the hydrophilic strengths to the PVA sponge [24]. The methyl groups' asymmetrical and symmetrical C-H stretching vibration modes could be perceived in the PVA spectrum at 2926 cm^{-1}. Moreover, the band at 2845 cm^{-1} corresponds to a methylene vibration band, whereas the distinct band at 1710 cm^{-1} is imputed to the stretching vibration band of the remaining acetyl carbonyl groups. The appearance of the band at 1450 cm^{-1} is attributed to asymmetrical and symmetrical CH bending vibrations of methyl groups [24].

Furthermore, a prominent band at 1118 cm^{-1} is the main indicator of the PVA structure [25], while a band at 1085 cm^{-1} corresponds to C–O–C. Incorporation of marjoram into PVA generated a new band at 1660 cm^{-1}, which interacted with the stretching vibration band of the remaining acetyl carbonyl groups in the PVA. Evidently, this band turned out to be clearer at high concentrations of marjoram. On the other hand, the addition of kaolin to PVA/marjoram sponges led to the appearance of peaks at 920 to 940 cm^{-1} related to

Al-OH vibration. Additionally, the bands at 530 and 789 cm^{-1} correspond to the Si-O-Al bond vibration band.

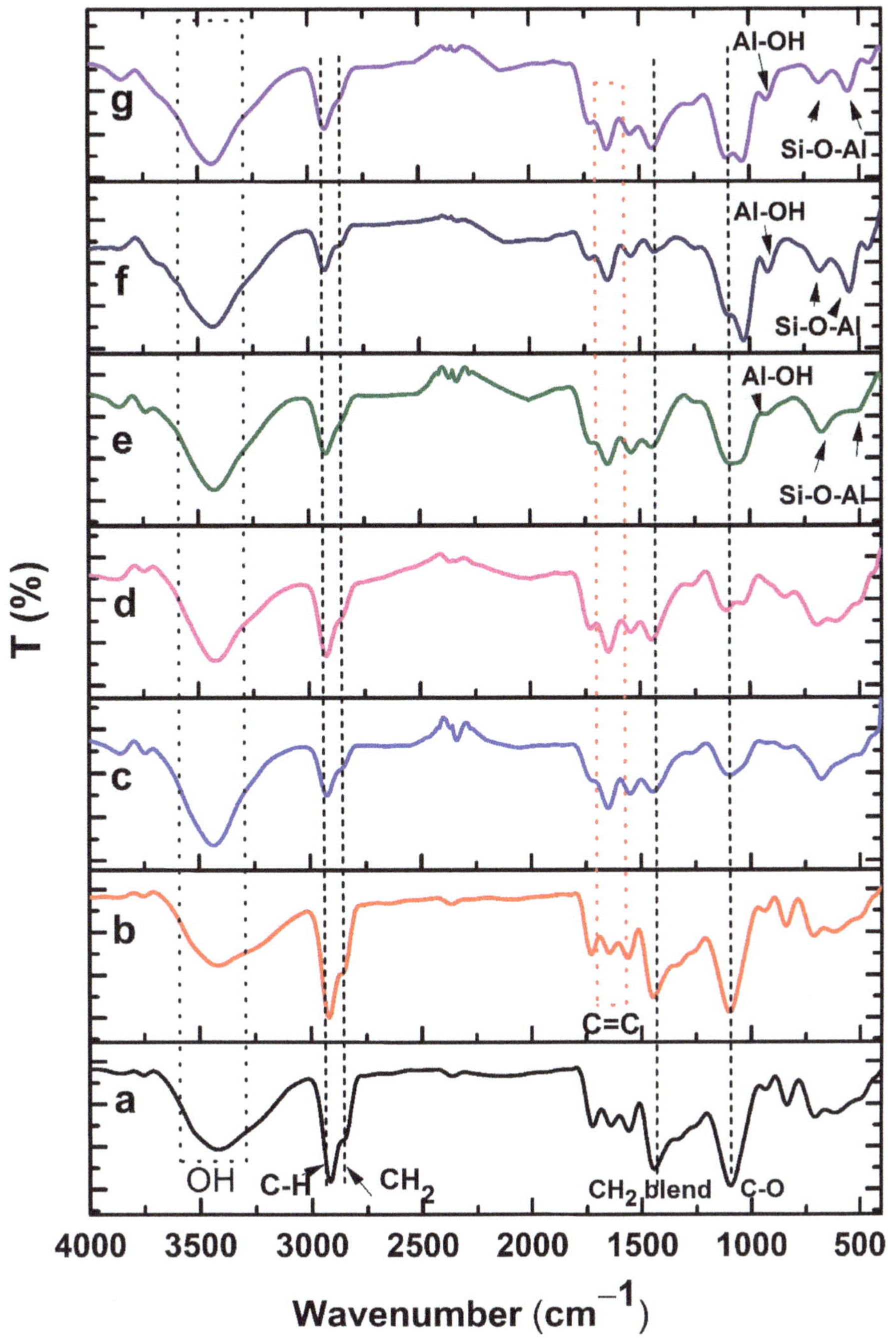

Figure 2. FT-IR spectra of (**a**) PVA, (**b**) PVA-M0.1, (**c**) PVA-M0.25, (**d**) PVA-M0.5, (**e**) PVA-M0.5-K0.1, (**f**) PVA-M0.5-K0.25, and (**g**) PVA-M0.5-K0.5 composite sponges.

2.2.2. SEM Analysis

Microstructures of the synthesized sponges were probed by means of SEM. The morphological surface of the pure PVA sponges exhibited fewer pores than the PVA sponge composites. However, the PVA sponge composites amalgamated with various ratios of either marjoram (PVA/marjoram) or marjoram and kaolin (PVA/marjoram/kaolin) revealed three-dimensional structures interconnected with varied pore sizes in asymmetric arrangements as illustrated in Figure 3a–g. Furthermore, cross-sectional micrographs of the PVA/marjoram and PVA/marjoram/kaolin sponges exposed asymmetric neat structures with prominent three-dimensional interconnected networks as portrayed in Figure 4a–g. Furthermore, the sponge composites showed porous sponge layers with evident lamellar structures comparable to previously applied PVA sponge composites in wound healing with remarkable performances [13,26]. The sponge with a dense structure, on the other hand, is not favorable for hemostatic wound dressings, which necessitates high conductivity to the wound for absorbing wound surplus exudate and interacting with blood to enhance the hemostatic reaction [26,27].

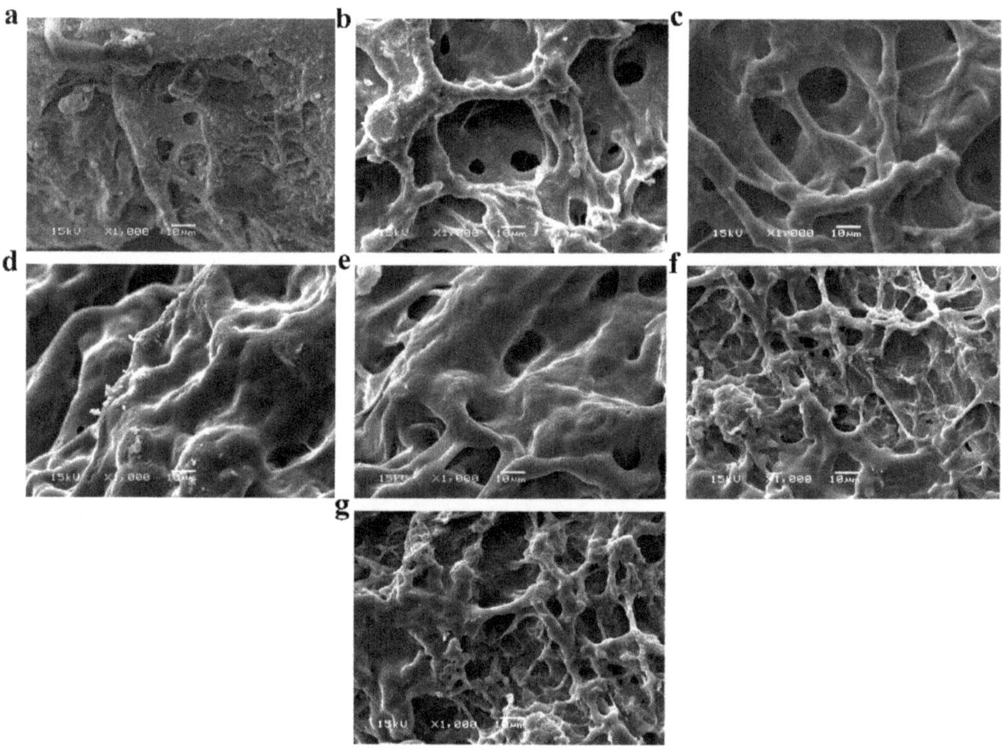

Figure 3. SEM images reveal surface morphologies of (**a**) PVA, (**b**) PVA-M0.1, (**c**) PVA-M0.25, (**d**) PVA-M0.5, (**e**) PVA-M0.5-K0.1, (**f**) PVA-M0.5-K0.25, and (**g**) PVA-M0.5-K0.5 composite sponges at a magnification of 1000×.

Moreover, antibacterial and antioxidant wound dressings should firmly seal the wound to stall the bacterial invasion and mitigate the free radicals generated during the wound healing process [28]. Given these requisites, the development of highly porous sponges is indispensable to promote the attachment and proliferation of keratinocytes and fibroblasts during epithelialization, thereby stimulating angiogenesis at the wound bed [29]. Besides, such sponges prevent the dehydration of the wound, permitting the exchange of fluids and gases toward the wound and furnishing the corresponding cells with nutrients required for its propagation [27].

From the data in Figure 4a′–d′, it can be perceived that the extent of pore sizes of the developed sponges increased with the rise of marjoram ratios, recording 12.9 ± 3.4 μm, 21.2 ± 6 μm, 22.7 ± 4.7 μm, 24.6 ± 5.2 μm for PVA, PVA-M0.1, PVA-M0.25, and PVA-M0.5, respectively. These findings are most likely due to the lessening of molecular crystallizations of PVA, which act on the cross-linking sites of marjoram extract, leading to the diminution in crosslinking density in the case of PVA/marjoram sponges [30,31]. Furthermore, the amalgamation of various kaolin concentrations into PVA/marjoram augmented the pore sizes, reporting 34 ± 5.8 μm, 29.5 ± 7.1 μm, 25.1 ± 4.9 μm for PVA-M0.5-K0.1, PVA-M0.5-K0.25, and PVA-M0.5-K0.5, respectively, as presented in Figure 4e′,f′. As delineated in Figure 4h, the maximum pore size was determined for PVA-M0.5-K0.1, while the increase in kaolin ratios lowered the pore size. This might be attributed to the aggregation of kaolin with marjoram extract as a consequence of the emulsification of the marjoram oil.

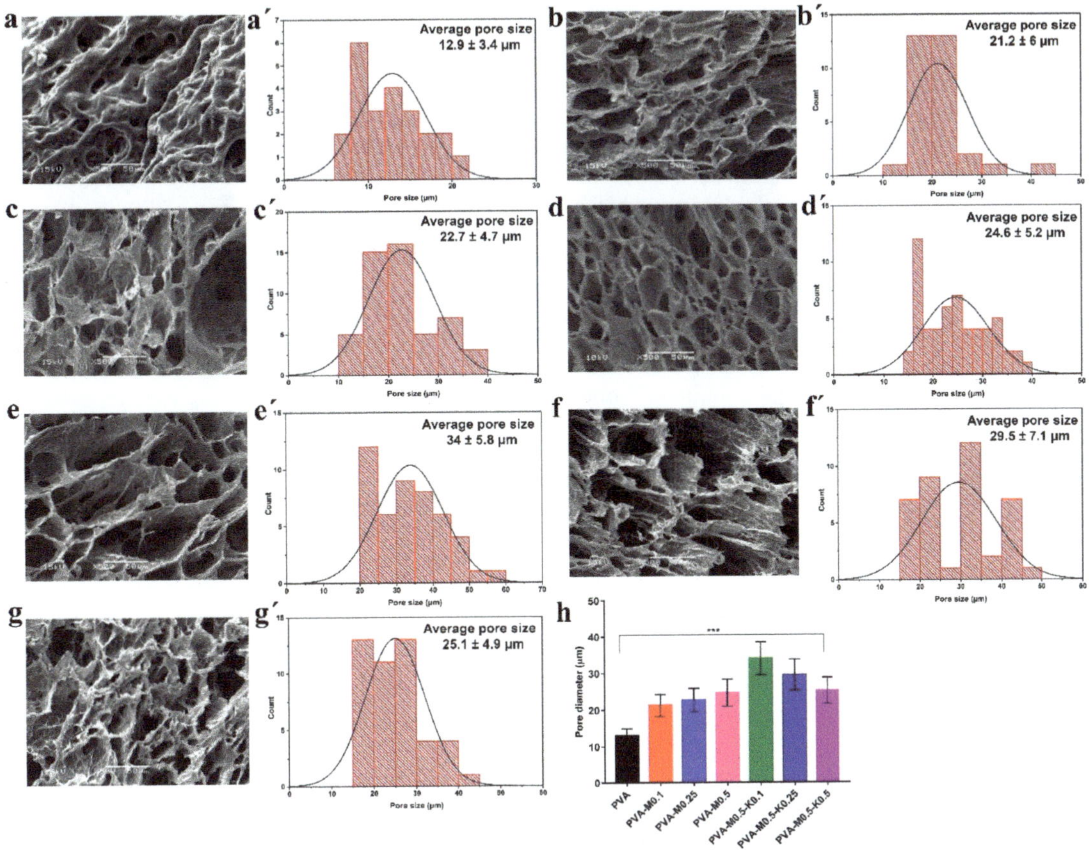

Figure 4. SEM images show cross-sectional and porous structures (**a**) PVA, (**b**) PVA-M0.1, (**c**) PVA-M0.25, (**d**) PVA-M0.5, (**e**) PVA-M0.5-K0.1, (**f**) PVA-M0.5-K0.25, and (**g**) PVA-M0.5-K0.5 composite sponges at a magnification of 500×. (**a′–g′**) illustrates the pore size distribution of PVA, PVA-M0.1, PVA-M0.25, PVA-M0.5, PVA-M0.5-K0.1, PVA-M0.5-K0.25, and PVA-M0.5-K0.5 composite sponges, respectively. (**h**) indicates the size of the average pores of PVA, PVA/marjoram, and PVA/marjoram/kaolin sponges. Results are presented as means ± SD (*** $p < 0.001$).

These findings are consistent with those of earlier studies [31], which demonstrated the outstanding impacts of the wound dressings with pore sizes in a range of 19.5–36.7 μm. Moreover, previously fabricated Poly(ionic liquid)/PVA hydrogels based on PVA with pore sizes ranging from 10 to 30 μm exhibited an influential role during the wound healing of rats [13]. Furthermore, previous reports showed that the wound dressings with pore sizes ranging from 20 to 125 μm had a decisive function in furnishing the dermal cells with oxygen and necessary nourishments, thereby ameliorating the regeneration of skin tissues [32]. Moreover, it has been reported that the measurements of significant human cells extend over a range from 2 to 120 μm [12]. Collectively, the PVA/marjoram and PVA/marjoram/kaolin with pore sizes in the range of 21–34 μm could be encouraging sponge composites for further implementation as wound dressings.

2.2.3. TGA

The thermal degradation behaviors of PVA and PVA/marjoram/kaolin sponges were investigated using TGA as shown in Figure 5. The figure charts represent several degradation steps of sponge composites. The initial stage started at ambient temperature, indicating the loss of piping water moisture content. Specifically, PVA neat sponges exhibited weight losses of 6.55% at 98.8 °C, PVA-M0.1 lost 6.76% at 73.84 °C, PVA-M0.25 lost 6.52% at 71.44 °C, PVA-M0.5 lost 6.16% at 67.67 °C, PVA-M0.5-K0.1 lost 5.98% at 67.5 °C, while PVA-M0.5-K0.25 and PVA-M0.5-K0.5 weight losses were 4.64% and 4.86%, respectively at 67.9 °C. Marjoram oil and kaolin had significant effects on the moisture contents of sponges, in which increase in marjoram or kaolin concentrations exhibited a significant decrease in the trapped solvent or water molecules. This might be attributed to the hydrophobic nature of marjoram oil, whereas in the case of kaolin, it could be explained by a decrease in the PVA content in the composite ratio, in which hydroxyl groups of PVA have a significant role in trapping moisture content. The combination of marjoram oil in the blend sponges resulted in new degradation characters perceived in peak between 109–200 °C with a magnitude weight loss of 3.89% for PVA-M0.1, 7.56% for PVA-M0.25 and 9.4% for PVA-M0.5. This could be attributed to the loss of volatile components of marjoram oil. It could also be recognized that this peak was absent in the sponge containing kaolin, which may be explained by the role of kaolin for stabilizing marjoram oil in the blend structure. Second weight loss for PVA sponges was observed from 226 °C to 314 °C with a weight loss of 62.5% corresponding to thermal decomposition of PVA backbone and eliminated water and carbon dioxide. The weight loss of this degradation stage was decreased by the addition of marjoram and more significantly by adding kaolin. Formulation of PVA blends with marjoram and kaolin stimulated the formations of the internal micro pore that can act as an interior atmosphere and trapped degraded byproduct carbon dioxide gases. The third degradation stage until 600 °C could be ascribed to the decomposition of organic residues. The remaining weight over 600 °C refers to the inorganic remains of kaolin, which did not degrade at this temperature.

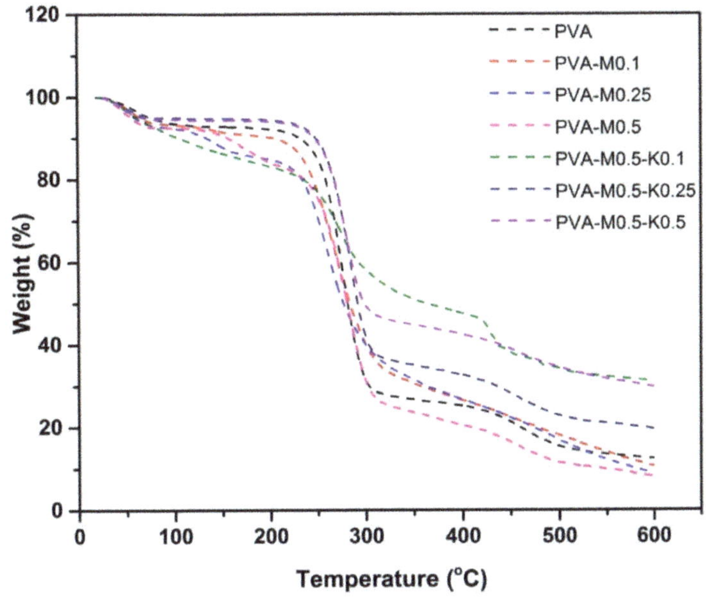

Figure 5. TGA charts of PVA, PVA/marjoram, and PVA/marjoram/kaolin composite sponges.

2.2.4. Gel Fraction, Swelling Profile, Hydrodegradation In Vitro, and Porosity Analyses

Gel fraction properties of PVA/marjoram and PVA/marjoram/kaolin sponges were inspected to explore the influence of marjoram oil and marjoram alongside kaolin particles on gel formation, respectively. Figure 6a demonstrates that the introduction of marjoram oil into the PVA significantly dropped the gel fraction percentages, recording 87.5 ± 3.4%, 80.3 ± 2.4%, 79.2 ± 2.9%, and 76.8 ± 2.5% for PVA, PVA-M0.1, PVA-M0.25, and PVA-M0.5, respectively. However, the results showed no significant differences correlated with the increase of marjoram ratios. On the other hand, the gel fraction ratios were insignificantly lowered for the PVA-M0.5-K0.1 (74.9 ± 2.1%) compared to the entire PVA/marjoram groups. At the same time, the additional growth in kaolin concentrations significantly amplified the gel fraction ratios, reporting 82 ± 3.8% and 83 ± 2.9% for PVA-M0.5-K0.25, PVA-M0.5-K0.5, respectively. Therefore, it seems possible that these results are due to the manipulation influence of marjoram and kaolin particles on the structure of PVA. Additionally, these outcomes indicate that incorporating marjoram and kaolin into the PVA weakened the crosslinking density, which might further enhance the swelling behavior of PVA/marjoram/kaolin dressings [33]. It could thus reinforce the absorption aptitude of excess wound exudates by these respective sponges, hampering adverse complications.

The swelling capability of hemostatic dressings acts as a key factor for hemostasis and wound healing downstream by governing the bleeding, drugs' release, degradation, and biological fluid absorption [34]. Accordingly, the 3-dimensional polymeric structures alongside the hydrophilic groups associated with the backbone of polymer chains improve the capacity of sponges to absorb great amounts of water without disintegration [35]. Therefore, the in vitro swelling characteristics of the developed sponges were analyzed, as displayed in Figure 6b. Swelling performances showed significant statistical differences ($p < 0.001$) in the swelling percentages for the entire sponges in relation to the PVA sponge. Furthermore, upon immersing the sponges in the water, quick swelling ratios were recorded for PVA/marjoram and PVA/marjoram/kaolin sponges with regard to the PVA. This is a very important property for hemostatic dressings in order to concentrate blood clotting determinants, thereby stimulating hemostasis [36]. Specifically, in the case of PVA/marjoram, the greatest water retention of 310 ± 9% was reported for PVA-M0.1

after 1.5 h among the other respective sponges. Furthermore, the increase in the ratio of marjoram oils resulted in significant reductions in the swelling manners, reporting 238 ± 11% and 236 ± 8% for PVA-M0.25, and PVA-M0.5 groups, respectively. This could be linked to the hydrophobic effect of oil ingredients [37].

On the other hand, the addition of kaolin particles at the lowest concretion for the PVA-M0.5-K0.1 sponge slightly influenced the swelling capacity, recording 233 ± 5%. Moreover, the statistical analyses exposed no significant variances in the PVA-M0.5-K0.1 sponge compared with the PVA-M0.25 and PVA-M0.5 groups. Nevertheless, the swelling ratios significantly decreased with the growth in kaolin levels. These findings could be attributed to the difference in the pore sizes, consistent with the data obtained from SEM analyses. Furthermore, the swelling ratios of the whole sponges levelled off after 4 h, accomplishing an equilibrium state. Additionally, the PVA-M0.1 group achieved the highest swelling ratio of 315 ± 12% after 4 h among the PVA/marjoram groups, while the greatest swelling ratio of 239 ± 7% was found for the PVA-M0.5-K0.1 sponge among the PVA/marjoram/kaolin sponges. Additionally, the graph shows that the similar results remained steady after 8 h. Previous studies reported the promotion of wound healing in vivo by applying wound dressings formulated on the basis of PVA and other biopolymers with different swelling percentages of 20%, 102% and 130% [13,38,39], highlighting the potential use of the PVA-M0.5-K0.1 sponge to ameliorate wound recovery.

The in vitro degradation of the PVA/marjoram and PVA/marjoram/kaolin sponges was studied by submerging them in PBS at 37 °C for predetermined times. As can be seen in Figure 6c, after 72 h of incubation, the PVA/marjoram sponges showed detectable weight losses of 23.7 ± 0.9%, 26.2 ± 1%, and 28.5 ± 0.6% for PVA-M0.1, PVA-M0.25, and PVA-M0.5, respectively. In contrast, the PVA group exhibited a weight loss of 20.5 ± 0.7%. These outcomes imply that the addition of marjoram oil promoted the degradation rate of sponges. On the other hand, the incorporation of kaolin alongside marjoram oil reduced the weight loss ratios, reporting 28.2 ± 0.8%, 23% ± 1, and 21.3 ± 0.8% for PVA-M0.5-K0.1, PVA-M0.5-K0.25, and PVA-M0.5-K0.5, respectively. This could be related to the aggregation of kaolin and marjoram, which could be perceived with the high concentration of kaolin. It is presumed that the in vitro degradation could impact the drug release represented by marjoram in this study, which is consistent with earlier studies [40,41]; thus, PVA-M0.5-K0.1 could be considered as the ideal wound dressing.

In aiming to appraise the water holding capacity of PVA/marjoram and PVA/marjoram/kaolin sponges, the porosity was tested. Figure 6d delineates the positive influence of marjoram oil on the porosity ratios for PVA/marjoram groups. The porosity for the PVA was 54.2 ± 2.4%, whereas PVA-M0.1, PVA-M0.25, and PVA-M0.5 reported porosities of 59.1 ± 2%, 65.8 ± 3.6%, and 74.4 ± 2.1%, respectively. This might be a result of the distortion impact of the marjoram oil on the internal structure of the polymer. Nonetheless, the introduction of kaolin particles into PVA-M0.5 lessened the porosity, and this decrease is correlated with the growth of the supplemented kaolin ratio. For PVA/marjoram/kaolin sponges, the porosity was determined to be 62.8 ± 3.9%, 60.5 ± 2.4%, and 55.8 ± 3.1% for PVA-M0.5-K0.1, PVA-M0.5-K0.25, and PVA-M0.5-K0.5, in respective order. These findings are in line with those of previous studies [42]. This performance might be related to the frequency of kaolin particles within the pores of sponges consuming some hydrogen bonds. Thus, the sponges turned out to be more compressed, and this obstructs the utilization of previously available pores [22]. The high porosity and water uptake competency of wound dressings are vital features for precluding microorganism invasion, expanding drug loading, and promoting dermal cell adherence and propagation. It could therefore ameliorate wound healing through preventing the prolongation of the inflammation phase.

In summary, swelling capacity, porosity, and biodegradation characteristics point to PVA-M0.5-K0.1 among the tested sponges for wound dressing applications.

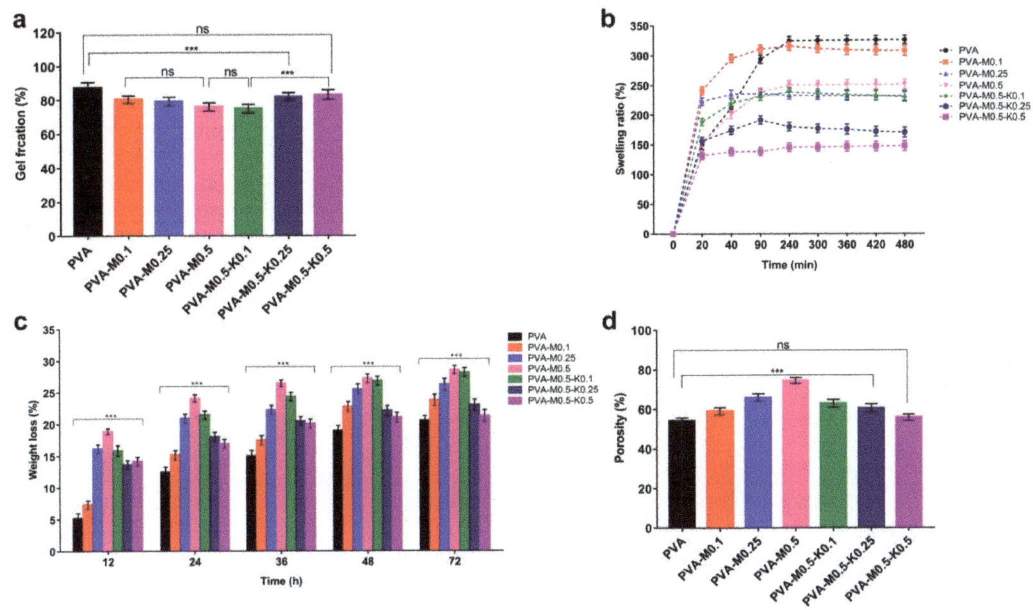

Figure 6. (**a**) Gel fractions, (**b**) Swelling characteristics, (**c**) In vitro weight loss, and (**d**) Porosity evaluations PVA/marjoram, and PVA/marjoram/kaolin composite sponges compared with the pure PVA sponge. Results are stated as means ± SD (n = 5) (*** $p < 0.001$, and (ns) points to non-significant difference).

2.3. Antibacterial Evaluation

The antibacterial activity of wound dressing is paramount to preclude the growth of pathogenic microorganisms on the wound bed and even on the surface of the wound dressing itself, which retards the rejuvenation of skin tissues and might incite tissue maceration [39,43–45]. To this end, we bolstered the devised sponges with marjoram oil as one of the natural antibacterial products as an alternative to conventional antibiotics to prevent the emergence of multi-drug resistant bacteria [21]. The antibacterial performances of PVA/marjoram and PVA/marjoram/kaolin sponges against *B. cereus* and *E. coli* were estimated by the growth turbidity method, as shown in Figure 7a. A significant positive correlation between the antibacterial capacity of PVA/marjoram sponges and the increase in the marjoram ratios is discernible. Specifically, the pure PVA sponges revealed no activity with regard to the tested bacteria. Remarkably, the addition of marjoram bestowed the antibacterial potency on the PVA-M0.1, PVA-M0.25, and PVA-M0.5 sponges, recording 31%, 65%, and 85% with regard to *B. cereus*, respectively. Moreover, PVA-M0.1, PVA-M0.25, and PVA-M0.5 sponges exerted growth inhibitions of 63%, 87%, and 90% in relation to *E. coli*, respectively. The variations in antibacterial capacity could be ascribed to the dissimilarity of cell walls for Gram-positive and Gram-negative bacteria.

The introduction of kaolin into the sponges in terms of PVA/marjoram/kaolin sponges showed no substantial differences in antibacterial activity for the PVA-M0.5-K0.1 and PVA-M0.5-K0.25 groups compared with the PVA-M0.5 group. By contrast, the antibacterial activity of the PVA-M0.5-K0.5 group was significantly diminished. This phenomenon could stem from the aggregation of kaolin with marjoram, hampering the release of marjoram oil into the medium.

The antibacterial features of PVA/marjoram and PVA/marjoram/kaolin sponges were further investigated using a colony-forming unit. As illustrated in Figure 7b, comparable trends to the previous results could be perceived. Significantly, PVA-M0.5 sponges could inhibit 95% and 97% of *B. cereus* and *E. coli*, respectively. Furthermore, the antibacterial behaviors of the PVA-M0.5-K0.1 group revealed no statistical differences in comparison with the PVA-M0.5 group, reporting growth inhibition ratios of 95% and 97% toward *B. cereus* and *E. coli*, respectively. In contrast to these findings, the inhibition ratios of the examined bacteria significantly lessened for PVA-M0.5-K0.25 and PVA-M0.5-K0.5 sponges. The variance of these results compared to the previous approach for PVA-M0.5-K0.25 could be explained by the inevitable measurement of viable and dead bacterial cells by the spectrophotometer in terms of the growth turbidity method. In summary, these findings suggest that the PVA-M0.5-K0.1 sponge could be implemented to frustrate microbial infections and further ameliorate wound healing.

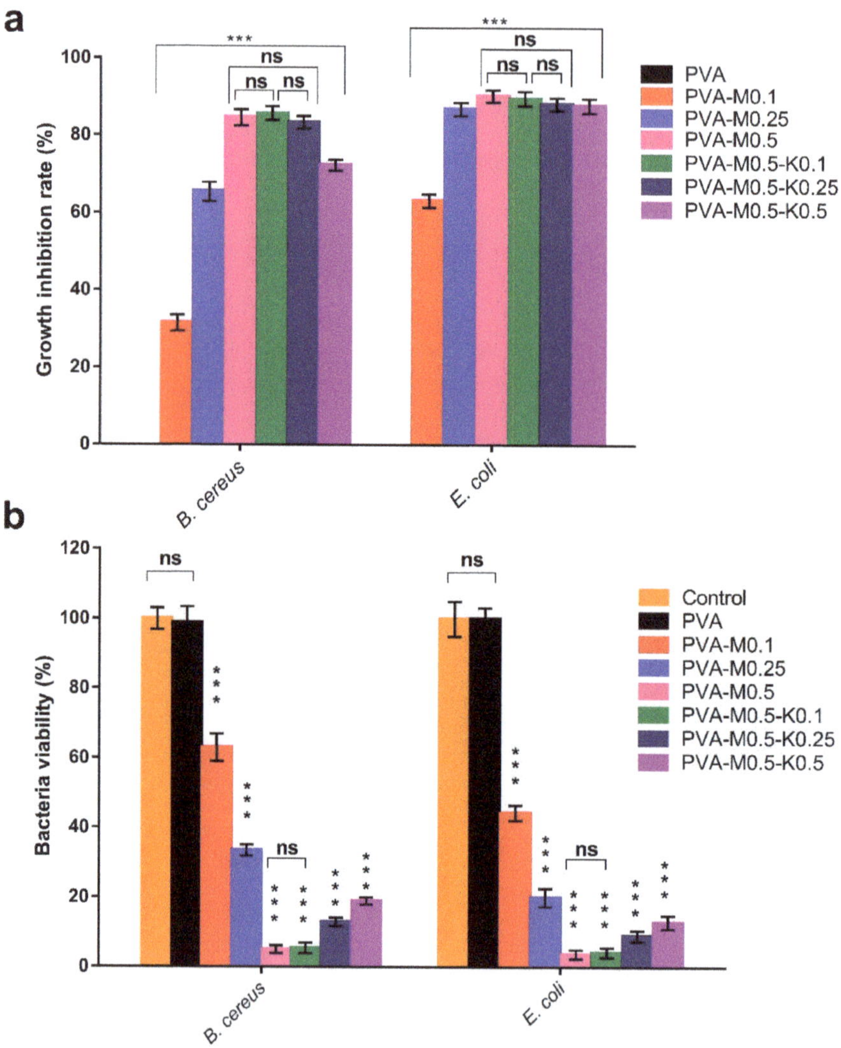

Figure 7. Antibacterial evaluations of PVA/marjoram and PVA/marjoram/kaolin composite sponges against *B. cereus* and *E. coli* compared to the pure PVA sponge adopting (**a**) growth turbidity method and (**b**) plate count method to determine the percentage of bacteria viability. Results are depicted as means ± SD (n = 5) (*** $p < 0.001$, and (ns) points to non-significant difference).

2.4. Total Phenolic Content and Antioxidant Evaluation

One of the detrimental bioburdens during wound healing is the overabundance of reactive oxygen species (ROS), provoking oxidative stress as a result of the phagocytosis mechanism [39,46]. This might incite lipid peroxidation, deactivation of vitals enzymes in addition to the damage of DNA, leading to impairment of wound healing and adjacent skin tissues [47]. Thus, the boost of wound dressing by antioxidant compounds is crucial to modulate the overabundance of ROS [29,48].

A striking attribute of essential oils is their containing various phenolic compounds, which exclusively endow them with vital biological activities, such as antioxidant characteristics, in order to scavenge reactive oxygen species (ROS). In this regard, it has been reported that marjoram oil contains phenolic acids and terpenoids [21]. To explore the efficiency of PVA/marjoram and PVA/marjoram/kaolin sponges to release the marjoram oil into the surrounding medium represented by the phenolic compounds, the total phenolic contents were determined subsequent to immersing the tested sponges in ethanol. Accordingly, the entire sponges exploited their structural integrity and released the corresponding phenolic mixtures. As can be seen from Figure 8a, phenolic compounds were not detected for the pure PVA sponge, which served as a negative control. For the composite sponges, there is a clear trend of increasing the phenolic contents with the rise in the marjoram oil for PVA-M0.1, PVA-M0.25, and PVA-M0.5 groups. The incorporation of kaolin into the sponges altered the release profile of marjoram oil; however, no significant reduction in the release profile was found for the PVA-M0.5-K0.1 sponge compared with the PVA-M0.5 sponge.

On the other hand, in comparison with the PVA-M0.5 sponge, the release profiles of phenolic contents were significantly diminished for PVA-M0.5-K0.25 and PVA-M0.5-K0.5 sponges. This manner could be ascribed to the adsorption of marjoram oil on the surface of kaolin particles. Concurrently, it is worth mentioning that there were no statistically significant differences in the release profile of phenolic contents for PVA-M0.5-K0.25 and PVA-M0.5-K0.5 groups in relation to the PVA-M0.5-K0.1 group.

Figure 8b displays the time-dependent decolorization of the $ABTS^{\bullet+}$ cationic radical by ethanol extracts of PVA, PVA/marjoram and PVA/marjoram/kaolin sponges. It could be discerned that the PVA sponge (control) exhibited slight $ABTS^{\bullet+}$ radical scavenging activity, which might be ascribed to the presence of hydroxyl groups along the PVA backbone. However, $ABTS^{\bullet+}$ radical scavenging activity was significantly enhanced by adding marjoram oil to PVA/marjoram groups. Besides, the amalgamation of kaolin in terms of PVA/marjoram/kaolin groups reduced the $ABTS^{\bullet+}$ radical scavenging capacity without a significant difference for PVA-M0.5-K0.1 compared to PVA-M0.5. Moreover, the statistical analyses demonstrated no significant differences in $ABTS^{\bullet+}$ radical scavenging capacity for PVA-M0.5-K0.25 and PVA-M0.5-K0.5 sponges in relation to the PVA-M0.5-K0.1 sponge. These results are in agreement with those obtained by measuring the total phenolic content in the previous section. Explicitly, the incidence of phenolic compounds in marjoram oil imparted an electron to $ABTS^{\bullet+}$, which further decolorized and converted to a neutral form [49–51].

To further evince the antioxidant potency of PVA/marjoram and PVA/marjoram/kaolin groups, an in vitro design system has been applied to estimate the studied materials' capacity to eradicate free radicals based on DPPH assay. The mechanism of this assay depends on the scavenging of the stable free radical 1,1-diphenyl-2-picrylhydrazyl (DPPH) by reducing the DPPH violet color into yellow-colored diphenyl-picrylhydrazine as a consequence of accepting an electron from antioxidant compounds [52]. Figure 8c delineates the scavenging activity of the DPPH dye by PVA/marjoram and PVA/marjoram/kaolin sponges. The results are consistent with those presented in the $ABTS^{\bullet+}$ assay. Furthermore, the pure PVA sponges showed a mild scavenging ratio of the DPPH dye on account of hydroxyl groups. At the same time, there are positive associations between the scavenging ratios of DPPH and the increase in marjoram oil contents. Moreover, the introduction of kaolin exposed no significant difference of DPPH scavenging for the entire PVA/marjoram/kaolin groups with regard to the PVA-M0.5 sponge.

Together, these results evidently indicate the potential application of PVA/marjoram/kaolin sponges in wound healing, particularly the PVA-M0.5-K0.1 sponge, without any significant influence on the emancipation of phenolic compounds from marjoram oil.

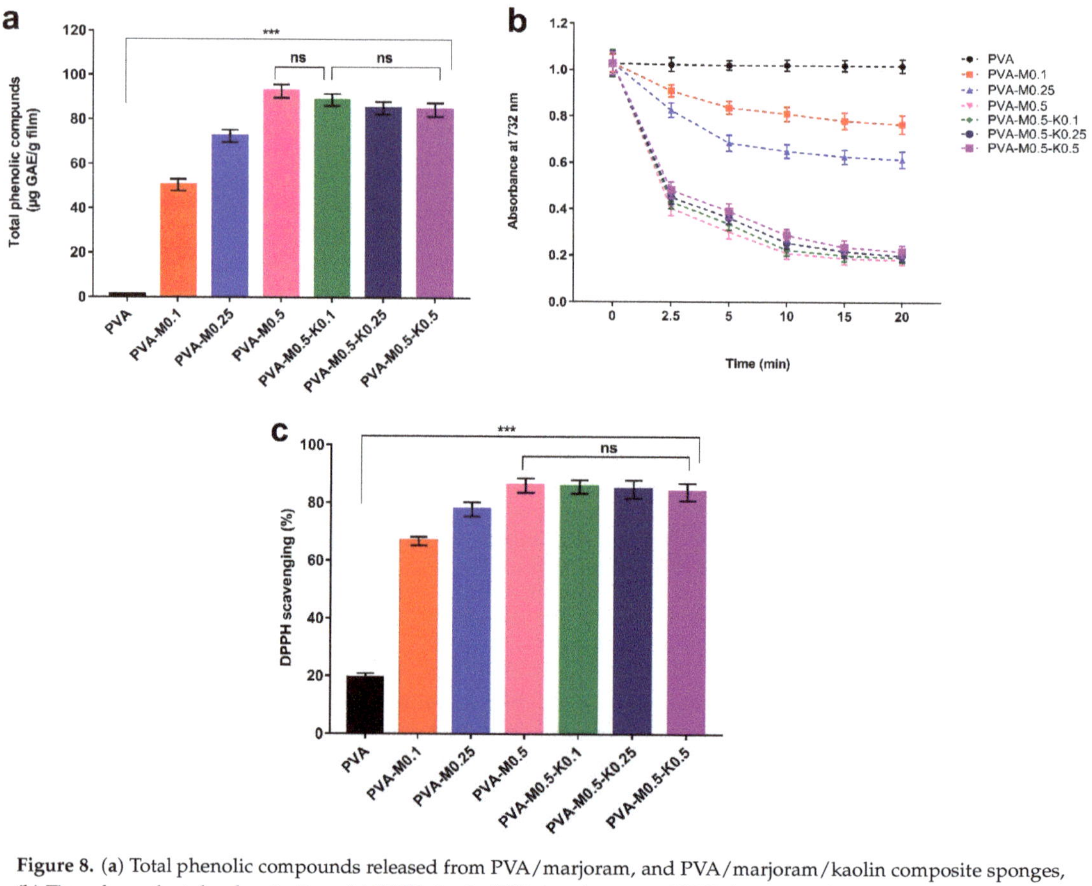

Figure 8. (**a**) Total phenolic compounds released from PVA/marjoram, and PVA/marjoram/kaolin composite sponges, (**b**) Time-dependent decolourization of $ABTS^{\bullet+}$ dye by PVA/marjoram, and PVA/marjoram/kaolin composite sponges, and (**c**) Scavenging competency of DPPH free radical by PVA/marjoram, and PVA/marjoram/kaolin composite sponges. Results are presented as means ± SD ($n = 5$) (*** $p < 0.001$, and (ns) points to a non-significant difference).

2.5. Hemocompatibility, Thrombogenicity and Cytotoxicity Evaluations

Hemocompatibility of biomaterials tailored to wound dressings, particularly hemostatic dressings, is an intrinsic property due to the unavoidable interactivity between blood and applied dressings [53,54]. The macroscopic photo in Figure 9a revealed the obvious variance in color between the seven sponge groups, the positive control (distilled water), and the negative control (PBS). Clearly, the entire sponge groups and the negative control emerged in yellow color without a significant difference. Conversely, the positive control tube appeared in red, implying the complete lysis of erythrocytes. The quantitative data of the hemolysis ratios for sponges PVA, PVA-M0.1, PVA-M0.25, PVA-M0.5, PVA-M0.5-K0.1, PVA-M0.5-K0.25 and PVA-M0.5-K0.5 were 1.9%, 1.84%, 1.65%, 1.45%, 1.56%, 1.79%, 1.98%, respectively, as provided in Figure 9b. It could be extrapolated that the tested sponges showed non-hemolytic activities (<2%) according to the American Society for

Testing and Materials (ASTM F 756-00, 2000), implying the good hemocompatibility of the PVA/marjoram/kaolin sponges.

The thrombogenicity test was conducted to appraise the PVA/marjoram/kaolin sponge groups' capacity to clot the blood, as illustrated in Figure 9c. Compared to the positive control, the thrombus formation was decreased for the PVA sponge; however, this diminution was not statistically significant. Therefore, this manner could be attributed to the hydrophilic trait of PVA. Nevertheless, it could be discovered that the introduction of marjoram oil substantially lessened thrombus formation. These results agree with those observed in prior studies [55,56], which elucidated this action by the presence of active phenolic compounds in marjoram oil, which hinder the aggregation of platelets. On the other hand, the supplementation of PVA/marjoram with different ratios of kaolin significantly escalated the weight of thrombus, which could be imputed to the influential blood clotting function of kaolin particles. Altogether, hemocompatibility and thrombogenicity findings suggest that PVA/marjoram/kaolin sponges could be significantly utilized as hemostatic wound dressings.

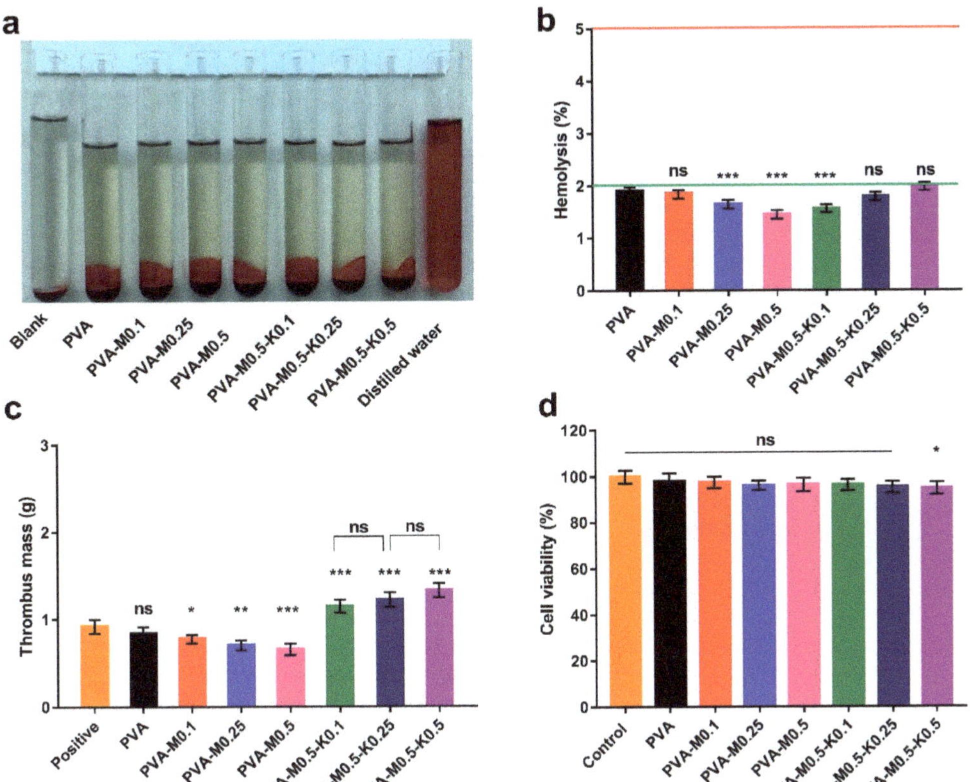

Figure 9. Photographs for the hemolysis test (**a**) and hemolytic percentages (**b**) reveal the hemocompatibility of PVA/marjoram and PVA/marjoram/kaolin composite sponges. (**c**) Thrombogenicity assessment for PVA/marjoram and PVA/marjoram/kaolin composite sponges demonstrate their capability to clot the blood through the formation of thrombus mass. (**d**) Cytotoxicity of PVA/marjoram and PVA/marjoram/kaolin composite sponges toward fibroblast cells show the safe behavior of the developed sponges. Results are presented as means ± SD (n = 6) (*** $p < 0.001$, ** $p < 0.01$, * $p < 0.05$, and (ns) points to a non-significant difference).

Cell compatibility test of wound dressings is crucial in order to assess the extent to which the fabricated sponges possess favorable biocompatibility in response to prevalent dermal cells, such as fibroblast, keratinocyte, and epithelial cells. During wound recovery, fibroblasts play a pivotal role in the construction of connective tissues, giving rise to the granulation of skin tissues and promoting skin remodeling [12,27,57]. Given this fact, alongside the cell interactions with wound dressings, we performed cytotoxicity investigations on fibroblast cells by the MTT assay. The cytotoxicity results exhibited no significant difference among the entire studied sponges, except for the sponge PVA-M0.5-K0.5 compared to untreated cells, as illustrated in Figure 9d. Thus, the feasible application of PVA-M0.5-K0.1 and PVA-M0.5-K0.25 sponges in wound dressings can be deduced from these findings since they could sustain the dermal cells in healthy behavior for achieving their functions properly.

Overall, these salient results advocate future in vivo wound healing studies for the PVA-M0.5-K0.1 sponge composite as an antibacterial and hemostatic wound dressing.

3. Materials and Methods

3.1. Materials and Bacterial Strains

PVA (Mw = 72 kDa) was purchased from ACROS Organics™, Carlsbad, CA, USA. Chinese marjoram oil and absolute ethanol were provided from Sinopharm Chemical Reagent Co., Ltd. (Beijing, China). Kaolin (hydrated aluminium silicate), MTT 3-(4,5-dimethylthiazol-2-yl)-2,5-diphenyltetrazolium bromide, sodium hydroxide, dimethyl sulfoxide (DMSO), and acid citrate dextrose solution (ACD) were obtained from Sigma–Aldrich (Chemie GmbH, Steinheim, Germany). Folin–Ciocalteu, gallic acid, 2,2-diphenyl-1-picrylhydrazyl (DPPH) and 2,2′azino-bis(3-ethylbenzothiazoline-6-sulphonic acid, ABTS) were purchased from Sigma-Aldrich Co., Ltd., St. Louis, MO, USA. Dulbecco's Modified Eagle Medium (DMEM) and trypsin were purchased from Gibco (ThermoFisher Scientific, Waltham, MA, USA). Yeast extract, tryptone, and sodium chloride were received from Bioshop (Canada Inc., Ontario, CA, Canada).

Bacillus cereus (*B. cereus*) and *Escherichia coli* (*E. coli*), representing Gram-positive and Gram-negative bacteria, respectively, were used to study the antibacterial activities of the fabricated dressings. Prior to performing the antibacterial assay, the bacterial strains were refreshed via inoculating into LB medium containing g/L: NaCl 10 g, peptone 10 g, and yeast extract 5 g, and then incubated at 37 °C and 150 rpm for 18 h.

3.2. Methodology

3.2.1. Preparation of Sponges

PVA/marjoram/kaolin sponge composites were fabricated adopting a freezing-thawing cycle approach in accordance with previously described procedures [58]. Briefly, mixtures of 5% (*w*/*v*) PVA, various concentrations of marjoram extract oil (0.1, 0.25, and 0.5 mL), and kaolin (0.1, 0.25, and 0.5 g) were thoroughly mixed prior to sonication and vortexing for 1 h. After that, the blends were poured into Petri dishes, followed by five successive cycles of freezing at −20 °C for 18 h and then thawing for 6 h at 25 °C. The sponges with different volumes of marjoram (0.1, 0.25, and 0.5 mL) were designated as PVA-M0.1, PVA-M0.25, and PVA-M0.5, respectively, in addition to a control sponge of PVA. Moreover, the sponge PVA-M0.5 was supplemented with various kaolin contents (0.1, 0.25, and 0.5 g) and labelled as PVA-M0.5-K0.1, PVA-M0.5-K0.25, and PVA-M0.5-K0.5, respectively. The developed sponges were frozen in liquid nitrogen for 10 min and then lyophilized for further examinations.

3.2.2. Characterization of the Sponges

FT-IR Analysis

The alterations of the chemical structures of the designed sponges were investigated using a Fourier transform infrared spectrophotometer after thoroughly mixing a weight of 5 mg of each sponge with potassium bromide (KBr). The FT-IR equipment (Shimadzu 8400S, Kyoto, Japan) was programmed to probe each sponge 40 times at a range of 400–4000 cm^{-1} for 40 scans.

Morphological Examination and Thermal Analysis

To scrutinize the formulated sponges' morphological variations, each sample was overlaid with a thin film of gold under vacuum conditions prior to surveying by scanning electron microscope (SEM, Joel Jsm 6360LA, Tokyo, Japan).

For thermal characterization, ~5 mg of each film in a sealed aluminium pan was analyzed employing a thermal gravimetric analyzer (TGA, Shimadzu 50/50H, Kyoto, Japan) within a temperature range of 20–600 °C with a heating rate (10 °C/min) and a nitrogen flow of 30 mL/min.

Gel Fraction, Swelling Profile, In Vitro Degradation, and Porosity of Sponges

For gel fraction determination, the sponges were placed in a vacuum oven for 24 h at 50 °C until dried and then weighed. Thereafter, the sponges were re-swollen for 24 h in distilled water until an equilibrium swelling point for eliminating soluble PVA was obtained. The sponges were subsequently dried in a vacuum oven at 50 °C and weighed [59]. The experiments were conducted in five replicates, and the gel fractions were estimated using Equation (1) as follows:

$$\text{Gel fraction (\%)} = (We/Wi) \times 100 \quad (1)$$

where We and Wi refer to the weights of the dried sponge and swollen sponge, respectively.

The swelling capacities of sponges were appraised via the determination of their weights after immersing them into the water for time intervals. Approximately 1 g of a dried sponge was weighed and then dipped into 500 mL of distilled water. Dynamic Swelling was performed at 25 °C until reaching an equilibrium state. Each swollen sample was withdrawn at predetermined time points, and the water-adhered onto the surface was gently blotted using filter papers before weighing. Each experiment was replicated five times, and the swelling ratios were estimated using Equation (2) as follow:

$$\text{Swelling ratio (\%)} = [(Ws - Wd)/Wd] \times 100 \quad (2)$$

where (Ws) denotes the weight of the swollen sponge, while (Wd) refers to the weight of the sponge at the initial time.

To evaluate the porosity of the sponges, the measurements were performed following the procedure used by Yin et al. [60]. First, sponges were dried at 50 °C for 2 h in a vacuum oven and then the dried weights were estimated. Next, the samples were plunged for 4 h in absolute ethanol. Subsequently, the swollen sponges were blotted to eliminate the extra ethanol using filter paper and then weighed. Finally, the porosity analyses were conducted in five replicates and calculated according to the Equation (3):

$$\text{Porosity (\%)} = [(W2 - W1)/pV] \times 100 \quad (3)$$

where W1 and W2 refer to the weight of the sponge prior to and after being immersed in absolute ethanol, respectively, "V" represents the volume of the sponge, and "p" denotes the density of absolute ethanol.

To perceive the hydrolytic degradation of the devised sponges, dried sponges were weighed, immersed into 3 mL PBS (0.1 M, pH 7.4), and maintained at 37 °C. The sponges were then taken out at different time points and mildly wiped with soft papers to eliminate the excess water on the surface of sponges. Following this, the samples were dried under vacuum conditions and weighed. All experiments were conducted in five replicates.

3.2.3. Antibacterial Analysis

The antibacterial evaluation of the prepared sponges was conducted adopting two approaches to determine optical densities and the colony-forming unit (CFU/mL). First, overnight bacterial cultures of *B. cereus* and *E. coli* were diluted in LB medium prior to adapting their turbidities were adapted following the McFarland 0.5 standard at 625 nm with 2×10^8 CFU/mL [61,62]. Subsequently, 100 µL of the diluted bacterial cultures were inoculated into 10 mL LB medium, including 50 mg of tested sponges, before being incubated for 18 h at 37 °C and 150 rpm. In contrast, the bacterial cultures without sponges were set out as controls. After incubation, the antibacterial capacity was estimated by two approaches; first by estimating the inhibition of bacterial growth employing a spectrophotometer at 600 nm and calculating the ratio of bacterial growth inhibition were determined using Equation (4):

$$\text{Bacterial growth inhibition (\%)} = [(\text{ODc} - \text{ODi})/\text{ODc})] \times 100 \quad (4)$$

where ODc and ODi are the optical densities of bacterial cultures untreated and treated with a tested sponge, respectively.

For the second evaluation method, 50 µL of bacterial cultures were spread over LB agar plates and then maintained for 24 h at 37 °C before ascertaining the colony-forming unit (CFU/mL). Thus, both antibacterial tests were accomplished in five replicates.

3.2.4. Total Phenolic Content

The phenolic contents of the formulated films were assessed based on reducing the Folin-Ciocalteu reagent from yellow to blue colored compound. First, 50 mg of each membrane was submerged into 5 mL of ethanol to extract the marjoram oil content in the membrane [63]. Afterwards, 0.5 mL of sponge supernatant was put into 2.0 mL of Folin–Ciocalteu reagent (10%, *v*/*v*), followed by adding 2 mL sodium carbonate solution (7.5%, *w*/*v*). The blend was maintained at 50 °C for 5 min, and the absorption was then gauged at 760 nm by means of a spectrophotometer. The determinations were replicated five times and estimated in relation to the stranded curve for gallic acid solutions (0–100 µg).

3.2.5. Antioxidant Activity Determination

$ABTS^{\bullet+}$ Radical Scavenging Assay

For the ABTS radical scavenging assessment, the radical cations were prompted by the reaction of an aqueous solution of $K_2S_2O_8$ (3.30 mg) in water (5 mL) with ABTS (17.2 mg). Then, the resultant bluish-green radical cation solution was kept overnight below 0 °C under dark conditions. Later, 1 mL of the solution was diluted to a final volume of 60 mL with distilled water and marked as the $ABTS^{\bullet+}$ solution. The samples were extracted as presented above in the estimation of total phenolic content. Following this, 0.1 mL of each sponge leachate was put into 2.0 mL of $ABTS^{\bullet+}$ solution. The $ABTS^{\bullet+}$ evaluation was implemented five times, and the absorption was appraised at 730 nm at various time points.

DPPH Radical Scavenging Activity

The antioxidant properties of the sponge leachates were assessed by adapting the 2,2-diphenyl-1-picrylhydrazyl (DPPH) approach [43,64]. Accordingly, 6 mg of DPPH were dissolved in 50 mL methanol (0.3 mM), and then a volume of 2.5 mL of each sponge extract and 2.5 mL of the DPPH solution was thoroughly mixed. Next, the tube was incubated at room temperature for 20 min under dark conditions. Afterwards, the decolorization of the dye was quantified at 517 nm using a spectrophotometer. The reactions were replicated five times, and inhibition percentages of radicals were computed by the following Equation (5):

$$\text{DPPH scavenging (\%)} = [(\text{Ac} - \text{As})/\text{Ac})] \times 100 \quad (5)$$

where Ac is the absorbance of the control DPPH solution and As is the absorbance of the sponge extract after reaction with the DPPH solution.

3.2.6. Hemocompatibility of the Sponges

In order to investigate the hemocompatibility of the formulated sponges, the hemolysis tests were executed as previously demonstrated with minor adaptations [65]. Anticoagulated blood was prepared for this determination by adding 1 mL of anticoagulant acid citrate dextrose solution (ACD) to 9 mL of blood. Prior to commencing the direct contact between blood and the tested membranes, about 1 cm^2 of each film was plunged in phosphate buffer solution (PBS, pH 7.0) for 72 h at 37 °C. Following that, the PBS was poured out before immersing the sponges in 1 mL of ACD blood and keeping the tubes at 37 °C for 3 h. For preparing negative and positive control tubes, the equivalent volumes of the ACD blood were added to 7 mL of PBS and water, respectively. The tubes were carefully inverted three times every 30 min to preserve the appropriate contact of the films with the blood. Subsequently, each liquid was carefully moved to new tubes and clarified via centrifugation for 15 min at 200 rpm. The hemoglobin released by hemolysis was estimated at 540 nm employing a spectrophotometer (Model Ultrospec 2000). All determinations were implemented in five replicates, and the hemolysis ratio was computed using Equation (6):

$$\text{Hemolysis}\ (\%) = [(\text{ODm} - \text{ODn})/(\text{ODp} - \text{ODn})] \times 100 \quad (6)$$

where ODm is the absorbance value of a tested sponge, ODn is the absorbance value of the negative control, and ODp is the absorbance value of the positive control.

3.2.7. Thrombogenicity Test

A gravimetric method was applied, as described earlier, to ascertain the amounts of formed thrombus over the surface of the fabricated sponges [57]. ACD blood was prepared as demonstrated above. Membranes were plunged into PBS for 48 h at 37 °C. On completion of incubation time, the PBS was then poured out, and the ACD blood was positioned over the examined materials. At the same time, positive control was set out by applying the equivalent amount of ACD blood to an empty Petri dish.

For prompting the clotting reaction of blood, 20 µL of a 10 M calcium chloride solution was put onto the sponges. After 45 min, the reactions were terminated by adding 5 mL H_2O. Subsequently, the clots were firmly attached with an additional 5 mL of a 36% formaldehyde solution and dried with tissue papers before weighing. Thrombogenicity examinations were repeated five times.

3.2.8. Cytotoxicity Test of the Sponges

The cellular toxicity of the membranes on NIH 3T3 (mouse fibroblast cells) was appraised adopting MTT [3-(4,5-Dimethythiazol-2-yl)-2,5-Diphenyltetrazolium Bromide] method as demonstrated earlier with some adaptations [66,67].

The NIH 3T3 cells were cultivated in Dulbecco's modified Eagle's medium (DMEM), dissolved in 10% fetal bovine serum and fostered at 5% CO_2 and 37 °C with a humidity of 85% in a CO_2 incubator. Then, 0.5% trypsin was applied to detach the fibroblasts, and the cells were thereafter seeded at 5×10^3 cells/well in a 96-well plate. The plate was then incubated in the CO_2 incubator as mentioned above for 24 h.

Concurrently, 30 mg of each sponge was sterilized by immersing in 70% ethanol, subjected to UV for 45 min, and transferred into a 24-well plate containing 1 mL of DMEM at 37 °C for 24 h. Subsequently, the medium was discarded from the plate containing fibroblast cells and replaced with 100 µL of the membrane's leach, while the control cells were supplied with 100 µL of standard DMEM medium. After that, the 96-well plate was incubated for 24 h before washing the cells with PBS three times. Then, the MTT test was commenced by adding 20 µL of MTT solution (5 mg/mL in serum-free medium) to each well, and the plate was then maintained in the CO_2 incubator for 3 h at 37 °C.

Subsequently, the MTT solution was replaced with 200 μL of dimethylsulfoxide (DMSO) for each well. Finally, the plate was shaken at 100 rpm for 5 min, and the absorbance values were estimated at 570 nm by means of a microtiter plate reader. The investigation was carried out in six replicates for each sponge, and the viable ratio of fibroblast cells was evaluated following Equation (7):

$$\text{Cell viability (\%)} = (\text{Am}/\text{Ac}) \times 100 \quad (7)$$

where (Am) refers to the absorbance of cells doped with tested membrane, while (Ac) points to the absorbance of untreated cells.

3.2.9. Statistical Analysis

GraphPad Prism software (V. 5) was employed to analyze the statistical significance of the entire data. One-way and two-way analyses of variance (ANOVA) with Tukey's multiple comparison tests were thus applied. The entire determination values are expressed as means ± SD, and they were significantly considered at p-value < 0.05, where $n = 5$, except for the cytotoxicity studies ($n = 6$).

4. Conclusions

In summary, novel sponges based on PVA boosted by marjoram essential oil and kaolin were successfully designed to frustrate massive bleeding and bacterial infection, which could further accelerate full-thickness wound healing. PVA/marjoram/kaolin sponges exhibited noticeable porous and lamellar structures. The amalgamation of marjoram and kaolin into PVA augmented the pore size of the devised sponges, thus encouraging cell attachment and proliferation. Moreover, they demonstrated great water absorption, which supports their competency to govern the hemorrhage quickly. PVA/marjoram/kaolin sponges presented an outstanding performance in scavenging free radicals as antioxidant sponges and revealed high antibacterial activity in relation to pathogenic bacteria. Furthermore, manifested thrombogenicity, hemocompatibility, and cellular compatibility were corroborated for the developed sponges. Thus, the results clearly indicate the PVA-M0.5-K0.1 sponge for future consideration as a hemostatic and antibacterial wound dressing. Therefore, future in vivo studies are justified to determine the extent of PVA-M0.5-K0.1 to enhance cutaneous wound restoration in cases of bleeding and microbial infections.

Author Contributions: T.M.T. and M.H.A. conceived the project; T.M.T., M.M.S. and M.H.A. conducted the experiments; T.M.T. and M.H.A. analyzed and interpreted the data, performed the statistical analyses, and designed the figures; T.M.T. and M.A.H. wrote the draft of the manuscript; T.M.T., M.H.A., A.M.O., T.H.A., M.S.M.-E. and M.H.A. revised the manuscript; T.M.T. and M.H.A. finalized the final version of the manuscript. All authors have read and agreed to the published version of the manuscript.

Funding: This research received no external funding.

Institutional Review Board Statement: Not applicable.

Informed Consent Statement: Informed consent was obtained from all subjects involved in the study.

Data Availability Statement: The datasets generated during the current study are available from the corresponding authors upon request.

Conflicts of Interest: The authors declare that they have no conflict of interest.

References

1. Berthet, M.; Gauthier, Y.; Lacroix, C.; Verrier, B.; Monge, C. Nanoparticle-Based Dressing: The Future of Wound Treatment? *Trends Biotechnol.* **2017**, *35*, 770–784. [CrossRef] [PubMed]
2. Tamer, T.M.; Valachová, K.; Hassan, M.A.; Omer, A.M.; El-Shafeey, M.; Eldin, M.S.M.; Šoltés, L. Chitosan/hyaluronan/edaravone membranes for anti-inflammatory wound dressing: In vitro and in vivo evaluation studies. *Mater. Sci. Eng. C* **2018**, *90*, 227–235. [CrossRef]

3. Pereira, R.F.; Bártolo, P.J. Traditional Therapies for Skin Wound Healing. *Adv. Wound Care (New Rochelle)* **2016**, *5*, 208–229. [CrossRef]
4. Gurtner, G.C.; Werner, S.; Barrandon, Y.; Longaker, M.T. Wound repair and regeneration. *Nature* **2008**, *453*, 314–321. [CrossRef]
5. King, D.R. Initial Care of the Severely Injured Patient. *N. Engl. J. Med.* **2019**, *380*, 763–770. [CrossRef] [PubMed]
6. Alarhayem, A.Q.; Myers, J.G.; Dent, D.; Liao, L.; Muir, M.; Mueller, D.; Nicholson, S.; Cestero, R.; Johnson, M.C.; Stewart, R.; et al. Time is the enemy: Mortality in trauma patients with hemorrhage from torso injury occurs long before the "golden hour". *Am. J. Surg.* **2016**, *212*, 1101–1105. [CrossRef] [PubMed]
7. Ma, Y.; Yao, J.; Liu, Q.; Han, T.; Zhao, J.; Ma, X.; Tong, Y.; Jin, G.; Qu, K.; Li, B.; et al. Liquid Bandage Harvests Robust Adhesive, Hemostatic, and Antibacterial Performances as a First-Aid Tissue Adhesive. *Adv. Funct. Mater.* **2020**, *30*, 2001820. [CrossRef]
8. Glick, J.B.; Kaur, R.R.; Siegel, D. Achieving hemostasis in dermatology-Part II: Topical hemostatic agents. *Indian Dermatol. Online J.* **2013**, *4*, 172–176. [PubMed]
9. Shefa, A.A.; Amirian, J.; Kang, H.J.; Bae, S.H.; Jung, H.-I.; Choi, H.-J.; Lee, S.Y.; Lee, B.-T. In vitro and in vivo evaluation of effectiveness of a novel TEMPO-oxidized cellulose nanofiber-silk fibroin scaffold in wound healing. *Carbohydr. Polym.* **2017**, *177*, 284–296. [CrossRef]
10. Zhao, X.; Guo, B.; Wu, H.; Liang, Y.; Ma, P.X. Injectable antibacterial conductive nanocomposite cryogels with rapid shape recovery for noncompressible hemorrhage and wound healing. *Nat. Commun.* **2018**, *9*, 2784. [CrossRef] [PubMed]
11. Liu, Y.; Xu, K.; Chang, Q.; Darabi, M.A.; Lin, B.; Zhong, W.; Xing, M. Highly Flexible and Resilient Elastin Hybrid Cryogels with Shape Memory, Injectability, Conductivity, and Magnetic Responsive Properties. *Adv. Mater.* **2016**, *28*, 7758–7767. [CrossRef] [PubMed]
12. Sultana, T.; Hossain, M.; Rahaman, S.; Kim, Y.S.; Gwon, J.-G.; Lee, B.-T. Multi-functional nanocellulose-chitosan dressing loaded with antibacterial lawsone for rapid hemostasis and cutaneous wound healing. *Carbohydr. Polym.* **2021**, *272*, 118482. [CrossRef] [PubMed]
13. Fang, H.; Wang, J.; Li, L.; Xu, L.; Wu, Y.; Wang, Y.; Fei, X.; Tian, J.; Li, Y. A novel high-strength poly(ionic liquid)/PVA hydrogel dressing for antibacterial applications. *Chem. Eng. J.* **2019**, *365*, 153–164. [CrossRef]
14. Qi, X.; Hu, X.; Wei, W.; Yu, H.; Li, J.; Zhang, J.; Dong, W. Investigation of Salecan/poly(vinyl alcohol) hydrogels prepared by freeze/thaw method. *Carbohydr. Polym.* **2015**, *118*, 60–69. [CrossRef] [PubMed]
15. Chen, Y.-N.; Peng, L.; Liu, T.; Wang, Y.; Shi, S.; Wang, H. Poly(vinyl alcohol)–Tannic Acid Hydrogels with Excellent Mechanical Properties and Shape Memory Behaviors. *ACS Appl. Mater. Interfaces* **2016**, *8*, 27199–27206. [CrossRef] [PubMed]
16. Awad, M.E.; López-Galindo, A.; Setti, M.; El-Rahmany, M.M.; Iborra, C.V. Kaolinite in pharmaceutics and biomedicine. *Int. J. Pharm.* **2017**, *533*, 34–48. [CrossRef] [PubMed]
17. Liang, Y.; Xu, C.; Li, G.; Liu, T.; Liang, J.F.; Wang, X. Graphene-kaolin composite sponge for rapid and riskless hemostasis. *Colloids Surf. B Biointerfaces* **2018**, *169*, 168–175. [CrossRef] [PubMed]
18. Hajlaoui, H.; Mighri, H.; Aouni, M.; Gharsallah, N.; Kadri, A. Chemical composition and in vitro evaluation of antioxidant, antimicrobial, cytotoxicity and anti-acetylcholinesterase properties of Tunisian *Origanum majorana* L. essential oil. *Microb. Pathog.* **2016**, *95*, 86–94. [CrossRef]
19. Arranz, E.; Jaime, L.; López de las Hazas, M.C.; Reglero, G.; Santoyo, S. Supercritical fluid extraction as an alternative process to obtain essential oils with anti-inflammatory properties from marjoram and sweet basil. *Ind. Crop. Prod.* **2015**, *67*, 121–129. [CrossRef]
20. Vági, E.; Simándi, B.; Suhajda, Á.; Héthelyi, É. Essential oil composition and antimicrobial activity of *Origanum majorana* L. extracts obtained with ethyl alcohol and supercritical carbon dioxide. *Food Res. Int.* **2005**, *38*, 51–57. [CrossRef]
21. Almasi, H.; Azizi, S.; Amjadi, S. Development and characterization of pectin films activated by nanoemulsion and Pickering emulsion stabilized marjoram (*Origanum majorana* L.) essential oil. *Food Hydrocoll.* **2020**, *99*, 105338. [CrossRef]
22. Tamer, T.M.; Sabet, M.M.; Omer, A.M.; Abbas, E.; Eid, A.I.; Mohy-Eldin, M.S.; Hassan, M.A. Hemostatic and antibacterial PVA/Kaolin composite sponges loaded with penicillin–streptomycin for wound dressing applications. *Sci. Rep.* **2021**, *11*, 3428. [CrossRef] [PubMed]
23. Aslam, M.; Kalyar, M.A.; Raza, Z.A. Polyvinyl alcohol: A review of research status and use of polyvinyl alcohol based nanocomposites. *Polym. Eng. Sci.* **2018**, *58*, 2119–2132. [CrossRef]
24. Mansur, H.S.; Oréfice, R.L.; Mansur, A.A.P. Characterization of poly(vinyl alcohol)/poly(ethylene glycol) hydrogels and PVA-derived hybrids by small-angle X-ray scattering and FTIR spectroscopy. *Polymer* **2004**, *45*, 7193–7202. [CrossRef]
25. Kamoun, E.A.; Kenawy, E.-R.S.; Tamer, T.M.; El-Meligy, M.A.; Eldin, M.S.M. Poly (vinyl alcohol)-alginate physically crosslinked hydrogel membranes for wound dressing applications: Characterization and bio-evaluation. *Arab. J. Chem.* **2015**, *8*, 38–47. [CrossRef]
26. Zhou, T.; Chen, S.; Ding, X.; Hu, Z.; Cen, L.; Zhang, X. Fabrication and Characterization of Collagen/PVA Dual-Layer Membranes for Periodontal Bone Regeneration. *Front. Bioeng. Biotechnol.* **2021**, *9*, 437. [CrossRef] [PubMed]
27. Fan, X.; Li, Y.; Li, N.; Wan, G.; Ali, M.A.; Tang, K. Rapid hemostatic chitosan/cellulose composite sponge by alkali/urea method for massive haemorrhage. *Int. J. Biol. Macromol.* **2020**, *164*, 2769–2778. [CrossRef]
28. Zhang, B.; He, J.; Shi, M.; Liang, Y.; Guo, B. Injectable self-healing supramolecular hydrogels with conductivity and photo-thermal antibacterial activity to enhance complete skin regeneration. *Chem. Eng. J.* **2020**, *400*, 125994. [CrossRef]

29. Wang, X.; Qi, J.; Zhang, W.; Pu, Y.; Yang, R.; Wang, P.; Liu, S.; Tan, X.; Chi, B. 3D-printed antioxidant antibacterial carboxymethyl cellulose/ε-polylysine hydrogel promoted skin wound repair. *Int. J. Biol. Macromol.* **2021**, *187*, 91–104. [CrossRef]
30. Dong, R.; Zhao, X.; Guo, B.; Ma, P.X. Self-Healing Conductive Injectable Hydrogels with Antibacterial Activity as Cell Delivery Carrier for Cardiac Cell Therapy. *ACS Appl. Mater. Interfaces* **2016**, *8*, 17138–17150. [CrossRef]
31. Qu, J.; Zhao, X.; Liang, Y.; Xu, Y.; Ma, P.X.; Guo, B. Degradable conductive injectable hydrogels as novel antibacterial, anti-oxidant wound dressings for wound healing. *Chem. Eng. J.* **2019**, *362*, 548–560. [CrossRef]
32. Tan, H.B.; Wang, F.Y.; Ding, W.; Zhang, Y.; Ding, J.; Cai, D.X.; Yu, K.F.; Yang, J.; Yang, L.; Xu, Y.Q. Fabrication and Evaluation of Porous Keratin/chitosan (KCS) Scaffolds for Effectively Accelerating Wound Healing. *Biomed. Environ. Sci.* **2015**, *28*, 178–189. [PubMed]
33. Sung, J.H.; Hwang, M.-R.; Kim, J.O.; Lee, J.H.; Kim, Y.I.; Kim, J.H.; Chang, S.W.; Jin, S.G.; Kim, J.A.; Lyoo, W.S.; et al. Gel characterisation and in vivo evaluation of minocycline-loaded wound dressing with enhanced wound healing using polyvinyl alcohol and chitosan. *Int. J. Pharm.* **2010**, *392*, 232–240. [CrossRef] [PubMed]
34. Adeli-Sardou, M.; Yaghoobi, M.M.; Torkzadeh-Mahani, M.; Dodel, M. Controlled release of lawsone from polycaprolactone/gelatin electrospun nano fibers for skin tissue regeneration. *Int. J. Biol. Macromol.* **2019**, *124*, 478–491. [CrossRef] [PubMed]
35. He, J.; Shi, M.; Liang, Y.; Guo, B. Conductive adhesive self-healing nanocomposite hydrogel wound dressing for photothermal therapy of infected full-thickness skin wounds. *Chem. Eng. J.* **2020**, *394*, 124888. [CrossRef]
36. Landsman, T.L.; Touchet, T.; Hasan, S.M.; Smith, C.; Russell, B.; Rivera, J.; Maitland, D.J.; Cosgriff-Hernandez, E. A shape memory foam composite with enhanced fluid uptake and bactericidal properties as a hemostatic agent. *Acta Biomater.* **2017**, *47*, 91–99. [CrossRef]
37. Ruiz-Navajas, Y.; Viuda-Martos, M.; Sendra, E.; Perez-Alvarez, J.A.; Fernández-López, J. In vitro antibacterial and antioxidant properties of chitosan edible films incorporated with Thymus moroderi or Thymus piperella essential oils. *Food Control.* **2013**, *30*, 386–392. [CrossRef]
38. Zhou, M.; Lin, F.; Li, W.; Shi, L.; Li, Y.; Shan, G. Development of nanosilver doped carboxymethyl chitosan-polyamideamine alginate composite dressing for wound treatment. *Int. J. Biol. Macromol.* **2021**, *166*, 1335–1351. [CrossRef]
39. He, J.; Liang, Y.; Shi, M.; Guo, B. Anti-oxidant electroactive and antibacterial nanofibrous wound dressings based on poly(ε-caprolactone)/quaternized chitosan-graft-polyaniline for full-thickness skin wound healing. *Chem. Eng. J.* **2020**, *385*, 123464. [CrossRef]
40. Liu, X.; You, L.; Tarafder, S.; Zou, L.; Fang, Z.; Chen, J.; Lee, C.H.; Zhang, Q. Curcumin-releasing chitosan/aloe membrane for skin regeneration. *Chem. Eng. J.* **2019**, *359*, 1111–1119. [CrossRef]
41. Omer, A.M.; Ziora, Z.M.; Tamer, T.M.; Khalifa, R.E.; Hassan, M.A.; Mohy-Eldin, M.S.; Blaskovich, M.A.T. Formulation of Quaternized Aminated Chitosan Nanoparticles for Efficient Encapsulation and Slow Release of Curcumin. *Molecules* **2021**, *26*, 449. [CrossRef] [PubMed]
42. Ninan, N.; Muthiah, M.; Park, I.-K.; Elain, A.; Thomas, S.; Grohens, Y. Pectin/carboxymethyl cellulose/microfibrillated cellulose composite scaffolds for tissue engineering. *Carbohydr. Polym.* **2013**, *98*, 877–885. [CrossRef]
43. Hassan, M.A.; Tamer, T.M.; Valachová, K.; Omer, A.M.; El-Shafeey, M.; Mohy Eldin, M.S.; Šoltés, L. Antioxidant and antibacterial polyelectrolyte wound dressing based on chitosan/hyaluronan/phosphatidylcholine dihydroquercetin. *Int. J. Biol. Macromol.* **2021**, *166*, 18–31. [CrossRef]
44. Liang, Y.; Zhao, X.; Hu, T.; Chen, B.; Yin, Z.; Ma, P.X.; Guo, B. Adhesive Hemostatic Conducting Injectable Composite Hydrogels with Sustained Drug Release and Photothermal Antibacterial Activity to Promote Full-Thickness Skin Regeneration During Wound Healing. *Small (Weinh. der Bergstr. Ger.)* **2019**, *15*, e1900046. [CrossRef] [PubMed]
45. Tamer, T.M.; Hassan, M.A.; Omer, A.M.; Baset, W.M.A.; Hassan, M.E.; El-Shafeey, M.E.A.; Eldin, M.S.M. Synthesis, characterization and antimicrobial evaluation of two aromatic chitosan Schiff base derivatives. *Process. Biochem.* **2016**, *51*, 1721–1730. [CrossRef]
46. Dunnill, C.; Patton, T.; Brennan, J.; Barrett, J.; Dryden, M.; Cooke, J.; Leaper, D.; Georgopoulos, N.T. Reactive oxygen species (ROS) and wound healing: The functional role of ROS and emerging ROS-modulating technologies for augmentation of the healing process. *Int. Wound J.* **2017**, *14*, 89–96. [CrossRef] [PubMed]
47. Akbik, D.; Ghadiri, M.; Chrzanowski, W.; Rohanizadeh, R. Curcumin as a wound healing agent. *Life Sci.* **2014**, *116*, 1–7. [CrossRef]
48. Qu, J.; Zhao, X.; Liang, Y.; Zhang, T.; Ma, P.X.; Guo, B. Antibacterial adhesive injectable hydrogels with rapid self-healing, extensibility and compressibility as wound dressing for joints skin wound healing. *Biomaterials* **2018**, *183*, 185–199. [CrossRef] [PubMed]
49. Moteriya, P.; Padalia, H.; Chanda, S. Characterization, synergistic antibacterial and free radical scavenging efficacy of silver nanoparticles synthesized using Cassia roxburghii leaf extract. *J. Genet. Eng. Biotechnol.* **2017**, *15*, 505–513. [CrossRef] [PubMed]
50. Tamer, T.M.; Valachová, K.; Mohyeldin, M.S.; Soltes, L. Free radical scavenger activity of chitosan and its aminated derivative. *J. Appl. Pharm. Sci.* **2016**, *6*, 195–201. [CrossRef]
51. Abdelrazik, T.M.; Valachová, K.; Mohyeldin, M.S.; Soltes, L. Free radical scavenger activity of cinnamyl chitosan schiff base. *J. Appl. Pharm. Sci.* **2016**, *6*, 130–136. [CrossRef]
52. Gharibi, R.; Yeganeh, H.; Rezapour-Lactoee, A.; Hassan, Z.M. Stimulation of Wound Healing by Electroactive, Antibacterial, and Antioxidant Polyurethane/Siloxane Dressing Membranes: In Vitro and in Vivo Evaluations. *ACS Appl. Mater. Interfaces* **2015**, *7*, 24296–24311. [CrossRef] [PubMed]

53. Mao, L.; Hu, S.; Gao, Y.; Wang, L.; Zhao, W.; Fu, L.; Cheng, H.; Xia, L.; Xie, S.; Ye, W.; et al. Biodegradable and Electroactive Regenerated Bacterial Cellulose/MXene (Ti3C2Tx) Composite Hydrogel as Wound Dressing for Accelerating Skin Wound Healing under Electrical Stimulation. *Adv. Healthc. Mater.* **2020**, *9*, 2000872. [CrossRef] [PubMed]
54. Tamer, T.M.; Collins, M.N.; Valachová, K.; Hassan, M.A.; Omer, A.M.; Mohy-Eldin, M.S.; Švík, K.; Jurčík, R.; Ondruška, L.; Biró, C.; et al. MitoQ Loaded Chitosan-Hyaluronan Composite Membranes for Wound Healing. *Materials* **2018**, *11*, 569. [CrossRef] [PubMed]
55. Yazdanparast, R.; Shahriyary, L. Comparative effects of Artemisia dracunculus, Satureja hortensis and Origanum majorana on inhibition of blood platelet adhesion, aggregation and secretion. *Vasc. Pharmacol.* **2008**, *48*, 32–37. [CrossRef]
56. Okazaki, K.; Nakayama, S.; Kawazoe, K.; Takaishi, Y. Antiaggregant effects on human platelets of culinary herbs. *Phytother. Res.* **1998**, *12*, 603–605. [CrossRef]
57. Tamer, T.M.; Hassan, M.A.; Valachová, K.; Omer, A.M.; El-Shafeey, M.E.A.; Mohy Eldin, M.S.; Šoltés, L. Enhancement of wound healing by chitosan/hyaluronan polyelectrolyte membrane loaded with glutathione: In vitro and in vivo evaluations. *J. Biotechnol.* **2020**, *310*, 103–113. [CrossRef]
58. Peppas, N.A.; Stauffer, S.R. Reinforced uncrosslinked poly (vinyl alcohol) gels produced by cyclic freezing-thawing processes: A short review. *J. Control. Release* **1991**, *16*, 305–310. [CrossRef]
59. Yang, X.; Liu, Q.; Chen, X.; Yu, F.; Zhu, Z. Investigation of PVA/ws-chitosan hydrogels prepared by combined γ-irradiation and freeze-thawing. *Carbohydr. Polym.* **2008**, *73*, 401–408. [CrossRef]
60. Yin, L.; Fei, L.; Tang, C.; Yin, C. Synthesis, characterization, mechanical properties and biocompatibility of interpenetrating polymer network–super-porous hydrogel containing sodium alginate. *Polym. Int.* **2007**, *56*, 1563–1571. [CrossRef]
61. Hassan, M.A.; Tamer, T.M.; Rageh, A.A.; Abou-Zeid, A.M.; Abd El-Zaher, E.H.F.; Kenawy, E.-R. Insight into multidrug-resistant microorganisms from microbial infected diabetic foot ulcers. *Diabetes Metab. Syndr. Clin. Res. Rev.* **2019**, *13*, 1261–1270. [CrossRef] [PubMed]
62. Hassan, M.A.; Omer, A.M.; Abbas, E.; Baset, W.M.A.; Tamer, T.M. Preparation, physicochemical characterization and antimicrobial activities of novel two phenolic chitosan Schiff base derivatives. *Sci. Rep.* **2018**, *8*, 11416. [CrossRef] [PubMed]
63. El Fawal, G.F.; Omer, A.M.; Tamer, T.M. Evaluation of antimicrobial and antioxidant activities for cellulose acetate films incorporated with Rosemary and Aloe Vera essential oils. *J. Food Sci. Technol.* **2019**, *56*, 1510–1518. [CrossRef] [PubMed]
64. Tamer, T.M.; Hassan, M.A.; Omer, A.M.; Valachová, K.; Eldin, M.S.M.; Collins, M.N.; Šoltés, L. Antibacterial and antioxidative activity of O-amine functionalized chitosan. *Carbohydr. Polym.* **2017**, *169*, 441–450. [CrossRef] [PubMed]
65. Hassan, M.A.; Amara, A.A.; Haroun, B.M. Leucocytes show improvement growth on PHA polymer surface. *Pak. J. Pharm. Sci.* **2010**, *23*, 332–336. [PubMed]
66. Mosmann, T. Rapid colorimetric assay for cellular growth and survival: Application to proliferation and cytotoxicity assays. *J. Immunol. Methods* **1983**, *65*, 55–63. [CrossRef]
67. El-Fakharany, E.; Hassan, M.; Taha, T. Production and application of extracellular laccase produced by Fusarium oxysporum EMT. *Int. J. Agric. Biol.* **2016**, *18*, 939–947. [CrossRef]

International Journal of
Molecular Sciences

MDPI

Article

Ultrasound Imaging by Thermally Tunable Phononic Crystal Lens

Yuqi Jin [1,2] and Arup Neogi [1,3,*]

1 Department of Physics, University of North Texas, P.O. Box 311427, Denton, TX 76203, USA; yuqijin@my.unt.edu
2 Department of Mechanical Engineering, University of North Texas, 3940 North Elm Suite F101, Denton, TX 76207, USA
3 Center for Agile and Adaptive Additive Manufacturing, 3940 North Elm Suite, Denton, TX 76207, USA
* Correspondence: arupn@yahoo.com

Abstract: This work demonstrates the detections and mappings of a solid object using a thermally tunable solid-state phononic crystal lens at low frequency for potential use in future long-distance detection. The phononic crystal lens is infiltrated with a polyvinyl alcohol-based poly n-isopropyl acrylamide (PVA-PNIPAm) bulk hydrogel polymer. The hydrogel undergoes a volumetric phase transition due to a temperature change leading to a temperature-dependent sound velocity and density. The temperature variation from 20 °C to 39 °C changes the focal length of the tunable solid-state lens by 1 cm in the axial direction. This thermo-reversible tunable focal length lens was used in a monostatic setup for one- and two-dimensional mapping scans in both frequency domain echo-intensity and temporal domain time-of-flight modes. The experimental results illustrated $1.03 \pm 0.15\lambda$ and $2.35 \pm 0.28\lambda$ on the lateral and axial minimum detectable object size. The experiments using the tunable lens demonstrate the capability to detect objects by changing the temperature in water without translating an object, source, or detector. The time-of-flight mode modality using the tunable solid-state phononic lens increases the signal-to-noise ratio compared to a conventional phononic crystal lens.

Keywords: tunable lens; ultrasonic detection; deep detection; phononic crystal; hydrogel

Citation: Jin, Y.; Neogi, A. Ultrasound Imaging by Thermally Tunable Phononic Crystal Lens. *Int. J. Mol. Sci.* **2021**, *22*, 7966. https://doi.org/10.3390/ijms22157966

Academic Editor: Peter John Jervis

Received: 15 June 2021
Accepted: 22 July 2021
Published: 26 July 2021

1. Introduction

Phononic crystals are artificially engineered crystals with a periodic arrangement [1–4] of scatterers in an ambient medium. Based on incident wavelength, these phononic crystals can either cause Bragg scattering [5,6] or behave as a homogeneous material [7,8]. As in crystals with electronic bandgap, the transient behavior of phononic crystals can be modified by changing the lattice diameter, spacing, or arrangement as needed for an application. The flexibility of these phononic crystals has led to the design of new classes of lenses [6], filters [9], beam splitters [10], and decomposition devices [11]. A phononic crystal lens has been demonstrated to behave similarly to a homogeneous optical lens beyond the long-wavelength limit [12]. Phononic crystal-based lenses can also yield sub-wavelength resolution in the evanescent near-field by negative refraction and by breaking the diffraction limit using the meta-materials properties [6,13–16]. However, these artificially designed lenses have barely been used to image any real object for practical application. Moreover, these lenses are passive structures, and the operating wavelength is fixed for a designed structure.

There are three main categories of nondestructive imaging or mapping modality grouped in terms of the measurement techniques: echo-intensity [17,18], time-of-flight [19–21], and elastography [22,23]. Each of these techniques modalities has its advantages. The time-of-flight mode imaging provides temporal information about the size and the location of an object in an ambient medium [20]. Echo-intensity detection can distinguish the contrast between an object and ambient medium in terms of acoustic impedance [18] and is normally presented in the frequency domain. For a standardized ambient media, echo-intensity detection can also provide information about the location of the object in the axial direction based on the overall attenuation from the ambient material. Elastography is a technique that differentiates the object's elasticity contrast from its ambient environment using active [24] or passive [25–27] experimental methods.

Poly-vinyl alcohol-based poly n-isopropyl acrylamide (PVA-PNIPAm) hydrogel is a polymer that undergoes a volumetric phase transition at ~32–33 °C. The elasticity of the polymer changes due to the phase transition and can result in over a 10% change in the velocity of sound in the PVA-PNIPAm [23]. This variance in sound velocity can be used to thermally tune the effective mechanical properties of the phononic lens interstitially filled with PVA-PNIPAm, and consequently achieve a tunable focal length lens.

In this study, the thermally tunable solid-state phononic crystal lens (TSSL) (Figure 1) is utilized for both frequency domain echo-intensity and temporal domain time-of-flight based one- or two-dimensional scans in a monostatic setup. As the temperature of the lens is gradually raised from room temperature (20 °C), the focal length of the lens shifts from 10.29λ to 8.82λ from the transducer surface (Figure 2). This all-acoustic technique is utilized without any advanced signal processing technique and can laterally resolve 1.03 ± 0.15λ. in one/two-dimensional echo-intensity mode (Figure 3 and Figure 4) and 2.35 ± 0.28λ. in the axial direction using time-of-flight mode (Figure 5). Compared with a water-based focusing sonic lens (FSL), the polymer-based TSSL demonstrates its tunability in its focal length and far-field detection capability, which was characterized and explained in our previous work [28]. In this work, detections and mappings with the tunable solid-state lens are demonstrated, and the practical potential of the lens is investigated through the echo-intensity and time-of-flight measurements. The lateral and axial minimum detectable object size of the tunable solid-state lens was also experimentally characterized in hard material target samples.

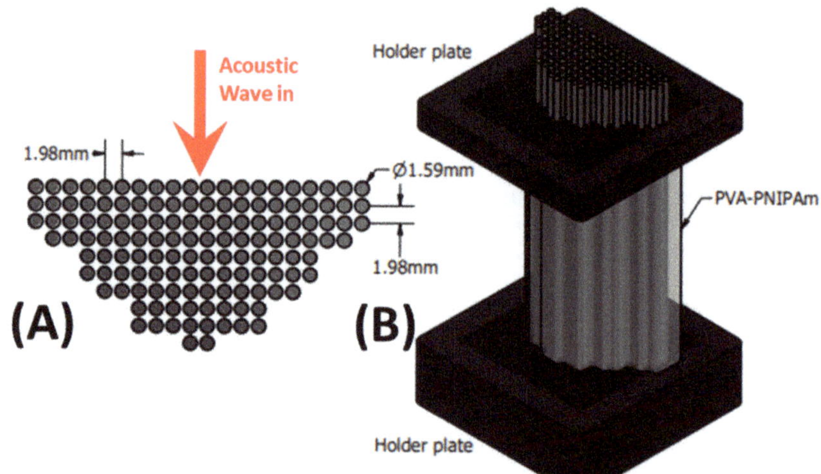

Figure 1. Design of the tunable solid-state lens. (**A**) The configuration of stainless-steel rods array as the scatterers in the phononic crystal structure. (**B**) The configuration of the lens including holder plates of the scatterer array and interstitially filled with thermally tunable hydrogel.

2. Results and Discussion

Focusing for the lens configurations both with and without hydrogel are shown in the intensity maps of Figure 2 at the operating frequency of 0.2 MHz, well within the homogenization zone as shown in the experimental section. For lens characterization, the full width at half the maximum of the transmitted signal is used for analysis. Objects are placed at the focal length of the lens and images are reconstructed from pulse-echo data. We observed from Figure 2E,F that the focal points of the water-based FSL are located at 45 mm away from the surface of the transducer. This position was located about 25 mm away from the apex of the phononic crystal lens. The FSL intensity maps show no appreciable effect of temperature on the lens's focal length as the water temperature is raised from 20 °C to 39 °C. The polymer's infiltration in the lens (TSSL) leads to an increase in the focal length of the lens to about 70 mm at room temperature (Figure 2A,C). Due to the change in the speed of sound and effective density of the polymer-filled phononic crystal, the lens's focal length increases. Figure 2B,D show that as the temperature is increased to 39 °C, the focal length of the TSSL lens reduces to about 60 mm in the axial direction. The numerical results (A and B) were qualitatively agreed with experimental results (C and D). The focal point of the lens at 39 °C is comparatively sharper than at 20 °C. The width of the focal point observed at 20 °C was clearly larger than the focal width at 39 °C. The enlarged focal point was expected to induce lower spatial resolution in detection.

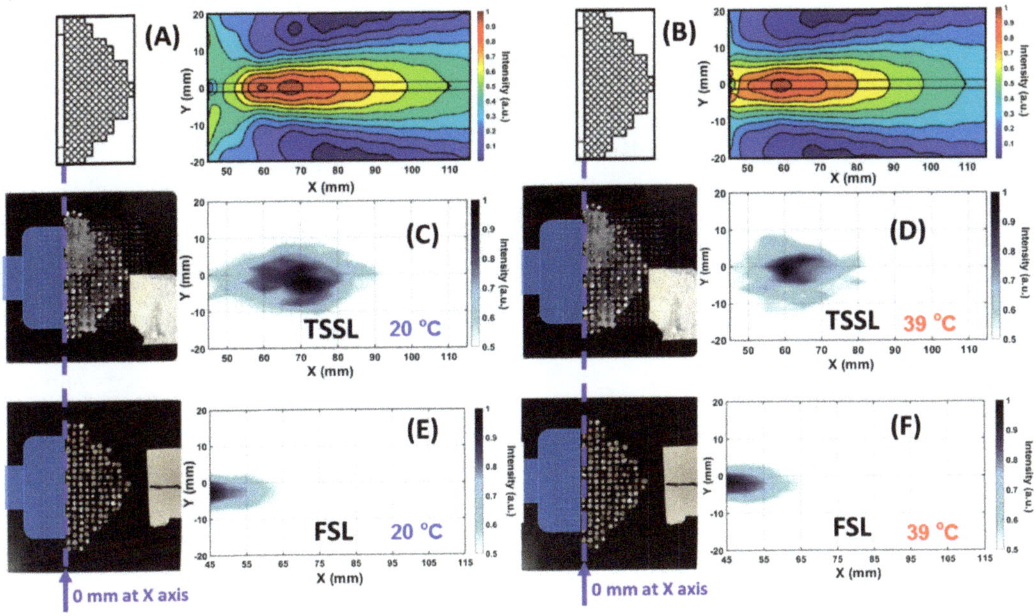

Figure 2. Sound intensity maps of TSSL (infilled with hydrogel) comparing with FSL (infilled with ambient water). Origin point of the *X*-axis in the contours was at the interface between transducer and lens. (**A**,**B**) illustrated numerical simulation of the focusing fields of TSSL at 20 °C and 39 °C. (**C**) Focusing point of TSSL at 20 °C. (**D**) Focusing point of TSSL at 39 °C. (**E**) Focusing point of FSL at 20 °C. (**F**) Focusing point of FSL at 39 °C.

The focalization mechanism illustrated in the Appendix A. The physical mechanism of focusing the ultrasonic was achieved differently from the acoustic meta-lens, which normally has a negative refraction index [29].

The temperature-dependent change in the lens's focal length property has been used to detect actual objects submerged in water. To test the detection feasibility of TSSL, two cylindrical acrylonitrile butadiene styrene (ABS) plastic rods with a diameter of 8 mm and 7 mm were 3D printed to test the lenses' detection capability. The structure is shown in the inset of Figure 3. The cylindrical rods were separated by a center-to-center distance of 10 mm, which corresponds to the tunable range of the TSSL (Figure 2C,D). These structures were placed along the lens's axial direction at the distance of 60 mm and 70 mm from the transducer surface as the blue dash lines indicated in Figure 2, which the one-dimensional scan along the Y direction. The 7-mm rod was placed closer to the TSSL than the 8 mm rod. Figure 3A illustrated the original reflected broadband pulse envelopes recorded at 0 mm from the 1D scan.

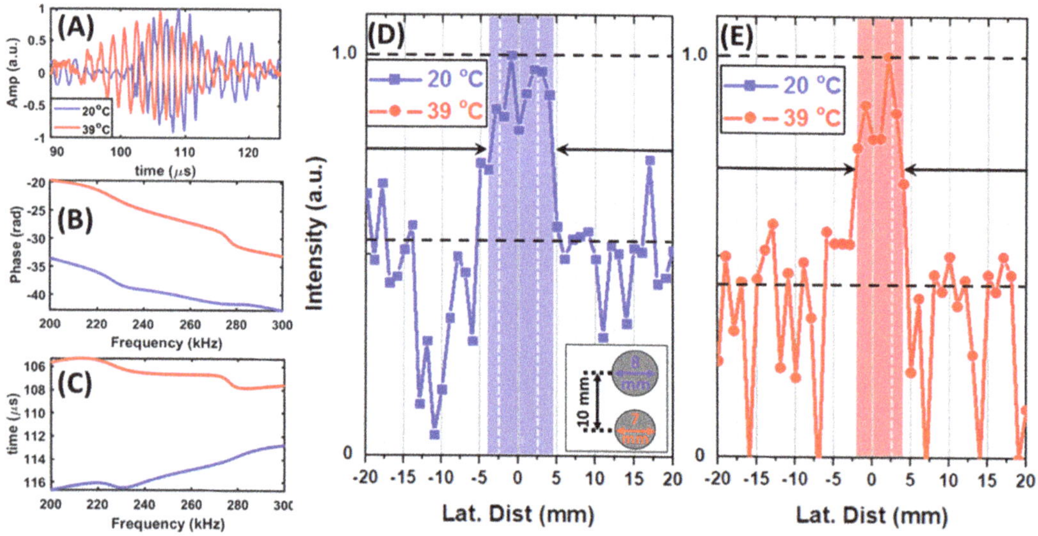

Figure 3. One-dimensional frequency-domain echo-intensity mode detection of two 3D printed ABS plastic rods in ambient water. The rods were located along the Y- (lateral) axis (Figure 2). The 7 mm diameter rod was 60 mm away from the transducer surface. The 8 mm diameter rod was 70 mm away from the transducer. (**A**) Temporal reflection waves detected from the target rods using TSSL at 20 °C and 39 °C. (**B**) The monochromatic components' phase delay in the range from 0.2 MHz to 0.3 MHz obtained by FFT of the signal showed in (**A**). (**C**) The arrival time of the monochromatic components in the range from 0.2 MHz to 0.3 MHz from the phase information illustrated in (**B**). The results for the one-dimensional echo-intensity mode detection was obtained by FFT of the temporal waves in (**A**) shown by the blue solid line with squares (TSSL at 20 °C) in (**D**) and red dash-dot line with circles (TSSL at 39 °C) in (**E**).

To determine the detailed temporal delay of the single-frequency component at 0.2 MHz, the wave envelopes were transformed to the frequency domain by FFT. The obtained phase of the frequency components in the range between 0.2 MHz and 0.3 MHz were illustrated in subfigure (B). In (C), the phase delay was translated in arriving time for a better illustration. The figures showed that the object detected at 39 °C was physical closer than the object found at 20 °C respected by the transducer. For clear estimations on the width difference between the two detected objects, the frequency domain single frequency component information at 0.2 MHz was sorted by the scanned direction along the lateral axis. Figure 3D,E show the transmitted ultrasonic wave's spatial profile through the lens at 20 °C and 39 °C. The wavelength of the emitting transducer used for the detection was about 6.8 mm. Our experimental observation showed that the TSSL detected the 8 mm diameter rod at 20 °C and the 7-mm diameter rod (located closer to the lens) at 39 °C without any axial translation of either the sample or the lens. Estimating the scanned object width values were from the intensity profiles with the averaged amplitude level between maximum and averaged noise. The object width accuracy of the TSSL operated in the 1D echo-intensity mode at 20 °C is 7% less than that at 39 °C. The loss of accuracy is due to the larger spatial dispersion of the TSSL at room temperature, as observed from Figure 2C,D.

Figure 4 shows the result of two-dimensional monostatic echo-intensity mode mapping of a 10 mm diameter aluminum rod in water ambient. In the experiments, the rod (sample) was translated within a 40 mm × 40 mm square area with a 2-mm interval on both the X- and Y-axes with the position of the source transducer and TSSL fixed. The center of the detection area was 80 mm away from the transducer surface on the axial (X) axis. Figure 4A,B show the reflection intensity profiles at 0.2 MHz of the object scanned by TSSL at 20 °C and 39 °C within identical scanning cross-section of the sample with motion. An increase in the temperature from 20 °C to 39 °C results in a scanned area shift backward by 10 mm on the X- (axial) axis. Both subfigures A and B showed an 11-mm wide object. Figure 4A depicts an ellipsoidal rod due to the focused beam's dispersion in the axial direction at 20 °C (Figure 2A).

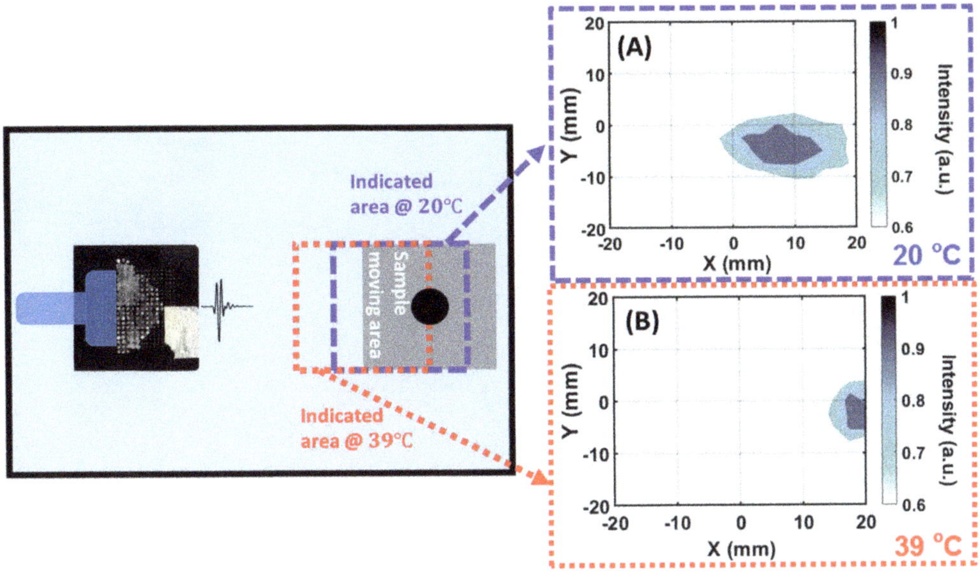

Figure 4. 2D frequency domain echo-intensity mode mappings by TSSL at 20 °C and 39 °C. A 10 mm diameter aluminum rod moved on the *X*- and *Y*-axis in a 40 mm by 40 mm area with a 2 mm interval on both axes. TSSL scanned data recorded at 20 °C (**A**) and 39 °C (**B**) with the sample moving in the same area.

Figure 5 shows a comparison of the far-field monostatic 1D time-of-flight mode mappings using the TSSL and the same structure without the PVA-PNIPAm polymer (an FSL). This modality uses the time of flight measurement of the reflected waves from the object. The results are presented in 3D surface plots (Figure 6) and 2D color contours (Figure A1) in terms of the lateral distance (*Y*-axis), the time-of-flight of the reflected wave with the normalized linear intensity scale. The scanned sample was a 16 mm diameter aluminum rod translated along the Y direction (laterally) by 60 mm in steps of 1-mm intervals. In Figure 5, the top four plots were obtained from the experiments when the sample was about 70 mm (farther) away from the transducer's surface on the *X*-axis. The bottom four plots were from the experiments when the sample was about 60 mm away from the transducer axially. Both TSSL and FSL were used for the temporal domain time-of-flight mode mapping under identical experimental conditions for comparison. The time-of-flight mode mapping was performed by lateral sweeping around the focal distance to detect the target object. The higher signal-to-noise ratio for the ultrasonic wave around the axial direction indicates the presence of an object. By detecting the temporal location and the reflected signal's length, the depth and axial size of the object could be calculated. These reflected signals are compared to a standard reference with a well-calibrated speed of sound and acoustic properties.

Figure 5 shows that the TSSL exhibits an improved signal-to-noise ratio when the object is at 70 mm for a lens temperature of 20 °C. At 39 °C, the object at 60 mm exhibits a higher signal-to-noise ratio. In order to have a better view of the contrast in signal-to-noise ratio under on-focusing and off-focusing conditions, the results were presented in 3D plots to show the amplitude difference. The conventional 2D contours versions of the TSSL results were posted in the Appendix A Figure A2.

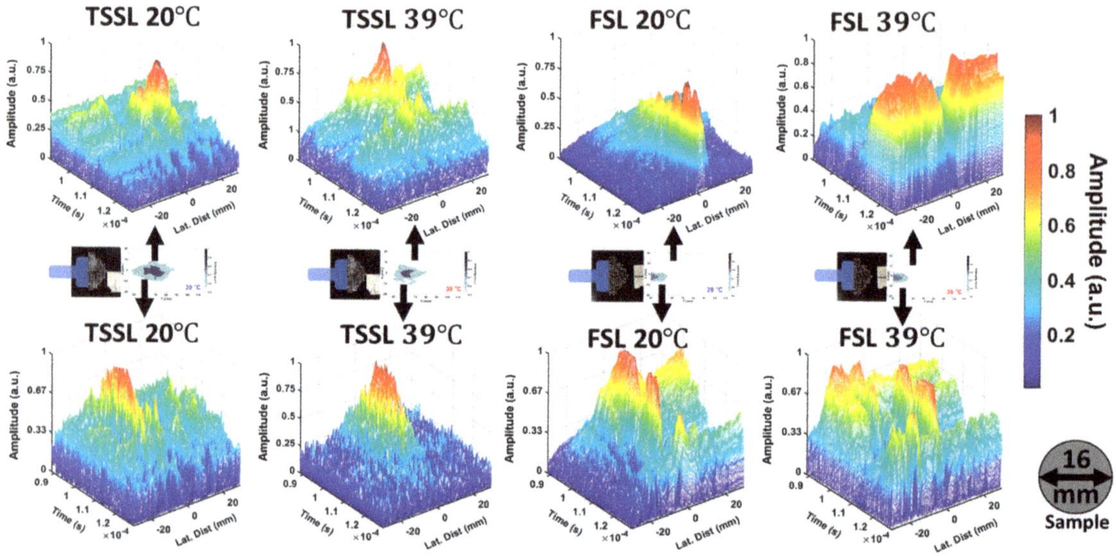

Figure 5. One-dimensional temporal domain time-of-flight mapping. A 16 mm diameter Aluminum rod translated on lateral direction (*Y*-axis) by a distance of 60 mm with 1 mm interval. The translating axis on *X*-axis was located 70 mm away from the transducer surface (Upper four subfigures (**A**,**B**,**E**,**F**)) and 60 mm away from the transducer surface (Lower four subfigures (**C**,**D**,**G**,**H**)).

From Figure 5A,D, the center of the reflected signal at 20 °C is observed at 107.1 μs and 93.8 μs (39 °C) from the sample to the TSSL which corresponds to the distances between the transducer surface and samples. The axial dimension of the scanned aluminum rod could be obtained from the half maximum of the temporal length of the reflection from the signals at 20 °C (farther sample) and 39 °C (closer sample), which were 7.27 μs and 5.52 μs respectively. As the sound speed in aluminum is 6400 m/s, the axial estimated values of the scanned sample were 23.2 mm and 17.6 mm.

The thermal sensitivity of the tunability of TSSL has been demonstrated in the above results. By changing the polymer phase to above the volume phase transition temperature, the focal length could be reversibly changed from 10.29λ to 8.82λ away from the transducer's surface. From the monostatic 1D frequency-domain echo-intensity mode detection, the lateral minimum detectable object size of the PVA-PNIPAm infilled TSSL was illustrated as detecting a 1.03λ wide object with a 0.15λ error. Since the operating frequency of the TSSL was in its first transmission band, a sub-wavelength minimum detectable object size was not expected. The lateral minimum detectable object size (1.03λ) of the TSSL was not compromised by the dispersion of the acoustic waves in the polymer and was close to the operating wavelength (6.8 mm). The axial minimum detectable object size of the TSSL that provides information about the rod's diameter was observed to be 2.35λ.

Compared to a tunable solid-state lens with a similar shape focusing on the sonic lens, the tunable solid-state lens focused the sound in the far-field which is more suitable for practical detections. Biomedical ultrasound imaging and detection are mostly operating in high frequency, such as 10 MHz for the smaller lateral minimum detectable object size. Higher frequency introduces larger attenuation and dispersion effects, reducing detection depth and penetration. In this study, a low operating frequency (0.2 MHz) tunable solid-state lens detections and mappings demonstrated about one wavelength lateral minimum detectable object size, which showed the potential possibility of clear, long-distance detection.

Further study on this novel type of phononic crystal lens can focus on increasing the lens's tunability and increasing the scanning speed between the variable focal points. The use of the different types of phase transition hydrogel [30], even microgel [31], or viscoelastic polymer with larger modulation of its physical properties, can be applied in TSSL to obtain a larger tunable focusing distance. Instead of conventional heating of the TSSL, electromagnetic waves such as infrared radiation [9] and radio-frequency electromagnetic field [32] can be a viable option.

3. Material and Methods

The phononic crystal structure used in this study was designed and fabricated to have a 50.1% filling fraction of stainless-steel scatterers in water (FSL) or PVA-PNIPAm (TSSL) in the square lattice configuration. The diameter of the stainless-steel rods was 1/16-inch (1.59 mm) with a lattice constant of 5/64-inch (1.98 mm) periodic arrangement. Based on our previously reported temperature-dependent density and sound speed in PVA-PNIPAm hydrogel [33], the temperature dependence of the dispersion relation of the first transmission band for the hydrogel infilled phononic crystal was calculated and presented in Figure 6A. The actual phononic crystal structure infiltrated with hydrogel before and after phase transition is shown in Figure 6B. The stainless-steel rod has the elastic modulus 195 GPa, Poisson's ratio 0.29, and density 7500 kg/m^3 from the datasheet. The dynamic bulk modulus of the PVA-PNIPAm hydrogel was calculated as a product of density (ρ) and square of speed of sound ($K = \rho c^2$), which was 1.65 GPa at room temperature and 2.42 GPa at 39 °C [34]. The effective sound velocity of the hydrogel infilled PnC lens is 2150 m/s at room temperature and 2289 m/s at 39 °C.

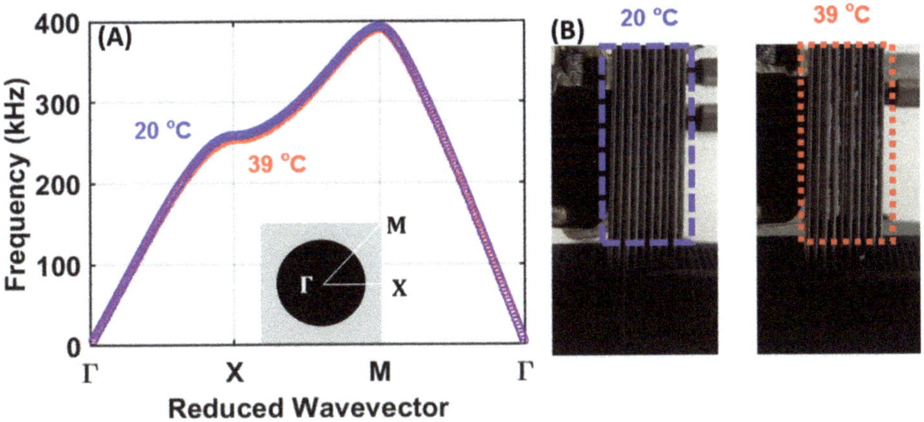

Figure 6. (**A**) Band structure of the first band of the PVA-PNIPAm filled phononic crystal at 20 °C (blue) and 39 °C (red) using the measured hydrogel density and speed of sound. Γ, X, and M points refer to the location of (0,0), (0,1) and (1,1) in the unit reciprocal unit cell. (**B**) Photographs of the TSSL at 20 °C and 39 °C ambient temperatures.

The numerically simulated results of sound intensity fields illustrated in Figure 2A,B were performed by finite element analysis (FEA) based COMSOL Multiphysics software using pressure acoustic and solid mechanics module. The acoustic transducer was defined as a plane wave source. The properties of ambient water were selected from the COMSOL library with the speed of sound of 1480 m/s at 295 K ambient, and density of 998 kg/m^3. The Young's modulus, speed of sound, and a density of stainless-steel rods were set to 200 GPa, 5800 m/s, and 7850 kg/m^3 respectively. The speed of sound and density values of PVA-PNIPAm hydrogel were 1345.4 m/s and 1048.9 kg/m^3 at 20 °C 1425.5 m/s 1269.3 kg/m^3 at 39 °C which were experimental measured in our previous works [33,34].

Figure 7A shows the bistatic experiment setup of sound intensity raster focusing point maps (Figure 2). A one-inch (25.4 mm) Olympus diameter planar transducer V301 and the lens in a deionized water tank were fully emerged and aligned for the transmission measurements. The transducer-generated broadband pulses at 0.5 MHz center frequency with effective bandwidth from 0.1 MHz to 0.9 MHz at 500 Hz repetition rate were focused using the lens. The 0.5 MHz center frequency immersion transducer was used as the acoustic source. Our target frequency of 0.2 MHz in frequency domain experiments was located close to the Gaussian-liked distribution bandwidth edge from 0.1 MHz to 0.9 MHz, which led to a relatively low initial signal-to-noise ratio.

Other energy loss sources also exist in the system, such as impedance mismatch between ambient water and TSSL and the hydrogel's attenuation. A needle hydrophone (Müller-Platte) with a tip diameter of 1 mm was used to measure the outcoming signal from the lens. The tip was used to map the acoustic pressure within a 40 mm (Y) by 70 mm (X) area with a 2-mm interval along the *Y*-axis and a 5-mm interval along the *X*-axis. The needle was raster-scanned using a custom MATLAB code and controlled by the computer. The program-controlled the Newmark NSG-G2-X2 translation stage to linearly translate with a step size of 0.5 mm/s and a time delay of 15 s to facilitate data accumulation. It also controlled the two-dimensional Newmark LC-200-11 linear translation stage holding the needle hydrophone. The temporal signal from the hydrophone was averaged (512 time windows) and acquired in the time domain channel on a Tektronix MDO 3024 oscilloscope. In the frequency domain echo-intensity mode detection results, reflection intensity was described as a single frequency component at only 0.2 MHz. The monochromatic information at 0.2 MHz was obtained from a time domain to frequency domain conversion of the recorded time-domain reflection signal by fast Fourier transformation (FFT).

With the single-frequency information at 0.2 MHz, the minimum detectable object size of the echo-intensity mode detection results (Figures 3 and 4) can be analyzed based on its wavelength of 6.8 mm in ambient water. Figure 7B illustrates the monostatic setup of the experiments which produced results in Figures 3–5. In the one-dimensional experiments, the sample was translated along arrow 1 (*Y*-axis) with 1-mm interval. In the two-dimensional experiment, the sample was translated in a 40 mm × 40 mm area along the direction of the arrows 1 and 2 (*Y*- and *X*-axis) with a 2-mm interval as indicated in Figure 7B.

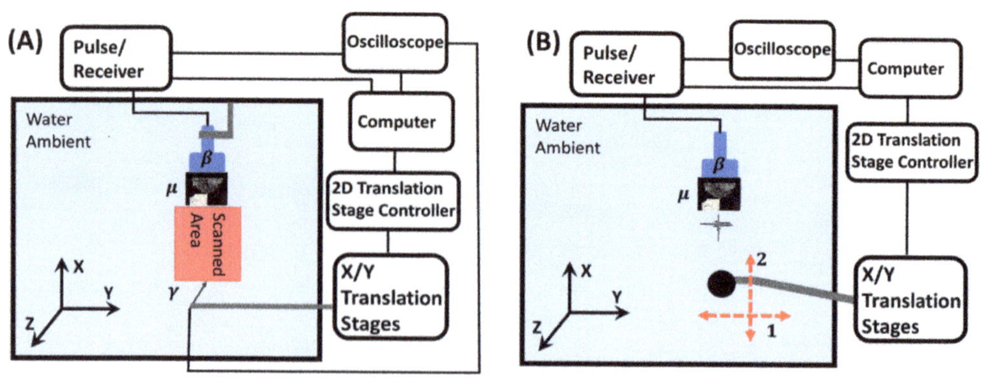

Figure 7. (**A**) Experimental setup of sound pressure field raster scan. (**B**) Experimental setup of monostatic frequency echo-intensity and time domain time-of-flight modes mappings in both 1D and 2D. One-dimensional detection was performed by moving the sample along arrow 1 (*X*-axis). 2D mapping was achieved by translating the object in both arrow 1 and 2 (*X*- and *Y*-axis). In the illustrations, β was transducer. μ indicated lens. Arrow γ represented needle hydrophone.

4. Conclusions

This work demonstrated the first practical experimental study of a thermal tunable solid-state phononic crystal focusing lens prototype. The results demonstrated the feasibility of applying a hydrogel infilled tunable solid-state lens in frequency domain echo-intensity and temporal domain time-of-flight modes monostatic detections and mappings in simple experimental environments. Comparing with the same shape water-infilled FSL, the tunability of the focusing capability of the TSSL were remarkable advantages.

Author Contributions: Conceptualization, Y.J.; methodology, Y.J.; software, Y.J.; validation, A.N. and A.N.; formal analysis, A.N.; resources, A.N.; data curation, Y.J.; writing—original draft preparation, Y.J. and A.N.; writing—review and editing, A.N. and A.N.; supervision, A.N.; funding acquisition, A.N. All authors have read and agreed to the published version of the manuscript.

Funding: This work is supported by an Emerging Frontiers in Research and Innovation (EFRI) grant from the National Science Foundation (NSF Grant No. 1741677), and the support from the infrastructure and support of Center for Agile & Adaptive and Additive Manufacturing (CAAAM) funded through State of Texas Appropriation #190405-105-805008-220 is also greatly acknowledged.

Institutional Review Board Statement: Not applicable.

Informed Consent Statement: Not applicable.

Data Availability Statement: Data available for requesting to the corresponding author.

Acknowledgments: We deeply appreciate the language checking and nice suggestions from Mathew Sofield and Ezekiel Walker.

Conflicts of Interest: The authors declare no conflict of interest.

Appendix A

In Figure A1, we posted the focusing mechanism of the TSSL in more detail. As the experimental sound intensity field (subfigure A) illustrated, with a planar ultrasound transducer source (at 0 mm on *Y*-axis), the near field effect provided slight convergence wave vectors as the source to the TSSL. In subfigure (B) and (C), we showed the 0.2 MHz equi-frequency contours of TSSL at 20 °C and 39 °C. On the lens's closer edge respected to the transducer, the wave was incident on the normal direction. Since the farther curve edge of the lens has an approximately 45° slope. After 45° clockwise rotation, 0° downward incidence in the EFCs described the emission vector from a prefect collimated source. In EFCs (subfigures B and C), we used approximately normal incident angle in the 45° rotated reciprocal unit cell to represent the around 45° edge lens curvature. Due to the near-field effect, we used a 5° incident angle as the source vector from the transducer to the lens in water. The red dash circle was indicated to the EFC of the ambient water. The dash black lines were indicated the momentum-matched conditions of $k_i = nG + k_r$, where k_i and k_r were the incidence vector from transducer in water and resultant vector from the lens in water. G was the reciprocal lattice vector. And n was the integral numbers. The dash black line was the momentum-matched condition when n equals 1. The k_x axis location of the black dash line was from the projection of the incidence vector k_i. From (B) and (C), the EFCs predicted the refraction angle results vectors in water at the outcoming region which equals to the refraction angle incidence vector.

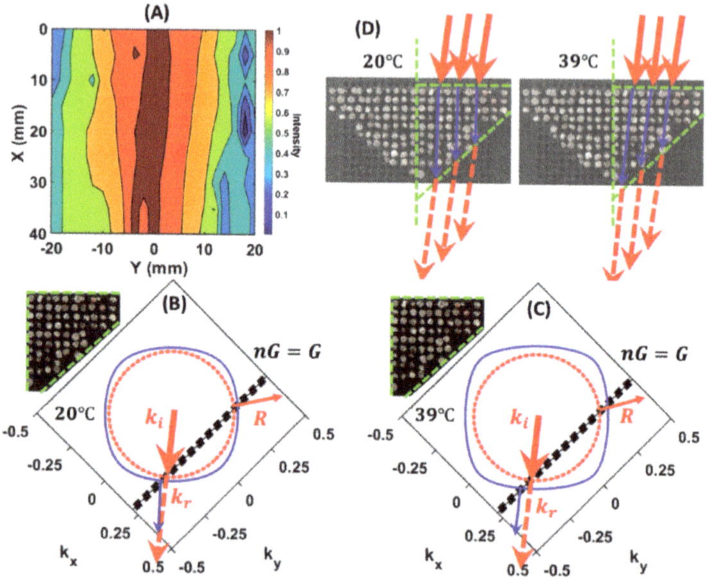

Figure A1. (**A**) Sound intensity field of the source planar transducer. The value of 0 mm on the *Y*-axis refers to the surface of the transducer. (**B**) Equi-frequency contour (EFC) of TSSL at 20 °C. (**C**) EFC of TSSL at 39 °C. (**D**) Illustration of the focusing mechanism based on the refracted wave vectors from (**B**,**C**) The unit of k_x, k_y and G was $2\pi/a$. The red arrows referred the wave direction in water. The blue arrows referred the wave direction in lens.

However, the energy propagation direction in the TSSL was different at 20 °C and 39 °C as the blue arrows in subfigure (B) and (C) showed. Vectors R were the backward propagation side beams. At 20 °C (39 °C), the group velocity inside the TSSL propagated more (less) comparable with 0° vertical direction. As the illustration in subfigure (D) showed, at 20 °C (39 °C), the smaller (larger) angle of internal refraction in TSSL resulted in a wider (narrower) outcoming converged beam and a farther (closer) focal length. In detail, at 20 °C, the source beam from the 1-inch diameter planar transducer had the beam radius around 11 mm and a slight convergence direction (near-field effect) at 10° as the red shadow and arrows illustrated. The wave propagated into the lens with the refracted angle around 2° respected to the horizontal direction which was obtained from the blue solid arrow in Figure A1B. The beam radius inside the lens was reduced to 10.1 mm at the curved lens edge around 13 mm on the axial axis respected to the edge close to the source transducer. The outcoming beam from the lens to ambient water had a refraction angle at 10° which was also showed in Figure A1B with the red dash arrow.

Based on the reduced beam radius (10.1 mm) and angle (10°), the calculated focal length was 70 mm as the following figure (A) showed. At 39 °C, the source beam from the transducer also had the radius around 11 mm and the slight convergence direction at 10° as the red shadow and arrows illustrated. The wave propagated into the lens with the refracted angle around 9°, which was obtained from the blue arrow in Figure A1C. Around the axial distance 15 mm apart from the flat edge of the lens, the beam radius reduced to around 7.9 mm at the curved lens edge. The outcoming beam from the lens to ambient water had a refraction angle at 10° which was also showed in Figure A1C. Based on the reduced beam radius (7.9 mm) and angle (10°), the calculated focal length was about 60 mm. The analytical EFC contours of the TSSL agreed with the behavior compared with the numerical and experimental determined temperature-dependent focal length in Figure 2.

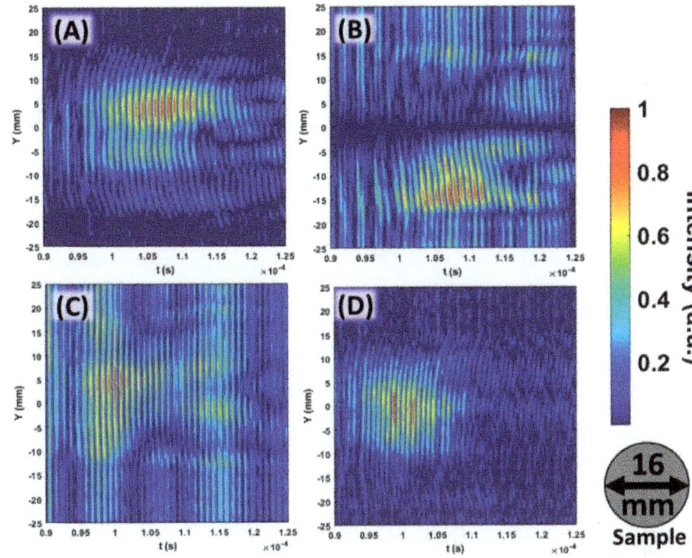

Figure A2. Two-dimensional contour version of the time-of-flight mode detection (Figure 5A–D) using TSSL at near and far distance at 20 °C and 39 °C. (**A**). Far distance (70 mm) between transducer and sample scanned by TSSL at 20 °C. (**B**). Far distance (70 mm) between transducer and sample scanned by TSSL at 39 °C. (**C**). Near distance (60 mm) between transducer and sample scanned by TSSL at 20 °C. (**D**). Near distance (60 mm) between transducer and sample scanned by TSSL at 39 °C.

References

1. Kushwaha, M.S.; Halevi, P.; Dobrzynski, L.; Djafari-Rouhani, B. Acoustic band structure of periodic elastic composites. *Phys. Rev. Lett.* **1993**, *71*, 2022. [CrossRef] [PubMed]
2. Liu, Z.; Chan, C.T.; Sheng, P.; Goertzen, A.L.; Page, J.H. Elastic wave scattering by periodic structures of spherical objects: Theory and experiment. *Phys. Rev. B* **2000**, *62*, 2446. [CrossRef]
3. Kushwaha, M.S.; Djafari-Rouhani, B.; Dobrzynski, L.; Vasseur, J.O. Sonic stop-bands for cubic arrays of rigid inclusions in air. *Eur. Phys. J. B Condens. Matter Complex Syst.* **1998**, *3*, 155. [CrossRef]
4. Yang, S.; Page, J.H.; Liu, Z.; Cowan, M.L.; Chan, C.T.; Sheng, P. Ultrasound tunneling through 3D phononic crystals. *Phys. Rev. Lett.* **2002**, *88*, 104301. [CrossRef] [PubMed]
5. Vasseur, J.O.; Deymier, P.A.; Chenni, B.; Djafari-Rouhani, B.; Dobrzynski, L.; Prevost, D. Experimental and theoretical evidence for the existence of absolute acoustic band gaps in two-dimensional solid phononic crystals. *Phys. Rev. Lett.* **2001**, *86*, 3012. [CrossRef] [PubMed]
6. Yang, S.; Page, J.H.; Liu, Z.; Cowan, M.L.; Chan, C.T.; Sheng, P. Focusing of sound in a 3D phononic crystal. *Phys. Rev. Lett.* **2004**, *93*, 024301. [CrossRef]
7. Krokhin, A.A.; Arriaga, J.; Gumen, L.N. Speed of sound in periodic elastic composites. *Phys. Rev. Lett.* **2003**, *91*, 264302. [CrossRef] [PubMed]
8. Sánchez-Dehesa, J.; Krokhin, A. *Introduction to Acoustics of Phononic Crystals. Homogenization at Low Frequencies*; Springer Science and Business Media LLC: Berlin/Heidelberg, Germany, 2016; Volume 1, pp. 1–21.
9. Walker, E.; Reyes, D.; Rojas, M.M.; Krokhin, A.; Wang, Z.; Neogi, A. Tunable ultrasonic phononic crystal controlled by infrared radiation. *Appl. Phys. Lett.* **2014**, *105*, 143503. [CrossRef]
10. Li, J.; Wu, F.; Zhong, H.; Yao, Y.; Zhang, X. Acoustic beam splitting in two-dimensional phononic crystals using self-collimation effect. *J. Appl. Phys.* **2015**, *118*, 144903. [CrossRef]
11. Jin, Y.; Zubov, Y.; Yang, T.; Choi, T.Y.; Krokhin, A.; Neogi, A. Spatial Decomposition of a Broadband Pulse Caused by Strong Frequency Dispersion of Sound in Acoustic Metamaterial Superlattice. *Materials* **2021**, *1*, 125.
12. Håkansson, A.; Cervera, F.; Sánchez-Dehesa, J. Sound focusing by flat acoustic lenses without negative refraction. *Appl. Phys. Lett.* **2005**, *86*, 054102. [CrossRef]
13. Li, J.; Liu, Z.; Qiu, C. Negative refraction imaging of acoustic waves by a two-dimensional three-component phononic crystal. *Phys. Rev. B* **2006**, *73*, 054302. [CrossRef]

14. Cheng, Y.; Zhou, C.; Wei, Q.; Wu, D.; Liu, X. Acoustic subwavelength imaging of subsurface objects with acoustic resonant metalens. *Appl. Phys. Lett.* **2013**, *103*, 224104. [CrossRef]
15. Li, Y.; Yu, G.; Liang, B.; Zou, X.; Li, G.; Cheng, S.; Cheng, J. Three-dimensional ultrathin planar lenses by acoustic metamaterials. *Sci. Rep.* **2014**, *4*, 6830. [CrossRef] [PubMed]
16. Chen, S.; Fan, Y.; Fu, Q.; Wu, H.; Jin, Y.; Zheng, J.; Zhang, F. A review of tunable acoustic metamaterials. *Appl. Sci.* **2018**, *8*, 1480. [CrossRef]
17. Tu, H.; Varghese, T.; Madsen, E.L.; Chen, Q.; Zagzebski, J.A. Ultrasound attenuation imaging using compound acquisition and processing. *Ultrason. Imaging* **2003**, *25*, 245. [CrossRef]
18. Passmann, C.; Ermert, H. A 100-MHz ultrasound imaging system for dermatologic and ophthalmologic diagnostics. *IEEE Trans. Ultrason. Ferroelectr. Freq. Control* **1996**, *43*, 545. [CrossRef]
19. Bevan, P.D.; Sherar, M.D. B-scan ultrasound imaging of thermal coagulation in bovine liver: Frequency shift attenuation mapping. *Ultrasound Med. Biol.* **2001**, *27*, 809. [CrossRef]
20. Fatemi, M.; Kak, A.C. Ultrasonic B-scan imaging: Theory of image formation and a technique for restoration. *Ultrason. Imaging* **1989**, *2*, 1. [CrossRef]
21. Gan, W.S. *Acoustical Imaging: Techniques and Applications for Engineers*; John Wiley & Sons: Hoboken, NJ, USA, 2012.
22. Righetti, R.; Ophir, J.; Srinivasan, S.; Krouskop, T.A. The feasibility of using elastography for imaging the Poisson's ratio in porous media. *Ultrasound Med. Biol.* **2004**, *30*, 215. [CrossRef]
23. Sarvazyan, A.P.; Rudenko, O.V.; Swanson, S.D.; Fowlkes, J.B.; Emelianov, S.Y. Shear wave elasticity imaging: A new ultrasonic technology of medical diagnostics. *Ultrasound Med. Biol.* **1998**, *24*, 1419. [CrossRef]
24. Dhyani, M.; Anvari, A.; Samir, A.E. Ultrasound elastography: Liver. *Abdom. Imaging* **2015**, *40*, 698. [CrossRef]
25. Jin, Y.; Walker, E.; Krokhin, A.; Heo, H.; Choi, T.Y.; Neogi, A. Enhanced instantaneous elastography in tissues and hard materials using bulk modulus and density determined without externally applied material deformation. *IEEE Trans. Ultrason. Ferroelectr. Freq. Control* **2019**, *67*, 624. [CrossRef]
26. Jin, Y.; Walker, E.; Heo, H.; Krokhin, A.; Choi, T.Y.; Neogi, A. Nondestructive ultrasonic evaluation of fused deposition modeling based additively manufactured 3D-printed structures. *Smart Mater. Struct.* **2020**, *29*, 045020. [CrossRef]
27. Jin, Y.; Yang, T.; Heo, H.; Krokhin, A.; Shi, S.; Dahotre, N.; Choi, T.-Y.; Neogi, A. Novel 2D dynamic elasticity maps for inspection of anisotropic properties in fused deposition modeling objects. *Polymers* **2020**, *12*, 1966. [CrossRef] [PubMed]
28. Walker, E.L.; Reyes-Contreras, D.; Jin, Y.; Neogi, A. Tunable hybrid phononic crystal lens using thermo-acoustic polymers. *ACS Omega* **2019**, *4*, 16585. [CrossRef] [PubMed]
29. Walker, E.L.; Jin, Y.; Reyes, D.; Neogi, A. Sub-wavelength lateral detection of tissue-approximating masses using an ultrasonic metamaterial lens. *Nat. Commun.* **2020**, *11*, 5967. [CrossRef]
30. Walker, E.; Reyes, D.; Krokhin, A.; Neogi, A. Anomalous temperature dependence of speed of sound of bulk poly (N-isopropylacrylamide) hydrogels near the phase transition. *Ultrasonics* **2014**, *54*, 1337. [CrossRef]
31. Städele, V.; Gasser, U.; Dietsch, H. Ellipsoidal hybrid magnetic microgel particles with thermally tunable aspect ratios. *Soft Matter* **2012**, *8*, 4427. [CrossRef]
32. Walker, E.L.; Wang, Z.; Neogi, A. Radio-frequency actuated polymer-based phononic meta-materials for control of ultrasonic waves. *NPG Asia Mater.* **2017**, *9*, e350. [CrossRef]
33. Jin, Y.; Heo, H.; Walker, E.; Krokhin, A.; Choi, T.Y.; Neogi, A. The effects of temperature and frequency dispersion on sound speed in bulk poly (Vinyl Alcohol) poly (N-isopropylacrylamide) hydrogels caused by the phase transition. *Ultrasonics* **2019**, *104*, 105931. [CrossRef] [PubMed]
34. Jin, Y.; Yang, T.; Ju, S.; Zhang, H.; Choi, T.Y.; Neogi, A. Thermally Tunable Dynamic and Static Elastic Properties of Hydrogel Due to Volumetric Phase Transition. *Polymers* **2020**, *12*, 1462. [CrossRef] [PubMed]

International Journal of
Molecular Sciences

MDPI

Article

Bio-Functionalized Chitosan for Bone Tissue Engineering

Paola Brun [1,*], Annj Zamuner [2], Chiara Battocchio [3], Leonardo Cassari [2], Martina Todesco [2], Valerio Graziani [3], Giovanna Iucci [3], Martina Marsotto [3], Luca Tortora [3], Valeria Secchi [3] and Monica Dettin [2]

1 Department of Molecular Science, University of Padua, Via A. Gabelli 63, 35121 Padua, Italy
2 Department of Industrial Engineering, University of Padua, Via F. Marzolo 9, 35131 Padua, Italy; annj.zamuner@unipd.it (A.Z.); leonardo.cassari@phd.unipd.it (L.C.); martina.todesco@unipd.it (M.T.); monica.dettin@unipd.it (M.D.)
3 Department of Science, Roma Tre University of Rome, Via della Vasca Navale 79, 00146 Rome, Italy; chiara.battocchio@uniroma3.it (C.B.); valerio.graziani@roma3.infn.it (V.G.); giovanna.iucci@uniroma3.it (G.I.); martina.marsotto@uniroma3.it (M.M.); luca.tortora@uniroma3.it (L.T.); valeria.secchi@uniroma3.it (V.S.)
* Correspondence: paola.brun.1@unipd.it; Tel.: +39-04-9827-2343

Abstract: Hybrid biomaterials allow for the improvement of the biological properties of materials and have been successfully used for implantology in medical applications. The covalent and selective functionalization of materials with bioactive peptides provides favorable results in tissue engineering by supporting cell attachment to the biomaterial through biochemical cues and interaction with membrane receptors. Since the functionalization with bioactive peptides may alter the chemical and physical properties of the biomaterials, in this study we characterized the biological responses of differently functionalized chitosan analogs. Chitosan analogs were produced through the reaction of GRGDSPK (RGD) or FRHRNRKGY (HVP) sequences, both carrying an aldehyde-terminal group, to chitosan. The bio-functionalized polysaccharides, pure or "diluted" with chitosan, were chemically characterized in depth and evaluated for their antimicrobial activities and biocompatibility toward human primary osteoblast cells. The results obtained indicate that the bio-functionalization of chitosan increases human-osteoblast adhesion ($p < 0.005$) and proliferation ($p < 0.005$) as compared with chitosan. Overall, the 1:1 mixture of HVP functionalized-chitosan:chitosan is the best compromise between preserving the antibacterial properties of the material and supporting osteoblast differentiation and calcium deposition ($p < 0.005$ vs. RGD). In conclusion, our results reported that a selected concentration of HVP supported the biomimetic potential of functionalized chitosan better than RGD and preserved the antibacterial properties of chitosan.

Keywords: chitosan; functionalization; Chit-HVP; Chit-RGD; XPS; NEXAFS; h-osteoblasts

Citation: Brun, P.; Zamuner, A.; Battocchio, C.; Cassari, L.; Todesco, M.; Graziani, V.; Iucci, G.; Marsotto, M.; Tortora, L.; Secchi, V.; et al. Bio-Functionalized Chitosan for Bone Tissue Engineering. *Int. J. Mol. Sci.* **2021**, *22*, 5916. https://doi.org/10.3390/ijms22115916

Academic Editor: Peter John Jervis

Received: 28 April 2021
Accepted: 27 May 2021
Published: 31 May 2021

1. Introduction

Natural biopolymers have attracted more and more attention as tissue engineering scaffolds because of their inherent biophysical and biochemical properties, such as renewability, biocompatibility, biodegradability, and targeting ability. Biopolymers present many advantages over synthetic material-based scaffolds in terms of half-life, stability, safety, and ease of manufacture. Among them, polysaccharide-based scaffolds are non-toxic, non-immunogenic, and display good targeting ability and biocompatibility. They can undergo in vivo enzymatic digestion generating metabolites characterized by low toxicity, too [1]. Moreover, functional groups of the polysaccharide backbone serve as anchoring sites for chemical modifications, generating versatile scaffolds of great significance in the biomedical field [2].

Chitosan is a natural polysaccharide composed of glucosamine and *N*-acetyl-glucosamine, produced by deacetylation of chitin. The amine groups in the glucosidic residue are positively charged at physiological pH. The presence of amino groups assures reactive sites for conjugation. Chitosan is insoluble in water at neutral pH and soluble in acidic conditions when the amino groups are protonated. As it is non-toxic, biodegradable, and biocompatible, the Food and Drug Administration has already approved chitosan for use in wound dressings. Several papers reported on chitosan and its potential application in biomedical fields, including wound healing, tissue regeneration, and drug delivery [3,4]. Indeed, because of its hydrophilic surface, biocompatibility, and biodegradability, chitosan has found significant application in promoting cell adhesion, proliferation, and differentiation [5]. Since chitosan has poor mechanical properties, hybrid materials and copolymers have been evaluated [5]. Sequences of short amino acids of the extracellular matrix (ECM) proteins have been used to guide interactions with specific cell receptors and improve the biological targets of biopolymers by inducing selected cellular responses. Indeed, developing hybrid biomaterials by including peptide sequences into biopolymers is an attractive alternative to confer functional cell-biomaterial interaction. The amino acid sequence RGD (Arginine-Glycine-Aspartic Acid) is a well-known tripeptide that induces cell adhesion and migration by interacting with integrin family receptors. The conjugation of GRGDS to carboxymethyl-trimethyl chitosan, proposed by Hansson A et al. [5], produced a scaffold for dermal healing that increases the initial adhesion 3–5 times and the cell spreading 12-fold compared with unfunctionalized scaffolds. In the article of Wang C et al. [6], a cyclic RGD peptide was conjugated to chitosan by a thiolation reaction and a cross-linking agent and used in addition to graphene oxide in drug delivery applications. Because of its osteoconductive properties and structural homology to proteoglycans of extracellular matrices, chitosan has been proposed as a coating in bone implants or a component in composite materials [7]. So far, experimental studies have reported that alloy surfaces, layer assembly, and the degree of deacetylation n chitosan affect the promotion of osseointegration, bone remodeling, and delivery of growth factors at the implant-tissue interface [8].

In addition to its advantages in tissue regeneration and biomedical applications, chitosan is endowed with intrinsic antimicrobial properties with no toxic effects on host cells [9]. Despite considerable progress, orthopedic device-related infections are clinically relevant problems whose incidence is estimated to be in the range of 0.5–5% for total joint replacements [10]. Infections of endosseous implants are related to pathogenic bacteria usually localized in the skin, such as *Staphylococcus aureus*, that contaminate the wound during a surgical procedure and are attracted by nutrients concentrated in the foreign biomaterial surface of the implant [11]. Adhesion to surface and proliferation induce the formation of bacterial aggregates, namely biofilms, that are highly resistant to antibiotics and human immune responses [12]. The antimicrobial activity of chitosan relies on positive charges of its protonated amino groups, which interact with molecules at the surfaces of bacteria and fungi, resulting in the leakage of intracellular content and cell death [13]. Therefore, the degree of acetylation, molecular weight, and conjugation with functional groups is critical in maintaining chitosan's antimicrobial properties. Moreover, it has been reported that solid chitosan (i.e., nanoparticles or hydrogels) has a negligible antibacterial effect as compared with soluble chitosan, since nanoparticles expose a reduced number of positive charge available for the binding with the bacterial surface [14].

In this work, the effort to confer new and effective bioactive properties to chitosan through the covalent conjugation with cell adhesive peptides was carried out. The chemistry here proposed for selective covalent conjugation consists of Schiff bases production, successively reduced between chitosan amino groups and aldehyde groups appositely inserted in the peptide backbone during solid phase peptide synthesis. In addition to the RGD peptide, the conjugation with an adhesive peptide named HVP, mapped on the human vitronectin and able to interact with osteoblasts through a proteoglycan-mediated mechanism, was investigated [15,16]. Indeed, previous studies demonstrated that the HVP peptide selects the attachment and supports the differentiation of osteoblasts while promoting cellular migration [17]. The functionalized chitosan matrices (named Chit-RGD and Chit-HVP) have been assessed as scaffolds for osteoblast adhesion, growth, and differentiation. The antibacterial properties have been tested, taking into account the substitutions induced by derivatization. The chemical characterization of Chit-RGD and Chit-HVP was performed through FT-IR, XPS, and NEXAFS analyses.

2. Results and Discussion

2.1. Chemical Composition and Biomolecule Stability: X-Ray Photoelectron Spectroscopy Investigation

For all samples, SR-induced XPS measurements were performed at the C1s, N1s, and O1s core levels. C1s, O1s, and N1s spectra measured for RGD and Chit-RGD are shown in Figure 1. The RGD C1s core level spectrum can be resolved by the curve-fitting analysis into four components, associated with chemically inequivalent carbon and labeled 1−4 in Figure 1. Peak 1 (BE = 285.0 eV) is attributed to aliphatic (C−C) carbons of the side chains and partially to carbon contamination; peak 2 (286.5 eV) is related to O=C−C̲−N carbons of the peptide backbone, with a contribution from C−N carbons of the lysine pending groups; peak 3 (288.0 eV) is due to O=C−N peptide carbons and $O=C-O^-$ carbons of the aspartate pending group; peak 4 (289.3 eV) is due to O=C−OH carbon sin protonated aspartate and $HN=C\text{-}NH_2$ carbons in arginine.

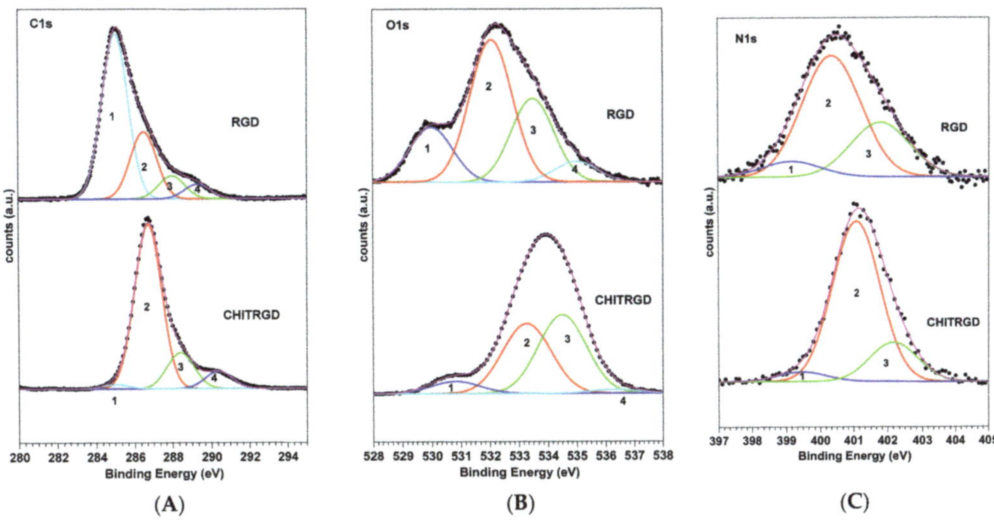

Figure 1. C1s (**A**), O1s (**B**) and N1s (**C**) spectra of RGD peptide and functionalized chitosan matrix Chit-RGD deposited on titania and curve-fitting analysis of the experimental spectra.

The O1s spectra show four-component peaks, labeled 1−4 in Figure 1. Peak 1 (530.0 eV) is assigned to the oxygens of TiO_2 in the titania substrate; peak 2 (532.1 eV) to O=C oxygens of the peptide backbone and of the carboxylate group of aspartates; peak 3 (533.5eV) to C-O oxygens; peak 4 (535.0 eV) is related to physisorbed water and possibly to C−O−H groups in protonated aspartate. The N1s spectrum results from peak 1 (399.17 eV) due to C=N nitrogens of the arginine pending group, a main peak at 400.4 eV related to peptide nitrogens and peak 3 (402.8 eV), due to protonated nitrogens of the lysine and arginine pending groups [17,18].

The Chit-RGD C1s, O1s, and N1s spectra present a similar shape and the same curve-fitting components of the RGD spectra but different peak relative intensities. For example, the intensity of peak 2 in the C1s spectrum of Chit-RGD shows an evident increase in comparison to peak 3 and a slight shift to higher BE (286.7 eV). Both effects are related to the contributions from C−O carbons of chitosan located at 286.6 eV [19].

Some changes are also present in the O1s spectrum. Peak 1, related to the titania substrate, decreases because of the presence of the chitosan film; peak 3 increases in intensity with respect to peak 2 due to contributions from O−C oxygens of chitosan [20].

Similar results were obtained by the analysis and comparisons of the C1s, O1s, and N1s spectra of HVP and Chit-HVP in Figure 2.

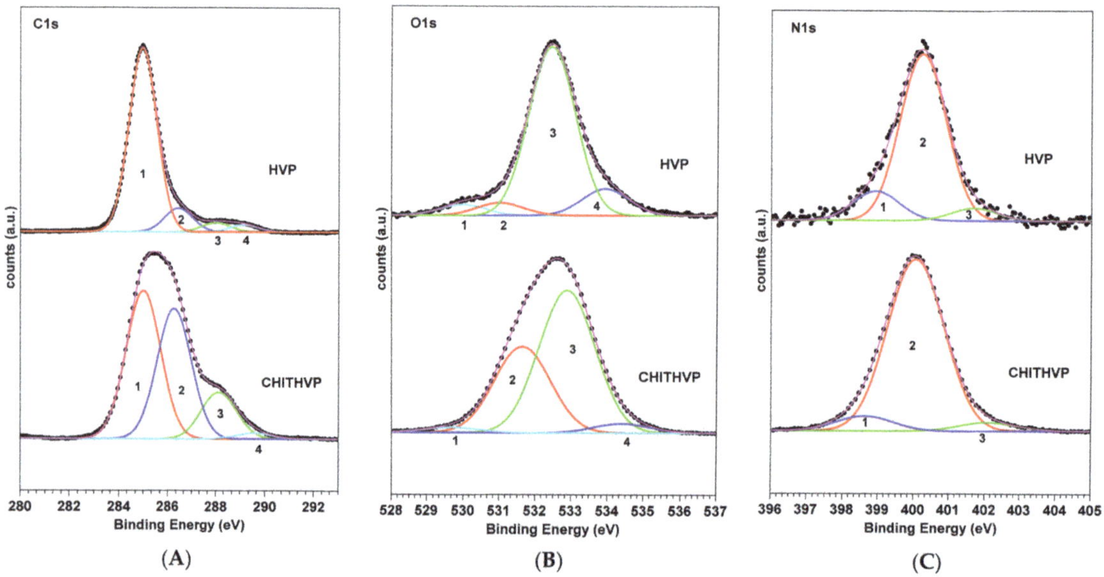

Figure 2. C1s (**A**), O1s (**B**) and N1s (**C**) spectra of HVP peptide and Chit-HVP functionalized chitosan matrix deposited on titania and curve-fitting analysis of the experimental spectra.

Table S1 in the Supplementary Materials lists the Binding Energy (BE), Full Width Half Maxima (FWHM), experimental atomic ratios, and assignments determined from the XPS spectra analysis of all four samples.

The SR-induced XPS data analysis confirms the successful anchoring of both peptides to titania surfaces, as well as their chemical stability. What is more, the substantial modifications arising in the C1s and O1s spectra of the peptides functionalized with chitosan are indicative of a successful peptide-chitosan conjugation, as expected from the synthetic path described in Section 3.2.2.

2.2. Molecular Structure and Organization: Near Edge X-ray Absorption Fine Structure Studies

First, we will analyze NEXAFS C and N K-edge spectra collected at the magic incidence angle (54.7° of incidence of the X-ray radiation on the sample surface, ensuring that no dichroic contributions arise from the sample's spatial orientation) on RGD, the complex Chit-RGD, HVP, and the complex Chit-HVP.

The carbon K-edge spectra of RGD and Chit-RGD are reported in Figure 3A. Those of HVP and Chit-HVP are shown in Figure 3B. The feature at about 288.7 eV is associated with a C 1s→π* transition of a C=O molecular orbital, the right shoulder at about 289.0 eV to a C 1s→π* transition of a C=N molecular orbital in the arginine pending group, and the left shoulder at about 288.0 eV to the histidine pending group just in the case of HVP and Chit-HVP [21]. The feature at 292 eV can be assigned to 1s→σ* transitions by the C−N and C-O molecular groups present in the amino acid pending groups and in chitosan in the case of Chit-RGD and Chit-HVP, so that this feature appears more intense in the Chit-RGD and Chit-HVP spectra. Additional features at 294 and 303 eV can be associated with 1s→σ* transitions by C−C and C=O molecular groups, respectively [22].

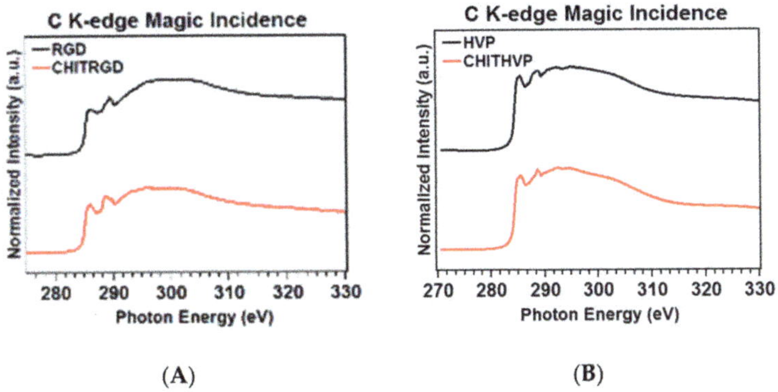

Figure 3. Carbon K-edge spectra of RGD and Chit-RGD (**A**) and of HVP and Chit-HVP (**B**).

As for N K-edge spectra (reported in Figure 4A,B), all samples show a sharp peak at 402 eV assigned to N 1s→π^* transitions of the peptide bonds, one peak at 399.5eV due to N 1s→π^* transitions of C=N arginine [23] and two broadbands at 406 and 413 eV to N 1s→σ^*N−H and N 1s→σ^*N−C resonances, respectively [22]. The spectra of HVP and Chit-HVP also show a peak at 400.4 eV related to C=N histidine [24].

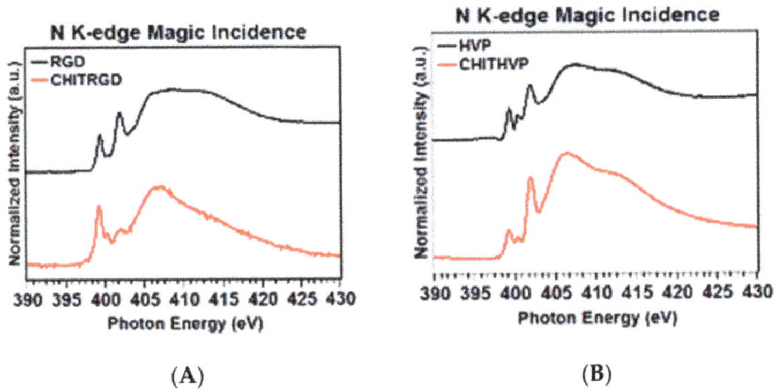

Figure 4. Nitrogen K-edge spectra of RGD and Chit-RGD (**A**) and of HVP and Chit-HVP (**B**).

To evaluate if a molecular preferential orientation occurs, we performed NEXAFS measurements at both C K-edge and N K-edge by varying the photon beam incidence angle on the sample surface from grazing (20°) to normal (90°). To highlight the most indicative dichroic effects, in Figure 5, angular dependent N K-edge spectra collected on RGD (a), HVP (b), Chit-RGD (c), and Chit-HVP (d) anchored onto titania surfaces are reported. As can be seen, HVP and Chit-HVP show a strong dichroic effect in the π^* region; the ratio between the peak intensities at normal (90°) and grazing (20°) incidence, determined for the selected resonance by peak fitting of the experimental data, can be inserted in the equation reported by Stöhr [25] for the 3-fold or higher symmetry substrates, assuming the polarization factor $p = 0.95$, allowing to calculate the average tilt angle between the π^* vector orbital for the C=O bond and the normal to the surface. For HVP (c) and Chit-HVP (d), the tilt angle of the peptide chains is about 65.5° and 67.5°, respectively. Interestingly, no dichroic effects or minimal effects are observed for RGD and Chit-RGD.

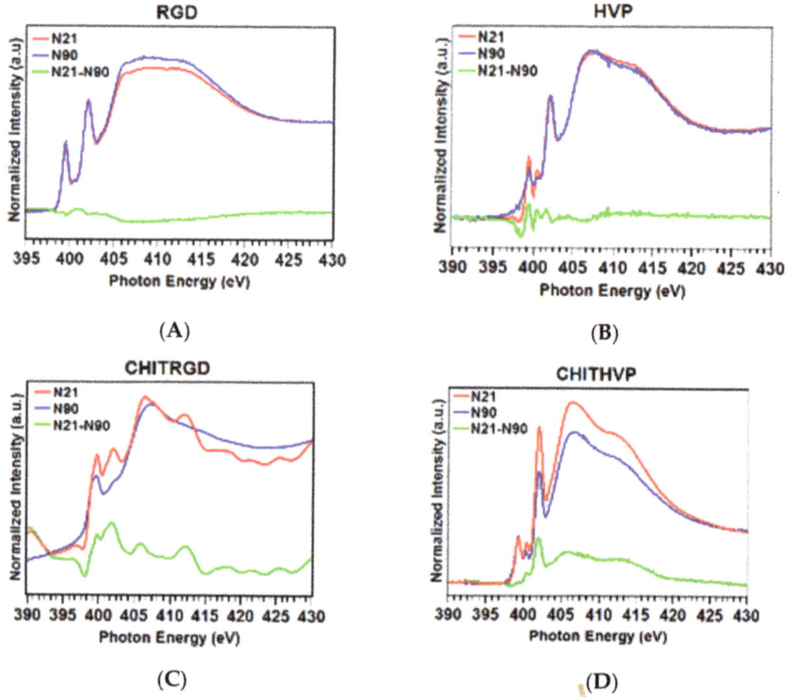

Figure 5. Angular dependent N K-edge NEXAFS spectra collected at normal (90°) and grazing (20°) incidence angles of the impinging X-ray beam on (**A**) RGD; (**B**) HVP; (**C**) Chit-RGD; (**D**) Chit-HVP. The difference spectra (grazing-normal), evidencing dichroic effects, are also shown (green lines).

2.3. FTIR Spectroscopy

FTIR spectra of pristine peptides HVP and RGD are shown in Figure 6. In the spectra of the samples, at high wavenumbers, the N-H stretching vibrations produce the broad peak at about 3300 cm^{-1}, while the two sharp peaks localized at 2930 and 2850 cm^{-1} can be ascribed to C-H stretching vibrations. At lower wavenumbers, the spectra evidence peaks related to the peptide bond, namely the C=O stretching vibration, or amide (I), located at 1660–1630 cm^{-1} and the N-H bending band, or amide (II), located at 1550–1530 cm^{-1} for both peptides. The position of amide (I) can be used to determine the peptide's secondary structure; a wavenumber of about 1660 cm^{-1} can be ascribed to an alpha-helix type conformation of the peptide backbone for most proteins and peptides, according to the literature [26]. The position of amide (II) is also in accordance with this evidence [27]. However, contributions to the IR spectra could also be due to the amino acids pending groups; the peptide sequence is GRGDSPK for RGD and FRHRNRKGY for HVP. In particular, the HVP peptide contains three arginine residues; however, the main contribution to the IR spectrum is the C=N stretching band of the guanidine moiety, located at about 1640 cm^{-1}, and therefore superimposed with the main amide (I) band [28]. The infrared spectra of the Chit-HVP and Chit-RGD assemblies, shown at the bottom of Figure 6, result from contributions of chitosan and of the immobilized peptides. In the chitosan spectrum [29–31], the most intense peaks are the broadband located at 3550 cm^{-1} and the intense peak located at 1080 cm^{-1}, attributed to O-H and C-O stretching vibrations, respectively. These peaks are evident in the Chit-HVP and Chit-RGD spectra of Figure 6. Contributions due to O-H bending vibrations around 1400 cm^{-1} are also evident in both spectra. For sample Chit-HVP, the position of the amide (I) and amide (II) bands reproduces the same features in the spectrum of the pristine HVP peptide. It is worth noticing that, since chitosan is obtained by the partial hydrolysis of chitin, weak residual amide (I) and amide (II) bands can be found in the spectrum of chitosan due to the incomplete hydrolysis of the amide functions. However, the position and intensity of the peaks in the Chit-HVP spectrum are a clear indication of successful peptide grafting to chitosan and of the retention of the peptide's secondary structure after immobilization.

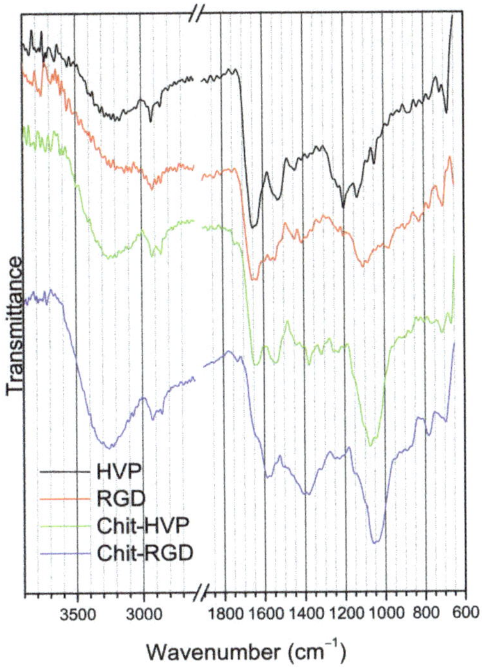

Figure 6. FTIR spectra of the pristine peptides (HVP, RGD) and the corresponding chitosan-peptide assemblies (Chit-HVP, Chit-RGD) in the 3900–2600 and 1900–600 cm^{-1} wavenumber ranges.

For the Chit-RGD assembly, the situation appears more complex. The two bands are less intense and appear shifted with respect to the pristine RGD spectrum. The amide (I) band appears as a shoulder at about 1640 cm^{-1}, on the high wavenumber side of the amide (I) band, which seems to be shifted to about 1580 cm^{-1}. These data could indicate that a lower peptide amount has been grafted to chitosan, in agreement with the XPS results.

2.4. *Functionalization of Chitosan with Biomimetic Peptides Affects Human Osteoblast Survival and Adhesion*

The biomimetic properties of chitosan matrices functionalized with HVP and RGD peptides were assessed in primary human osteoblast cells by the MTT test. As reported in Figure 7A, following 2 h in culture, chitosan matrices functionalized with only HVP (Chit-HVP 100% in concentration) reported an increased number of attached osteoblast cells compared with non-functionalized chitosan (Chit). Chitosan functionalized with RGD 100% (Chit-RGD 100%) did not report a significant increase in cell adhesion compared with Chit-HVP 100% or Chit.

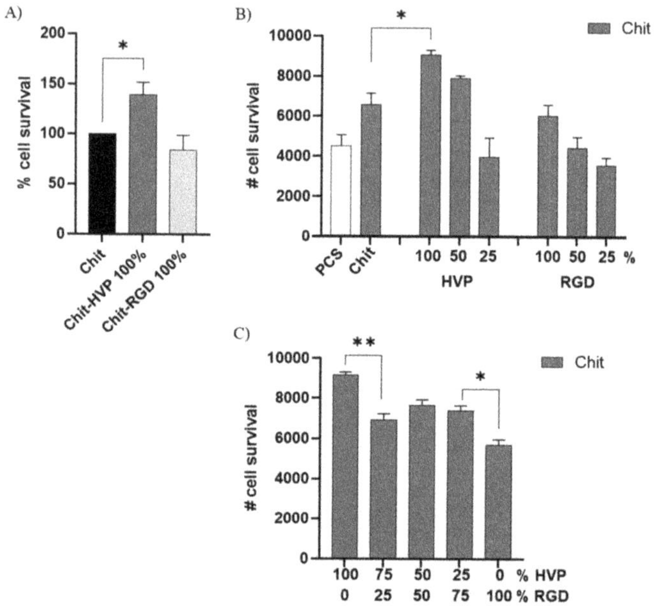

Figure 7. MTT assay performed in primary human osteoblast cells seeded for 2 h on different functionalized chitosan matrices. The cells were quantified by setting a standard curve for each experiment obtained by planting a known number of osteoblast cells. In (**A**), the percentage of cell survival was calculated on the number of cells retrieved in chitosan. (**B**,**C**) number of cells adhering to different functionalized chitosan. Data are reported as mean ± st err of three independent experiments, each performed in duplicate. * indicates $p < 0.005$. ** indicates $p < 0.001$. PCS: plastic culture surface; Chit: chitosan. % indicates the concentration of HVP or RGD.

Different combinations of HVP and RGD with Chit or the two peptide concentrations were also tested to identify the best-performing blend. After 2 h in culture, we observed that the number of viable cells was reduced with the decreasing concentrations of the biomimetic peptides used for functionalized chitosan (Figure 7B). Human osteoblast cells preferentially attach to, and survive in, Chit-HVP 100%. Considering that HVP and RGD interact with the same integrin receptor family on cells [5,15,16], we combined HVP and RGD at different concentrations to functionalized chitosan matrices. As reported in Figure 7C, the addition of 25% HVP improved the biomimetic properties of Chit-RGD as reported by the increased number of viable cells in Chit-HVP25%RGD75% ($p < 0.05$ vs. Chit-RGD 100%).

On the other hand, Chit-HVP 100% is the best performing functionalization, since the replacement of 25% of HVP with RGD significantly worsens the ability to attract osteoblast cells (Figure 7C). Overall, by testing different combinations of peptide concentrations, we did not observe an increase in the number of viable cells compared with Chit-HVP 100%, ruling out a possible synergistic effect of the two adhesive sequences. Since it has been reported that a positive surface charge, the pH, and the molecular weight of chitosan impact its stability and biological activity [5], our data reveal that the molecular structure of Chit-HVP (Figures 4 and 5) supports the biomimetic potential of chitosan better than RGD functionalization.

The adhesion of osteoblast cells on differentially functionalized chitosan matrices was evaluated by immunocytochemistry, analyzing the expression and distribution of the smooth muscle actin (α-SMA), a marker of cytoplasmic fibers involved in stress tensor and mechanotransduction during cell adhesion [32,33]. As reported in Figure 8, after 24 h in culture, osteoblast cells adhered to the chitosan matrix and reported a blunt expression of α-SMA compared with cells cultured on unfunctionalized glass coverslips. Osteoblast cells cultured on Chit-HVP 100% or Chit-HVP 50% are elongated and pointed on both ends, exhibiting a spindle shape. Moreover, in cells cultured on Chit-HVP 100%, actin fibers are structurally organized in a filament network, and the expression of α-SMA increased. Cells cultured on Chit- RGD 100% or Chit-RGD 50% reported an elongated shape, but actin fibers accumulate in the cytoplasm, resulting in a less organized distribution of α-SMA.

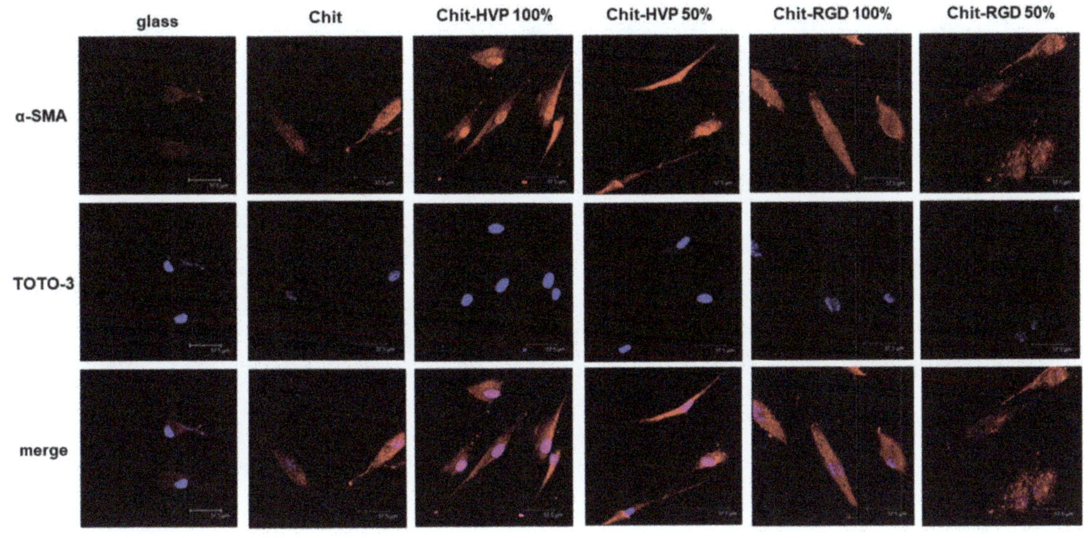

Figure 8. Immunofluorescence on human primary osteoblast cells cultured for 24 h on differently functionalized chitosan matrices. Cells were stained with anti-αSMA antibody (red). Nuclei were stained with TOTO-3 (blue). The images were acquired using a Leica TCSNT/SP2 confocal microscope (Leica Microsystems). Scale bar: 37.5 μm. The images are representative of three independent experiments.

2.5. HVP Functionalization of Chitosan Increases Proliferation of Human Osteoblast Cells

Cell proliferation was evaluated using the intracellular dye CFSE and a cytofluorimetric analysis on human osteoblast cells cultured 72 h on different functionalized chitosan matrices. As reported in Figure 9, in cells cultured on a plastic culture surface or non-functionalized chitosan matrices, the fluorescent intensity relative to cell proliferation was recorded in 3.17 ± 1% and 4.3 ± 1.1% of cells, respectively. Cells cultured on Chit-HVP 100% reported a significant increase in cell proliferation ($p < 0.05$) compared with Chit. However, cell proliferation decreased in Chit-HVP 50% to levels comparable to those of chitosan matrices functionalized with RGD.

Several studies reported chitosan ability in sustaining osteoblast proliferation when chitosan is used as a coating in titanium implants or alone in the form of devices for bone regeneration. Indeed, chitosan nanofiber scaffolds increased DNA replication and cell proliferation in a time-dependent manner, up to 5 days [34]. In our study, cell proliferation was evaluated following 72 h of culture, the time at which we observed the maximal proliferative effect in a preliminary set of experiments. Even if our study did not consider chitosan organized in nanofibers, unlike chitosan or RGD, HVP functionalization was the only one able to augment osteoblast proliferation, a critical event in bone tissue regeneration.

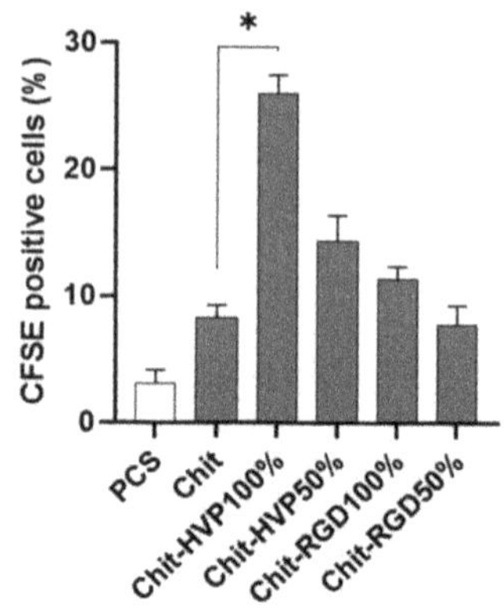

Figure 9. Cell proliferation evaluated in primary human osteoblast cells loaded with CFSE intracellular fluorescent dye. The cells were cultured for 72 h on chitosan matrices and then analyzed by cytofluorimetric analysis. Data are reported as mean ± st err of three independent experiments, each performed in duplicate. * indicates $p < 0.005$. PCS: plastic culture surface; Chit: chitosan. % indicates the concentration of HVP or RGD.

2.6. Chitosan Functionalization with HVP Supports Human Osteoblast Differentiation

To further evaluate the role of functionalization of chitosan matrices in sustaining the differentiation of primary human osteoblast cells, the expression of genes involved in bone formation was assessed by a quantitative RT-PCR. As reported in Figure 10, cells cultured on Chit-HVP 100% and Chit-HVP 50% increased the mRNA specific transcript levels of *Spp1* gene-coding osteopontin involved in bone homeostasis [35]. At the same time, Chit-HVP increased the expression of *Runx2* mRNA (Figure 10), a key transcription factor associated with increased levels of osteocalcin and sialoprotein involved in osteoblast differentiation [35].

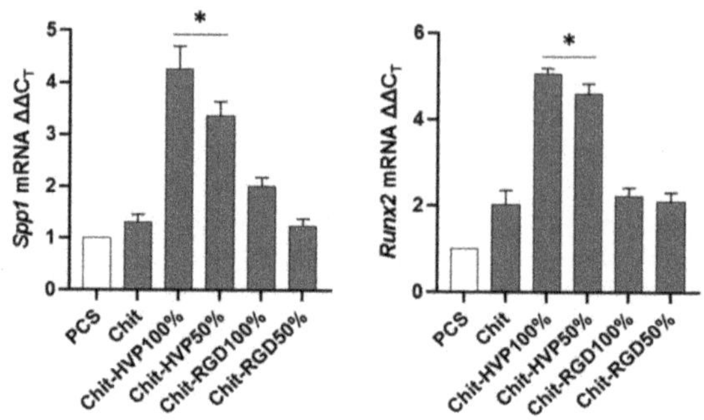

Figure 10. Quantitative RT-PCR performed on human osteoblast cells cultured for 24 h on functionalized chitosan matrices. Data were quantified by the $\Delta\Delta C_T$ method using hGAPDH as a reference gene. Data are reported as mean ± st err of three independent experiments, each performed in triplicate. * indicates $p < 0.005$ vs. Chit. PCS: plastic culture surface; Chit: chitosan. % indicates the concentration of HVP or RGD.

To confirm osteoblast differentiation, we cultured cells on functionalized chitosan matrices for seven days. The cells were then stained with Alizarin red. As reported in Figure 11, osteoblast cells cultured for 7 days on Chit-HVP 100% and Chit-HVP 50% reported strong signals relative to Alizarin staining, demonstrating increased calcium deposition. Calcium deposition was evident to a less extent in cells cultured on matrices functionalized with RGD. In a parallel set of experiments, the dye was extracted from cells, and optical density (O.D.) was determined to quantify calcium deposition. As reported in Figure 11B), we observed an increase in O.D. values in cells cultured in Chit-HVP, confirming the increased formation of calcium deposition.

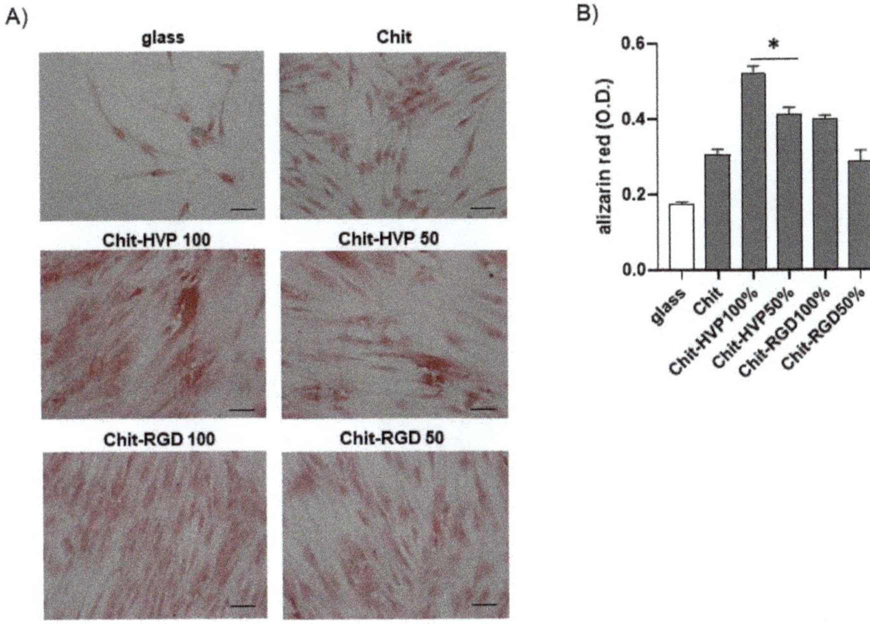

Figure 11. Osteoblast cells were cultured for 7 days on functionalized chitosan matrices and then fixed and stained with Alizarin red to visualize calcium deposition. (**A**) The images were acquired using a Leica microscope equipped with a digital camera. (**B**) The cells were lysed in acetic acid, and the optical density (O.D.) was recorded at 405 nm. Data are reported as the mean ± st err of three independent experiments, each performed in duplicate. * indicates $p < 0.005$ vs. Scale bar: 75 μm. Chit. PCS: plastic culture surface; Chit: chitosan. % indicates the concentration of HVP or RGD.

2.7. *Chitosan Functionalization with Biomimetic Peptides Reduces the Antibacterial Activity*

As reported, the antimicrobial activity of chitosan is linked with the conjugation to functional groups and the degree of acetylation [13]. In the context of endosseous implants, devices naturally endowed with antimicrobial effects confer important advantages.

Therefore, we evaluated whether the functionalization of chitosan with HVP or RGD alters the antimicrobial activity of chitosan matrices against the biofilm of *S. aureus*. To this aim, cultures of *S. aureus* were grown for 48 h on different functionalized chitosan matrices and then stained using a LIVE/DEAD BacLight Bacterial Viability Kit. As reported in Figure 12A, the antibacterial effect of chitosan was confirmed against *S. aureus*, as very few bacterial cells were retrieved compared to the glass surface. Functionalization with HVP or RGD reduced chitosan's antibacterial effect, as more bacterial cells were visualized in Chit-RGD 100 and Chit-RGD 50 or Chit-HVP 100 and Chit-HVP 50 as compared to the unfunctionalized chitosan. Live and dead bacteria were enumerated and, considering the different number of bacterial cells in functionalized surfaces, viable *S. aureus* increased as the concentrations of the peptides increased (Figure 12B), suggesting that structural characteristics of functionalized chitosan guide the antibacterial effect. Generally, chitosan reported a stronger bactericidal effect against Gram-positive bacteria than against Gram-negative bacteria [36]. In chitosan functionalized with HVP, we should expect a similar trend. However, by tuning the concentration of HVP, and thus changing the chemical charges at the surface of Chit-HVP, we envisage the possibility to improve the antimicrobial effects of chitosan against Gram-negative bacteria.

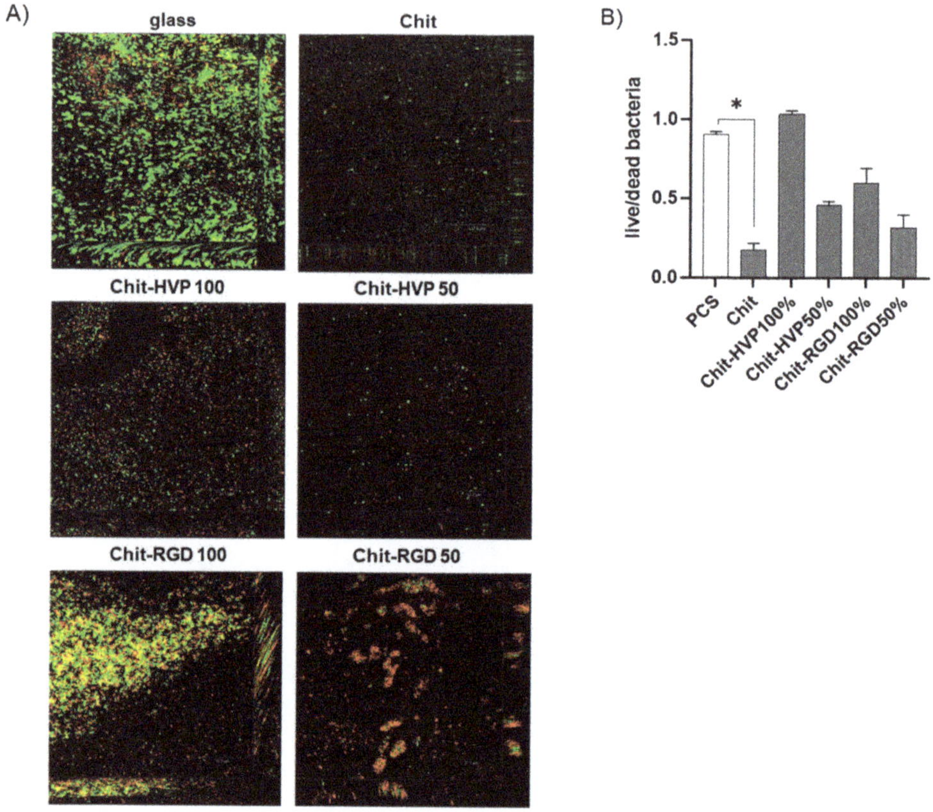

Figure 12. Cultures of *S. aureus* were grown on different functionalized chitosan matrices for 48 h, resulting in a mature biofilm in glass surfaces. Bacterial cultures were stained with a LIVE/DEAD BacLight Bacterial Viability Kit and (**A**) images were acquired using a Leica TCSNT/SP2 confocal microscope. Scale bar: 37.5 μm. The images are representative of three independent experiments, each performed in duplicate. (**B**) Data were analyzed using ImageJ and the ratio of live (bacteria with SYTO9 green fluorescence)/dead (bacteria with a high propidium iodide fluorescence and a low SYTO9 green fluorescence) bacteria was recorded. Data are reported as mean ± st err. * indicates $p < 0.005$ vs. glass surfaces. Chit: chitosan. % indicates the concentration of HVP or RGD.

Overall, our data indicate that chitosan bio-functionalization with HVP confers beneficial features on endosseous devices but also lead to the loss of the natural antibacterial effect of chitosan. The reported results showed the different biomimetic properties of the two adhesive sequences resulting in different yields of functionalization, as demonstrated by XPS and FT-IR data, and correlated to different bioactivity of synthetic peptides. On the other hand, it is well known that the cellular response depends on a peculiar pattern and concentration of peptides that need to be optimized. Actually, several circulating proteases are responsible for biomimetic peptides' degradation in vivo, thus complicating the performance of functionalized biomaterials. Therefore, as in vitro experiments do not take into account the in vivo variables, more tunable systems should be set to calibrate the concentration of biomimetic peptides on hybrid biomaterials. Indeed, given the results reported in this study, the best performing matrix for bone tissue application results in Chit-HVP 100%. However, the analysis of cell and bacterial adhesion reveals that Chit-HVP

50% should preserve both the antibacterial feature of the biomaterial and the beneficial cellular features.

3. Materials and Methods

3.1. Materials

Acetic acid (AcOH), NaOH, Methanol (MeOH), trifluoroacetic acid (TFA), Triethoxysilane (TES), *N,N*-dimethylformamide (DMF), and Pyridoxal 5-phosphate (PLP) were purchased from Sigma-Aldrich. 2-(1H-benzotriazol-1-il)-1,1,3,3-tetramethyluronium exafluorophosphate (HBTU), 1-hydroxybenzotrazole (HOBt), H-Phe Nova Syn Tg resin, and all Fmoc-protected amino acids were purchased from Merck (Darmstadt, Germany). *N*-methyl-2-pyrrolidone (NMP), dichloromethane (DCM), *N,N*-Diisopropylethylamine (DIEA), and piperidine were from Biosolve (Leenderweg, Valkenswaard, The Netherlands). Ethanol (EtOH) was provided by VWR Chemicals (Milan, Italy). Chitosan 70/1000 was purchased from Heppe Medical Chitosan GmbH (Halle, Germany).

3.2. Methods

3.2.1. Peptide Synthesis

(1) GRGDSPK-Aldehyde

The GRGDSPK peptide, named RGD for simplicity (sequence H-Gly-Arg-Gly-Asp-Ser-Pro-Lys-NH_2) was synthesized with Fmoc Chemistry and a solid phase strategy using a Syro I synthetizer (MultiSynTech, Witten, Germany). Double coupling was carried out for loading. The peptide was cleaved from the resin with TFA:H_2O:TES (95:2.5:2.5) for 1.5 h under stirring. The peptide was purified by RP-HPLC. The conversion of the N terminal group NH_2-CH_2- into an aldehyde group was achieved by treating the peptide with 20 molar equivalents of PLP in a 25 mM phosphate buffer at pH = 6.5 for 18h at 37 °C. Finally, the GRGDSPK-aldehyde was purified by RP-HPLC up to 98% homogeneity (Figure S1 in the Supplementary Materials). The identity of the target product was determined through mass spectrometry (Figure S2 in the Supplementary Materials).

(2) HVP-Aldehyde

The HVP-aldehyde peptide (sequence H-Phe-Arg-His-Arg-Asn-Arg-Lys-Gly-Tyr-7-aminoheptanoic acid-Phe-H) was synthesized with Fmoc Chemistry and a solid phase strategy using a Syro I synthesizer (MultiSynTech, Witten, Germany). H-Phe Nova Syn Tg resin (0.2 mmoles/g) was used as solid support. The 7-aminoheptanoic acid, introduced as a spacer, was conjugated using a single coupling, whereas the following residues were introduced with double couplings. The side-chain deprotection was carried out by treating the peptide-resin with 4 mL TFA for 30 min. The resin was washed with DCM and dried under vacuum for 30 min, and then the cleavage of the peptide from the solid support was obtained with 9.9 mL of an AcOH:H_2O:DCM:MeOH (10:5:63:21) solution treatment for 1h under magnetic stirring. The crude peptide was purified by RP-HPLC and characterized by analytical RP-HPLC and mass spectrometry as reported in the Supplementary Materials (Figures S3 and S4).

3.2.2. Functionalization of Chitosan

Chitosan (100 mg) dissolved in 5.5 mL AcOH 0.2M. EtOH (3.75 mL) was added and the pH was adjusted to 5.1 with NaOH 1N. GRGDSPK-aldehyde (11.54 mg) was summed. Finally, 100 mg of $NaCNBH_3$ were added. The solution was stirred for 24 h at room temperature. After adjusting the pH to 7 with NaOH 1N, the addition of EtOH caused the product precipitation. Chit-RGD was filtered with gooch G3 and dried under vacuum for 1 h. The same protocol was used for the preparation of Chit-HVP, starting from 14 mg of chitosan and treating with 12.87 mg of HVP-aldehyde.

3.2.3. Chemical Characterization

Samples for chemical analysis were prepared by incubation of TiO_2/Si(111) surfaces for about 16 h with a mother solution of the selected biomolecules (0.4 mg of Chit-RGD or, respectively, Chit-HVP dissolved in 120 μL of an AcOH 0.2M water solution).

(1) XPS

XPS experiments have been performed at MSB, Materials Science Beamline, at ELETTRA-Sincrotrone Trieste, using an ultra-high vacuum experimental system. MSB is placed at a bending magnet exit and, using a plane grating monochromator, can provide photon energy ranging from ultraviolet to soft X-rays (20 to 1000 eV). The X-ray spectrometer operates at the background pressure of low 10^{-10} mbar, and the emitted photoelectrons are analyzed by a SPECS Phoibos 150 electron energy analyzer.

For SR XPS measurements, photoelectrons have been collected normally to the sample surface, while the radiation impinges with 60° with respect to the sample surface.

All spectra were energy-referenced to the C1s signal of aliphatic carbon located at 285.0 eV. A curve-fitting analysis of the experimental spectra was performed using Gaussian curves as fitting functions. Atomic ratios were calculated from peak intensities using Scofield's cross-section values.

(2) Near-Edge X-ray Absorption Fine Structure (NEXAFS)

NEXAFS spectra were acquired at the BEAR beamline (bending magnet for emission absorption and reflectivity), at the ELETTRA storage ring. BEAR is installed at the left exit of the 8.1 bending magnet exit. The apparatus is based on a bending magnet as a source and beamline optics delivering photons from 5 eV up to about 1600 eV with a selectable degree of ellipticity. The UHV end station is equipped with a movable hemispherical electron analyzer and a set of photodiodes to collect angle-resolved photoemission spectra, optical reflectivity, and fluorescence yield. In these experiments, we used ammeters to measure the drain current from the sample. C and N K-edge spectra were collected at normal (90°), grazing (20°), and magic (54.7°) incidence angles of the linearly polarized photon beam with respect to the sample surface. In addition, our carbon and nitrogen K-edge spectra have been further calibrated using the resonance at 288.70 eV, assigned to the C=O $1s \rightarrow \pi^*$ transition, and the resonance at 402.00 eV, assigned to the $1s \rightarrow \pi^*$ transition of the peptide bonds, respectively. The raw C and N K-edge NEXAFS spectra were normalized to the incident photon flux by dividing the sample spectrum by the spectrum collected on a freshly sputtered gold surface. The spectra were then normalized by subtracting a straight line that fits the part of the spectrum below the edge and setting to 1 the value at 330.00 and 430.00 eV for carbon and nitrogen, respectively.

(3) FT-IR

A Continuum FT-IR microscope mounted on a Nicolet iS50 FT–IR spectrometer (ThermoFisher®, Waltham, MA, USA) working with a liquid-nitrogen-cooled MCT (mercury-cadmium telluride) detector was employed for acquiring FT-IR spectra of the investigated samples. The spectra were collected in μ–ATR mode employing a Ge tip, in the region of 400–4000 cm^{-1}, with a resolution 8 cm^{-1} and 256 scans per spectrum. The optical bench apertures were set to 100 and the gain to 8.

3.2.4. In Vitro ASSAYS

(1) Human Osteoblast Cells' Isolation and Culture

Human (h) osteoblast cells were obtained from explants of cortical mandible bone collected during a surgical procedure. After collection, the bone fragments were cultured at 37 °C in 5% CO_2 and 95% humidity in a complete medium (DMEM supplemented with 10% *v*/*v* heat-inactivated fetal bovine serum, and penicillin-streptomycin solution, all provided by ThermoFischer) and incubated until the cells migrated from the bone fragments. At the cell confluence, the bone fragments were removed and the cells were detached using trypsin EDTA (Gibco) and cultured in a complete medium supplemented with 50 mg/mL ascorbic acid, 10 nM dexamethasone, and 10 mM β-glycerophosphate (all purchased from Merck). The osteoblast phenotype was confirmed by the von Kossa staining. The study was approved by the Ethical Committee of the University Hospital of Padova. The patients were informed of the study's aims and protocol and gave their written informed consent.

(2) Cell Adhesion

Osteoblast cells were seeded and cultured for 2 h on different functionalized chitosan matrices. Cell viability was assessed using the MTT (3-(4,5-dimethylthia-zole-2-yl)-2,5-diphenyl tetrazoliumbromide; Merck) assay. Briefly, at the end of incubation, the cells were rinsed three times with PBS to remove non-adherent cells and then incubated with MTT (5 mg/mL in 100 μL final volume) at 37 °C for 4 h. The reaction was stopped by adding 0.01 N HCl in 10% *v*/*v* SDS. The cells were quantified by setting a standard curve for each experiment, obtained by seeding a known number of h-osteoblast cells. The absorbance of cell lysates was recorded at 620 nm.

(3) Immunofluorescence

Human osteoblast cells were cultured for 24 h on chitosan matrices layered on glass coverslips. The cells were fixed in 4% *w*/*v* paraformaldehyde (PFA) for 10 min and then washed three times (5 min each) in PBS. The cells were permeabilized with 0.1% Triton X-100, and nonspecific binding sites were then blocked by incubation with 2% bovine serum albumin (BSA) in PBS for 30 min. The cells were incubated with an anti-αSmooth Muscle Actin (α-SMA) antibody conjugated with tetramethylrhodamine B isothiocyanate (TRITC; 50 mg/mL, Merck). For nuclear counterstaining, the samples were incubated with TOTO-3 conjugated with Alexa Fluor 647. After extensive washing, the samples were mounted and analyzed using a Leica TCSNT/SP2 confocal microscope. The images were digitally stored using the Leica software.

(4) Proliferation Assay

The cells were incubated at 37 °C for 10 min in a prewarmed PBS containing 0.1% *v*/*v* BSA and a 25 mM carboxyfluorescein diacetate succinimidyl ester (CFSE, Molecular Probe, Invitrogen). Staining was quenched by adding five volumes of ice-cold culture media. The cells were then washed, counted using Trypan blue, seeded at 8×10^4 cells/mL, and incubated in a fresh culture medium at 37 °C for 72 h. Cell proliferation was evaluated by the partitioning of a fluorescent dye between daughter cells using a BD FACS-Calibur flow cytometer.

(5) Quantitative Real-Time Polymerase Chain Reaction

Specific mRNA transcript levels coding human Secreted Phosphoprotein 1 (*Spp1*) and Runt-related transcription factor 2 (*Runx2*) were quantified in osteoblast cells cultured 24 h on differently functionalized chitosan matrices. At the end of incubation, the total RNA was extracted using the EZN.A lysis buffer (Total R.N.A. Kit I (Omega Bio-Tek, Milan, Italy). Contaminating DNA was removed by DNase I treatment (Omega Bio-Tek). cDNA synthesis and subsequent polymerization were performed in a one-step using the iTaq Universal SYBR Green One-Step Kit (Bio-Rad). The reaction mixture contained a 200 nM forward primer, 200 nM reverse primer, iTaq universal SyBR Green reaction mix, and iScript reverse transcriptase, and a 200 ng total RNA Real-time PCR was performed using the ABI PRISM 7700 Sequence Detection System (Applied Biosystems, Milan, Italy). Data were quantified by the $\Delta\Delta C_T$ method using hGAPDH as the reference gene. Target and reference genes were amplified with efficiencies near 100%. The oligonucleotides used for PCR are listed in Table 1.

Table 1. Oligonucleotide sequences.

Gene (Accession #)	Sequence	
Gapdh (NM_001289726)	Fw: 5′-agtgccagcctcgtcccgta-3′	Rv: 5′-caggcgcccaatacggccaa-3′
Spp1 (NM_000582)	Fw: 5′-aagtttcgcagacctgacatc-3′	Rv: 5′-ggctgtcccaatcagaagg-3′
Runx2 (NM_001024630)	Fw: 5′-cagtgacaccatgtcagcaa-3′	Rv: 5′-gctcacgtcgctcattttg-3′

(6) Alizarin Staining

Osteoblast cells cultured on differently functionalized chitosan matrices were washed with PBS and fixed using 10% PFA for 30 min. The cells were then stained with 40 mM Alizarin red (pH 4.2) for 40 min in the dark at room temperature. The cells were washed in distilled water and mounted. The images were obtained using a Leica microscope equipped with a digital camera. In a parallel set of experiments, cells fixed in PFA were incubated at −20 °C for 30 min and then lysed in acetic acid 10% v/v. The samples were incubated at 85 °C for 10 min and centrifuged, and the supernatants' pH was neutralized before reading the absorbance of Alizarin red at 405 nm using a microplate reading (Tecan).

(7) Bacterial Cultures and Biofilm Staining

Staphylococcus aureus was purchased from CCUG (Culture Collection University of Gothenburg, Gothenburg, Sweden) and maintained in a Trypticase Soy (TS) agar or broth. Bacterial cells were harvested from overnight cultures, and 10^2 colony forming units (CFU)/mL were cultured on differently functionalized chitosan matrices layered on glass coverslips. The bacteria were cultured in a total volume of 2 mL of TS broth for 48 h at 37 °C without agitation. At the end of incubation, the glasses were washed in 0.85% w/v NaCl to remove loosely bound bacterial cells. The biofilms were stained with the LIVE/DEAD BacLightTM Bacterial Viability Kit (Thermo Fisher, Milan, Italy) for 15 min at room temperature in the dark. Briefly, the samples were incubated with 5 µM SYTO9 (a green fluorescent nucleic acid dye labeling both live and dead bacteria) and 30 µM propidium iodide, penetrating only bacteria with damaged membranes. Unlabeled dyes were removed by washing, and the samples were visualized using confocal microscopy. Data were analyzed using ImageJ, and the ratio of live/dead bacteria (i.e., bacteria with a high propidium iodide fluorescence and a low SYTO9 green fluorescence) was recorded.

(8) Statistical Analysis

Biological data are reported as mean ± standard error. The statistical analysis was performed using the One-way ANOVA test followed by Bonferroni's multicomparison test, using Graph-Pad Prism 3.03. p-values < 0.05 were considered statistically significant.

4. Conclusions

Our results reported that the molecular structure of Chit-HVP supported the biomimetic potential of chitosan better than RGD functionalization. In fact, after 2 h of culture, Chit-HVP resulted in the best performing functionalization in osteoblast cell adhesion, whereas even a 25% replacement of Chit-HVP content with Chit-RGD significantly decreased cell adhesion. Furthermore, at 72 h of incubation, osteoblasts cultured on Chit-HVP reported a significant increase in cell proliferation compared with Chit, and cell proliferation decreased with the decreasing percentage of HVP content. Chit-HVP 100% and Chit-HVP 50% supported the adhesion of osteoblasts, the formation of elongated cell shapes, and increased osteoblast differentiation. Even if chitosan bio-functionalization with HVP conferred beneficial features on endosseous devices, it reduced the antibacterial effect of chitosan against biofilms of *S. aureus*. While Chit-HVP 100% showed the best results in osteoblast adhesion, proliferation, and differentiation, only Chit-HVP 50% preserved the antibacterial effect and supported cell adhesion.

Supplementary Materials: The following are available online at https://www.mdpi.com/article/10.3390/ijms22115916/s1.

Author Contributions: Conceptualization, P.B. and M.D.; methodology, P.B., M.D. and G.I.; formal analysis, P.B.; Peptides preparation and chitosan functionalization, A.Z., L.C. and M.T.; chemical characterization: C.B., V.G., M.M., L.T. and V.S., biological investigation, P.B.; resources, M.D. and G.I.; data curation, P.B. and G.I.; writing—Original draft preparation, P.B., A.Z., G.I. and M.D.; supervision, M.D.; project administration, M.D.; funding acquisition, M.D. All authors have read and agreed to the published version of the manuscript.

Funding: This work was supported by the University of Padova (www.unipd.it, accessed on 29 May 2021) SID 2017 (DETT_SID17_01).

Institutional Review Board Statement: Not applicable.

Informed Consent Statement: Not applicable.

Data Availability Statement: Not applicable.

Conflicts of Interest: The authors declare no conflict of interest.

References

1. Bagheri-Khoulenjani, S.; Taghizadeh, S.; Mirzadeh, H. An Investigation on the Short-Term Biodegradability of Chitosan with Various Molecular Weights and Degrees of Deacetylation. *Carbohydr. Polym.* **2009**, *78*, 773–778. [CrossRef]
2. Zhang, Y.; Sun, T.; Jiang, C. Biomacromolecules As Carriers in Drug Delivery and Tissue Engineering. *Acta Pharm. Sin. B* **2018**, *8*, 34–50. [CrossRef] [PubMed]
3. Palma, P.J.; Ramos, J.; Martins, J.B.; Diogenes, A.; Figueiredo, M.H.; Ferreira, P.; Viegas, C.; Santos, J.M. Histologic Evaluation of Regenerative Endodontic Procedures with the Use of Chitosan Scaffolds in Immature Dog Teeth with Apical Periodontitis. *J. Endod.* **2017**, *43*, 1279–1287. [CrossRef] [PubMed]
4. Huang, Y.-M.; Lin, Y.-C.; Chen, C.-Y.; Hsieh, Y.-Y.; Liaw, C.-K.; Huang, S.-W.; Tsuang, Y.-H.; Chen, C.-H.; Lin, F.-H. Thermosensitive Chitosan–Gelatin–Glycerol Phosphate Hydrogels as Collagenase Carrier for Tendon–Bone Healing in a Rabbit Model. *Polymers* **2020**, *12*, 436. [CrossRef] [PubMed]
5. Hansson, A.; Hashom, N.; Falson, F.; Rousselle, P.; Olivier, J.; Borchard, E.G. In Vitro Evaluation of an RGD-Functionalized Chi-Tosan Derivative for Enhanced Cell adhesion. *Carbohydr. Polym.* **2012**, *90*, 1494–1500. [CrossRef]
6. Wang, C.; Chen, B.; Zou, M.; Cheng, G. Cyclic RGD-Modified chitosan/Graphene Oxide Polymers for Drug Delivery and Cellular Imaging. *Colloids Surf. B* **2014**, *122*, 332–340. [CrossRef]
7. Leedy, M.R.; Martin, H.J.; Andrew Norowski, P.; Jennings, J.A.; Haggard, W.O.; Bumgardner, J.D. Use of Chitosan as a Bioactive Implant Coating for Bone-Implant applications. *Chitosan Biomater.* **2011**, *244*, 129–165.
8. Kjalarsdóttir, L.; Dýrfjörd, A.; Dagbjartsson, A.; Laxdal, E.H.; Örlygsson, G.; Gíslason, J.; Einarsson, J.M.; Ng, C.-H.; Jónsson, H. Bone Remodeling Effect of a Chitosan and Calcium Phosphate-Based Composite. *Regen. Biomater.* **2019**, *6*, 241–247. [CrossRef]
9. Vaz, J.M.; Pezzoli, D.; Chevallier, P.; Campelo, C.S.; Candiani, G.; Mantovani, D. Antibacterial Coatings Based on Chitosan for Pharmaceutical and Biomedical Applications. *Curr. Pharm. Des.* **2018**, *24*, 866–885. [CrossRef]
10. Morgenstern, M.; Kuehl, R.; Zalavras, C.G.; McNally, M.; Zimmerli, W.; Burch, M.A.; Vandendriessche, T.; Obremskey, W.T.; Verhofstad, M.H.J.; Metsemakers, W.J. The Influence of Duration of Infection on Outcome of Debridement and Implant Retention in Fracture-Related Infection. *Bone Jt. J.* **2021**, *103-B*, 213–221. [CrossRef]
11. Schierholz, J.; Beuth, J. Implant Infections: A Haven for Opportunistic Bacteria. *J. Hosp. Infect.* **2001**, *49*, 87–93. [CrossRef]
12. Verlee, S.; Mincke, E.C.; Stevens, V. Recent Developments in Antibacterial and Antifungal Chitosan and Its derivatives. *Carbohydr. Polym.* **2017**, *164*, 268–283. [CrossRef]
13. Sadeghi, A.; Dorkoosh, F.; Avadi, M.; Saadat, P.; Rafiee-Tehrani, M.; Junginger, H. Preparation, Characterization and Antibacterial Activities of Chitosan, N-Trimethyl Chitosan (TMC) and N-Diethylmethyl Chitosan (DEMC) Nanoparticles Loaded with Insulin Using Both the Ionotropic Gelation and Polyelectrolyte Complexation Methods. *Int. J. Pharm.* **2008**, *355*, 299–306. [CrossRef] [PubMed]
14. Brun, P.; Zamuner, A.; Peretti, A.; Conti, J.; Messina, G.M.L.; Marletta, G.; Dettin, M. 3D Synthetic Peptide-Based Architectures for the Engineering of the Enteric Nervous System. *Sci. Rep.* **2019**, *9*, 5583. [CrossRef]
15. Zamuner, A.; Brun, P.; Scorzeto, M.; Sica, G.; Castagliuolo, I.; Dettin, M. Smart Biomaterials: Surfaces Functionalized with Proteo-Lytically Stable Osteoblast-Adhesive peptides. *Bioact. Mater.* **2017**, *2*, 121–130. [CrossRef] [PubMed]
16. Brun, P.; Scorzeto, M.; Vassanelli, S.; Castagliuolo, I.; Palù, G.; Ghezzo, F.; Messina, G.M.; Iucci, G.; Battaglia, V.; Sivolella, S.; et al. Mechanisms Underlying the Attachment and Spreading of Human Osteoblasts: From Transient Interactions to Focal Adhesions on Vitronectin-Grafted Bioactive Surfaces. *Acta Biomater.* **2013**, *9*, 6105–6115. [CrossRef]

17. Secchi, V.; Franchi, S.; Santi, M.; Dettin, M.; Zamuner, A.; Iucci, G.; Battocchio, C. Self-Assembling Behavior of Cysteine-Modified Oligopeptides: An XPS and NEXAFS Study. *J. Phys. Chem. C* **2018**, *122*, 6236–6239. [CrossRef]
18. Secchi, V.; Franchi, S.; Santi, M.; Vladescu, A.; Braic, M.; Skála, T.; Nováková, J.; Dettin, M.; Zamuner, A.; Iucci, G.; et al. Biocompatible Materials Based on Self-Assembling Peptides on Ti25Nb10Zr Alloy: Molecular Structure and Organization in-Vestigated by Synchrotron Radiation Induced techniques. *Nanomaterials* **2018**, *8*, 148. [CrossRef]
19. Battocchio, C.; Concolato, S.; De Santis, S.; Fahlman, M.; Iucci, G.; Santi, M.; Sotgiu, G.; Orsini, M. Chitosan Functionalization of Titanium and Ti6Al4V Alloy with Chloroacetic Acid as Linker Agent. *Mater. Sci. Eng. C* **2019**, *99*, 1133–1140. [CrossRef]
20. Skwarczynska, A.; Kamińska, M.; Owczarz, P.; Bartoszek, N.; Walkowiak, B.; Modrzejewska, Z. The Structural (FTIR, XRD, and XPS) and Biological Studies of Thermosensitive Chitosan Chloride Gels With β-Glycerophosphate Disodium. *J. Appl. Polym. Sci.* **2018**, *135*. [CrossRef]
21. Solomon, D.; Lehmann, J.; Kinyangi, J.; Liang, B.; Heymann, K.; Dathe, L.; Hanley, K.; Wirick, S.; Jacobsen, C. Carbon (1s) NEXAFS Spectroscopy of Biogeochemically Relevant Reference Organic Compounds. *Soil Sci. Soc. Am. J.* **2009**, *73*, 1817–1830. [CrossRef]
22. Polzonetti, G.; Battocchio, C.; Iucci, G.; Dettin, M.; Gambaretto, R.; Di Bello, C.; Carravetta, V. Thin Films of a Self-Assembling Peptide on TiO2 and Au Studied by NEXAFS, XPS and IR Spectroscopies. *Mater. Sci. Eng. C* **2006**, *26*, 929–934. [CrossRef]
23. Li, H.; Jiang, J.; Luo, Y. Identification of the Smallest Peptide with a Zwitterion as the Global Minimum: A First-Principles Study on Arginine-Containing Peptides. *Phys. Chem. Chem. Phys.* **2017**, *19*, 12117–12126. [CrossRef]
24. Feyer, V.; Plekan, O.; Ptasińska, S.; Iakhnenko, M.; Tsud, N.; Prince, K.C. Adsorption of Histidine and a Histidine Tripeptide on Au(111) and Au(110) from Acidic Solution. *J. Phys. Chem. C* **2012**, *116*, 22960–22966. [CrossRef]
25. Stöhr, J. *NEXAFS Spectroscopy*; Springer Series in Surface Sciences; Springer: Berlin/Heidelberg, Germany, 1992. [CrossRef]
26. Haris, P.I.; Chapman, D. The Conformational Analysis of Peptides Using Fourier Transform IR Spectroscopy. *Biopolymers* **1995**, *37*, 251–263. [CrossRef] [PubMed]
27. Nevskaya, N.; Chirgadze, Y. Infrared Spectra and Resonance Interactions of amide-I and II Vibrations of α-helix. *Biopolymers* **1976**, *15*, 637–648. [CrossRef] [PubMed]
28. NIST Database. Available online: https://webbook.nist.gov/Chemistry/ (accessed on 10 December 2020).
29. Tortora, L.; Concolato, S.; Urbini, M.; Giannitelli, S.M.; Basoli, F.; Rainer, A.; Trombetta, M.; Orsini, M.; Mozetic, P. Functionali-Zation of poly(ε-Caprolactone) Surface with Lactose-Modified Chitosan via Alkaline Hydrolysis: ToF-SIMS characterization. *Biointerphases* **2016**, *11*, 02A32. [CrossRef]
30. Kumirska, J.; Czerwicka, M.; Kaczyński, Z.; Bychowska, A.; Brzozowski, K.; Thöming, J.; Stepnowski, E.P. Application of Spec-Troscopic Methods for Structural Analysis of Chitin and chitosan. *Mar. Drugs* **2010**, *8*, 1567–1636. [CrossRef]
31. Secchi, V.; Franchi, S.; Ciccarelli, D.; Dettin, M.; Zamuner, A.; Serio, A.; Iucci, G.; Battocchio, C. Biofunctionalization of TiO_2 Surfaces with Self-Assembling Layers of Oligopeptides Covalently Grafted to Chitosan. *ACS Biomater. Sci. Eng.* **2019**, *5*, 2190–2199. [CrossRef] [PubMed]
32. Hinz, B.; Dugina, V.; Ballestrem, C.; Wehrle-Haller, B.; Chaponnier, E.C. α-Smooth Muscle Actin Is Crucial for Focal Adhesion Maturation in myofibroblasts. *Mol. Biol. Cell* **2003**, *14*, 2508–2519. [CrossRef]
33. Ho, M.-H.; Liao, M.-H.; Lin, Y.-L.; Lai, C.-H.; Lin, P.-I.; Chen, R.-M. Improving Effects of Chitosan Nanofiber Scaffolds on Osteoblast Proliferation and maturation. *Int. J. Nanomed.* **2014**, *9*, 4293–4304.
34. Si, J.; Wang, C.; Zhang, D.; Wang, B.; Hou, W.; Zhou, Y. Osteopontin in Bone Metabolism and Bone Diseases. *Med. Sci. Monit.* **2020**, *26*, e919159-1. [CrossRef] [PubMed]
35. Paredes, R.; Arriagada, G.; Cruzat, F.; Villagra, A.; Olate, J.; Zaidi, K.; Van Wijnen, A.; Lian, J.B.; Stein, G.S.; Stein, J.L.; et al. Bone-Specific Transcription Factor Runx2 Interacts with the 1α,25-Dihydroxyvitamin D3 Receptor To Up-Regulate Rat Osteocalcin Gene Expression in Osteoblastic Cells. *Mol. Cell. Biol.* **2004**, *24*, 8847–8861. [CrossRef] [PubMed]
36. No, H.K.; Park, N.Y.; Lee, S.H.; Meyers, S.P. Antibacterial Activity of Chitosans and Chitosan Oligomers with Different Molecular Weights. *Int. J. Food Microbiol.* **2002**, *74*, 65–72. [CrossRef]

International Journal of
Molecular Sciences

MDPI

Article

Accelerated and Improved Vascular Maturity after Transplantation of Testicular Tissue in Hydrogels Supplemented with VEGF- and PDGF-Loaded Nanoparticles

Federico Del Vento [1], Jonathan Poels [1], Maxime Vermeulen [1], Bernard Ucakar [2], Maria Grazia Giudice [1,3], Marc Kanbar [1], Anne des Rieux [2] and Christine Wyns [1,3,*]

1 Gynecology-Andrology Unit, Institute of Experimental and Clinical Research, Medical School, Catholic University of Louvain, UCLouvain, 1200 Brussels, Belgium; federico.delvento@uclouvain.be (F.D.V.); jonathan.poels@uclouvain.be (J.P.); vermeulen.maxime@live.be (M.V.); maria.giudice@uclouvain.be (M.G.G.); marc.kanbar@uclouvain.be (M.K.)

2 Advanced Drug Delivery and Biomaterials Unit, Louvain Drug Research Institute, Catholic University of Louvain, UCLouvain, 1200 Brussels, Belgium; bernard.ucakar@uclouvain.be (B.U.); anne.desrieux@uclouvain.be (A.d.R.)

3 Department of Gynecology-Andrology, Saint-Luc University Hospital, 1200 Brussels, Belgium

* Correspondence: christine.wyns@uclouvain.be

Citation: Del Vento, F.; Poels, J.; Vermeulen, M.; Ucakar, B.; Giudice, M.G.; Kanbar, M.; des Rieux, A.; Wyns, C. Accelerated and Improved Vascular Maturity after Transplantation of Testicular Tissue in Hydrogels Supplemented with VEGF- and PDGF-Loaded Nanoparticles. *Int. J. Mol. Sci.* **2021**, *22*, 5779. https://doi.org/10.3390/ijms22115779

Academic Editor: Peter John Jervis

Received: 13 May 2021
Accepted: 24 May 2021
Published: 28 May 2021

Abstract: Avascular transplantation of frozen–thawed testicular tissue fragments represents a potential future technique for fertility restoration in boys with cancer. A significant loss of spermatogonia was observed in xeno-transplants of human tissue most likely due to the hypoxic period before revascularization. To reduce the effect of hypoxia–reoxygenation injuries, several options have already been explored, like encapsulation in alginate hydrogel and supplementation with nanoparticles delivering a necrosis inhibitor (NECINH) or VEGF. While these approaches improved short-term (5 days) vascular surfaces in grafts, neovessels were not maintained up to 21 days; i.e., the time needed for achieving vessel stabilization. To better support tissue grafts, nanoparticles loaded with VEGF, PDGF and NECINH were developed. Testicular tissue fragments from 4–5-week-old mice were encapsulated in calcium-alginate hydrogels, either non-supplemented (control) or supplemented with drug-loaded nanoparticles (VEGF-nanoparticles; VEGF-nanoparticles + PDGF-nanoparticles; NECINH-nanoparticles; VEGF-nanoparticles + NECINH-nanoparticles; and VEGF-nanoparticles + PDGF-nanoparticles + NECINH-nanoparticles) before auto-transplantation. Grafts were recovered after 5 or 21 days for analyses of tissue integrity (hematoxylin–eosin staining), spermatogonial survival (immuno-histo-chemistry for promyelocytic leukemia zinc finger) and vascularization (immuno-histo-chemistry for α-smooth muscle actin and CD-31). Our results showed that a combination of VEGF and PDGF nanoparticles increased vascular maturity and induced a faster maturation of vascular structures in grafts.

Keywords: testicular tissue transplantation; fertility preservation; VEGF; PDGF; vascular maturity; necrosis inhibitor; spermatogonia stem cells; nanoparticles; tissue engineering

1. Introduction

Fertility impairment induced by anticancer treatments in the pediatric population [1–5] represents a major concern for the quality of life of oncological survivors [6,7], and several fertility preservation (FP) programs have been set up worldwide to preserve patients' reproductive potential [8–10]. While for female and post-pubertal male patients FP is at a clinical stage [11–13], for pre-pubertal boys it is still considered experimental.

As spermatogonia are the only germ cells present in the testes before puberty [14], their cryopreservation is the only option for FP [15,16], with the perspective of either cell/tissue auto-transplantation [17] or in vitro maturation using different culture systems [18–20].

Although there is so far no reported outcome for clinical human prepubertal testicular cells or tissue auto-transplantation, several preclinical investigations in animal species [17], including non-human primates [21,22], pave the way for future clinical implementation. As a proof of concept, auto-transplantation of frozen–thawed prepubertal rhesus macaques testicular tissue fragments of large size (9–20 mm^3) allowed complete spermatogenesis and the generation of offspring [23]. With regard to human tissue, frozen–thawed immature testicular tissue (ITT) xeno-transplanted to nude mice allowed preservation of spermatogonia able to proliferate and initiate differentiation [15,16,24]. However, as testicular sampling aimed at FP is usually kept at a minimum to avoid potential harm caused by the procedure, only small amounts of testicular tissue are available for cryostorage and future use [4].

A persistent observation when small ITT fragments (1–10 mm^3) were xenografted was the loss of a high proportion of spermatogonia [24,25], as high as 67% after 5 days of transplantation [26]. This could be explained by the phylogenetic distance between the mouse and human, but also by the avascular transplantation procedure as there is no surgical anastomosis to the host´s vascular system [15]. Tissue revascularization has been shown to occur spontaneously in the grafts where small capillaries connect to the recipient's larger blood vessels assuring tissue survival [27]. However, as observed in ovarian tissue transplantation [28], before the development of a neo-vascularization, the graft is exposed to a period of hypoxia, which might be responsible for the reduction in germ cell numbers.

Hence, improving or accelerating the development of a mature vascular network in grafts could increase both tissue and spermatogonia survival. In this regard, enhancing tissue vascularization with vascular growth factors has been the goal of several studies involving testicular tissue transplantation [29,30].

Vascular endothelial growth factor (VEGF) is a polypeptide that initiates angiogenesis by inducing chemotaxis and proliferation of endothelial cells [31,32]. It has been used to support bovine ITT xeno-transplantation, increasing both the weight of the recovered grafts and the number of seminiferous tubules containing elongated spermatids [33]. More recently, human ITT fragments cultured with VEGF 5 days before being xeno-transplanted to nude mice also showed a better seminiferous tubule integrity in grafts [30].

Unlike ex vivo supplementation of vascular growth factors, systemic administration of molecules that increase angiogenesis calls for caution because of severe possible collateral effects, such as hypotension [34] or promotion of tumor angiogenesis [35]. To overcome these limitations, localized delivery using tissue engineering techniques has been considered [36,37].

In our previous study, embedding of mice testicular tissue in an alginate hydrogel loaded with dextran-chitosan nanoparticles (NPs) delivering VEGF supported short-term angiogenesis, leading to a significantly increased vascular surface 5 days post-transplantation. However, such beneficial effect disappeared after 21 days, suggesting that the newly generated vascular structures might lack vessel stability [29]. These results motivated follow-up investigations on novel approaches aiming at decreasing the hypoxic damage to testicular cells and tissue. In this regard, controlled local drug delivery with NPs loaded with a necrosis inhibitor improved spermatogonial survival in mice testicular tissue auto-grafts [38]. Accelerating revascularization and improving vascular maturity in grafts is another strategy. Amongst the several growth factors involved in the generation and maturation of blood vessels [39], platelet-derived growth factor (PDGF) is a candidate molecule to optimize angiogenesis, as it was shown to play a role in the stabilization of vascular structures [40]. However, it is still unclear which approach is superior, either acting on graft revascularization or protecting cells and tissue from hypoxic damage, or a combination of approaches, and which molecules are the most effective.

Therefore, the objective of this study was to explore and compare the effects of sustained and localized delivery of two vascular growth factors and a necrosis inhibitor on mouse testicular tissue grafts.

Poly(D,L-lactide-co-glycolide)/poly(ethylene glycol (PLGA/PLGA-PEG) was used to form nanoparticles containing either one vascular growth factor (VEGF or PDGF) or a necrosis inhibitor (NECINH). Each molecule was encapsulated individually in separate nanoparticles that were then combined in the alginate hydrogel.

Testicular tissue from 4–5-week-old mice was obtained after bilateral orchidectomy, and fragments of 1 mm^3 were used for orthotopic transplantation experiments. Before grafting, tissue fragments were encapsulated in a matrix composed of 1% calcium-alginate hydrogel (control) or in a 1% calcium-alginate hydrogel supplemented with one or a combination of drug-loaded nanoparticles, including five different conditions, namely, VEGF-NPs (V group); VEGF-NPs and PDGF-NPs (V+P group); NECINH-NPs (N group); VEGF-NPs and NECINH-NPs (V+N group); and VEGF-NPs, PDGF-NPs and NECINH-NPs (V+P+N group). Grafts were recovered after 5 or 21 days for analyses on tissue integrity, spermatogonial survival and vascularization.

2. Results

2.1. Nanoparticle Characterization

The physicochemical characteristics of the nanoparticles are summarized in Table 1.

Table 1. Physicochemical characteristics of the NPs.

	Size (nm)	PDI	Z-Potential (mV)	Encapsulation Efficiency
VEGF-NPs	199 ± 18	0.18	−37 ± 7	90% ± 4%
PDGF-NPs	191 ± 8	0.14	−42 ± 9	94% ± 2%
NECINH-NPs	310 ± 13	0.26	−27 ± 5	65% ± 5%

The objective of this part of the study was to define how the different molecules were released from the NPs incorporated in a 1% alginate hydrogel and whether the combination of bioactive cues in the same hydrogel will impact their release.

VEGF release was very slow (less than 2.5% of the encapsulated dose in 21 days) and was not significantly impacted by the presence of other factors (Figure 1A). PDGF release in the same conditions was faster than VEGF (12% after 21 days) and was not affected by the presence of PDGF- or/and NECINH-loaded NPs (Figure 1B).

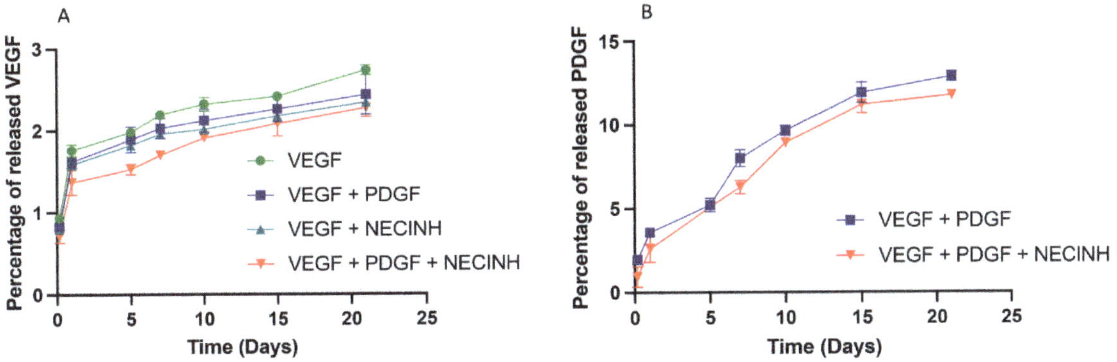

Figure 1. In vitro release of vascular growth factors from NPs incorporated alone or in combination with other NPs, in an alginate hydrogel. (**A**) VEGF-loaded PLGA:PLGA-PEG NPS and (**B**) PDGF-loaded PLGA:PLGA-PEG NPs were embedded in a 1% alginate matrix and VEGF and PDGF were quantified in the release buffer (40 mM $CaCl_2$ PBS) at 34 °C for 21 days (n = 4).

2.2. Tissue Integrity

The goal here was to evaluate the impact of NP supplementation on the integrity of an immature testicular tissue fragment orthotopically auto-transplanted after embedding in an alginate hydrogel.

Seminiferous tubule (ST) integrity was evaluated on hematoxylin–eosin-stained slides of grafts recovered after 5 (3231 STs) and 21 days (2181 STs).

After 5 days of transplantation, the majority of the STs were classified as satisfactory (Score 2) based on morphological evaluation. No statistically significant difference between supplementations was observed when intact (Score 1) STs or intact and satisfactory (Score 1 + 2) STs were taken in account (Figure 2A and Table 2).

Table 2. Impact of NP supplementation on seminiferous tubule integrity 5 days after grafting.

Condition	1% Alginate (Control)	V	V+P	N	V+N	V+P+N
Intact (Score 1)	0.65% ± 0.71%	0.10% ± 0.17%	0.58% ± 0.61%	0.14% ± 0.25%	0.13% ± 0.23%	0%
Satisfactory (Score 2)	59.51% ± 15.91	50.70% ± 10.4%	67.21% ± 6.00%	62.60% ± 11.10	68.31% ± 3.30%	62.11% ± 3.81%
Damaged (Score 3)	38.70% ± 16.52%	48.79% ± 10.43%	32.11% ± 5.71%	37.19% ± 10.91	31.53% ± 3.60%	37.90% ± 3.81%

Results are expressed as the mean percentage of STs ± standard deviation.

With regard to grafts recovered after 21 days, supplementation with NECINH-NPs (N) and the combination of VEGF and NECINH-NPs (V+N) were the most efficient at supporting the preservation of ST integrity compared to the other conditions (Figure 2A,B). Addition of PDGF-NPs had no additional effect when combined with VEGF-NPs alone (V+P) or with the combination of VEGF and NECINH-NPs (V+N+P).

2.3. Spermatogonial Survival

We focused on a specific population of undifferentiated spermatogonia that includes the spermatogonial stem cells using immunohistochemistry against promyelocytic leukemia zinc finger (PLZF) for their quantification in grafts (Figure 3A,B).

The number of PLZF-positive cells per ST after 21 days of transplantation was the highest when NECINH-NPs were added to the graft and was significantly higher than in grafts supplemented with VEGF-NPs. No significant difference was observed between the other conditions.

2.4. Vascularization

As neovascularization is crucial for graft survival and functionality, the impact of NP drug delivery on α-smooth muscle actin (α-SMA)-positive vessels was evaluated, as well as on stable blood vessels (identified by CD-31, a cell marker of mature endothelial structures). The vascular surface was quantified by morphometry using ImageScope on digitalized images of tissue recovered after 5 and 21 days of transplantation. Results are expressed as the measured surface of α-SMA-positive or CD-31-positive blood vessels per total graft surface ± standard deviation.

The α-SMA-positive vascular surface per graft increased when the graft was supplemented with VEGF-NPs, either alone or combined with other NPs, 5 days after transplantation (Figures 4A and 5A). Combination with PDGF and NECINH had no significant effect. However, after 21 days, the surface of the α-SMA-positive blood vessels declined, except when the VEGF and PDGF NPs were combined (Figures 4C and 5C).

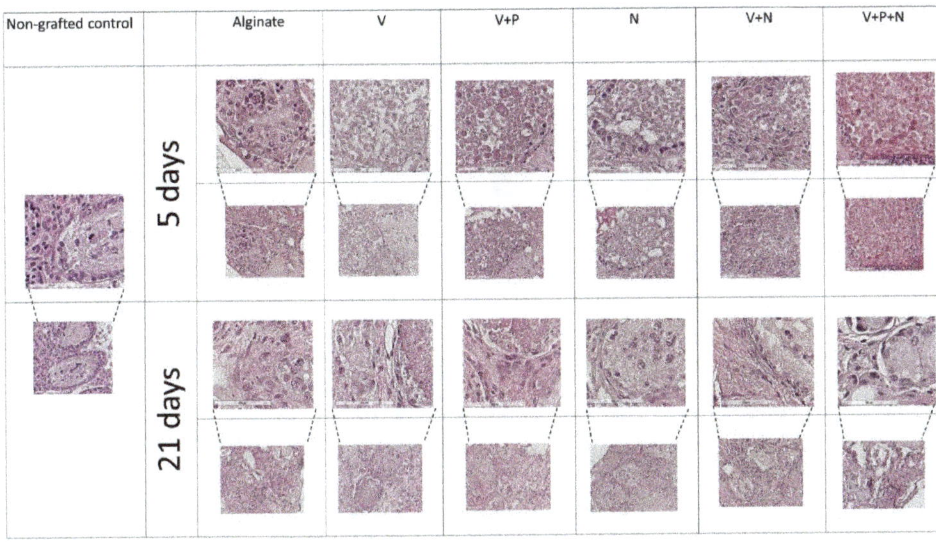

(A)

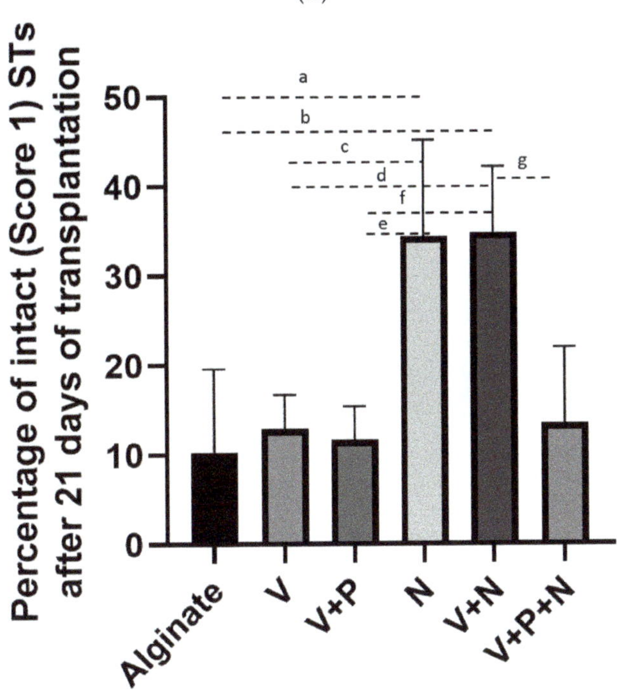

(B)

Figure 2. Impact of NP supplementation on seminiferous tubule integrity. Mouse testicular tissues were encapsulated in alginate supplemented with five different combinations in addition to a negative control (Alginate, V, V+P, N, V+N and V+P+N) for 21 days. Seminiferous tubule integrity was (**A**) visualized by hematoxylin–eosin-staining 5 and 21 days after orthotopic auto-transplantation and (**B**) quantified after 21 days. Results are expressed as the mean percentage of intact (Score 1) STs ± standard deviation. a: $p = 0.02$; b: $p = 0.02$; c: $p = 0.04$; d: $p = 0.04$; e: $p = 0.03$, f: $p = 0.03$; g: $p = 0.04$ ($n = 3$).

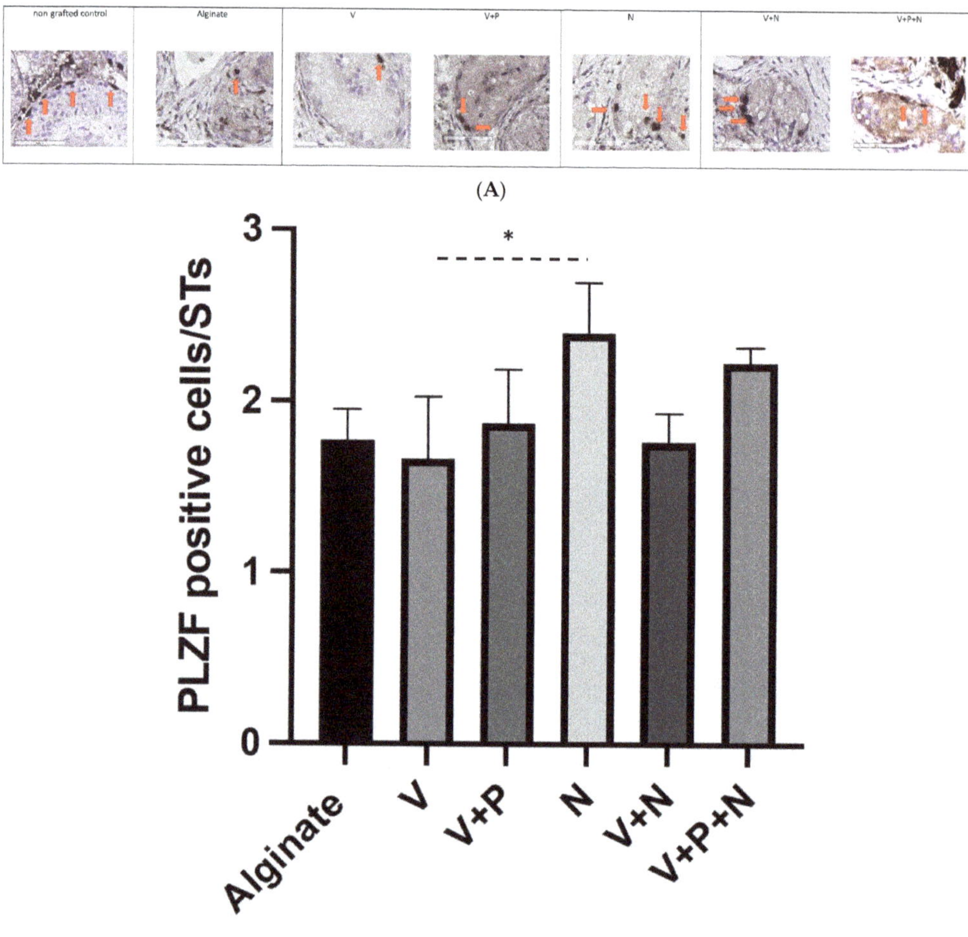

Figure 3. Impact of NP supplementation on germ cell survival 21 days after grafting. Mouse testicular tissue fragments were encapsulated in alginate supplemented with five different combinations in addition to a negative control (Alginate, V, V+P, N, V+N and V+P+N). (**A**) Immunostaining against PLZF was used to identify undifferentiated spermatogonia, (Red arrows highlight positive cells, scale bar = 60 µm) and (**B**) to quantify them. Results are expressed as the mean PLZF positive cells/STs (Alginate 1.77 ± 0.17; 1.79 ± 0.46; V+P: 1.87 ± 0.31; N: 2.4 ± 0.29; V+N. 1.76 ± 0.16; and V+P+N: 2.23 ± 0.08) (n = 3). * p = 0.03.

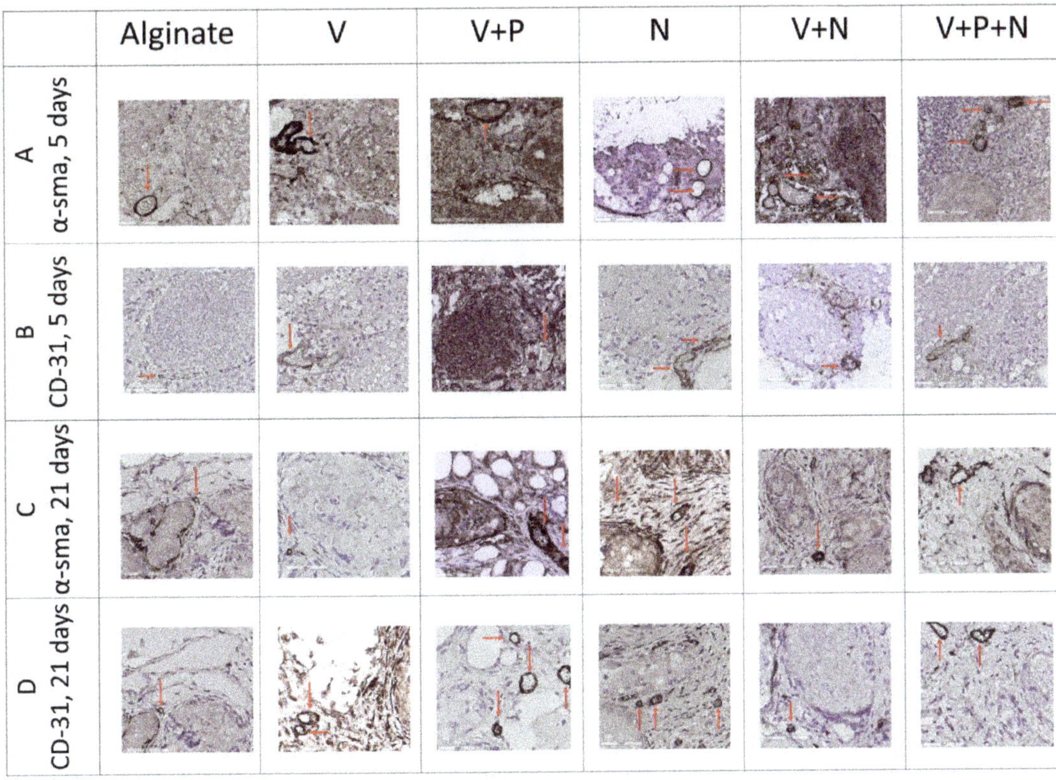

Figure 4. Impact of NP supplementation on the vascularization of mice testicular tissue auto-transplanted for 5 and 21 days. Mouse testicular tissue fragments were encapsulated in alginate supplemented with five different combinations in addition to a negative control (Alginate, V, V+P, N, V+N and V+P+N) and the grafts were stained after 5 and 21 days for α-SMA and CD31. Red arrows highlight positive vascular structures. Scale bar = 60 μm. $n = 3$.

A combination of VEGF- and PDGF-NPs, NECINH-NPs, and a combination of NECINH-NPs and VEGF-NPs led to a significant increase in CD31-positive surfaces in grafts (Figures 4B and 5B) 5 days after transplantation. Only the addition of PDGF-NPs allowed to maintain a significantly larger surface of CD31 staining over a longer period (21 days) (Figures 4D and 5D).

When looking at the evolution over time of these vascular markers, an increased α-SMA-positive surface was recorded between 5 and 21 days when VEGF-NPs and PDGF-NPs were employed together (V+P group) (Figure 6A). The CD-31-positive vascular surface was increased in the V+P+N group between 5 and 21 days and reduced when the PDGF-NPs were not employed (N and V+N) (Figure 6B).

The ratio between the CD-31/α-SMA-positive vascular surface, representing the proportion of stabilized vessels, was not different between groups after 5 days (Alginate: 2.44 ± 0.38; V: 0.42 ± 0.12; V+P: 1.14 ± 0.41; N: 1.84 ± 0.43; V+N: 0.79 ± 0.33; V+P+N: 0.61 ± 0.14. After 21 days, the proportion of stabilized vessels was statistically higher when the V+P, N and V+P+N groups were compared to the alginate group (Figure 7).

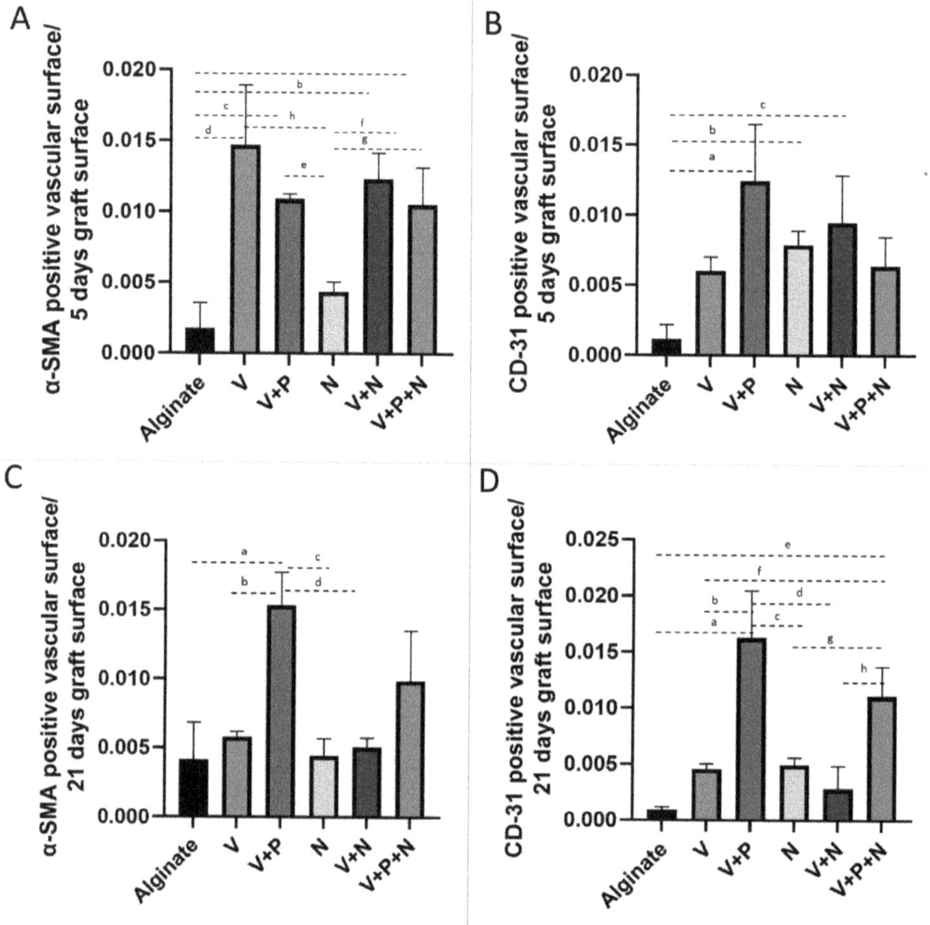

Figure 5. Impact of NP supplementation on the vascularization of mice testicular tissue auto-transplanted for 5 and 21 days. Mouse testicular tissue fragments were encapsulated in alginate supplemented with five different combinations in addition to a negative control (Alginate, V, V+P, N, V+N and V+P+N) and the positive surface for α-SMA (A and C, 5 and 21 days, respectively) and CD31 (B and D, 5 and 21 days, respectively) was quantified using VisioPharm (n = 36). (**A**) IHC for α-SMA, 5 days. Encapsulation with a matrix containing VEGF-NPs (alone or combined with PDGF-NPs or NECINH-NPs) increased vascular surfaces. Results are expressed as the mean positive vascular surface per total graft surface ± standard deviation: Alginate ($1.74 \times 10^{-3} \pm 1.81 \times 10^{-3}$); V ($1.47 \times 10^{-2} \pm 4.21 \times 10^{-3}$); V+P ($1.09 \times 10^{-2} \pm 3.29 \times 10^{-4}$); N ($4.63 \times 10^{-3} \pm 6.85 \times 10^{-4}$); V+N ($1.23 \times 10^{-2} \pm 1.80 \times 10^{-3}$); V+P+N ($1.05 \times 10^{-2} \pm 2.64 \times 10^{-3}$). a: $p < 0.01$; b: $p < 0.01$; c: $p < 0.01$; d: $p < 0.01$; e: $p < 0.01$; f: $p = 0.01$; g: $p = 0.05$; h: $p < 0.01$. (**B**) IHC for α-SMA, 21 days. Encapsulation with a matrix containing VEGF-NPs and PDGF-NPs increased vascular surface: alginate ($4.13 \times 10^{-3} \pm 2.69 \times 10^{-3}$); V ($5.79 \times 10^{-3} \pm 3.92 \times 10^{-4}$); V+P ($1.53 \times 10^{-2} \pm 2.38 \times 10^{-3}$); N ($4.42 \times 10^{-3} \pm 1.25 \times 10^{-3}$); V+N ($5.06 \times 10^{-3} \pm 6.78 \times 10^{-4}$); V+P+N ($9.89 \times 10^{-3} \pm 3.62 \times 10^{-3}$). a: $p < 0.01$; b: $p < 0.01$; c: $p < 0.01$; d: $p < 0.01$. (**C**) IHC for CD-31, 5 days. Supplementation with V+P, N and V+N improved vascular maturity compared to alginate encapsulation only. Alginate ($1.04 \times 10^{-4} \pm 1.17 \times 10^{-3}$); V ($7.15 \times 10^{-3} \pm 6.00 \times 10^{-3}$); V+P ($1.20 \times 10^{-2} \pm 1.24 \times 10^{-2}$); N ($6.83 \times 10^{-3} \pm 7.9 \times 10^{3}$); V+N ($6.06 \times 10^{-3} \pm 9.48 \times 10^{-3}$); V+P+N ($8.63 \times 10^{-3} \pm 6.37 \times 10^{-3}$). a: $p < 0.01$; b: $p = 0.0467$; c: $p = 0.01$. (**D**) IHC for CD-31, 21 days. Supplementation with VEGF-NPs and PDGF-NPs increased vascular maturity regardless of NECINH-NPs (V+P and V+P+N). Alginate ($9.27 \times 10^{-4} \pm 2.61 \times 10^{-4}$); V ($4.56 \times 10^{-3} \pm 4.82 \times 10^{-4}$); V+P ($1.63 \times 10^{-2} \pm 4.19 \times 10^{-3}$); N ($4.97 \times 10^{-3} \pm 6.27 \times 10^{-4}$); V+N ($2.84 \times 10^{-3} \pm 2.02 \times 10^{-3}$); V+P+N ($1.11 \times 10^{-2} \pm 2.64 \times 10^{-3}$). a: $p < 0.0001$; b: $p < 0.01$; c: $p < 0.01$; d: $p < 0.01$; e: $p < 0.01$; f: $p = 0.03$; g: $p = 0.04$; h: $p < 0.01$. $n = 3$.

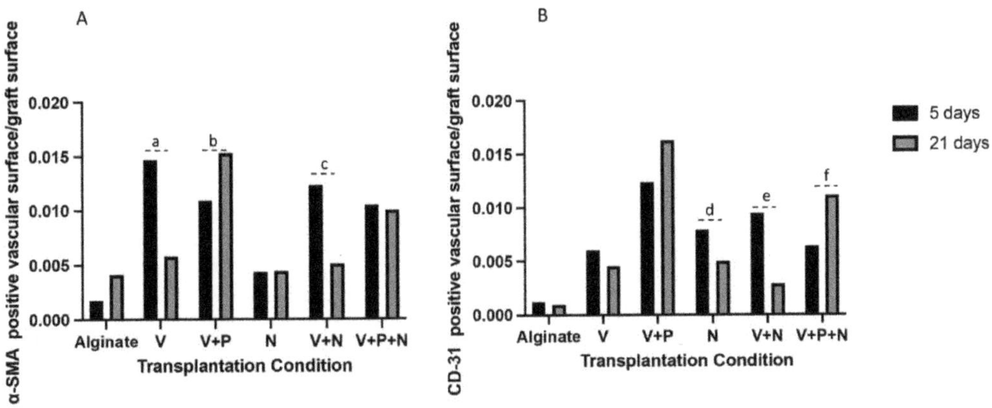

Figure 6. Impact of NP supplementation on the blood vessels' evolution between 5 and 21 days in mice testicular tissue auto-transplants. (**A**) α-SMA- and (**B**) CD31-positive surfaces in the grafts were compared after 5 and 21 days of transplantation. a: $p = 0.03$; b: $p = 0.04$; c: $p < 0.01$; d: $p = 0.01$; e: $p = 0.02$; f: $p = 0.03$. $n = 3$.

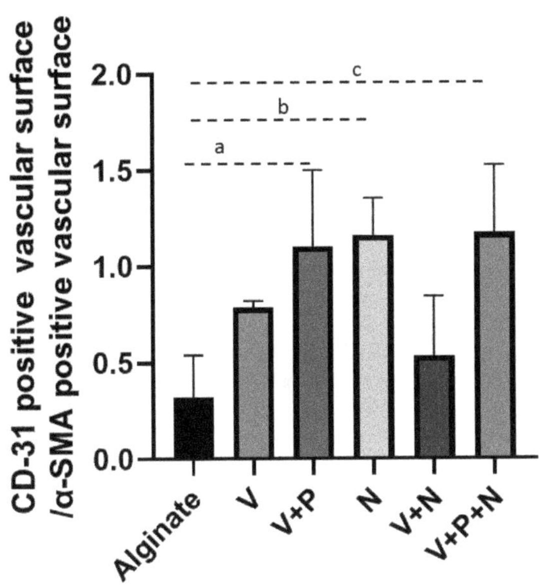

Figure 7. Impact of NP supplementation on neo-vascularization stabilization. The CD-31/α-SMA ratio was calculated for grafts retrieved after 21 days a: $p = 0.04$; b: $p = 0.02$; c: $p = 0.02$. $n = 18$.

3. Discussion

Auto-transplantation of frozen–thawed testicular tissue is a potential option to restore a patient's fertility after gonadotoxic treatment and could therefore play a role in improving the quality of life of prepubertal cancer survivors. Based on short- and long-term human ITT xeno-transplantation experiments, the reduction in spermatogonia numbers appeared to be the highest during the first three weeks following transplantation [15,24,26], a period of time that has been proven to be necessary for the generation of stable vascular structures [41] and that has been characterized by hypoxic damage to tissue and cells [42]. The optimization of the grafting technique, targeting an increased spermatogonial survival, represents an open challenge. Delivery of drugs directly to the transplanted tissue could reduce or avoid the potential collateral effects of a systemic administration and enhance the graft's efficiency by improving tissue integrity and increasing the number of surviving spermatogonial stem cells able to differentiate into mature spermatozoa. Tissue engineering techniques that employ vascular growth factors and necrosis inhibitors could provide an essential support to small testicular fragments grafts.

Here, we showed that localized and sustained delivery of vascular growth factors from polymeric nanoparticles improved vascularization of mouse testicular tissue grafts.

Analysis of IHC for α-SMA on tissue recovered after 5 days of transplantation showed that VEGF-NPs increased the vascular surface in the graft. This result was not influenced by simultaneous supplementation with other types of NPs and it was not maintained after 21 days of grafting. While this is in agreement with previously published results, linking an increased short-term vascularization to NPs delivering VEGF [29], it motivated the addition in this study of a second growth factor known for its ability to stabilize neo-vascularization (PDGF) [40]. This led to further progress in terms of graft angiogenesis enhancement. Indeed, after 21 days of transplantation, both the vascular surface and vessel maturity were significantly increased by simultaneous addition of VEGF-NPs and PDGF-NPs, compared to alginate hydrogel and to supplementation with VEGF-NPs.

The reduction in vascular surface evidenced by α-SMA immunostaining observed after 3 weeks of grafting, when the only vascular growth factor administered was VEGF (V and V+N groups), can be linked to insufficient vessel stabilization. In support of this theory, addition of NPs delivering PDGF to VEGF-NPs (V+P group) was associated with an increased graft vascular surface at 21 days. Improvement in the CD-31/α-SMA ratio observed after 21 days with PDGF-NPs application (V+P and V+P+N groups) further corroborated our hypothesis that the increased vascular surface was due to an improved maturation of the vessels. Interestingly, vascular maturity also improved after 5 days of transplantation (CD-31 IHC) in the presence of PDGF, which suggests that adding PDGF-loaded NPs accelerated the revascularization process in grafts.

Two of the transplantation conditions involving NECINH-NPs (V+N and N group) were also associated with an improvement in short-term vascular stability. However, this effect was not maintained after 21 days unless PDGF-NPs were present.

Moreover, we observed a significant improvement in the CD-31-positive vascular surface between 5 and 21 days when NECINH was added to vascular growth factors (V+P+N group versus V+P group), which suggests that NECINH might support the effect of PDGF. A higher CD-31/α-SMA ratio in the V+P+N group after 21 days of grafting provided further support to this hypothesis. No previous report has linked NECINH to pericytes chemotaxis. However, necrosis inhibition and reduction of the inflammation process associated with NECINH administration [43] could exert a yet partially unknown effect on angiogenesis that could explain the observed improvement in vascular maturity. This lack of information on the potential role of NECINH in neovascularization therefore calls for further studies.

Simultaneous use of NPs containing vascular growth factors and NECINH (V+P+N group) did not result in an improvement in the intact (Score 1) seminiferous tubule sections number, nor in an improved spermatogonial survival compared to the other groups employing NECINH-NPs. Unknown interactions between these molecules or the negative effects of vascular growth factors on the STs' integrity are therefore not excluded.

No previous study investigated the concurrent use of these bioactive molecules. PDGF plays a role in the stabilization of vascular structures, both in physiological situations and pathological processes, such as wound healing and cancer [44]. The effect on vessel maturity has been linked to the PDGF chemotactic properties on pericytes [40,45]. Potentially, PDGF could be responsible for increased ST fibrosis [46], which counteracts the beneficial effects of NECINH. This could explain our observations on tissue integrity after concomitant supplementation of the three types of NPs (V+P+N group), while this parameter was still improved in the N and V+N conditions.

Moreover, cell signal pathways involving PDGF have also been identified in several tissues of the human body, including the testis, with effects displayed especially during testicular embryonic development [47]. Hence, while the repercussion of PDGF delivery to support testicular tissue grafts has never been studied, its impact could likely go beyond angiogenesis.

It is important to mention that the in vitro release of vascular growth factors is not to be held accountable for any of the differences between groups, as the drug-delivery profiles were similar for all matrixes applied and were not influenced by the concurrent supplementation of multiple nanoparticles. NECINH release from NPs and alginate hydrogels already has been reported in a previous publication [38]. As combining different nanoparticles within the same gel did not impact their release, we assumed it would be the same for NECINH.

In addition, modification in the release profile of VEGF did not seem to change the vascular effects of the VEGF-NPs. Indeed, while the VEGF concentration in the alginate hydrogel was the same as the one used in our previous transplantation experiment [29], the rate of VEGF release was slower in this study (2.5% after 21 days with PLGA/PLGA-PEG NPs compared to 7.8% with Dextran-chitosan-NPs, reported by Poels et al. in 2016); still, the VEGF effect on neo-vascularization was similar.

ST integrity and spermatogonial survival after 21 days of transplantation were increased when nanoparticles containing NECINH were administered alone, corroborating previous results [38], although no further benefit was obtained by improving the vascular network in grafts with vascular growth factors. VEGF-regulated pathways within the testes could be implicated. Indeed, VEGF receptors have been found on testicular cells, including spermatogonia, Sertoli cells and Leydig cells [48], and an impact of VEGF on spermatogonial self-renewal and differentiation has been suggested, of specific importance being the balance between the different isoforms of VEGF [49]. We used VEGF-164 in our experiments based on the long-known role of this isoform in the angiogenesis process [31]. While VEGF-164 was shown to participate in germ cells proliferation [50], it has also been linked to increased germ cell differentiation [51]. Hence, the number of PLZF-positive cells could have been influenced by a triggered differentiation of germ cells within grafts. Further studies on the short-term impact of vascular growth factors on germ cells should be conducted to fully elucidate the underlying mechanisms.

Altogether, our results represent a step towards a betterment of the testicular tissue transplantation technique with the potential for clinical application. So far, use of vascular growth factors in clinical trials led to encouraging results when administered via a local injection, e.g., in the infarcted heart [52], although application outside of experimental settings is not available yet. Considering unwanted systemic effect in the context of cancer patients, sustained and localized drug delivery seems to be a better option for administration of such molecules. Delivery of VEGF through nanoparticles has also been explored in vivo in rodents, with promising results for pathological conditions such as limb ischemia [53]. As PLGA is an FDA-approved excipient [54] with possible applications in humans [55], the strategy explored in these experiments could potentially be applied in clinical trials. Moreover, other types of avascular transplantation procedures, involving, for example, skin [56] or ovarian tissue [57], could benefit from accelerated vascular maturity.

Overall, the role of locally controlled vascular growth factor administration and the effects of simultaneous multiple bioactive molecule delivery needs to be further elucidated, especially with regard to longer-term grafting of testicular tissue and the effects on germ-cell differentiation.

4. Materials and Methods

4.1. Encapsulation of Bioactive Factors in Polymeric Nanoparticles

Recombinant murine VEGF-164 was produced as previously described [58], while PDGF-BB (101–14 B Peprotech, London, UK) and necrosis inhibitor NecroX-5 were purchased (ALX-430–167-M005, Enzo life science, New York, NY, USA).

VEGF, PDGF and Necrox were encapsulated as previously described [38,59]. Briefly, for vascular growth factors, 50 mg of polymers (25 mg of Poly(D,L-lactide-co-glycolide) (PLGA) (Resomer® RG 502H, Sigma-Aldrich-719897, Darmstadt, Germany) and 25 mg of PLGA/poly(ethylene glycol (PLGA-PEG)) (Resomer® RGP d 50155, Boehringer Ingelheim, Ingelheim, Germany) were dissolved in 1 mL of dichloromethane. In total, 50 µg of VEGF (1 mg/mL) or PDGF (1 mg/mL) were incorporated into the formulation and encapsulated by double emulsion.

NECINH nanoparticles were produced by the single emulsion technique, as previously reported [38]. In total, 250 µg of NECINH were used for this formulation.

4.1.1. Nanoparticle Characterization

Nanoparticles were characterized in terms of size, PDI (polydispersity index) and zeta-potential using a Zetasizer (Malvern Panalytical, Malvern, UK) [38].

To calculate the encapsulation efficiency, non-encapsulated VEGF and PDGF concentrations were determined by the ELISA (enzyme-linked immunosorbent assay) sandwich test (900-K99 and 900-K04, Peprotech, respectively).

4.1.2. In Vitro Drug Release Profile

In vitro release experiments were performed for each formulation and different combinations.

For each condition, NPs were mixed with a solution of 1% alginate (SLM100, FMC BioPolymers, NovaMatrix™, Sandvika, Norway) in 3-(*n*-morpholino) propanesulfonic acid, 4-morpholinepropanesulfonic acid (MOPS buffer) (M3183; Sigma-Aldrich, Darmstadt, Germany). Then, 5 µL of $CaCl_2$ (50 mM solution in MOPS) were added to 45 uL of alginate to form the gel. Hydrogels were covered with 250 µL of 40 mM $CaCl_2$ in PBS and incubated at 34 °C for 21 days. Release media were collected and replaced at 4 h, and 1, 5, 7, 10, 15 and 21 days. The VEGF and PDGF concentrations were determined by ELISA. Results were expressed as the percentage of released growth factor compared to the total amount of growth factor initially introduced into the hydrogel.

4.2. Testicular Tissue Collection

Ethics Review Board and the Committee on Animal Research of UCLouvain (project 2018/UCL/MD/20, approval date: 22 June 2018) approved all experiments involving animals, and the laws currently in force in Belgium (Royal Decree on the Protection of Experimental Animals 29 May 2013) were respected.

Thirty-six NMRI (Naval Medical Research Institute) male mice were purchased from Janvier Labs (Saint Berthevin Cedex, France) and kept in individual cages in the animal facility of UCLouvain (Linné building, UCLouvain, Woluwe Saint-Lambert, Belgium), where they received animal care according to university guidelines. Room temperature and humidity were checked every day and a usual night/day cycle was applied. Food and water were available ad libitum.

Surgery was performed in an equipped area of the animal house facility of the UCLouvain University. Intraperitoneal injection of medetomidine (1 mg/kg) (Domitor, Pfizer, Cambridge, MA, USA) and ketamine (75 mg/kg) (Anesketin, Eurovet, Heusden-Zolder, Belgium) was used to provide anesthesia, while buprenorphine (0.1 mg/kg) (Temgesic, Schering Plough, Kenilworth, NJ, USA) provided analgesia.

Through a scrotal incision, we performed a bilateral orchidectomy. For each animal, one fragment of 1 mm^3 of the testicular tissue was used for transplantation and one fragment of the same size was kept as the non-grafted control.

4.3. Encapsulation of Tissue and NPs

Fragments and NPs were encapsulated in alginate hydrogels, as previously described [29,38]. Briefly, we added NP solutions to the 1% alginate hydrogels to explore the impact of six different types of supplementation (Alginate, V, V+P, V+N and V+P+N). The concentrations of the bioactive molecules in the hydrogels were 0.06 μg/μL for VEGF [29,60] and PDGF [61], and 0.13 μg/μL for NECINH [38]. Mice testicular tissue fragments were placed in 45 μL of the hydrogel and 5 μL of $CaCl_2$ were added to obtain gelation.

4.4. Testicular Tissue Transplantation

Hydrogels containing tissue fragments were orthotopically autografted and we reversed anesthesia with intraperitoneal injection of atipamezole (1 mg/kg) (Antisedan, Pfizer, Cambridge, MA, USA). After 2 different timings (5 and 21 days), mice were euthanized by cervical dislocation to allow graft recovery.

4.5. Tissue Analyses

Recovered grafts were kept in modified Davidson's fluid (mDF) for one night before fixation in paraffin. Blocks were cut in 5 μm-thick sections that were placed on Superfrost Plus slides (VWR, Leuven, Belgium). Immunohistochemistry analysis were performed according to common protocols. After dewaxing and rehydration, slides were incubated for 30 min at room temperature (RT) in 0.3% H_2O_2 to block endogenous peroxidase activity and soaked in citrate buffer at 98 °C 60 min for heat-induced antigen retrieval. Then, 10% normal goat serum (NGS, Invitrogen, Merelbeke, Belgium) and 1% bovine serum albumin (BSA) (Invitrogen, Merelbeke, Belgium) were used (30 min at RT) to block non-specific binding sites.

Incubation with primary antibody followed indications and concentrations of the manufacturer's datasheet. Anti-promyelocytic leukemia zinc finger (PLZF) (rabbit anti-PLZF antibody, 1/400, Sigma-Aldrich, HPA001499, St. Louis, MO, USA) was employed to identify undifferentiated spermatogonia [62]. Vascular structures and mature vascular structures were labeled, applying antibodies directed, respectively, towards α-SMA (anti-alpha smooth muscle actin, ab5694, Abcam, Cambridge, UK) [63] and CD31 (Cluster of differentiation-31) (PECAM-1/D8V9E- XP Rabbit, Cell Signaling, Danvers, MA, USA) [64]. Slides were incubated for 60 min at RT with a secondary anti-rabbit antibody (Envision+ system-labeled polymer-horseradish peroxidase (HRP); DAKO, K4003, Agilent, Santa Clara, CA, USA). Diaminobenzidine (DAKO K3468) was employed as chromogen, and Mayer's hematoxylin was applied to counterstain the nuclei. Slides were washed with tap water and dehydrated before being sealed with Entellan mounting medium (Sigma-Aldrich, St. Louis, MO, USA) and coverslips. A Leica SCN400 slide scanner (Leica Biosystems, WETZLAP, Nußloch, Germany) was used to obtain digital images of the whole sections for tissue analyses.

The tissue integrity score was assessed on hematoxylin–eosin (HE)-stained slides, applying the following scoring criteria: STs were classified as intact (Score 1, cells adhere to the basement membrane, no signs of necrosis were spotted and cells cohesion was present), satisfactory (Score-2, intra-tubular cells could still be spotted in spite of presence of intra-tubular necrosis) or damaged (Score 3, STs showed complete necrosis). For PLZF IHC, results were expressed as the mean number of positive cells per seminiferous tubule (ST) section.

The α-SMA and CD31 results were expressed as the identified vascular surface per total graft surface

4.6. Statistical Analysis

Statistical analysis was performed using JMP Software (version JPM-pro 15, Cary, NC, USA).

Normal distribution of the data was evaluated with a Shapiro–Wilk test, and a one-way ANOVA with a Tukey post-hoc test was applied to compare the effects of the six encapsulation conditions. Student's t-test was applied to assess the difference for each group at the two time points. Results are expressed as the mean ± standard deviation and p values < 0.05 were considered significant.

5. Conclusions

Supplementation of alginate hydrogels with nanoparticles delivering PDGF induced improvement in terms of vascularization and vascular maturity in testicular tissue grafts compared to supplementation with VEGF alone, while potential interactions with NECINH deserve further investigations.

Our study shows how support to cells and tissue can be provided by simultaneous local delivery of multiple bioactive molecules that target different biological processes. The results obtained with NP drug-delivery systems offer new perspectives for further application in regenerative medicine, especially in the field of fertility preservation.

Author Contributions: F.D.V. was responsible for performing the experiments, analysis of the results, and drafting of the manuscript. J.P. participated in the project design and contributed to the analysis of the results. B.U. produced the nanoparticles and provided technical support for their characterization. M.V., M.G.G. and M.K. provided assistance during the experiments and reviewed the manuscript. A.d.R. participated in the project design, interpretation of results and reviewed the manuscript. C.W. was responsible for the project concept, interpretation of results, critical discussion, and review of the manuscript. All authors have read and agreed to the published version of the manuscript.

Funding: This project was supported by grants from the Fondation contre le cancer (2016-141 FAF-C/2016/745) and the Fondation Salus Sanguinis. Anne des Rieux is a research associate from the FRS-FNRS (Fonds de la Recherche Scientifique, Belgique) and is a recipient of subsidies from the Fonds Spéciaux de Recherche Scientifique (FSR, UCL).

Institutional Review Board Statement: The study was conducted according to the guidelines of the Declaration of Helsinki, and approved by the Ethics Review Board and the Committee on Animal Research of UCLouvain (project 2018/UCL/MD/20, approval date: 22 June 2018).

Informed Consent Statement: Not applicable.

Conflicts of Interest: The authors declare no conflict of interest.

References

1. Wyns, C.; Curaba, M.; Vanabelle, B.; Van Langendonckt, A.; Donnez, J. Options for fertility preservation in prepubertal boys. *Hum. Reprod. Updat.* **2010**, *16*, 312–328. [CrossRef]
2. Stukenborg, J.-B.; Jahnukainen, K.; Hutka, M.; Mitchell, R.T. Cancer treatment in childhood and testicular function: The importance of the somatic environment. *Endocr. Connect.* **2018**, *7*, R69–R87. [CrossRef]
3. Pampanini, V.; Wagner, M.; Asadi-Azarbaijani, B.; Oskam, I.C.; Sheikhi, M.; Sjödin, M.O.D.; Lindberg, J.; Hovatta, O.; Sahlin, L.; Björvang, R.D.; et al. Impact of first-line cancer treatment on the follicle quality in cryopreserved ovarian samples from girls and young women. *Hum. Reprod.* **2019**, *34*, 1674–1685. [CrossRef] [PubMed]
4. Wyns, C.; Kanbar, M.; Giudice, M.G.; Poels, J. Fertility preservation for prepubertal boys: Lessons learned from the past and update on remaining challenges towards clinical translation. *Hum. Reprod. Updat.* **2021**, *27*, 433–459. [CrossRef] [PubMed]
5. Kanbar, M.; De Michele, F.; Giudice, M.G.; Desmet, L.; Poels, J.; Wyns, C. Long-term follow-up of boys who have undergone a testicular biopsy for fertility preservation. *Hum. Reprod.* **2020**, *36*, 26–39.
6. Thomson, A.B.; Campbell, A.J.; Irvine, D.C.; Anderson, R.A.; Kelnar, C.J.H.; Wallace, W.H.B. Semen quality and spermatozoal DNA integrity in survivors of childhood cancer: A case-control study. *Lancet* **2002**, *360*, 361–367. [CrossRef]
7. Picton, H.M.; Wyns, C.; Anderson, R.A.; Goossens, E.; Jahnukainen, K.; Kliesch, S.; Mitchell, R.T.; Pennings, G.; Rives, N.; Tournaye, H.; et al. A European perspective on testicular tissue cryopreservation for fertility preservation in prepubertal and adolescent boys. *Hum. Reprod.* **2015**, *30*, 2463–2475. [CrossRef]
8. Ming, J.M.; Chua, M.E.; Lopes, R.I.; Maloney, A.M.; Gupta, A.A.; Lorenzo, A.J. Cryopreservation of testicular tissue in pre-pubertal and adolescent boys at risk for infertility: A low risk procedure. *J. Pediatr. Urol.* **2018**, *14*, 274.e1–274.e5. [CrossRef] [PubMed]
9. Valli-Pulaski, H.; Peters, K.A.; Gassei, K.; Steimer, S.R.; Sukhwani, M.; Hermann, B.P.; Dwomor, L.; David, S.; Fayomi, A.P.; Munyoki, S.; et al. Testicular tissue cryopreservation: 8 years of experience from a coordinated network of academic centers. *Hum. Reprod.* **2019**, *34*, 966–977. [CrossRef] [PubMed]

10. Goossens, E.; Jahnukainen, K.; Mitchell, R.T.; van Pelt, A.; Pennings, G.; Rives, N.; Poels, J.; Wyns, C.; Lane, S.; Rodriguez-Wallberg, K.A.; et al. Fertility Preservation in Boys: Recent Developments and New Insights (Dagger). *Hum. Reprod. Open* **2020**, *2020*, hoaa016. [CrossRef]
11. Demeestere, I.; Simon, P.; Dedeken, L.; Moffa, F.; Tsépélidis, S.; Brachet, C.; Delbaere, A.; Devreker, F.; Ferster, A. Live birth after autograft of ovarian tissue cryopreserved during childhood. *Hum. Reprod.* **2015**, *30*, 2107–2109. [CrossRef]
12. Jadoul, P.; Guilmain, A.; Squifflet, J.; Luyckx, M.; Votino, R.; Wyns, C.; Dolmans, M. Efficacy of ovarian tissue cryopreservation for fertility preservation: Lessons learned from 545 cases. *Hum. Reprod.* **2017**, *32*, 1046–1054. [CrossRef]
13. Lambertini, M.; Peccatori, F.A.; Demeestere, I.; Amant, F.; Wyns, C.; Stukenborg, J.B.; Paluch-Shimon, S.; Halaska, M.J.; Uzan, C.; Meissner, J.; et al. Fertility Preservation and Post-Treatment Pregnancies in Post-Pubertal Cancer Patients: Esmo Clinical Practice Guidelines(Dagger). *Ann. Oncol.* **2020**, *31*, 1664–1678. [CrossRef] [PubMed]
14. Clermont, Y. Kinetics of spermatogenesis in mammals. *Arch. D'anatomie Microsc. Et De Morphol. Exp.* **1967**, *56*, 7–60.
15. Wyns, C.; Curaba, M.; Martinez-Madrid, B.; van Langendonckt, A.; François-Xavier, W.; Donnez, J. Spermatogonial Survival after Cryopreservation and Short-Term Orthotopic Immature Human Cryptorchid Testicular Tissue Grafting to Immunodeficient Mice. *Hum. Reprod.* **2007**, *22*, 1603–1611. [CrossRef] [PubMed]
16. Poels, J.; Van Langendonckt, A.; Many, M.-C.; Wese, F.-X.; Wyns, C. Vitrification preserves proliferation capacity in human spermatogonia. *Hum. Reprod.* **2013**, *28*, 578–589. [CrossRef]
17. Del Vento, F.; Vermeulen, M.; De Michele, F.; Giudice, M.G.; Poels, J.; Rieux, A.D.; Wyns, C. Tissue Engineering to Improve Immature Testicular Tissue and Cell Transplantation Outcomes: One Step Closer to Fertility Restoration for Prepubertal Boys Exposed to Gonadotoxic Treatments. *Int. J. Mol. Sci.* **2018**, *19*, 286. [CrossRef]
18. De Michele, F.; Vermeulen, M.; Wyns, C. Fertility restoration with spermatogonial stem cells. *Curr. Opin. Endocrinol. Diabetes Obes.* **2017**, *24*, 424–431. [CrossRef]
19. De Michele, F.; Poels, J.; Vermeulen, M.; Ambroise, J.; Gruson, D.; Guiot, Y.; Wyns, C. Haploid Germ Cells Generated in Organotypic Culture of Testicular Tissue From Prepubertal Boys. *Front. Physiol.* **2018**, *9*, 1413. [CrossRef]
20. Vermeulen, M.; Del Vento, F.; Kanbar, M.; Ruys, S.P.D.; Vertommen, D.; Poels, J.; Wyns, C. Generation of Organized Porcine Testicular Organoids in Solubilized Hydrogels from Decellularized Extracellular Matrix. *Int. J. Mol. Sci.* **2019**, *20*, 5476. [CrossRef]
21. Liu, Z.; Nie, Y.H.; Zhang, C.C.; Cai, Y.J.; Wang, Y.; Lu, H.P.; Li, Y.Z.; Cheng, C.; Qiu, Z.L.; Sun, Q. Generation of Macaques with Sperm Derived from Juvenile Monkey Testicular Xenografts. *Cell Res.* **2016**, *26*, 139–142. [CrossRef]
22. Ntemou, E.; Kadam, P.; Van Saen, D.; Wistuba, J.; Mitchell, R.T.; Schlatt, S.; Goossens, E. Complete spermatogenesis in intratesticular testis tissue xenotransplants from immature non-human primate. *Hum. Reprod.* **2019**, *34*, 403–4133. [CrossRef]
23. Fayomi, A.P.; Peters, K.; Sukhwani, M.; Valli-Pulaski, H.; Shetty, G.; Meistrich, M.L.; Houser, L.; Robertson, N.; Roberts, V.; Ramsey, C.; et al. Autologous grafting of cryopreserved prepubertal rhesus testis produces sperm and offspring. *Science* **2019**, *363*, 1314–1319. [CrossRef] [PubMed]
24. Wyns, C.; Van Langendonckt, A.; Wese, F.-X.; Donnez, J.; Curaba, M. Long-term spermatogonial survival in cryopreserved and xenografted immature human testicular tissue. *Hum. Reprod.* **2008**, *23*, 2402–2414. [CrossRef] [PubMed]
25. Goossens, E.; Geens, M.; De Block, G.; Tournaye, H. Spermatogonial survival in long-term human prepubertal xenografts. *Fertil. Steril.* **2008**, *90*, 2019–2022. [CrossRef] [PubMed]
26. Poels, J.; Abou-Ghannam, G.; Herman, S.; Van Langendonckt, A.; Wese, F.-X.; Wyns, C.; Abou-Ghannam, G.; Wese, F.-X. In Search of Better Spermatogonial Preservation by Supplementation of Cryopreserved Human Immature Testicular Tissue Xenografts with N-acetylcysteine and Testosterone. *Front. Surg.* **2014**, *1*, 47. [CrossRef]
27. Schlatt, S.; Westernströer, B.; Gassei, K.; Ehmcke, J. Donor-Host Involvement in Immature Rat Testis Xenografting into Nude Mouse Hosts1. *Biol. Reprod.* **2010**, *82*, 888–895. [CrossRef]
28. Van Eyck, A.-S.; Jordan, B.F.; Gallez, B.; Heilier, J.-F.; Van Langendonckt, A.; Donnez, J. Electron paramagnetic resonance as a tool to evaluate human ovarian tissue reoxygenation after xenografting. *Fertil. Steril.* **2009**, *92*, 374–381. [CrossRef]
29. Poels, J.; Abou-Ghannam, G.; Decamps, A.; Leyman, M.; Rieux, A.D.; Wyns, C. Transplantation of testicular tissue in alginate hydrogel loaded with VEGF nanoparticles improves spermatogonial recovery. *J. Control. Release* **2016**, *234*, 79–89. [CrossRef]
30. Ntemou, E.; Kadam, P.; Van Laere, S.; Van Saen, D.; Vicini, E.; Goossens, E. Effect of recombinant human vascular endothelial growth factor on testis tissue xenotransplants from prepubertal boys: A three-case study. *Reprod. Biomed. Online* **2019**, *39*, 119–133. [CrossRef]
31. Ferrara, N.; Davis-Smyth, T. The Biology of Vascular Endothelial Growth Factor. *Endocr. Rev.* **1997**, *18*, 4–25. [CrossRef]
32. Yancopoulos, G.D.; Davis, S.; Gale, N.W.; Rudge, J.S.; Wiegand, S.J.; Holash, J. Vascular-specific growth factors and blood vessel formation. *Nat. Cell Biol.* **2000**, *407*, 242–248. [CrossRef] [PubMed]
33. Schmidt, J.A.; De Avila, J.M.; McLean, D.J. Effect of Vascular Endothelial Growth Factor and Testis Tissue Culture on Spermatogenesis in Bovine Ectopic Testis Tissue Xenografts. *Biol. Reprod.* **2006**, *75*, 167–175. [CrossRef] [PubMed]
34. Hariawala, M.D.; Horowitz, J.R.; Esakof, D.; Sheriff, D.D.; Walter, D.H.; Keyt, B.; Isner, J.M.; Symes, J.F. VEGF Improves Myocardial Blood Flow but Produces EDRF-Mediated Hypotension in Porcine Hearts. *J. Surg. Res.* **1996**, *63*, 77–82. [CrossRef]
35. Folkman, J. Role of Angiogenesis in Tumor Growth and Metastasis. *Semin. Oncol.* **2002**, *29* (Suppl. 16), 15–18. [CrossRef]
36. Eiselt, P.; Kim, B.-S.; Chacko, B.; Isenberg, B.; Peters, M.; Greene, K.; Roland, W.; Loebsack, A.; Burg, K.; Culberson, C.; et al. Development of Technologies Aiding Large-Tissue Engineering. *Biotechnol. Prog.* **1998**, *14*, 134–140. [CrossRef] [PubMed]

37. Strobel, H.A.; Qendro, E.I.; Alsberg, E.; Rolle, M.W. Targeted Delivery of Bioactive Molecules for Vascular Intervention and Tissue Engineering. *Front. Pharmacol.* **2018**, *9*, 1329. [CrossRef]
38. Del Vento, F.; Vermeulen, M.; Ucakar, B.; Poels, J.; Rieux, A.D.; Wyns, C. Significant Benefits of Nanoparticles Containing a Necrosis Inhibitor on Mice Testicular Tissue Autografts Outcomes. *Int. J. Mol. Sci.* **2019**, *20*, 5833. [CrossRef]
39. Folkman, J.; Klagsbrun, M. Angiogenic factors. *Science* **1987**, *235*, 442–447. [CrossRef] [PubMed]
40. Lindahl, P.; Johansson, B.R.; Leveen, P.; Betsholtz, C. Pericyte Loss and Microaneurysm Formation in Pdgf-B-Deficient Mice. *Science* **1997**, *277*, 242–245. [CrossRef]
41. Benjamin, L.; Hemo, I.; Keshet, E. A plasticity window for blood vessel remodelling is defined by pericyte coverage of the preformed endothelial network and is regulated by PDGF-B and VEGF. *Development* **1998**, *125*, 1591–1598. [CrossRef] [PubMed]
42. Israely, T.; Nevo, N.; Harmelin, A.; Neeman, M.; Tsafriri, A. Reducing ischaemic damage in rodent ovarian xenografts transplanted into granulation tissue. *Hum. Reprod.* **2006**, *21*, 1368–1379. [CrossRef] [PubMed]
43. Nam, S.-Y.; Shin, B.-H.; Lee, M.; Lee, S.; Heo, C.Y. NecroX-5 ameliorates inflammation by skewing macrophages to the M2 phenotype. *Int. Immunopharmacol.* **2019**, *66*, 139–145. [CrossRef] [PubMed]
44. Wang, C.; Liu, Y.; He, D. Diverse effects of platelet-derived growth factor-BB on cell signaling pathways. *Cytokine* **2019**, *113*, 13–20. [CrossRef] [PubMed]
45. Soriano, P. Abnormal kidney development and hematological disorders in PDGF beta-receptor mutant mice. *Genes Dev.* **1994**, *8*, 1888–1896. [CrossRef]
46. Wick, G.; Grundtman, C.; Mayerl, C.; Wimpissinger, T.-F.; Feichtinger, J.; Zelger, B.; Sgonc, R.; Wolfram, D. The Immunology of Fibrosis. *Annu. Rev. Immunol.* **2013**, *31*, 107–135. [CrossRef]
47. Basciani, S.; Mariani, S.; Spera, G.; Gnessi, L. Role of Platelet-Derived Growth Factors in the Testis. *Endocr. Rev.* **2010**, *31*, 916–939. [CrossRef]
48. Ergün, S.; Kiliç, N.; Fiedler, W.; Mukhopadhyay, A. Vascular endothelial growth factor and its receptors in normal human testicular tissue. *Mol. Cell. Endocrinol.* **1997**, *131*, 9–20. [CrossRef]
49. Caires, K.C.; De Avila, J.M.; Cupp, A.S.; McLean, D.J. VEGFA Family Isoforms Regulate Spermatogonial Stem Cell Homeostasis in Vivo. *Endocrinology* **2012**, *153*, 887–900. [CrossRef]
50. Baltes-Breitwisch, M.M.; Artac, R.A.; Bott, R.C.; McFee, R.M.; Kerl, J.G.; Clopton, D.T.; Cupp, A.S. Neutralization of vascular endothelial growth factor antiangiogenic isoforms or administration of proangiogenic isoforms stimulates vascular development in the rat testis. *Reproduction* **2010**, *140*, 319–329. [CrossRef]
51. Caires, K.C.; De Avila, J.; McLean, D.J. Vascular endothelial growth factor regulates germ cell survival during establishment of spermatogenesis in the bovine testis. *Reproduction* **2009**, *138*, 667–677. [CrossRef]
52. Simon-Yarza, T.; Formiga, F.R.; Tamayo, E.; Pelacho, B.; Prosper, F.; Blanco-Prieto, M.J. Vascular Endothelial Growth Factor-Delivery Systems for Cardiac Repair: An Overview. *Theranostics* **2012**, *2*, 541–552. [CrossRef]
53. Zhang, Z.D.; Xu, Y.Q.; Chen, F.; Luo, J.F.; Liu, C.D. Sustained delivery of vascular endothelial growth factor using a dextran/poly(lactic-co-glycolic acid)-combined microsphere system for therapeutic neovascularization. *Hear. Vessel.* **2019**, *34*, 167–176. [CrossRef] [PubMed]
54. Zhang, Q.; Hubenak, J.; Iyyanki, T.; Alred, E.; Turza, K.C.; Davis, G.; Chang, E.I.; Branch-Brooks, C.D.; Beahm, E.K.; Butler, C.E. Engineering vascularized soft tissue flaps in an animal model using human adipose–derived stem cells and VEGF+PLGA/PEG microspheres on a collagen-chitosan scaffold with a flow-through vascular pedicle. *Biomaterials* **2015**, *73*, 198–213. [CrossRef] [PubMed]
55. Matsuki, K.; Sugaya, H.; Takahashi, N.; Kawasaki, T.; Yoshimura, H.; Kenmoku, T. Degradation of Cylindrical Poly-Lactic Co-Glycolide/Beta-Tricalcium Phosphate Biocomposite Anchors After Arthroscopic Bankart Repair: A Prospective Study. *Orthopedics* **2018**, *41*, e348–e353. [CrossRef] [PubMed]
56. Jiang, R.; Lin, C.; Jiang, C.; Huang, Z.; Gao, W.; Lin, D. Nobiletin enhances the survival of random pattern skin flaps: Involvement of enhancing angiogenesis and inhibiting oxidative stress. *Int. Immunopharmacol.* **2020**, *78*, 106010. [CrossRef] [PubMed]
57. Kong, H.S.; Lee, J.; Youm, H.W.; Kim, S.K.; Lee, J.R.; Suh, C.S.; Kim, S.H. Effect of Treatment with Angiopoietin-2 and Vascular Endothelial Growth Factor on the Quality of Xenografted Bovine Ovarian Tissue in Mice. *PLoS ONE* **2017**, *12*, e0184546. [CrossRef] [PubMed]
58. Rieux, A.D.; Ucakar, B.; Mupendwa, B.P.K.; Colau, D.; Feron, O.; Carmeliet, P.; Préat, V. 3D systems delivering VEGF to promote angiogenesis for tissue engineering. *J. Control. Release* **2011**, *150*, 272–278. [CrossRef] [PubMed]
59. Rieux, A.D.; De Berdt, P.; Ansorena, E.; Ucakar, B.; Damien, J.; Schakman, O.; Audouard, E.; Bouzin, C.; Auhl, D.; Simón-Yarza, T.; et al. Vascular endothelial growth factor-loaded injectable hydrogel enhances plasticity in the injured spinal cord. *J. Biomed. Mater. Res. Part A* **2013**, *102*, 2345–2355. [CrossRef]
60. Silva, E.A.; Mooney, D.J. Effects of VEGF temporal and spatial presentation on angiogenesis. *Biomaterials* **2010**, *31*, 1235–1241. [CrossRef]
61. Sun, Q.; Silva, E.A.; Wang, A.; Fritton, J.C.; Mooney, D.; Schaffler, M.B.; Grossman, P.M.; Rajagopalan, S. Sustained Release of Multiple Growth Factors from Injectable Polymeric System as a Novel Therapeutic Approach Towards Angiogenesis. *Pharm. Res.* **2009**, *27*, 264–271. [CrossRef] [PubMed]
62. Costoya, J.A.; Hobbs, R.M.; Barna, M.; Cattoretti, G.; Manova, K.; Sukhwani, M.; Orwig, K.E.; Wolgemuth, D.J.; Pandolfi, P.P. Essential role of Plzf in maintenance of spermatogonial stem cells. *Nat. Genet.* **2004**, *36*, 653–659. [CrossRef] [PubMed]

63. Bai, H.; Wang, Z.; Li, M.; Sun, P.; Wei, S.; Wang, Z.; Xing, Y.; Dardik, A. Adult Human Vein Grafts Retain Plasticity of Vessel Identity. *Ann. Vasc. Surg.* **2020**, *68*, 468–475. [CrossRef]
64. Onufer, E.J.; Aladegbami, B.; Imai, T.; Seiler, K.; Bajinting, A.; Courtney, C.; Sutton, S.; Bustos, A.; Yao, J.; Yeh, C.-H.; et al. EGFR in enterocytes & endothelium and HIF1α in enterocytes are dispensable for massive small bowel resection induced angiogenesis. *PLoS ONE* **2020**, *15*, e0236964.

International Journal of
Molecular Sciences

MDPI

Article

Synthesis and Evaluation of AlgNa-g-Poly(QCL-co-HEMA) Hydrogels as Platform for Chondrocyte Proliferation and Controlled Release of Betamethasone

Jomarien García-Couce [1,2], Marioly Vernhes [3], Nancy Bada [1], Lissette Agüero [1], Oscar Valdés [4], José Alvarez-Barreto [5], Gastón Fuentes [1,2,*], Amisel Almirall [1,2] and Luis J. Cruz [2]

1. Centro de Biomateriales, Universidad de La Habana, La Habana 10400, Cuba; jgcouce@gmail.com (J.G.-C.); bada@biomat.uh.cu (N.B.); lissetteaguerol@gmail.com (L.A.); amisel.almirall@gmail.com (A.A.)
2. TNI Group, Department of Radiology, Leiden University Medical Center (LUMC), 2333 ZA Leiden, The Netherlands; L.J.Cruz_Ricondo@lumc.nl
3. Laboratorio de Biología Molecular, Departamento de Radiobiología, Centro de Aplicaciones Tecnológicas y Desarrollo Nuclear (CEADEN), La Habana 11300, Cuba; mariolys@ceaden.edu.cu
4. Centro de Investigación de Estudios Avanzados del Maule (CIEAM), Vicerrectoría de Investigación y Posgrado, Universidad Católica del Maule, Talca 3460000, Chile; ovaldes@ucm.cl
5. Department of Chemical Engineering, Universidad de San Francisco de Quito, Quito 170901, Ecuador; jalvarezb@usfq.edu.ec

* Correspondence: gastonfe@biomat.uh.cu or gastonfe@gmail.com

Citation: García-Couce, J.; Vernhes, M.; Bada, N.; Agüero, L.; Valdés, O.; Alvarez-Barreto, J.; Fuentes, G.; Almirall, A.; Cruz, L.J. Synthesis and Evaluation of AlgNa-g-Poly(QCL-co-HEMA) Hydrogels as Platform for Chondrocyte Proliferation and Controlled Release of Betamethasone. *Int. J. Mol. Sci.* **2021**, 22, 5730. https://doi.org/10.3390/ijms22115730

Academic Editor: Peter John Jervis

Received: 31 March 2021
Accepted: 11 May 2021
Published: 27 May 2021

Abstract: Hydrogels obtained from combining different polymers are an interesting strategy for developing controlled release system platforms and tissue engineering scaffolds. In this study, the applicability of sodium alginate-g-(QCL-co-HEMA) hydrogels for these biomedical applications was evaluated. Hydrogels were synthesized by free-radical polymerization using a different concentration of the components. The hydrogels were characterized by Fourier transform-infrared spectroscopy, scanning electron microscopy, and a swelling degree. Betamethasone release as well as the in vitro cytocompatibility with chondrocytes and fibroblast cells were also evaluated. Scanning electron microscopy confirmed the porous surface morphology of the hydrogels in all cases. The swelling percent was determined at a different pH and was observed to be pH-sensitive. The controlled release behavior of betamethasone from the matrices was investigated in PBS media (pH = 7.4) and the drug was released in a controlled manner for up to 8 h. Human chondrocytes and fibroblasts were cultured on the hydrogels. The MTS assay showed that almost all hydrogels are cytocompatibles and an increase of proliferation in both cell types after one week of incubation was observed by the Live/Dead® assay. These results demonstrate that these hydrogels are attractive materials for pharmaceutical and biomedical applications due to their characteristics, their release kinetics, and biocompatibility.

Keywords: hydrogels; sodium alginate; betamethasone; drug delivery; cartilage tissue engineering

1. Introduction

Articular cartilage is a highly specific, avascular, connective tissue that lines the end of each bone and forms the joint, providing a smooth and gliding surface. It is predominantly formed of water, proteoglycans, and type II collagen. Chondrocytes are the only functional cells that compose articular cartilage [1,2]. Osteoarthritis (OA) is the most common condition that impacts cartilage. This is a disease characterized by the loss of the matrix and its functionality, producing inflammation and severe pain, which are symptoms usually treated with anti-inflammatory drugs. Corticosteroids (e.g., dexamethasone, triamcinolone, and betamethasone) are frequently used to treat OA due to their potent anti-inflammatory activity. It has been proven that a low dose exerts effects on down-regulating the gene expression level of MMPs (MMP-1, MMP-3, and MMP-13) in chondrocytes, reducing glycosaminoglycans loss [1]. However, their prolonged administration causes many side effects. In more advanced stages of the disease and due to the limited capacity for self-repair, replacement or induced regeneration of the damaged area is often required [3]. To date, different treatment methods have been designed, such as mosaicplasty, osteochondral allograft transplantation, and autologous chondrocyte implantation, among others. However, there are still limitations to producing cartilage with full biological activity [2,4]. Therefore, tissue engineering continues to be an interesting and promising alternative to repair cartilage defects in a functional way and to apply more localized treatments to reduce side effects [5,6].

Hydrogels are three-dimensional polymeric structures, capable of retaining large amounts of water or fluids without dissolving or losing their integrity. Due to these characteristics, they have a high similarity to living tissues, and they are suitable platforms for tissue engineering, drug-carrying matrices, artificial articular cartilage, and smart devices to external stimuli, such as pH and temperature [7]. To treat joint cartilage damage, hydrogels are created to be resident substitutes when replacing injured cartilage or to be used as cell-laden or drug-laden components that promote or stimulate tissue regeneration [8]. The use of natural polymers, such as polysaccharides, for developing these matrices has been frequently studied in past decades. Chitosan, hyaluronic acid, and sodium alginate are among the most widely used [3]. Sodium alginate (AlgNa) is a linear anionic polysaccharide, composed of β-D mannuronic acid and α-L-guluronic acid units, which are obtained mainly from brown marine algae [9,10]. Due to its properties (biocompatible, biodegradable, and non-toxic), several materials have been developed from alginate, such as microparticles, hydrogels, and films, for applications in the biomedical field [11]. Specifically, sodium alginate hydrogels have a similar structure to natural extracellular matrix, and, for that reason, their use in tissue engineering and drug delivery has grown exponentially [11,12], which previous studies carried out by different authors have demonstrated. Yamaoka et al. showed that the alginate matrices studied help chondrocytes reduce cell-to-cell contacts and maintain cell shape and function. These properties can improve the expression and accumulation of cartilaginous matrices, such as type II collagen and GAG [13]. In another work, Baghaban Eslaminejad et al. [14] observed the presence of abundant microvilli developed on the surfaces of chondrocytes seeded in alginate matrices, which is indicative of cellular participation in the secretion of the matrix by scanning electron microscopy. This morphology, in turn, would be the result of the active interaction of chondrocytes with alginate molecules. In another study, the incorporation of alginate in poly(2-hydroxyethyl methacrylate) hydrogels increased the amount of glycosaminoglycans and the proliferation of cultured chondrocytes [7]. However, hydrogels obtained only from natural polymers are mechanically weak due to the large amount of water they absorb. As a solution to this limitation, synthetic polymers are usually added to improve the physicochemical properties of hydrogels. Among the synthetic polymers, poly(2-hydroxyethyl methacrylate) (HEMA), which is a neutral hydrophilic polymer, has been one of the most used in tissue engineering due to its versatile properties, such as biocompatibility, good mechanical properties, and ease of synthesis [4,5,15]. Hydrogels for cartilage tissue engineering from the combination of polymers, natural (e.g., chitosan, collagen), and synthetic (polyethylene glycol and polyvinyl alcohol) have been reported by several authors [4,11,16]. 2-acryloyloxyethyl trimethyl ammonium chloride (QCL) is a positively charged monomer composed of a polymerizable double bond and a quaternary ammonium at the end of its structure. According to previous reports, compounds with quaternary ammonium groups in their structure have been used as vectors to target therapeutic agents (anti-inflammatory and osteoarthritis modifying drugs) toward the cartilage, and to interact with sulfate and carboxyl groups present in the proteoglycans of ECM [17–20]. It is known that QCL's homopolymers and copolymers are highly charged structures, capable of interacting through electrostatic attractions with other oppositely charged compounds [19–22]. The inclusion of hydrogels designed for cartilage tissue engineering provides an additional benefit because they improve the tissue-material interaction.

In this work, AlgNa-g-poly(QCL-co-HEMA) hydrogels are prepared by radical copolymerization of HEMA and QCL in the presence of AlgNa and *N,N*-methylene diacrylamide (MBA) as a crosslinking agent. Although there are some similar reports on systems based on one or more of these components [10,23–26], this specific composition has not been reported or studied as a drug carrier and platform for chondrocytes proliferation. The structure and morphology are analyzed by Fourier transform-infrared spectroscopy (FTIR) and scanning electron microscopy (SEM). The swelling capacity is measured as a function of time and pH. The betamethasone (BTM) release from the prepared matrices in phosphate buffered saline (PBS) is evaluated. Furthermore, the cytocompatibility in vitro with C-28 and 3T3 cells using MTS and Live/Dead® assays is evaluated. All analyses are carried out based on the composition of the hydrogels.

2. Results and Discussion

2.1. AlgNa-g-Poly(QCL-co-HEMA) Hydrogels

The hydrogels were obtained by a mechanism of copolymerization and simultaneous chemical crosslinking, as shown in Figure 1. Initially, the thermal decomposition of the potassium persulfate (KPS, initiator) occurs, generating the anionic sulfate radicals. Next, the radicals interact with the -OH group of AlgNa, extracting the hydrogen to form the corresponding macroradicals. These macroradicals act as active centers through which the grafting of QCL and HEMA monomer molecules in the main chains of AlgNa begins. Chains propagation and a simultaneous cross-linking process with MBA occur, forming the three-dimensional graft copolymer AlgNa-g-poly(QCL-co-HEMA). Mechanisms similar to the one proposed were reported in previous works and in other systems [24–27].

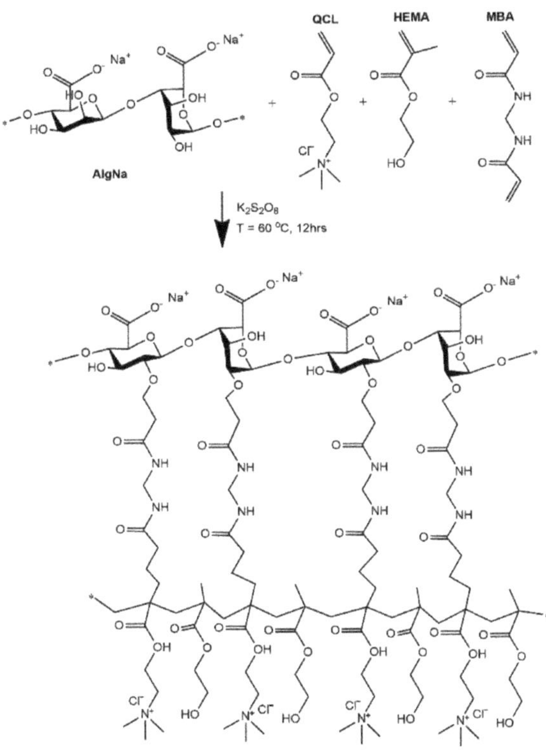

Figure 1. Schematic representation of chemical structures of polymer, monomers, and hypothetically cross-linked structures of AlgNa-g-poly(QCL-co-HEMA) hydrogels.

2.2. Hydrogel Characterizations

2.2.1. FTIR Spectroscopy Analysis

FT-IR spectroscopy was used to confirm the structure of the prepared hydrogels, the spectrums of the initial pure components, and some samples of hydrogels, which are shown in Figure 2. Figure 2a shows the spectrum of AlgNa. The broad band at 3258 cm^{-1} can be attributed to an -OH stretching vibration. Other characteristic signals observed at 1598 and 1403 cm^{-1} are attributed to asymmetric and symmetric stretching vibration bands of the -COO^{-} group [26]. Signals at 1080, 1025, and 813 cm^{-1} correspond to stretching vibrations of asymmetric and symmetric -C-O-C and Na-O, respectively [28]. Both monomers employed have C = O and C = C groups in their structure whose characteristic peaks are observed at 1718 and 1636 cm^{-1} in the QCL spectra (Figure 2b) [29,30] and at 1711 and 1636 cm^{-1} in the case of HEMA (Figure 2c) [31,32]. In the QCL spectrum, other characteristic peaks are also observed at 1478 and 1185 cm^{-1}, which are attributed to the bending band of quaternary ammonium groups (-$N^+(CH_3)_3$) and the C-O of ester, respectively [29,33]. At 950 cm^{-1}, a singular band related to the C–N stretching of quaternary ammonium groups appears [34].

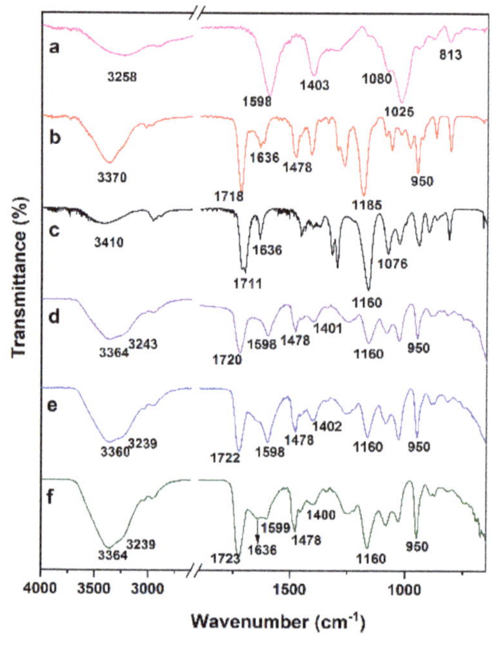

Figure 2. FTIR spectra of AlgNa (**a**), QCL (**b**), HEMA (**c**), hydrogel H2 (**d**), hydrogel H4 (**e**), and hydrogel H6 (**f**).

In addition to the C=O and C=C characteristic bands, HEMA shows intense signals at 940 and 813 cm^{-1} associated with the -C=C- group [35], and, at 1160 and 1076 cm^{-1}, corresponding to the C-O-C stretching vibrations of the ester group [25,31]. In the hydrogel spectrums (Figure 2d–f), all absorption bands mentioned above are present while some of them overlap. The signals between 1720 and 1724 cm^{-1} are due to the C=O stretching frequency, while those between 1070 and 1185 cm^{-1} are caused by the stretching frequencies of C-O-C bonds of ester groups, demonstrating the presence of the QCL and HEMA units in the gel network. Additionally, the intensity of bands at 1478 and 950 cm^{-1}, ascribed to QCL, increases between the spectrum 2d and 2f due to the growth of QCL in the hydrogel's composition. Signals at 1598 cm^{-1} are due to the C=O of the carboxylate group of alginates, whereas the bands at 1025 cm^{-1} are due to the symmetric C-O-C stretching frequencies. In addition, the asymmetric stretching vibration of the COO^- group of alginate at 1402 cm^{-1} remains nearly the same in the hydrogels, indicating a lack of chemical interactions occurs during the hydrogel's obtention process [24]. On the other hand, the bands at 1636, 940, and 813 cm^{-1} associated with the C=C stretching vibration were not found in the spectrum of the samples H2 and H4 (Figure 2d,e), while the spectrum 2f shows a small band at 1636 cm^{-1}. This result was only observed for sample H6, which contains the highest percentage of QCL. This is a charged monomer, and also has three methyl factors linked at the end of the chain, forming a bulky group. Both conditions can cause the approach between the monomer units to be limited, which implies that their inclusion in the growing polymer chains is reduced and, therefore, some unreacted monomers remained, causing the corresponding signal of methylene.

2.2.2. Morphology Characterization

Figure 3 (H1 to H6) show the SEM micrographs of the internal morphology of AlgNa-g-poly (QCL-co-HEMA) hydrogels. It is observed that all the samples have a porous structure, and, although the pore size varies, it does not show a marked difference with respect to variation of the sample's composition. In the SEM images, it is observed that hydrogels have interconnected pores through which the water molecules or fluids can easily spread and, therefore, act on the swelling degree and rate of the material, as well as constitute as sites of external stimuli interactions, such as pH or temperature. This porous structure is appropriate to allow attachment, growth, and proliferation of cells inside the matrix.

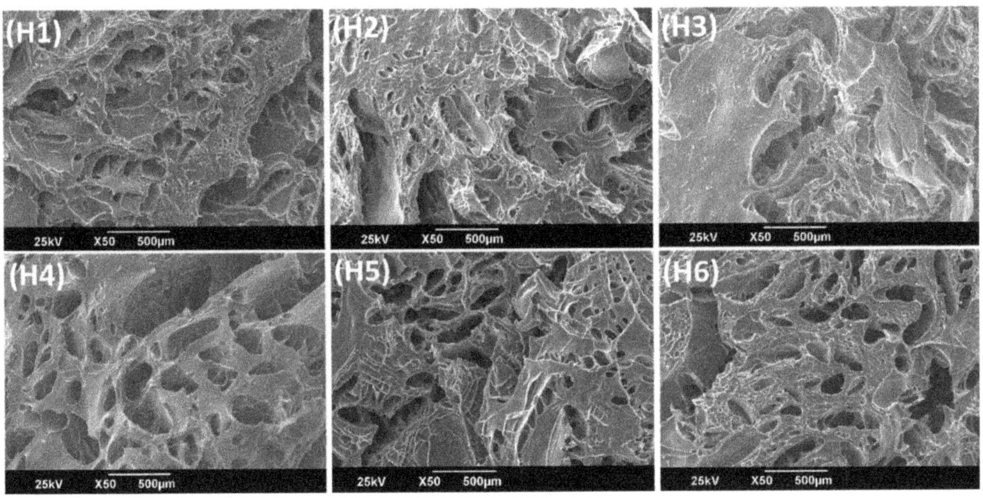

Figure 3. Scanning electron microscopy images of AlgNa-g-poly(QCL-co-HEMA) hydrogels.

2.2.3. Swelling Studies

The swelling process of hydrogels is conditioned by different factors, of which the matrix composition and pH of the medium have been extensively studied. The matrices developed in this research consist of three components of which each one contributes a functional group with hydrophilic characteristics (-COOH, $-N^+(CH_3)_3$, $-OH^-$). Now, these groups can, in turn, undergo interactions with each other depending on the pH of the medium. Futhermore, the swelling process in the hydrogels under study will be strongly influenced by their composition and the pH of the medium. Figure 4A,B show the swelling profiles of the hydrogels at pH 2.0 and 7.4, respectively, and Figure 4C shows the maximum swelling at equilibrium in more detail.

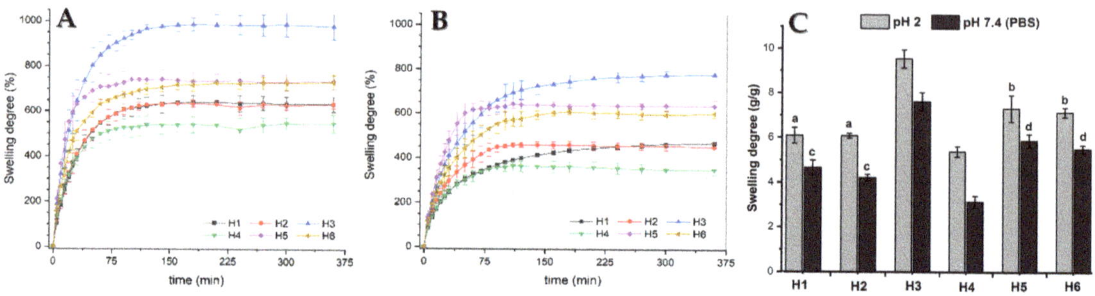

Figure 4. Swelling studies. (**A**) and (**B**) hydrogels' profiles at pH 2.0 and 7.4, respectively. (**C**) Maximum swelling at equilibrium in more detail. There are no significant differences between two compared samples at same pH, for a and b ($p \geq 0.05$), or for c and d ($p \geq 0.01$).

The results show that there is a significant difference in the swelling of the hydrogel samples at different pH values. It can be seen that the swelling degree is higher at pH 2.0 than at pH 7.4 in all compositions, which is contrary to what is described in the literature for matrices containing AlgNa in their structure [36,37]. This result is due to the fact that AlgNa has a pKa between 3.2–4.0, so, at pH 7.4, the carboxylic groups of its units are in a carboxylates form (-COO^-) and can be linked through electrostatic interactions with the $N^+(CH_3)_3$ groups of QCL (Figure 5), causing the chains to shrink and the matrix to become more compact, reducing the fluid rate entry. At pH 2.0, the carboxylate ions are protonated, and the carboxyl group is in acid form. Hence, electrostatic interactions do not occur. In addition, it can be observed that, at pH 7.4, the swelling degree of the hydrogel changes depending on the composition. Samples containing 20% AlgNa in their composition have less swelling when compared with samples that have 10% AlgNa. These results reaffirm that electrostatic interactions caused by the AlgNa structure have a major role in the swelling process.

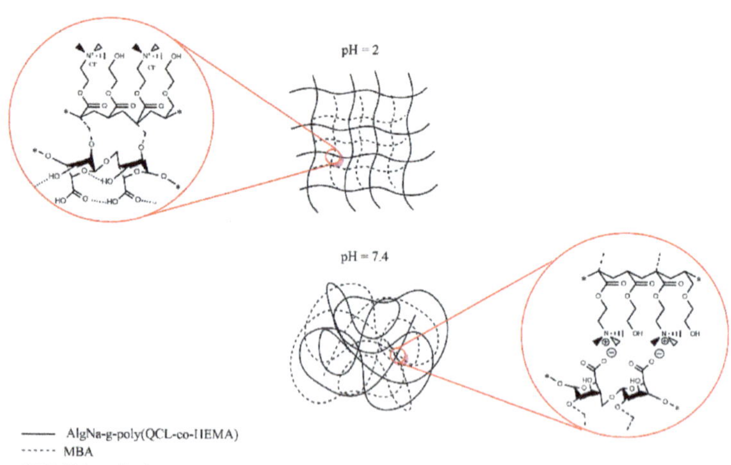

Figure 5. Schematic representation of the interactions between functional groups in the polymeric network chains at pH 2.0 and pH 7.4.

It is also valid to discuss the fact that, at pH 2.0, although electrostatic interactions do not occur, hydrogels show a similar swelling behavior as compared to the one observed at pH 7.4 regarding the composition of the materials. In this case, hydrogen bond formation among carboxylic groups with each other and with OH^- groups of the sugar moiety of the AlgNa structure are the predominant interactions produced, which act as a barrier that strengthens the polymeric network and hinders the entrance of water molecules inside the hydrogel network [36,38]. According to this, the samples containing 20% of AlgNa in its structure (H1, H2, and H4) present a greater number of groups capable of forming hydrogen bonding as compared to the samples with 10% of AlgNa in its composition. In the samples with 20% AlgNa, the chains interact with each other more. These chains are closer, and, therefore, the matrix becomes more compact, delaying the entry of fluid into the material and reducing the swelling, as can observed in Figure 4A.

2.2.4. Kinetic Swelling Study

To have a more detailed idea of the build of the swelling process matrix, the experimental data were analyzed using Equation (2). The results are shown in Table 1. It can be seen that the value of n for all samples, both in swelling at pH 2 and PBS, is between 0.5 and 1. Therefore, it can be deduced that the swelling occurs by an anomalous process according to the model proposed. That is, in the swelling process, the diffusion phenomenon is simultaneously accompanied by the viscoelastic relaxation of the polymer chains (both processes have similar rates). In a previous work published by Goel et al. [15], which evaluated the influence of the inclusion of a cationic monomer on the swelling process of hydrogels of 2-hydroxyethylmethacrylate-co-[2-(methacryloyloxy)ethyl] trimethyl ammonium chloride, values of n greater than 0.5 were reported with the addition of small quantities of the cationic monomer, and, when increased, n values also increased, reaching values close to 1. According to the research conclusions, this occurs because the charged segments present in the hydrogel will repel each other to open the matrix and cause a more rapid diffusion of the water into the interior. In our work, this approach is not totally valid because the matrix contains AlgNa, which is an anionic polysaccharide whose -COO^- groups, as mentioned above, will be linked by electrostatic interactions with the $N^+(CH3)_3$ groups of the QCL, causing a contraction of the polymeric network that reduces the rate of diffusion of the surrounding fluid into the matrix. Therefore, a relationship between the concentration of QCL and the value of n will not be as proportional as can be seen in the results obtained by Goel et al.

Table 1. Swelling kinetic constant (k), transport exponent (n), and coefficient of determination (R^2), according to Equation (2).

Samples	pH = 2.0			PBS, pH = 7.4		
	k	n	R^2	k	n	R^2
H1	38 ± 3	0.55 ± 0.02	99.67	33 ± 2	0.74 ± 0.02	99.90
H2	40 ± 4	0.60 ± 0.03	99.56	41 ± 3	0.69 ± 0.03	99.79
H3	52 ± 4	0.60 ± 0.03	99.55	65 ± 3	0.69 ± 0.02	99.93
H4	35 ± 1	0.59 ± 0.01	99.97	47 ± 2	0.64 ± 0.01	99.95
H5	45 ± 4	0.71 ± 0.03	99.80	68 ± 6	0.72 ± 0.03	99.84
H6	49 ± 3	0.59 ± 0.02	99.72	53 ± 2	0.69 ± 0.02	99.95

2.3. *Betamethasone Release Study*

2.3.1. FTIR Characterization of BTM Loaded Hydrogels

To identify the BTM in the hydrogels once charged, as well as to identify possible interactions between the matrix and the drug, a study was carried out using FTIR spectroscopy of the empty and loaded hydrogels. Figure 6 shows the FTIR spectrums of pure BTM, and three samples of AlgNa-g-poly(QCL-co-HEMA) hydrogels that are empty and loaded with BTM. In the FTIR spectra of pure BTM (Figure 6g), the characteristic absorption bands of this compound can be observed.

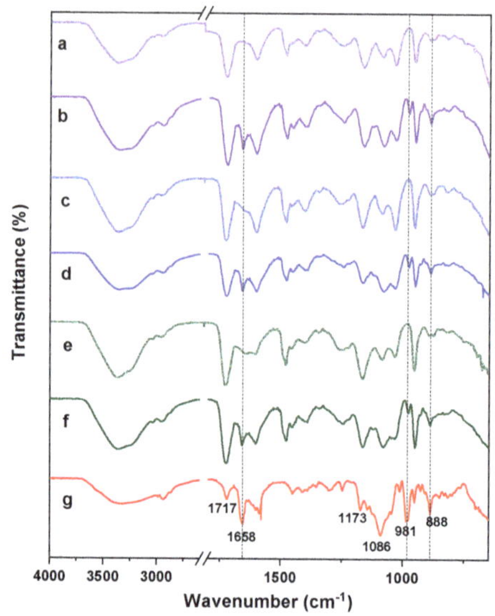

Figure 6. FTIR spectra of BTM (**g**), hydrogel H2 empty and loaded (**a**,**b**), hydrogel H4 empty and loaded (**c**,**d**), and hydrogel H6 empty and loaded (**e**,**f**).

The signals observed at 1717 and 1658 cm^{-1} correspond to the C=O stretching vibration in the carbonyl group of the chain and the ketonic group of ring # 1 in the cortisol structure [39,40]. Additionally, bands at 1173 and 1086 cm^{-1} are attributed to asymmetric and symmetric stretching vibration of -C-O. The signals seen at 981 and 888 cm^{-1} can be attributed to P–O stretching vibration in phosphate moiety and the C–C skeleton vibration in the cyclopentane ring, respectively [40,41].

In the spectra of the loaded hydrogels, the characteristic bands of the BTM present in the spectrum of pure BTM are observed. As can be seen, there are no significant changes in the peaks, which signals that there is no demonstrable chemical interaction among the hydrogel networks and the groups present in the drug molecule. The drug is then trapped in the matrix only by physical interactions.

2.3.2. Drug Release Study

One of the important parameters when designing a hydrogel-based, controlled delivery system for cartilage repair is its loading ability for the bioactive substance. The incorporation of the drug into the matrix can be performed by a swelling-diffusion process [42], as in our case. Drug retention can occur by physical or chemical immobilization in the polymeric network or through electrostatic interactions [43]. The drug load and loading efficiency of the studied hydrogels are presented in Table 2.

In previous works in which drug encapsulation studies have been carried out, the authors report that the swelling of the material [44,45], the concentration of the drug loading solution [46], and the interactions between the drug and the matrix [47] are the predominant factors on the loading of the drug into the matrix. The results observed in Table 2 show that the drug load in this work does not have a proportional relationship with the swelling degree, according to the results previously analysed in Section 2.2.3. However, it is observed that, when the concentration of HEMA increases in the formulation, the drug load also increases. In addition, for the same percentage of HEMA, the samples with 20% AlgNa have a higher drug load except for 10% of HEMA. Considering the described behavior and the molecular structure of BTM in which there are acceptor sites capable of forming hydrogen bonds (H-bond acceptor), such as the oxygens of the phosphate group located at the end of the molecule [48,49], it can be deduced that an interaction between the drug molecules and OH groups present in HEMA and AlgNa is taking place.

Table 2. Relation between hydrogels' composition and drug load, loading efficiency, and BTM released.

Samples	Composition (%)			Drug Load (µg of Drug/mg of Hydrogel)	Loading Efficiency (%)	BTM Released (%)
	HEMA	QCL	AlgNa			
H6	10	80	10	58.0	31.8	100 *
H4	10	70	20	57.8	27.9	100 **
H5	20	70	10	74.1	52.7	99.7
H2	20	60	20	99.8	37.6	95.5
H3	30	60	10	113.7	58.8	91.4
H1	30	50	20	137.4	59.6	83.2

* BTM released before 4 h. ** BTM released before 5 h.

After immersion of the BTM-loaded hydrogels in a PBS solution, the amount of BTM released gradually increased over time, as can be seen in Figure 7. A burst was not observed in any of the samples, which suggests that the drug is homogeneously distributed in the matrix and is not collected only on the material surface. The release study was carried out for 8 h and it was observed that the samples H6 and H4 released all the encapsulated content before the fifth hour of study. The sample H5 release reached almost 100% during the studied time, while the rest of the samples (H2, H3, and H1) released 95%, 91%, and 83%, respectively.

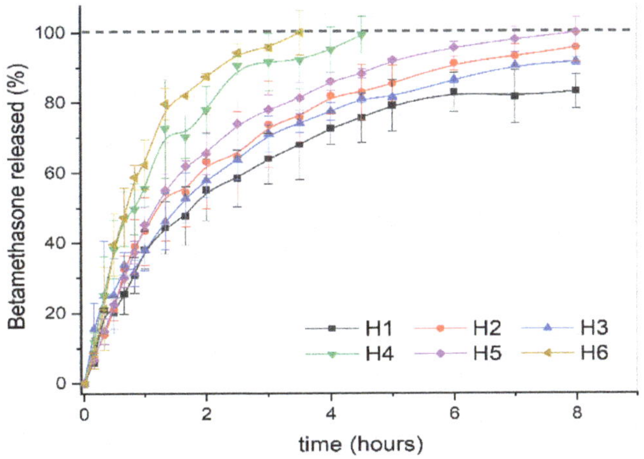

Figure 7. Cumulative in vitro release study of Betamethasone from AlgNa-g-poly(QCL-co-HEMA) hydrogels in PBS, pH = 7.4.

The release processes are influenced by different factors, such as solubility of the encapsulated drug, crosslinking, and material composition, external factors such as pH or temperature, and possible interactions between the encapsulated drug and components of the polymeric matrix. According to the observed results, it can be deduced that, in our case, the swelling degree is not the fundamental variable that affects the drug release process since there is no direct relationship between the swelling degree previously determined and the release rate obtained for each sample. On the other hand, taking into account that the encapsulated BTM is a phosphate sodium salt and it is totally soluble in PBS, we can also rule out this factor as a variable determining the behavior of the release process. Then, evaluating the composition of the material, we can observe (as summarized in Table 2) that, when the concentration of HEMA in the matrix increases, the release of BTM decreases. For the same concentration of HEMA, when the concentration of AlgNa increases, the drug release decreases as well. This result is in accordance with the previous observation in the hydrogel loading process. In this case, the interaction between AlgNa and HEMA with the drug via H bonding causes the release of the drug to be delayed from the matrices that have a higher content of HEMA and AlgNa.

2.3.3. Kinetic Drug Release Study

In order to know the kinetic mechanism of encapsulated BTM release, the in vitro experimental data were adjusted to the equation of the Korsmeyer-Peppas kinetic model, which is very useful when more than one mechanism is involved in the drug release or when accurate mechanisms are unknown [50]. As can be seen in Table 3, all the calculated n values are above 0.5 and the correlation coefficient with this model was high ($R^2 > 0.96$). For samples H1 to H5, the value of n is between 0.5–1, indicating that the BTM released follows a non-Fickian diffusion mechanism, in which the drug released by diffusion and relaxation of the polymer chains occurs simultaneously. However, in the sample H6, the value of n is slightly higher than 1, which implies that the release is controlled by a case II diffusion, where the diffusion of the drug toward the outside of the matrix occurs very quickly when compared to the chain relaxation processes, which explains the accelerated release suffered by the drug in this matrix.

Table 3. Drug release rate constant (k), diffusion exponent (n), and coefficient of determination (R^2), according to the Korsmeyer-Peppas model.

Samples	Korsmeyer-Peppas		
	k	n	R^2
H1	2.7 ± 0.6	0.63 ± 0.05	97.53%
H2	1.8 ± 0.6	0.76 ± 0.08	96.24%
H3	4.4 ± 0.7	0.54 ± 0.04	98.28%
H4	2.9 ± 0.9	0.73 ± 0.08	97.82%
H5	1.0 ± 0.1	0.93 ± 0.04	99.55%
H6	1.0 ± 0.4	1.0 ± 0.1	98.57%

2.4. Cytocompatibility Studies

For the practical application of hydrogels in different biomedical fields, the developed material does not have inherent cytotoxicity. In vitro cytotoxicity standard ISO 10993-5 states that "reduction of cell viability by more than 30% is considered a cytotoxic effect". In this study, the cytotoxicity of the blank hydrogels was investigated in C-28 and 3T3 cells and the results of the MTS assay are presented in Figure 8A,B respectively.

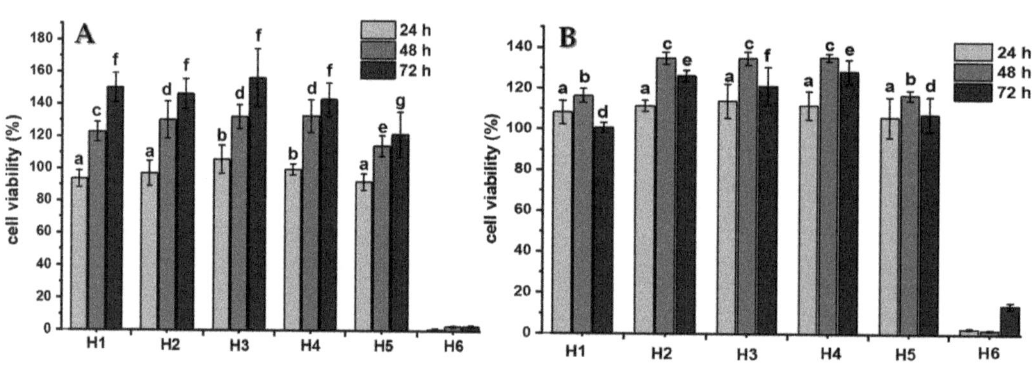

Figure 8. Cytotoxicity evaluation of the hydrogel's extracts exposed to C28 (**A**) and 3T3 (**B**) cells using the MTS assay method. There are no significant differences among several compared samples at the same time interval, for any letters, in each graph ($p \geq 0.05$).

The results show that the viability of the hydrogel extract-treated cells was higher than 90% after 24 h and even higher than 100% after 48 and 72 h of study in samples H1 to H5. The cell viability values obtained indicate that these materials are non-cytotoxic and also favor proliferation against the cells studied. Conversely, sample H6 shows high cytotoxicity for both cells evaluated. The number of viable cells are less than 20% in both kinds of cell cultures when they interact with the hydrogel extract in all evaluated times. This could be conditioned by the previous result obtained in the FTIR analysis, where the signal corresponding to the vinylic group is observed, indicating that an unreacted monomer is present in the hydrogel and known monomers are molecules with a high toxicity.

After quantitative analysis results were obtained by the MTS assay, a qualitative analysis was carried out via the Live/Dead® assay to visualize the distribution of living and dead cells after 7 days of being seeded in different samples of AlgNa-g-poly(QCL-co-HEMA) hydrogels. Figure 9A,B show the microphotographs of cells seeded on each hydrogel after the staining. In the microphotographs obtained for the samples from H1 to H5 (Figure 9A(a–e),B(a–e)), most cells observed are alive. Additionally, cells are spread on the hydrogel's surfaces and elongated morphologies are also observed, which is a sign that cells are healthy. In some images, it is possible to observe grouped cells forming colonies, which also shows that the environment where they are seeded is favorable for their proliferation. On the contrary, in sample H6, the number of dead cells is higher than the number of living cells, which is in agreement with the result obtained by the MTS assay discussed above.

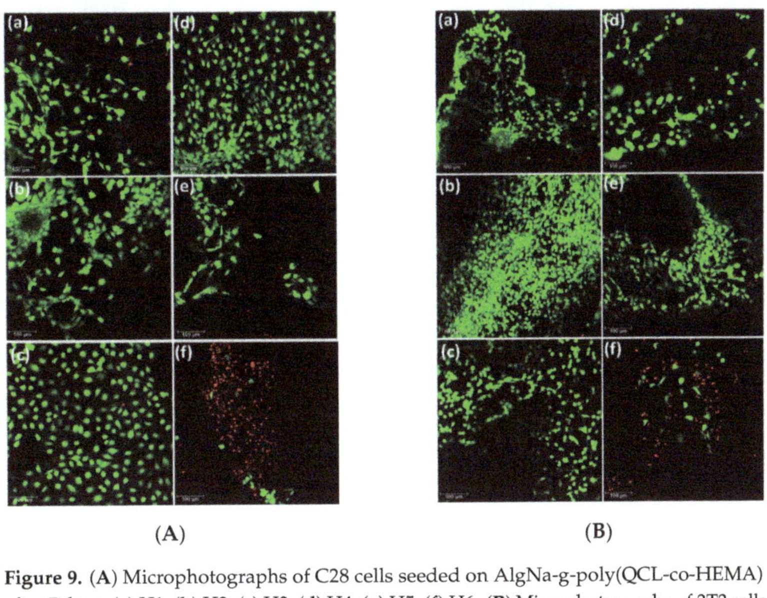

Figure 9. (**A**) Microphotographs of C28 cells seeded on AlgNa-g-poly(QCL-co-HEMA) hydrogels after 7 days, (**a**) H1, (**b**) H2, (**c**) H3, (**d**) H4, (**e**) H5, (**f**) H6. (**B**) Microphotographs of 3T3 cells seeded on AlgNa-g-poly(QCL-co-HEMA) hydrogels after 7 days, (**a**) H1, (**b**) H2, (**c**) H3, (**d**) H4, (**e**) H5, (**f**) H6.

According to the results obtained, it can be concluded that the AlgNa-g-poly(QCL-co-HEMA) hydrogels, of samples from H1 to H5, are biocompatible against C28 and 3T3 cells, demonstrating their potential as a scaffold for cartilage tissue engineering and a drug delivery system.

3. Materials and Methods

3.1. Materials

Commercial 2-hydroxyethylmethacrylate (HEMA, Sigma-Aldrich, Amsterdam, The Netherlands), 2-acryloxyethyl-trimethylammonium chloride (QCL, Sigma-Aldrich, Amsterdam, The Netherlands), potassium persulfate ($K_2S_2O_8$, Honeywell Fluka, Fisher Scientific, Madrid, Spain), and N, N methylenebisacrylamide (MBA, Merck BV, Amsterdam, The Netherlands) were used as purchased. Sodium alginate from *Macrocystis periferia* (viscosity-average molecular weight of 3.83×10^4 g/mol) was purchased from Sigma-Aldrich Chemie GmbH, Taufkirchen, Germany. Dulbecco's phosphate buffered saline (PBS), Dulbecco's Modified Eagles Medium (DMEM, high glucose, with Glutamax^{TM}), fetal bovine serum (FBS), penicillin, and streptomycin were purchased from Life Technologies (Breda, The Netherlands). 3-(4,5-dimethylthiazol-2-yl)-5-(3-carboxymethoxyphenyl)-2-(4-sulfophenyl)-2H-tetrazolium (MTS, Promega Benelux BV, Leiden, The Netherlands), Calcein-AM/ethidium homodimer-1 Live/Dead® assay kit (Invitrogen, Carlsbad, CA, USA) were used for cell studies.

3.2. *Preparation of AlgNa-g-Poly(QCL-co-HEMA) Hydrogels*

AlgNa-g-poly(QCL-co-HEMA) hydrogels were prepared by graft copolymerization of QCL and HEMA onto the sodium alginate (AlgNa) chain in the presence of potassium persulfate (1% wt) as a thermal initiator, and N, N methylenebisacrylamide (MBA, 2% wt) as a crosslinking agent. The composition of the hydrogels is listed in Table 4. The final volume of the mixture used was 10 mL. The calculated amount of AlgNa was dissolved in 5 mL of distilled water. KPS and MBA were added under constant stirring. Then, the monomer quantity was determined for each formulation. The final solution was placed in a glass tube, degassed for 30 min in an N_2 atmosphere to eliminate the dissolved oxygen in the system, and sealed under vacuum. The glass tubes were placed in a thermostatic water bath at 60 °C for 12 h. The hydrogels obtained were cut into discs for later studies.

Table 4. Composition in a reaction mix of different hydrogels (total mass 2 g).

Sample	QCL (% wt)	HEMA (% wt)	AlgNa (% wt)
H1	50	30	20
H2	60	20	20
H3	60	30	10
H4	70	10	20
H5	70	20	10
H6	80	10	10

3.3. *Hydrogel Characterizations*

3.3.1. Instrumental Analysis

Dried hydrogel samples were characterized by Fourier transform-infrared (FT-IR) spectroscopy and scanning electron microscopy (SEM). The FT-IR spectra were obtained on a Shimadzu IRSpirit-T spectrophotometer (Shimadzu Co., Kyoto, Japan), equipped with an attenuated total reflectance accessory (ATR). The absorption spectra were recorded in the spectral range of 4000–600 cm^{-1} with a resolution of 4 cm^{-1} for 32 scans. Scanning electron microscopic observations were performed to visualize pores and morphology of the hydrogels using a SEM-JEOL JSM-6360LV microscope (Jeol Co., Tokyo, Japan). Small pieces of hydrogels were sectioned and freeze-dried to observe the internal structure. The samples were placed on an aluminum mount, sputtered with gold palladium, and then scanned at an accelerating voltage of 25 kV.

3.3.2. Swelling Studies

The swelling kinetics of the terpolymeric hydrogels were studied in different media (HCl, pH 2.0 and PBS, pH 7.4). The samples previously weighed were immersed in 20 mL of each swelling medium at 37 °C in a thermostatic bath for 24 h to reach the swelling equilibrium. At established time intervals, the hydrogels were carefully taken out from the solution. The excess of water on the surface was removed with a filter paper, and then weighted. In all the experiments, the water uptake was determined by gravimetric measurements using an analytical balance. The swelling degree was calculated as follows.

$$swelling\ degree\ (\%) = \frac{w_s - w_d}{w_d} \times 100 \quad (1)$$

where w_s and w_d are the weights of the swollen and dry hydrogels, respectively. Experiments were carried out in triplicate.

In order to determine the mechanism of diffusion of the pH 2.0 solution and PBS into the hydrogels, the following equation was used.

$$\frac{w_t}{w_\infty} = Kt^n \quad (2)$$

where K is a kinetic constant characteristic of the polymer system, w_t is the mass of solvent absorbed at time t, w_∞ is the mass of solvent absorbed at equilibrium, and n is an empirical number called a transport exponent. Equation (2) is applied to the initial stages of swelling up to 60%. A value of $n = 0.5$ is taken as an indication that the process is diffusion controlled (Fickian diffusion) or Case I transport, whereas, for $n = 1.0$, the swelling is considered to be controlled by the chains relaxation (Case II transport). When the value of n lies between 0.5 and 1.0, the process is considered anomalous (non-Fickian diffusion, which is a combination of diffusion and relaxation) [15].

3.4. Betamethasone In Vitro Release Study

To study the behavior of the developed hydrogels as drug delivery systems, betamethasone as a model drug was loaded into them. The dry hydrogels (pieces of 40–70 mg) were immersed in 10 mL of 1 mg/mL betamethasone solution for 24 h. After incubation, the BTM-loaded hydrogels were taken out, rinsed with distilled water to remove the excess drug on the surface, and dried at room temperature for 3 days. The loaded hydrogels were characterized by FTIR spectroscopy in order to observe if they suffered any changes in the structure after incubation in BTM solution or if any chemical interaction between the drug and matrix occurred.

The amount of drug loaded (DL) in the hydrogel discs (ug of drug/mg of hydrogel) and the loading efficiency (LE) were calculated by the following equations [46,51].

$$DL\ (\mu g\ of\ drug/mg\ of\ hydrogel) = \frac{C_1V_1 - C_2V_2}{m_H} \quad (3)$$

$$LE\ (\%) = \frac{C_1V_1 - C_2V_2}{C_1V_1} \times 100 \quad (4)$$

where C_1 is the initial concentration of the loading BTM solution (mg/mL), V_1 is the initial volume of the BTM solution (mL), C_2 and V_2 are the final concentration and remaining volume of the BTM solution after the hydrogel's incubation including 5 mL of water used to rinse the loaded hydrogels. m_H is the weight of the hydrogel disc before being loaded. The amount of BTM was determined spectrophotometrically at 242 nm using a BTM standard calibration curve prepared at concentrations between 5 and 40 µg/mL.

The in vitro release studies were performed in PBS at pH 7.4 and 37 °C. Briefly, the loaded hydrogel samples were incubated in 5 mL of PBS and, at selected time intervals, 0.5 mL of medium was extracted and replaced with an equal volume of fresh PBS. The BTM concentration released was obtained by the same method mentioned above and the cumulative released amount (% BTM released) in each time interval was calculated according to Equation (5).

$$\%BTM_{released} = \frac{m_r}{m_t} * 100 \quad (5)$$

where m_r is the released amount of BTM at selected time intervals and m_t is the initial amount of BTM loaded in the dry hydrogels. Experiments were run in triplicate.

The release mechanism from AlgNa-g-poly(QCL-co-HEMA) hydrogels was investigated using released experimental data and the Korsmeyer-Peppas power law equation (Equation (2)) [15]. In this case, w_t is the mass released at time t and w_∞ is the total mass of BTM encapsulated in the matrix.

3.5. Cell Culture and Cytotoxicity Assay

Human chondrocyte C-28 and fibroblast 3T3 cells were cultured in DMEM containing 10% fetal bovine serum (FBS) and 1% antibiotics (penicillin-streptomycin) at 37 °C under a 5% CO_2 humidified atmosphere.

The in vitro cytotoxicity of the hydrogels was evaluated based on their impact on cell structures such as mitochondria by the MTS assay (3-(4,5-dimethylthiazol-2-yl)-5-(3-carboxymethoxyphenyl)-2-(4-sulfophenyl)-2H-tetrazolium). The indirect extraction method to evaluate hydrogel samples' cytotoxicity was used according to the ISO 10993-5 standard and previous literature [25,52]. The hydrogel extract solutions were prepared by incubating samples previously sterilized, in DMEM culture medium for 48 h at 37 °C. After incubation, the extract solutions were filtered through a 0.22-μm syringe filter. Alternatively, C-28 and 3T3 cells were seeded in 96-well plates at a density of 1×10^4 cells per well and cultured in 5% CO_2 at 37 °C for 24 h. Afterward, the medium in each well was replaced with 100 μL of the hydrogels' extract solutions and the treated cells were incubated for 24, 48, and 72 h. After the incubation time, the medium was discarded, the MTS solution was added, and plates were incubated for another 3 h in darkness. The absorbance of each well was measured by a micro-plate reader (VersaMax, Molecular Devices, San José, CA, USA, Program Softmax Pro) at 490 nm. Cells incubated only with DMEM culture medium were used as a negative control (100%), and the relative cell viability of the treated groups was calculated according to the next equation.

$$\%\ Cell\ viability = \frac{OD_{samples}}{OD_{control}} \times 100 \tag{6}$$

In order to evaluate the biocompatibility of materials after 7 days in contact with C28 and 3T3 cells, the Live/Dead® assay was made. Live/Dead® is a quick and easy two-color assay where Calcein AM fluoresces green upon the reaction of intracellular esterase and stains live cells. Ethidium homodimer-1, which binds to the DNA of dead membrane compromised cells, stains dead cells (red). First, small 2–3 mm thin disks of hydrogels were placed in 48-well plates overnight in DMEM medium. Then, the medium was removed, an aliquot of C28 human chondrocytes or C3T3 fibroblast suspension (2×10^5 cells) was seeded on the surface of the hydrogel disk and incubated for 6 h at 5% CO_2 and 37 °C to allow cell attachment. Subsequently, cell-seeded hydrogels were transferred into new 24-well plates. Then, 1 mL of DMEM media was added on each well and incubated for 7 days. After every 24 h, culture medium was replaced. At 7 days, the hydrogels were rinsed with PBS and stained with Calcein AM/ethidium homodimer-1 using the Live-Dead Assay Kit (Invitrogen), according to the manufacturers' instructions. The images of the C28 and 3T3 cells on the surface hydrogels were captured by a fluorescence microscope (Leica DM 5500 B, Leica Microsystems GmbH, Wetzlar, Germany).

3.6. Statistical Analysis

Graphs and statistics were performed with OriginPro 2021 (OriginLab Corp., Northampton, MA, USA). Data are reported as mean ± standard deviation (SD), unless stated otherwise. Error bars represent the SD calculated from tests of triplicate measurements for each scaffold. Statistical analysis was significant by a One-Way Analysis of Variance (ANOVA) for $p < 0.05$ or $p < 0.01$, according to the *t*-test for two samples or a multiple samples' comparison.

4. Conclusions

In this study, a series of AlgNa-g-poly(QCL-co-HEMA) hydrogels were successfully synthesized by graft co-polymerization of QCL and HEMA onto the sodium alginate chain. Through FTIR characterization, it was possible to confirm the formation of a graft co-polymer and the structural composition of these materials, constituted mainly by groups such as -OH, -COOH, -C=O and $N^+(CH_3)_3$. The morphological study shows that an internal structure is porous in all samples, which is an advantage in materials for biomedical applications. The swelling study showed that the pH of the medium and the composition of the hydrogels affect the swelling degree, and, based on the applied kinetic model, it was observed that the fluid penetrates the polymer network through an anomalous process. In vitro release of BTM loaded into the hydrogels were carried out for 8 h and the studies showed an effective BTM release from hydrogels in a controlled manner conditioned by a simultaneous process of diffusion and relaxation of the polymeric chains, according to the Korsmeyer-Peppas kinetic model used in the analysis. In the case of in vitro cytocompatibility studies, it was confirmed through MTS and Live/Dead® assays that samples from H1 to H5 are cytocompatible and allow the adhesion and cell proliferation against C28 and 3T3 cells, while, in the sample H6, the cell viability was less than 20%, which demonstrates that a QCL concentration higher than 70% is not suitable for biomedical materials. We can conclude that AlgNa-g-poly(QCL-co-HEMA) hydrogel matrices obtained (compositions from H1 to H5) are promising materials as platforms for chondrocyte proliferation and drug delivery.

Author Contributions: Conceptualization and Methodology, J.G.-C., M.V., L.A., G.F., A.A. and L.J.C.; investigation and data curation, J.G.-C., M.V., N.B. and L.A.; investigation and formal analysis, G.F., A.A., J.A.-B. and O.V.; writing—original draft preparation, J.G.-C., M.V., G.F. and O.V.; writing—review and editing, J.G.-C., G.F., L.J.C. and J.A.-B.; resources, supervision, project administration, and funding acquisition, L.J.C. and J.A.-B. All authors have read and agreed to the published version of the manuscript.

Funding: We would like to acknowledge the financial support from the European Union through Erasmus PLUS doctoral fellowship and mobility staff, project code 2015-1-NL01-KA 107-008639 (J.G.-C., A.A. and G.F. respectively). This work was also supported by project grants from the European Commission H2020-MSCA-RISE (644373—PRISAR), H2020-MSCA-RISE (777682—CANCER), H2020-WIDESPREAD-05-2017-Twinning (807281—ACORN), H2020-WIDESPREAD-2018-03 (852985—SIMICA), H2020-SCARISE-2016 (734684—CHARMED) and MSCA-ITN-2015-ETN (675743—ISPIC), 861190 (PAVE), 857894 (CAST), 859908 (NOVA-MRI), 872860 (PRISAR2). This work was also financially supported by the VIDI personal grant (project number 723.012.110) (Luis J. Cruz). Oscar Valdes gratefully acknowledges the financial support of Agencia Nacional de Investigación y Desarrollo (ANID, Chile) through projects FONDECYT REGULAR 1210107.

Institutional Review Board Statement: Not applicable.

Informed Consent Statement: Not applicable.

Acknowledgments: The authors acknowledge María Victoria Muñoz, MA in English Literature, for her invaluable efforts in the manuscript correction.

Conflicts of Interest: The authors declare no conflict of interest.

References

1. Kou, L.; Xiao, S.; Sun, R.; Bao, S.; Yao, Q.; Chen, R. Biomaterial-engineered intra-articular drug delivery systems for osteoarthritis therapy. *Drug Deliv.* **2019**, *26*, 870–885. [CrossRef]
2. Samuel, S.; Ahmad, R.E.; Ramasamy, T.S.; Karunanithi, P.; Naveen, S.V.; Kamarul, T. Platelet-rich concentrate in serum-free medium enhances cartilage-specific extracellular matrix synthesis and reduces chondrocyte hypertrophy of human mesenchymal stromal cells encapsulated in alginate. *Platelets* **2019**, *30*, 66–74. [CrossRef] [PubMed]
3. Kaviani, A.; Zebarjad, S.M.; Javadpour, S.; Ayatollahi, M.; Bazargan-Lari, R. Fabrication and characterization of low-cost freeze-gelated chitosan/collagen/hydroxyapatite hydrogel nanocomposite scaffold. *Int. J. Polym. Anal. Charact.* **2019**, *24*, 191–203. [CrossRef]
4. Silva, C.R.; Babo, P.S.; Gulino, M.; Costa, L.; Oliveira, J.M.; Silva-Correia, J.; Domingues, R.M.A.; Reis, R.L.; Gomes, M.E. Injectable and tunable hyaluronic acid hydrogels releasing chemotactic and angiogenic growth factors for endodontic regeneration. *Acta Biomater.* **2018**, *77*, 155–171. [CrossRef] [PubMed]
5. Urbanek, O.; Kołbuk, D.; Wróbel, M. Articular cartilage: New directions and barriers of scaffolds development—Review. *Int. J. Polym. Mater.* **2019**, *68*, 396–410. [CrossRef]

6. Wang, J.; Wang, Y.; Sun, X.; Liu, D.; Huang, C.; Wu, J.; Yang, C.; Zhang, Q. Biomimetic cartilage scaffold with orientated porous structure of two factors for cartilage repair of knee osteoarthritis. *Artif. Cells Nanomed. Biotechnol.* **2019**, *47*, 1710–1721. [CrossRef]
7. Passos, M.F.; Carvalho, N.M.S.; Rodrigues, A.A.; Bavaresco, V.P.; Jardini, A.L.; Maciel, M.R.W.; Maciel Filho, R. PHEMA Hydrogels Obtained by Infrared Radiation for Cartilage Tissue Engineering. *Int. J. Chem. Eng.* **2019**, *2019*, 10. [CrossRef]
8. Chuang, E.-Y.; Chiang, C.-W.; Wong, P.-C.; Chen, C.-H. Hydrogels for the Application of Articular Cartilage Tissue Engineering: A Review of Hydrogels. *Adv. Mater. Sci. Eng.* **2018**, *2018*, 14. [CrossRef]
9. Khalid, I.; Ahmad, M.; Minhas, M.U.; Barkat, K. Preparation and characterization of alginate-PVA-based semi-IPN: Controlled release pH-responsive composites. *Polym. Bull.* **2018**, *75*, 1075–1099. [CrossRef]
10. Torres, M.L.; Oberti, T.G.; Fernández, J.M. HEMA and alginate-based chondrogenic semi-interpenetrated hydrogels: Synthesis and biological characterization. *J. Biomater. Sci. Polym. Ed.* **2020**, *32*, 504–523. [CrossRef]
11. Hasnain, M.S.; Ray, P.; Nayak, A.K. Chapter 5—Alginate-based interpenetrating polymer networks for sustained drug release. In *Alginates in Drug Delivery*, 1st ed.; Nayak, A.K., Hasnain, M.S., Eds.; Academic Press: London, UK, 2020; pp. 101–128.
12. Farokhi, M.; Jonidi Shariatzadeh, F.; Solouk, A.; Mirzadeh, H. Alginate Based Scaffolds for Cartilage Tissue Engineering: A Review. *Int. J. Polym. Mater.* **2020**, *69*, 230–247. [CrossRef]
13. Yamaoka, H.; Asato, H.; Ogasawara, T.; Nishizawa, S.; Takahashi, T.; Nakatsuka, T.; Koshima, I.; Nakamura, K.; Kawaguchi, H.; Chung, U.I.; et al. Cartilage tissue engineering using human auricular chondrocytes embedded in different hydrogel materials. *J. Biomed. Mater. Res. Part A* **2006**, *78*, 1–11. [CrossRef] [PubMed]
14. Eslaminejad, M.B.; Taghiyar, L.; Falahi, F. Quantitative analysis of the proliferation and differentiation of rat articular chondrocytes in alginate 3D culture. *Iran. Biomed. J.* **2009**, *13*, 153–160.
15. Goel, N.K.; Kumar, V.; Bhardwaj, Y.K.; Chaudhari, C.V.; Dubey, K.A.; Sabharwal, S. Swelling Response of Radiation Synthesized 2-Hydroxyethylmethacrylate-co-[2-(methacryloyloxy)ethyl] Trimethylammonium Chloride Hydrogels Under Various In Vitro Conditions. *J. Biomater. Sci. Polym. Ed.* **2009**, *20*, 785–805. [CrossRef]
16. Bölgen, N.; Yang, Y.; Korkusuz, P.; Güzel, E.; El Haj, A.J.; Pişkin, E. 3D ingrowth of bovine articular chondrocytes in biodegradable cryogel scaffolds for cartilage tissue engineering. *J. Tissue Eng. Regen. Med.* **2011**, *5*, 770–779. [CrossRef] [PubMed]
17. Nicolas, C.; Verny, M.; Giraud, I.; Ollier, M.; Rapp, M.; Maurizis, J.-C.; Madelmont, J.-C. New Quaternary Ammonium Oxicam Derivatives Targeted toward Cartilage: Synthesis, Pharmacokinetic Studies, and Antiinflammatory Potency. *J. Med. Chem.* **1999**, *42*, 5235–5240. [CrossRef]
18. Giraud, I.; Rapp, M.; Maurizis, J.-C.; Madelmont, J.-C. Application to a Cartilage Targeting Strategy: Synthesis and in Vivo Biodistribution of 14C-Labeled Quaternary Ammonium−Glucosamine Conjugates. *Bioconjugate Chem.* **2000**, *11*, 212–218. [CrossRef]
19. Miot-Noirault, E.; Vidal, A.; Morlieras, J.; Bonazza, P.; Auzeloux, P.; Besse, S.; Dauplat, M.-M.; Peyrode, C.; Degoul, F.; Billotey, C.; et al. Small rigid platforms functionalization with quaternary ammonium: Targeting extracellular matrix of chondrosarcoma. *Nanomed. NBM* **2014**, *10*, 1887–1895. [CrossRef] [PubMed]
20. Rivero, R.; Alustiza, F.; Capella, V.; Liaudat, C.; Rodriguez, N.; Bosch, P.; Barbero, C.; Rivarola, C. Physicochemical properties of ionic and non-ionic biocompatible hydrogels in water and cell culture conditions: Relation with type of morphologies of bovine fetal fibroblasts in contact with the surfaces. *Colloids Surf. B* **2017**, *158*, 488–497. [CrossRef]
21. Zhu, Z.; Zhu, Z. Adhesion of starch-g-poly(2-acryloyloxyethyl trimethyl ammonium chloride) to cotton and polyester fibers. *Starch/Staerke* **2014**, *66*, 566–575. [CrossRef]
22. Shen, S.; Zhu, Z.; Liu, F. Introduction of poly[(2-acryloyloxyethyl trimethyl ammonium chloride)-co-(acrylic acid)] branches onto starch for cotton warp sizing. *Carbohydr. Polym.* **2016**, *138*, 280–289. [CrossRef] [PubMed]
23. Ozay, O.; Ilgin, P.; Ozay, H.; Gungor, Z.; Yilmaz, B.; Kıvanç, M.R. The preparation of various shapes and porosities of hydroxyethyl starch/p(HEMA-co-NVP) IPN hydrogels as programmable carrier for drug delivery. *J. Macromol. Sci. Part A* **2020**, *57*, 379–387. [CrossRef]
24. Mandal, B.; Ray, S.K. Synthesis of interpenetrating network hydrogel from poly(acrylic acid-co-hydroxyethyl methacrylate) and sodium alginate: Modeling and kinetics study for removal of synthetic dyes from water. *Carbohydr. Polym.* **2013**, *98*, 257–269. [CrossRef]
25. Hu, X.; Wang, Y.; Zhang, L.; Xu, M.; Dong, W.; Zhang, J. Redox/pH dual stimuli-responsive degradable Salecan-g-SS-poly(IA-co-HEMA) hydrogel for release of doxorubicin. *Carbohydr. Polym.* **2017**, *155*, 242–251. [CrossRef]
26. Clara, I.; Natchimuthu, N. Hydrogels of sodium alginate based copolymers grafted with sodium-2-acrylamido-2-methyl-1-propane sulfonate and methacrylic acid for controlled drug delivery applications. *J. Macromol. Sci. Pure Appl. Chem.* **2018**, *55*, 168–175. [CrossRef]
27. Bardajee, G.R.; Hooshyar, Z.; Rastgo, F. Kappa carrageenan-g-poly (acrylic acid)/SPION nanocomposite as a novel stimuli-sensitive drug delivery system. *Colloid Polym. Sci.* **2013**, *291*, 2791–2803. [CrossRef]
28. Jiang, H.; Yang, Y.; Lin, Z.; Zhao, B.; Wang, J.; Xie, J.; Zhang, A. Preparation of a novel bio-adsorbent of sodium alginate grafted polyacrylamide/graphene oxide hydrogel for the adsorption of heavy metal ion. *Sci. Total Environ.* **2020**, *744*, 10. [CrossRef]
29. Rivas, B.L.; Aguirre, M.D.C. Water-soluble polymers: Optimization of arsenate species retention by ultrafiltration. *J. Appl. Polym. Sci.* **2009**, *112*, 2327–2333. [CrossRef]
30. Torres, C.C.; Urbano, B.F.; Campos, C.H.; Rivas, B.L.; Reyes, P. Composite hydrogel based on surface modified mesoporous silica and poly[(2-acryloyloxy)ethyl trimethylammonium chloride]. *Mater. Chem. Phys.* **2015**, *152*, 69–76. [CrossRef]

31. Zhang, L.; Zheng, G.-J.; Guo, Y.-T.; Zhou, L.; Du, J.; He, H. Preparation of novel biodegradable pHEMA hydrogel for a tissue engineering scaffold by microwave-assisted polymerization. *Asian Pac. J. Trop. Med.* **2014**, *7*, 136–140. [CrossRef]
32. Lee, B.-S.; Chen, Y.-J.; Wei, T.-C.; Ma, T.-L.; Chang, C.-C. Comparison of Antibacterial Adhesion When Salivary Pellicle Is Coated on Both Poly(2-hydroxyethyl-methacrylate)- and Polyethylene-glycol-methacrylate-grafted Poly(methyl methacrylate). *Int. J. Mol. Sci.* **2018**, *19*, 2764. [CrossRef]
33. Yang, X.; Wang, B.; Qiao, C.; Li, Z.; Li, Y.; Xu, C.; Li, T. Molecular interactions in N-[(2-hydroxyl)-propyl-3-trimethyl ammonium] chitosan chloride-sodium alginate polyelectrolyte complexes. *Food Hydrocoll.* **2020**, *100*, 105400. [CrossRef]
34. Onder, A.; Ilgin, P.; Ozay, H.; Ozay, O. Removal of dye from aqueous medium with pH-sensitive poly[(2-(acryloyloxy)ethyl]trimethylammonium chloride-co-1-vinyl-2-pyrrolidone] cationic hydrogel. *J. Environ. Chem. Eng.* **2020**, *8*, 104436. [CrossRef]
35. Vargün, E.; Usanmaz, A. Degradation of Poly(2-hydroxyethyl methacrylate) Obtained by Radiation in Aqueous Solution. *J. Macromol. Sci. Pure Appl. Chem.* **2010**, *47*, 882–891. [CrossRef]
36. El-Sherbiny, I.M.; Abdel-Mogib, M.; Dawidar, A.-A.M.; Elsayed, A.; Smyth, H.D.C. Biodegradable pH-responsive alginate-poly (lactic-co-glycolic acid) nano/micro hydrogel matrices for oral delivery of silymarin. *Carbohydr. Polym.* **2011**, *83*, 1345–1354. [CrossRef]
37. Das, D.; Pham, H.T.T.; Lee, S.; Noh, I. Fabrication of alginate-based stimuli-responsive, non-cytotoxic, terpolymric semi-IPN hydrogel as a carrier for controlled release of bovine albumin serum and 5-amino salicylic acid. *Mater. Sci. Eng. C* **2019**, *98*, 42–53. [CrossRef]
38. Ghobashy, M.M.; Bassioni, G. pH stimuli-responsive poly(acrylamide-co-sodium alginate) hydrogels prepared by γ-radiation for improved compressive strength of concrete. *Adv. Polym. Technol.* **2018**, *37*, 2123–2133. [CrossRef]
39. Dou, W.-H.; Zhou, G.-M.; Kang, Q.-Q. Study of the Epimers of Dexamethasone Sodium Phosphate and Betamethasone Sodium Phosphate by FTIR, FT-Raman and SERS. *Spectrosc. Spectr. Anal.* **2012**, *32*, 2664–2668. [CrossRef]
40. Monajjemzadeh, F.; Gholizadeh, N.; Yousefzadeh Mobaraki, N.; Jelvehgari, M. Physicochemical and in vitro mucoadhesive properties of microparticles/discs of betamethasone for the management of oral lichen planus. *Pharm. Dev. Technol.* **2016**, *21*, 996–1005. [CrossRef] [PubMed]
41. Ghasemnejad, M.; Ahmadi, E.; Mohamadnia, Z.; Doustgani, A.; Hashemikia, S. Functionalized silica nanoparticles as a carrier for Betamethasone Sodium Phosphate: Drug release study and statistical optimization of drug loading by response surface method. *Mater. Sci. Eng. C* **2015**, *56*, 223–232. [CrossRef]
42. Qi, X.; Li, J.; Wei, W.; Zuo, G.; Su, T.; Pan, X.; Zhang, J.; Dong, W. Cationic Salecan-based hydrogels for release of 5-fluorouracil. *RSC Adv.* **2017**, *7*, 14337–14347. [CrossRef]
43. Rey-Rico, A.; Madry, H.; Cucchiarini, M. Hydrogel-Based Controlled Delivery Systems for Articular Cartilage Repair. *Biomed. Res. Int.* **2016**, *2016*, 1215263. [CrossRef]
44. Wei, W.; Qi, X.; Li, J.; Zuo, G.; Sheng, W.; Zhang, J.; Dong, W. Smart Macroporous Salecan/Poly(N,N-diethylacrylamide) Semi-IPN Hydrogel for Anti-Inflammatory Drug Delivery. *ACS Biomater. Sci. Eng.* **2016**, *2*, 1386–1394. [CrossRef]
45. Samanta, H.S.; Ray, S.K. Synthesis, characterization, swelling and drug release behavior of semi-interpenetrating network hydrogels of sodium alginate and polyacrylamide. *Carbohydr. Polym.* **2014**, *99*, 666–678. [CrossRef]
46. Ilgin, P.; Ozay, H.; Ozay, O. A new dual stimuli responsive hydrogel: Modeling approaches for the prediction of drug loading and release profile. *Eur. Polym. J.* **2019**, *113*, 244–253. [CrossRef]
47. Chen, J.; Liu, M.; Chen, W.; Zhang, N.; Zhu, S.; Zhang, S.; Xiong, Y. Synthesis, Swelling and Drug-Release Behaviour of a Poly(N,N-diethylacrylamide-co-(2-dimethylamino) ethyl methacrylate) Hydrogel. *J. Biomater. Sci. Polym. Ed.* **2011**, *22*, 1049–1068. [CrossRef]
48. Dalapati, S.; Alam, M.A.; Jana, S.; Guchhait, N. Reduced Schiff-base assisted novel dihydrogenphosphate–water polymer. *CrystEngComm* **2012**, *14*, 6029–6034. [CrossRef]
49. Basaran, I.; Emami Khansari, M.; Pramanik, A.; Wong, B.M.; Hossain, M.A. Binding and selectivity of dihydrogen phosphate by H-bond donors and acceptors in a tripodal-based thiourea receptor. *Tetrahedron Lett.* **2015**, *56*, 115–118. [CrossRef]
50. Kim, A.R.; Lee, S.L.; Park, S.N. Properties and in vitro drug release of pH- and temperature-sensitive double cross-linked interpenetrating polymer network hydrogels based on hyaluronic acid/poly (N-isopropylacrylamide) for transdermal delivery of luteolin. *Int. J. Biol. Macromol.* **2018**, *118*, 731–740. [CrossRef] [PubMed]
51. Cheng, C.; Zhang, X.; Meng, Y.; Chen, L.; Zhang, Q. Development of a dual drug-loaded hydrogel delivery system for enhanced cancer therapy: In situ formation, degradation and synergistic antitumor efficiency. *J. Mater. Chem. B* **2017**, *5*, 8487–8497. [CrossRef] [PubMed]
52. Roointan, A.; Farzanfar, J.; Mohammadi-Samani, S.; Behzad-Behbahani, A.; Farjadian, F. Smart pH responsive drug delivery system based on poly(HEMA-co-DMAEMA) nanohydrogel. *Int. J. Pharm.* **2018**, *552*, 301–311. [CrossRef] [PubMed]

International Journal of
Molecular Sciences

MDPI

Article

Bioprinting Via a Dual-Gel Bioink Based on Poly(Vinyl Alcohol) and Solubilized Extracellular Matrix towards Cartilage Engineering

Mohsen Setayeshmehr [1,2], Shahzad Hafeez [2], Clemens van Blitterswijk [2], Lorenzo Moroni [2], Carlos Mota [2,*] and Matthew B. Baker [2,*]

1 Biomaterials and Tissue Engineering Department, School of Advanced Technologies in Medicine, Isfahan University of Medical Sciences, Isfahan 81746-73461, Iran; setayeshmehr.m@gmail.com

2 MERLN Institute for Technology Inspired Regenerative Medicine, Complex Tissue Regeneration, Maastricht University, 6229 Maastricht, The Netherlands; s.hafeez@maastrichtuniversity.nl (S.H.); c.vanblitterswijk@maastrichtuniversity.nl (C.v.B.); l.moroni@maastrichtuniversity.nl (L.M.)

* Correspondence: c.mota@maastrichtuniversity.nl (C.M.); m.baker@maastrichtuniversity.nl (M.B.B.)

Abstract: Various hydrogel systems have been developed as biomaterial inks for bioprinting, including natural and synthetic polymers. However, the available biomaterial inks, which allow printability, cell viability, and user-defined customization, remains limited. Incorporation of biological extracellular matrix materials into tunable synthetic polymers can merge the benefits of both systems towards versatile materials for biofabrication. The aim of this study was to develop novel, cell compatible dual-component biomaterial inks and bioinks based on poly(vinyl alcohol) (PVA) and solubilized decellularized cartilage matrix (SDCM) hydrogels that can be utilized for cartilage bioprinting. In a first approach, PVA was modified with amine groups (PVA-A), and mixed with SDCM. The printability of the PVA-A/SDCM formulations cross-linked by genipin was evaluated. On the second approach, the PVA was functionalized with cis-5-norbornene-endo-2,3-dicarboxylic anhydride (PVA-Nb) to allow an ultrafast light-curing thiol-ene cross-linking. Comprehensive experiments were conducted to evaluate the influence of the SDCM ratio in mechanical properties, water uptake, swelling, cell viability, and printability of the PVA-based formulations. The studies performed with the PVA-A/SDCM formulations cross-linked by genipin showed printability, but poor shape retention due to slow cross-linking kinetics. On the other hand, the PVA-Nb/SDCM showed good printability. The results showed that incorporation of SDCM into PVA-Nb reduces the compression modulus, enhance cell viability, and bioprintability and modulate the swelling ratio of the resulted hydrogels. Results indicated that PVA-Nb hydrogels containing SDCM could be considered as versatile bioinks for cartilage bioprinting.

Keywords: poly(vinyl alcohol); decellularized cartilage matrix; bioprinting; thiol-ene cross-linking

Citation: Setayeshmehr, M.; Hafeez, S.; van Blitterswijk, C.; Moroni, L.; Mota, C.; Baker, M.B. Bioprinting Via a Dual-Gel Bioink Based on Poly(Vinyl Alcohol) and Solubilized Extracellular Matrix towards Cartilage Engineering. *Int. J. Mol. Sci.* **2021**, *22*, 3901. https://doi.org/10.3390/ijms22083901

Academic Editor: Peter John Jervis

Received: 28 January 2021
Accepted: 8 March 2021
Published: 9 April 2021

1. Introduction

Additive manufacturing techniques are increasingly used for biofabrication of three-dimensional (3D) scaffolds and constructs in tissue engineering (TE) [1]. Bioprinting offers controlled patterning and deposition of polymeric hydrogels or composites to fabricate well-defined constructs with the capability to combine various material and their compositions [2]. The bioprinting of cell-laden biomaterials, termed bioinks, allows the deposition of cells encapsulated in a defined 3D construct and can develop into living tissue-engineered constructs [3]. However, the available biomaterial inks and bioinks, which balance tailorability with desired performance for bioprinting and tissue growth, remains low [4].

A number of bioprinters with different dispensing principles have been commercialized, including inkjet or droplet-based, extrusion or pressure-based, and laser-assisted bioprinting [5]. All of these systems allow spatial control over bioink deposition while offering distinctive ranges of processable inks formulations. Extrusion bioprinting [6] is one of the most frequently used systems owing to its ease of processing, low cost of adoption, and ability to bioprint high cell densities. Extrusion systems are normally used to dispense in a layer-by-layer fashion filaments in a controlled way and the viscosity and gelation kinetics of the selected bioinks plays a key role in defining their performance [7]. High viscosity formulations normally offer structures with high shape fidelity, but upon gelation, high network density limits encapsulated cells in both mobility and the ability to reorganize the matrix. On the contrary, lower viscosity materials provide a less compact network and more permissive environment for cells but can suffer from low printability or structural integrity [7].

There are many efforts toward bridging this incompatibility gap, where systems not only meet the demands for good printability but also provide a suitable environment for cells [8]. Various hydrogel systems have been developed as biomaterial inks for bioprinting [9], including natural polymers [10] such as gelatin [11], collagen [12], alginate [13], and extracellular matrix (ECM)-derived materials [14] or synthetic polymers such as poly(ethylene glycol) [15], Pluronic [16], poly(vinylpyrrolidone) [17], and poly(vinyl alcohol) (PVA) [18]. Depending on the choice of polymer, different mechanisms can be employed for the cross-linking reactions. Photo-polymerizable systems have recently become a popular choice due to spatial and temporal control of cross-linking [19]. Free radical polymerization of acrylate derivatives has been extensively used in the design of photoreactive bioinks [2]; however, this free-radical chain-growth reaction characteristically yields all carbon kinetic-chain backbones, which greatly affect and tunability of mechanical properties. Photoinitiated thiol-ene cross-linking is well-matched as an alternative cross-linking mechanism and recently has reached utility in biofabrication [10,13,20,21]. The thiol-ene photo-cross-linking can be stoichiometrically controlled, chemospecific, and less sensitive to oxygen and is able to form homogeneous hydrogel networks when compared to free radical acrylate derivatives [22]. In addition, the same thiol-ene chemistry can allow the decoration of the hydrogel scaffold with numerous thiol-incorporating peptides and biomolecules, and dithiol (DT) cross-linkers can be designed to facilitate degradation. An example of thiol-ene chemistry for fabrication of cell-laden hydrogels include hyaluronic acid systems functionalized by both methacrylates and norbornenes to create dual cross-linking systems that allow bioprinting via cross-linking upon exposure to UV [23]. These tools facilitate the creation of uniform constructs with features not achievable with other techniques [13].

PVA is a water-soluble, odorless, tasteless, white polymer widely used in industry [18]. PVA is commonly considered as nontoxic polymer and has found use as an additive for food, cosmetic products, and in food packaging [24], but also for pharmaceutical and biomedical applications [25]. An additional advantage of PVA is the opportunity of post-polymerization modification due to its secondary hydroxyl groups, often accomplished via the formation of esters, ethers, and acetals [26]. Previously, PVA hydrogel scaffolds have been created by different cross-linking methods, via incorporation of polymerizable functionalities, whereas photo-cross-linking is of outstanding attention enabling in situ hydrogel formation [27–30]. Photo-reactive PVA had been investigated for usage as TE scaffolds, e.g., for in situ polymerization for minimally invasive implantation methods [27]. Furthermore, PVA is modified with allyl succinic anhydride and cis-5-norbornene-endo-2,3-dicarboxylic anhydride to produce macromers with reactive ene groups. The obtained macromers were photo-cross-linked via thiol-ene chemistry (via a thiol modified PVA) and resulted in mechanically tunable hydrogel formulations. Cell studies demonstrated that the resulted hydrogels exhibit low toxicity [18]. PVA hydrogels have been reported to resist protein adsorption and cell adhesion, allowing the incorporation of bioactivity for rationally tuning cell–matrix interactions [31].

ECM materials can either be harvested from cell-derived matrices from in vitro culture or can be obtained directly from native tissue. ECM from either source has to be decellularized to eliminate nucleic acids and cellular components, which may have the potential to cause adverse immunological reactions [32,33]. Some research groups have already established that decellularized cartilage has chondroinductive potential [34]. We recently reported the chondroinductive potential of genipin cross-linked PVA/devitalized cartilage (DC) matrix in vitro, where we observed chondroinductivity of human adipose-derived mesenchymal stromal cells cultured on ECM-based scaffolds [32,35].

In recent years, use of different ECM derivatives has been examined in culturing a variety of cell types, consisting of the usage of solubilized ECM and re-forming it as a hydrogel [36,37]. A key problem of bioprinting ECM-derived material is the high sensitivity of biomaterial ink concentration and viscosity during the bioprinting process, associated with the nozzle diameter selection, combined with the poor or slow gelation after extrusion [38]. However, by incorporation of cross-linkable synthetic materials into naturally derived ECM, one can modulate and improve the printability of the resultant hydrogel. Previous studies have established biomaterial inks based on PVA macromeres, yet lack of bioactivity is a problematic issue regarding to these synthesized materials [18]. Photoactive PVA hydrogels have been functionalized with the cell-adhesive peptide arginylglycylaspartic acid (RGD) and found to support the attachment and spreading of fibroblasts [27]. As also previously shown, incorporation of DC into PVA can achieve both advantage of ECM derivatives and synthetic polymers to consider as a strong vehicle for TE applications [35]. Incorporation of ECM into PVA hydrogel not only can improve biological characteristics of the resulting bioink but can also modulate the viscosity, which is critical parameter in bioprinting.

Hence, the aim of this study was to improve cell compatible and bioactive PVA/ECM bioinks, which can be suitable for cartilage bioprinting purposes. We explored two different formulations (Figure 1) in mixing modified PVA with decellularized ECM. First, amine modified PVA was tested and crosslinked with Genipin (Figure 1A), but poor printing performance was found. We then switched to thiol-ene crosslinking of the PVA (PVA-Nb, Figure 1B) in order to facilitate rapid formation of the multicomponent bioink. These PVA-Nb hydrogels with decellularized ECM were characterized for printing performance, mechanical properties, and the suitability for cartilage tissue engineering.

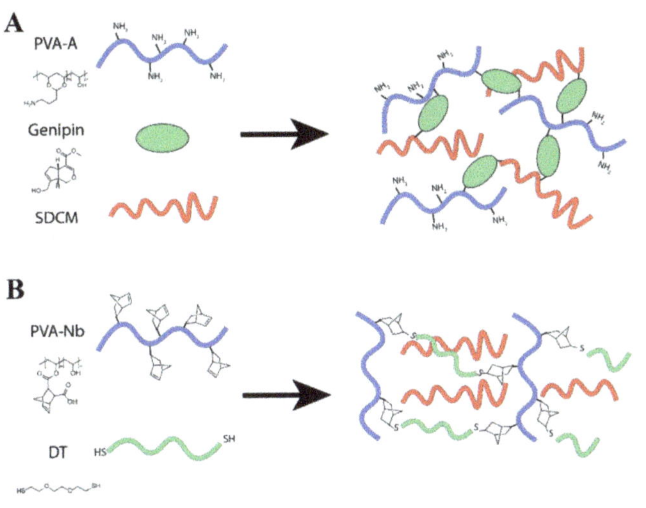

Figure 1. Schematic representations of the (**A**) PVA modified with amine groups (PVA-A)/solubilized decellularized cartilage matrix (SDCM) biomaterial inks formulated from 10 wt% (PVA-A + SDCM) and 0.1% wt/wt genipin and desired PVA-A/SDCM dual-gel. (**B**) PVA functionalized with cis-5-norbornene-endo-2,3-dicarboxylic anhydride (PVA-Nb)/SDCM biomaterial ink.

2. Results

2.1. Characterization of Cartilage Matrix

DNA, sGAG, and Hydroxyproline Content: Biochemical content analysis was performed on natural, devitalized, acellular, and solubilized cartilage matrices. The DNA, sGAG, and hydroxyproline contents of natural cartilage were determined to be 1.10 ± 0.20 μg, 6.97 ± 0.20 μg, and 8.70 ± 0.10 μg, respectively (Figure 2). Following devitalization and cryogrinding, there was a 67% reduction in DNA, negligible reduction in GAG, and an 18% reduction in hydroxyproline ($p < 0.05$). Following decellularization, there was a 95.5% reduction in DNA, negligible reduction in GAG, and a 19% reduction in hydroxyproline ($p < 0.05$) (Figure 2). After solubilizing and dialysis, the DNA content further reduced to 0.2% of that of the original lyophilized cartilage ($p < 0.05$). Although there were no significant reductions in GAG content through the devitalization and decellularization, after solubilizing and dialysis, the GAG content further reduced to 12% of that of the natural cartilage. Furthermore, following solubilization and dialysis, the hydroxyproline content was reduced by 35% compared to natural cartilage ($p < 0.05$).

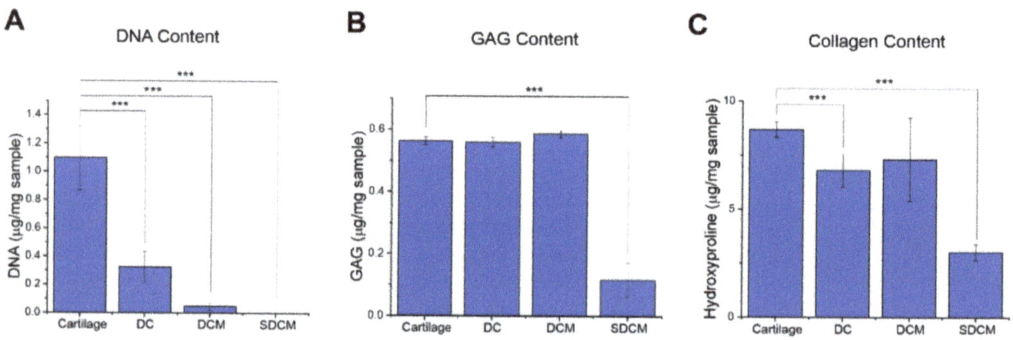

Figure 2. Analysis of key biomolecules during the divitalization process of the processed extracellular matrix (ECM). (**A**) DNA content; (**B**) glycosaminoglycans (GAG) content; and (**C**) collagen content of lyophilized cartilage, devitalized cartilage (DC), decellularized cartilage matrix (DCM), and solubilized decellularized cartilage matrix (SDCM) (Data presented as mean ± SD; *** $p < 0.001$; $n = 4$).

2.2. PVA Modification

Synthesis of Amine-Functionalized PVA

PVA was modified with 4-ABA under acidic conditions to target 10%, 50%, and 100% amination. After reaction and purification, ^{1}H NMR showed characteristic peaks near 2.9 ppm, confirming successful amination (Figure 3). The efficiency and amination percentage (relative to polymer subunits) were calculated from the ratio of the alpha methylene peak (2.9 ppm) to the -CH from the polymer backbone (3.9 ppm) in the spectra (Figure 3A–D). The result of different reaction lengths and equivalents of 4-ABA were briefly studied, with the results presented in Table 1. We observed that increasing the reaction time to 24 h led to the highest efficiency for attachment of the 4-ABA to PVA and that attempts at higher functional densities (i.e., 50% and 100%) led to lower functionalization efficiencies—up to 37% functionalization could be obtained in this study.

The cross-linking of the different PVA-A polymers using genipin (0.1% wt/wt) was investigated to evaluate the effect of amination percentage on gelation time (Table 1). The presence of pale blue color and a positive vial inversion test were considered to evaluate gelation time of the hydrogels at 37 °C. As expected, higher functional densities of amines grafted to the PVA resulted in faster gelation kinetics (Table 1). Based on the gelation kinetics, functionalization amount, and functionalization efficiency, PVA with 30% amine modification (PVA-A24-50) and 0.1% wt/wt genipin were chosen to carry forward for 3D bioprinting (referred hereafter as PVA-A).

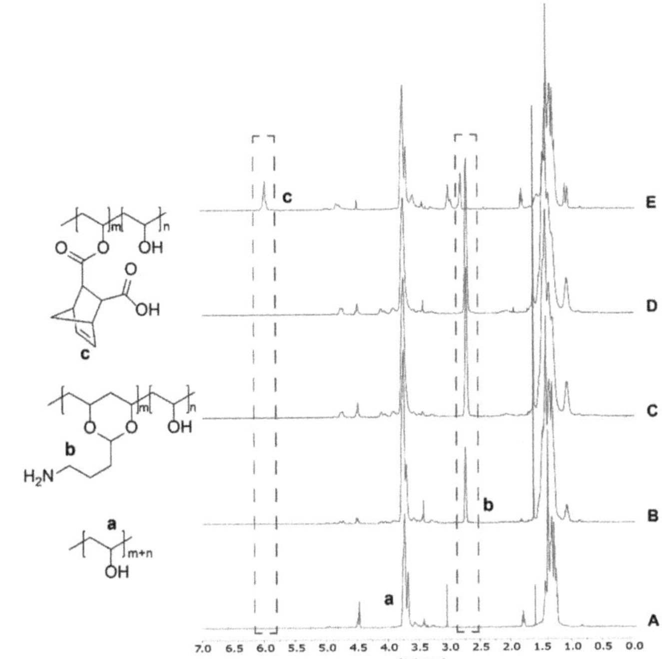

Figure 3. Representative ^{1}H NMR spectra of modified polymers used in this study. (**A–D**) ^{1}H NMR spectrum of different PVA amination for 24 h reaction time point (sample nomenclature PVA-AX-Y, where X corresponds to the reaction time (h) and Y to the maximum theoretical amination percentage). (**A**) Pure PVA, (**B**) PVA-A24-10, (**C**) PVA-A24-50, and (**D**) PVA-A24-100. Successful backbone modification was verified by the peak at 2.9 ppm corresponding to amine linker. (**E**) ^{1}H NMR spectrum of norbornene-modified PVA. Successful backbone modification was verified by vinyl peak at 6.25 ppm.

Table 1. Poly(vinyl alcohol) (PVA) amination in different concentrations and different time points. The results of the gelation time of the different PVA modified with amine groups (PVA-A) polymers using genipin (0.1% wt/wt) as a cross-linker.

Sample Name [a] PVA-AX-Y	Reaction Time (h)	Target Amination (mol%)	Calculated Amination (mol%)	Efficiency (%)	Gelation Time (h)
PVA-A1-10	1	10	3.7	37	72
PVA-A1-50	1	50	8.8	17.6	72
PVA-A1-100	1	100	18.0	18.0	48
PVA-A2-10	2	10	5.2	52	72
PVA-A2-50	2	50	14.9	29.8	48
PVA-A2-100	2	100	23.7	23.7	24
PVA-A24-10	24	10	9.7	97	72
PVA-A24-50	24	50	29.7	59.3	24
PVA-A24-100	24	100	37.4	37.4	24

[a] In the sample naming, X corresponds to the reaction time (h) and Y corresponds to the maximum theoretical amination percentage.

2.3. 3D Bioprinting of PVA-A/SDCM

Different biomaterial inks were formulated from 10 wt% (PVA-A + SDCM) and 0.1% wt/wt genipin to evaluate the printability of the synthesized materials and their suitability for bioprinting approaches. Three different compositions were prepared consisting of: (A) PVA-A, (B) PVA-A/SDCM30, and (C) PVA-A/SDCM50. Due to the low kinetics of the genipin cross-linking, suitable printability was only possible after 2 h once the cross-linking reaction started. After allowing the formulations to pre-cross-link, 3D hydrogels scaffolds were manufactured. Semi porous structures were observed in the PVA-A/SDCM30 and PVA-A/SDCM50 scaffolds (Figure S1). Of note, the printing conditions used to produce these scaffolds match those empirically optimized for best structure. Finally, 5 mm/s deposition speed and 30 kPa extrusion pressure were used to fabricate structures of the three biomaterial ink formulations, leading to poor feature reconstruction for the as-printed scaffold (Figure S1). These scaffolds have shown stability in PBS for over 1 month.

2.4. Synthesis of Norbornene-Functionalized PVA

PVA was modified using a norbornene anhydride under acidic conditions to obtain the norbornene ester-modified polymer (Figure 1). Peaks observed in the ^{1}H NMR spectrum at chemical shifts of: 6.22 (a, 2H, CH_5CH), 4.02 (1H, polym-CH-O), 3.28 and 3.09 (2H, CH[ring]), and 1.39 (2H, CH_2[ring]) confirmed successful norbornene modification [18]. According to the ^{1}H NMR spectrum, the calculated norbornene modification was 6.5 mol% with respect to monomer units of PVA (Figure 3E).

2.5. Volume Change Analysis and Swelling

The effect of ECM on the dimensional and structural stability of the composite hydrogels was evaluated, as these are the critical factors in the cartilage healing process. Hence, to evaluate the effect of incorporation of SDCM to PVA-Nb on dimensional stability, different PVA-Nb/SDCM formulations were cross-linked using 0.5 equivalent DT (equimolar functional groups) and LAP into cylindrical (bulk) hydrogels. Photographs of the hydrogels at 0, 2, and 28 days following cross-linking and soaking in PBS at 37 °C are shown in Figure 4A. The results of the diameter measurement showed that inclusion of SDCM to hydrogels decreased the amount of swelling. The average diameter of the PVA-Nb samples showed an increase in day 2 and 28 compared to day 1, which was 9.14 ± 0.04, 9.41 ± 0.09, and 8.00 mm, respectively. While, the average diameter of the PVA-Nb/SDCM50 samples showed a less pronounced increase in day 2 and 28 compared to day 1, which was 8.48 ± 0.16, 8.83 ± 0.23, and 8.00 mm, respectively (Figure 4B). The results of the swelling test during 48 h showed that the PVA-Nb group swelled significantly (168%) more than PVA-Nb/SDCM30 (122%) and PVA-Nb/SDCM50 (130%) groups (Figure 4C).

2.6. Mechanical Testing of Cross-Linked Hydrogels

The Young's modulus under compression after incubation in PBS was measured for the different bulk hydrogels after 2 and 28 days (Figure 5). A decrease in compressive mechanical properties was observed with the incorporation of larger amounts of SDCM in the formulation (50%), while the pure PVA-Nb gels and the 30% SDCM formulations had similar mechanical properties both at day 2 and 28. Two days after cross-linking, the compressive modulus of the PVA-Nb was 21.43 ± 1.16 kPa, whereas that of the PVA-Nb/SDCM30 and PVA-Nb/SDCM50 groups were 6% and 72% smaller, respectively ($p < 0.05$) (Figure 5). Furthermore, the compressive modulus of the PVA-Nb/SDCM30 group was 70% higher than that of the PVA-Nb/SDCM50 group ($p < 0.05$). Four weeks after incubation in PBS at 37 °C, the compressive modulus of the PVA-Nb was 8.07 ± 2.30 kPa, which was 62% smaller compared to day 2. Similarly, the PVA-Nb/SDCM30 group showed 53% reduction, while the PVA-Nb/SDCM50 group exhibited 30% reduction in elastic modulus compared to day 2 ($p < 0.05$). Over the 4 weeks of incubation, the PVA-Nb/SDCM50 showed the smallest reduction in mechanical properties (Figure 5).

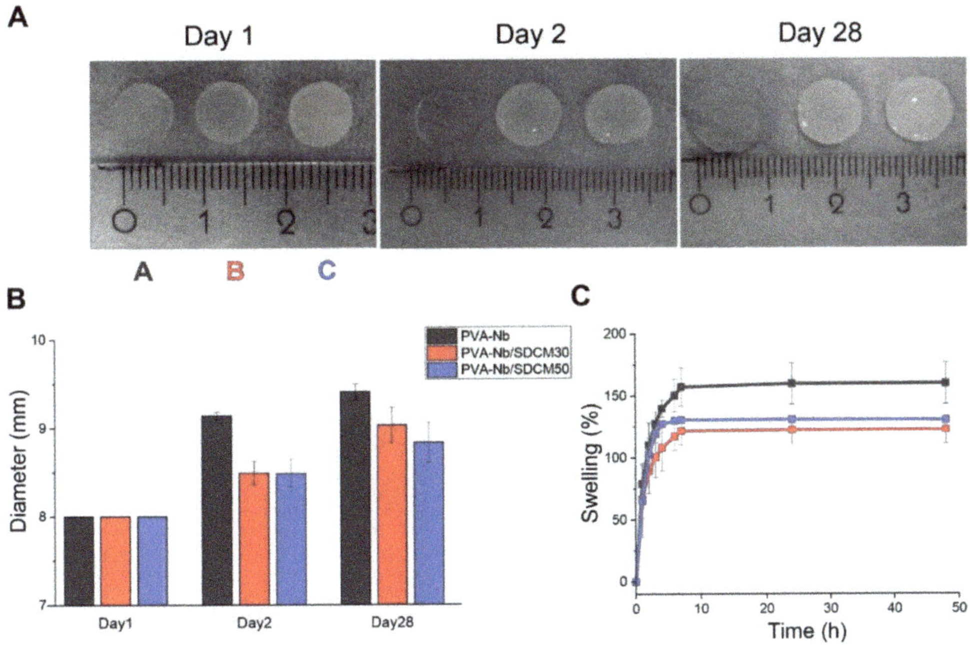

Figure 4. Swelling of norbornene-based hydrogels in phosphate-buffered saline (PBS). (**A**) The photograph of the hydrogels 0, 2, and 28 days after cross-linking and soaking in PBS at 37 °C. Morphology of PVA-Nb hydrogel (10 % wt/vol) cross-linked with 0.5 equivalent of thiol 2,2′-(Ethylenedioxy)diethanethiol (DT) and 2 mM lithium phenyl-2,4,6-trimethylbenzoylphosphinate (LAP). (**A**) PVA-Nb, (**B**) PVA-Nb/SDCM30, and (**C**) PVA-Nb/SDCM50 were successfully cross-linked into hydrogels. (**B**) The sizes of the hydrogels over the course of the 28-day experiment and the (**C**) swelling percentages are presented over the course of 48 h incubation in PBS.

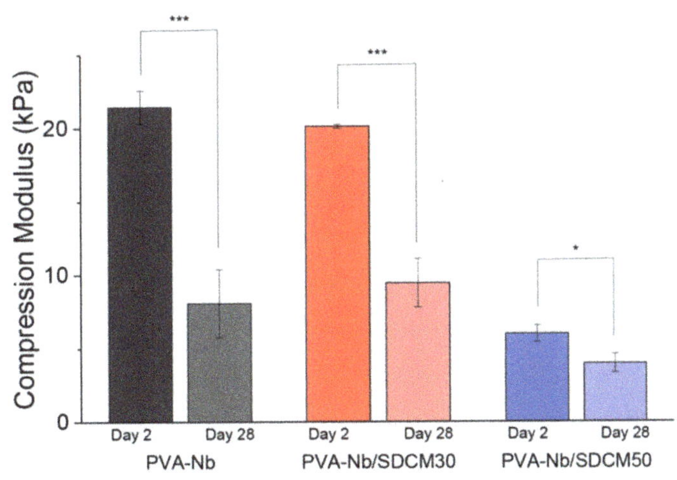

Figure 5. Compression modulus of PVA-Nb hydrogels (10 % wt/vol) cross-linked with 0.5 equivalent DT and 2 mM LAP. PVA-Nb, PVA-Nb/SDCM30, and PVA-Nb/SDCM50 after 2- and 28-days' incubation in PBS. * $p < 0.5$, *** $p < 0.01$.

2.7. *Live/Dead Assay of Bulk Hydrogels*

To evaluate the cytocompatibility of the hydrogels, ATDC5 viability in the bulk hydrogels was investigated at day 1 and 7 via Live/Dead staining (Figure S2). All hydrogel formulations showed over 70% cell viability at all time points measured. Both the PVA-Nb and the PVA-Nb/SDCM30 showed higher numbers of viable cells in day 1 (82% and 88%, respectively) with a slight decrease in the percentage of viable cells over the 7-day experiment (71% and 82%, respectively). The PVA-Nb/DSCM50 showed the lowest day 1 viability of the three samples (70%), yet the percentage of viable cells increased over the 7-day experiment (78%).

2.8. *Morphology and Distribution of the Chondrocytes*

To investigate cell remodeling and cell distribution of cell encapsulated hydrogels, Alcian blue staining was performed. Seven days after encapsulation the ATDC5 cells, hydrogels were stained with Alcian blue and Fast red. A round morphology, demonstrating a central core surrounded by a basophil transition zone stained with Alcian blue was observed (Figure S2-II). One of the main challenges in bioprinting is cell sedimentation and ensuring a homogenous distribution of cells within a printed hydrogel [39]; we observed homogeneously dispersed chondrocytes in all hydrogels at day 7.

2.9. *3D Bioprinting of PVA-Nb/SDCM*

Biomaterial inks were formulated from 10 wt% (PVA-Nb + SDCM), 2 mM LAP, and 0.5 equivalent DT. Three different compositions were prepared consisting: (A) PVA-Nb, (B) PVA-Nb/SDCM30, and (C) PVA-Nb/SDCM50. Scaffolds printed in the geometry of a cube. Porous-like structures can be seen in the PVA-Nb/SDCM50 scaffold (Figure S3C), but not in the printed PVA-Nb construct (Figure S3A). Of note, the printing conditions used to produce these scaffolds match those optimized for best structure. Finally, 5 mm/s deposition speed and 30 kPa extrusion pressure were used to produce constructs of the three biomaterial ink formulations.

2.10. *Bioink Bioprinting*

A syringe cartridge was loaded with PVA-Nb, SDCM, 2 mM LAP, and 0.5 equivalent DT and ATDC5 (1×10^7 cells/mL) in order to make the bioink formulation. Bioinks of two different formulations were tested (A) PVA-Nb and (B) PVA-Nb/SDCM50. Porous scaffolds were bioprinted in the geometry of a cube. Porous-like structures can be seen in the PVA-Nb/SDCM50 scaffold (Figure 6B1). Scaffolds were bioprinted with empirically optimized valued to obtain the best structure to match the theoretical dimensions (side length of 10 mm, 5 strands, 1.50 mm between strands, total height of 4 mm, 160 μm/layer, 25 layers). A 5 mm/s deposition speed with 30 kPa extrusion pressure was used to produce structures of the two bioink formulations.

Employing optimized conditions, multilayer constructions like a simple cubic structure were created with sufficient integrity over 25 layers of bioprinting. As shown in Figure 6, the structural integrity and initial 3D scaffolds of PVA-Nb/SDCM50 structure were well-preserved during culture. Some swelling slightly induced a decrease in the porosity of the constructs, yet porous-like structures were maintained and observed in the X-Y and Z planes as shown in Figure 6-B1,B7. The obtained scaffolds for the PVA-Nb-based bioinks showed limited resolution without a defined porous network. Furthermore, the swelling of this formulation was more pronounced when kept in culture during 7 days.

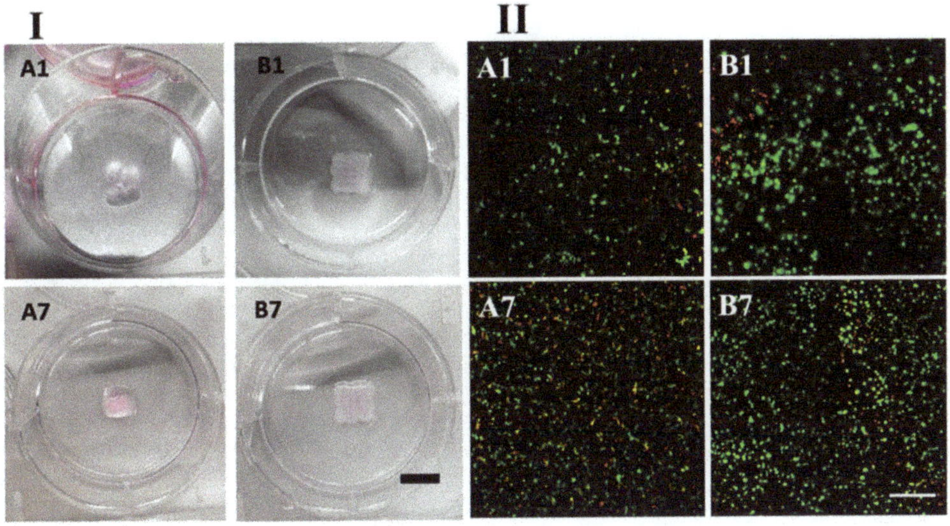

Figure 6. () Images of 3D bioprinted hydrogels, and () cell viability in bioprinted constructs. (**A**) PVA-Nb and (**B**) PVA-Nb/SDCM50 at day 1 and day 7. Scaffolds bioprinted in the geometry of a cube. Porous-like structures can be seen in the PVA-Nb/SDCM50 scaffold (scale bar 10: mm). Green stain represents live cells and red stain represents dead cells (scale bar 100 µm).

2.11. *Live/Dead Assay of Bioprinted Hydrogels*

To evaluate the cytocompatibility of the bioink and the bioprinting procedure, bioinks containing ATDC5 were bioprinted into 6-well plates and constructs were cultured in media for 1 week (Figure 6-I; A,B). To guarantee that the observed cytocompatibility was well-maintained within all depths of the printed constructs, Z-stack images for each sample were analyzed (Figure 6-II; A,B) and quantified (Figure S4). Florescent images of PVA-Nb constructs stained with Live/Dead assays exhibited high cell viabilities (~85%) at day 1, with a slight decrease in cell viability (~75%) at day 7 after printing (Figure 6-II; A1,A7). However, the PVA-NB/SDCM50 bioink showed suitable cell viabilities (~60%) at day 1 with a slight increase (~75%) in viability on day 7 (Figure 6-II; B1,B7). The results indicated that the viability increased in time in the PVA-NB/SDCM50 inks, while the PVA-Nb inks did not support an increase in viable cells. However, slightly more dead cells could be seen at day 1 in the PVA-Nb/SDCM50 scaffolds as compared to PVA-Nb (Figure 6-II; A1,B1).

3. Discussion

The aim of this study was to create a PVA/ECM formulation that can be utilized for bioprinting of cartilage constructs. By blending the bioactivity of solubilized and decellularized ECM with the tailorability and control of cross-linking of synthetically modified PVA, benefits from both biomaterials can be combined. In the present study, two approaches were investigated. First, PVA-A biomaterial ink containing SDCM and cross-linked with genipin was investigated, building upon previous promising results with this material as a chondroinductive scaffold [35]. However, the slow cross-linking limited the use of the formulations investigated as biomaterials inks or bioink for bioprinting. An alternative formulation, based on a norbornene-modified PVA (PVA-Nb) was developed in order to improve the gelation process with a rapid photo-induced thiol-ene cross-linking. PVA-Nb formulated with SDCM allowed to obtain bioprint 3D structures, and was proven suitable for bioink formulation showing high cell viability post-bioprinting up to 7 days. In a recently published study, a new class of cartilage ECM (cECM)-functionalized alginate bioink for the bioprinting of cartilaginous tissues was also investigated. The developed bioinks were 3D-printable, supported mesenchymal stromal cells viability post-printing and robust chondrogenesis in vitro in bioinks containing the higher percentage of cECM [40], in line with the similar findings in our current study.

The dense structure of cartilage presents exclusive challenges for decellularization in terms of tolerable elimination of cellular fragments while preserving non-collagenous components [41]. The most common chemicals to remove cell debris from the cartilage consists of SDS, Triton, trypsin, pepsin, hydrochloric acid, and guanidine hydrochloride, however, GAG loss during decellularization is known to occur using these materials [42]. Furthermore, retaining the GAG content through decellularization could potentially improve DCM properties including the elasticity, tensile, and compressive properties of DCM [43], as well as chondroinductivity effects in vitro and in vivo [44]. In a previous study, decellularization of bovine nucleus pulposus tissue was investigated using sodium deoxycholate, SDS, and Triton X-100, in combination with freeze-thaw cycles and DNase treatment [45]. The results showed no changes in collagen II maintenance, but there was significant GAG loss (80%) compared to the native cartilage. In another study, using the various detergents (Triton X-100, sulfobetaine-10, and sulfobetaine-16) in short cyclic intervals resulted in 49–55% GAG retention, while effectively removing cellular components [46]. In the present study, DC was decellularized using a modified method previously established via osmotic shock, detergent washes, and enzymatic treatment to preserve GAG and collagen content as much as possible (14). By documenting the biochemical makeup of the material during the decellularization process, it was observed that large decreases of DNA (95% removed) occurred during the decellularization process with good retention of GAGs (~100%) and hydroxyproline (82%) content. For consistency, both of our attempted approaches used identical DCM.

Various decellularized ECM-derived bioinks, such as liver, muscle, tendon, cardiac tissue, and cartilage, have been developed, which are used ECM in solubilized form [42]. The solubilization process occurred after lyophilization and destruction via pepsin digestion in the presence of acetic or hydrochloric acid solution following neutralization of the pH to 7.4. The resulted pregel was appropriate for combining with cells [47]. However, to use the DCM as a bioink, the decellularization process was followed by a solubilization and purification process via dialysis, which led to further decreases in DNA content (not detected), yet also decreased the GAG (to 12%) and hydroxyproline content (to 35%).

In our initial approach, PVA was successfully modified with amine functional groups. We observed that the degree of modification could be controlled via stoichiometry or reaction time. A decrease in reaction efficiency was observed when targeting higher levels of amination. Higher levels of functionalization led to faster genipin cross-linking speeds; however, these cross-linking kinetics within the hour range ultimately led to difficult optimization and a limited suitability for bioprinting. A recently developed approach for formulation of bioinks is the pre-cross-linking of precursor solutions to a state of higher viscosity, followed by final cross-linking after bioprinting [48]; a similar approach was employed here. Nevertheless, the 3D bioprinted structures from these formulations were not satisfactory (Figure S2A–C) and were only able to be manufactured after the pre-cross-linking reaction was initiated inside of the cartridge.

Noticing the poor cross-linking kinetics associated with genipin, in our second approach, we turned to the photoinitiated thiol-ene reaction chosen to introduce fast cross-linking in the 3D structure. Norbornene is one of the most reactive substrates for the radical thiol-ene reaction. PVA was successfully modified with cis-5-norbornene-endo-2,3-dicarboxylic anhydride (PVA-Nb, 7% functionalization). Different compositions of the PVA and SDCM were prepared with consistent solids content (10 wt%) and a varying composition (0%, 30%, and 50% SDCM). These hydrogel formulations showed rapid gelation upon exposure to UV light and were deemed promising for use as a bioink. Rathan et al. showed that alginate-SDCM bioinks containing 0.2% and 0.4% (wt/vol) SDCM and 2.45% alginate in DMEM had similar behavior to the formulations developed in this study in relation to the increase in resolution with the increase in SDCM and the overall cell viability observed [40].

Previous studies on formulation of bioinks based on solubilized ECM demonstrated that by manipulating the concentrations, molecular weights, and geometries, a range of shear elastic modulus values could be reached, spanning from 113.66 Pa to 19.798 $\pm$ 0.24 kPa, near 200-fold increase in stiffness [49]. In addition, the value of Young's modulus of pericellular matrix around chondrocytes reported in previous studies is in the order to 10–100 kPa [50]. Furthermore, the elastic modulus of healthy articular cartilage is in the range from 130 to 573 kPa [51]. However, comprehensive experiments with PVA/SDCM were conducted to evaluate the influence of the SDCM ratio in water uptake/swelling, mechanical properties, and bioprintability of the different formulations. The results showed that the mass swelling ratio decreased with increasing of SDCM content. Compression testing showed that both the pure PVA-Nb and the SDCM30 had similar mechanical properties over time (up to day 28 in PBS). Inclusion of higher amounts of SDCM (50%) led to a decrease in mechanical properties. These studies suggest that there is a delicate relationship between composition and cross-link density. As the PVA fraction is decreased (thereby decreasing the covalent cross-link density), SDCM can reinforce the hydrogel to a point; however, adding larger amounts of SDCM to the composition begins to compromise the compressive modulus. All in all, the PVA-Nb/SDCM50 was chosen as a bioink for bioprinting proposes, because of proper printability and higher cell compatibility compared to other formulations.

Pepsin and HCl solutions have been employed for ECM digestion in previous studies. These harsh environments might be predictable to result in undesirable degradation of biochemical content [49]. However, we have demonstrated the efficiency of this solubilizing process approach, and the resulted SDCM has shown to significantly increase ATDC5 cell viability when incorporated into the PVA hydrogels. We observed that the cytocompatibility of PVA-Nb formulations depended on the SDCM concentration with formulations containing 30 and 50 wt% SDCM exhibiting higher cell viability at day 7 compared to no SDCM incorporation. The applicability of the synthesized PVA-Nb/SDCM hydrogels for bioprinting was tested and results showed that this hydrogel exhibited a defined porous structure, which is dependent on the SDCM concentration. The structural integrity and initial 3D geometries of structure of PVA-Nb/SDCM50 hydrogels was well preserved after deposition. Results indicated that PVA-Nb/SDCM hydrogels can be considered as the versatile bioinks for cartilage bioprinting, as further adjustments to the DT cross-linking length and cross-linking density can be explored (via the PVA component) without drastically changing formulation viscosity.

4. Materials and Methods

4.1. Preparation of Solubilized Decellularized Cartilage

Three calf knees (males that were approximately 6–8 months old) were purchased from a local abattoir (Maastricht, Netherland). Articular cartilage from the knee joints was removed and collected using scalpels. The cartilage was then rinsed twice in PBS and stored at −20 °C. After freezing overnight, the cartilage was thawed (this was repeated three times) and then coarsely powdered with liquid nitrogen using a freeze-mill (SPEX SamplePrep, Metuchen, NJ, USA). This devitalized cartilage (DC) [14] was powdered to improve the diffusion of the solutions used for the decellularization process. The DC was placed into dialysis bags (3500 MWCO) and decellularized using a modified method previously established via osmotic shock, detergent washes, and enzymatic treatment [14]. The DC-containing bags were exposed in a hypertonic salt solution (HSS) under mild stirring overnight at room temperature. The bags were then placed in a stirrer at 220 rpm and washed twice with Triton X-100 (0.01% v/v) followed with HSS to permeabilize intact cellular membranes. The tissue was then exposed overnight with Benzonase (0.0625 KU mL^{-1}) at 37 °C, then, the tissue was further exposed overnight with sodium lauroyl sarcosine (SLS, 1% v/v), and finally, the tissue was washed with ethanol (40% v/v) at 50 rpm and then in PBS at 50 rpm followed by 24 h of rinsing with distillated water (dH_2O) at 220 rpm. The tissue was then removed from the bags, frozen, and lyophilized. Decellularized cartilage matrix (DCM) was solubilized via a modified protocol from a previously described method [14]. DCM powder was first mixed in 0.1 M HCl at a concentration of 10 mg DCM per mL HCl. Pepsin was then added at a concentration of 1 mg/mL and the solution was stirred at 200 rpm for 3 days at room temperature. The solution was then brought back to physiological pH, by adding 1 M NaOH. The solubilized DCM (SDCM) was then centrifuged at 10,000× *g* for 3 min and the supernatant was collected, frozen, lyophilized, cryo-milled (freeze-mill SPEX SamplePrep, Metuchen, NJ, USA) and stored at −20 °C until further use.

4.2. Biochemical Analysis

DNA Quantification: Tris-EDTA-buffered solution containing 1 mg/mL Proteinase K, 1 µg/mL iodoacetamide, and 18.5 µg/mL pepstatin A (all materials from Sigma-Aldrich, St. Louis, MO, USA) was used to digest each sample (n = 3) at 65 °C for 24 h. The CyQUANT DNA assay kit (Molecular Probes, Eugene, OR, USA) was used to quantify DNA content of samples according to the manufacturer's instructions, using a spectrofluorometer at 480/520 nm excitation/emission wavelength (CLARIOstar microplate reader; BMG Labtech, Cary, NC, USA) [52,53].

Sulfated Glycosaminoglycans (sGAG) Quantification: sGAG content was measured by following the method previously described on the Proteinase K-digested samples [53–55]. Briefly, sGAG content was determined spectrophotometrically with the 9-dimethylmethylene blue chloride (DMMB, Sigma-Aldrich) dye in PBE buffer (3.72 g/L Na_2EDTA and 14.2 g/L Na_2HPO4, pH 6.5) using a multiwell plate reader (Bio-TEK Instruments, Winooski, VT, USA) at 520 nm. The sGAG concentration of each sample was obtained by interpolation of its absorbance from the standard curve [53].

Collagen Quantification: The hydroxyproline assay kit (MAK008, Sigma-Aldrich) was used to quantify hydroxyproline content of samples according to the manufacturers' instructions. For this purpose, 100 µL of 12 M HCl (Sigma-Aldrich) was used to digest 10 mg of lyophilized samples (n = 3 for each group) at 120 °C for 3 h [54–56]. Finally, the absorbance of the hydroxyproline standard solution and digested samples was measured using multiwell plate reader (Bio-TEK Instruments) at 520 nm. After drawing the standard curve, hydroxyproline concentration of each sample was obtained by interpolation of its absorbance from the standard curve [35].

4.3. PVA Modifications

Synthesis of Amine-Functionalized PVA: PVA was modified with primary amine functional groups via a method previously described (Figure 7) [35,57]. Briefly, PVA was dissolved in dH_2O at 90 °C to prepare a 12% wt/vol solution. An equal volume of 10 mol% 4-aminobutyraldehyde diethyl acetal (4-ABA) (Sigma-Aldrich) was added dropwise to the solution (PVA was modified with three different target aminations, i.e., 10, 50, and 100 mol%). The pH was decreased close to zero by adding dropwise HCl 12 M, and then, the solution was mixed up to 24 h. Then, the pH was increased to 8.0 by adding an adequate quantity of NH_4OH in order to quench the reaction. The resulted solutions were purified by dialysis membrane (MWCO 3500, Sigma-Aldrich) in dH_2O and finally lyophilized. Samples were collected in three time points during reaction (1, 2, and 24 h) for amination analysis. Sample nomenclature was defined as PVA-A*-** where * corresponds to the reaction time (h) and ** the targeted amination percentage. NMRs are shown in Figures S5–S7.

Synthesis of Norbornene-Functionalized PVA: PVA was modified with cis-5-norbornene-endo-2,3-dicarboxylic anhydride (Nb) via a method previously described (Figure 7) [18]. Briefly, a 20% (wt/vol) solution of Nb (0.5 equivalent with respect to PVA subunits) in anhydrous DMSO was added dropwise to a stirred 4% (wt/vol) PVA solution in anhydrous DMSO, and 0.125 wt% p-toluene sulfonic acid (p-TsOH) was added as catalyst at 50 °C under nitrogen atmosphere and stirred during 24 h. After cooling to room temperature, the reaction mixture was dialyzed (MWCO 3500, Sigma-Aldrich) against dH_2O (7 cycles), sodium carbonate buffer (pH ≈ 8.2, overnight, 1 cycle), and again dH_2O (7 cycles). The aqueous solution was evaporated and dried in a rotary evaporator, to obtain functionalized PVA-Nb as a white powder. NMRs are shown in Figure S8.

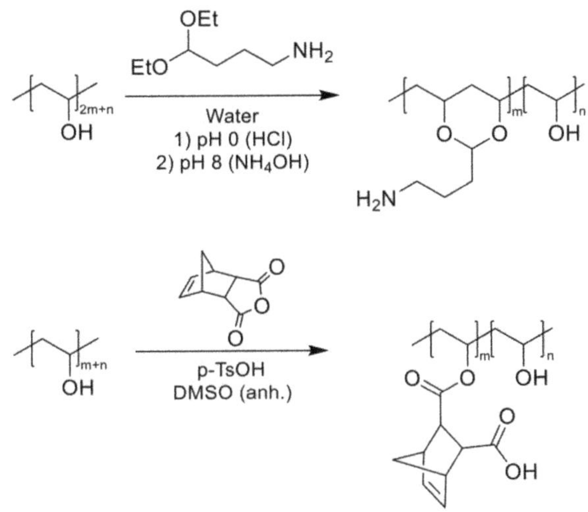

Figure 7. Schematic representation of the modification of poly(vinyl alcohol) (PVA) with amine groups (**top**) and 2-norbornene groups (**bottom**).

4.4. Nuclear Magnetic Resonance (^1H NMR) Spectroscopy

^1H NMR was used to confirm the modification of the PVA with amine and norbornene groups. For this, samples were dissolved in D_2O, and the spectra was acquired using a Bruker 700 spectrometer. The signal of deuterated solvent (D_2O at 4.79 ppm) was used as reference [13].

4.5. PVA-A/SDCM and PVA-Nb/SDCM Hydrogel Fabrication and Characterization

PVA-A/Genipin Gelation Time: To investigate the influence of the amination percentage with the gelation time, PVA-A hydrogels with different degrees of amination were cross-linked with genipin (0.1% dry weight of genipin/dry weight of sample (wt/wt)). The color change to pale blue and vial inversion test were used to show PVA-A cross-linking after 72 h at 37 °C. Briefly, different modified PVA-A formulations were dissolved in dH_2O at 60 °C, and nine different PVA-A solutions were prepared. In addition, genipin (sc-203057A, Santacruz, Santa Cruz, CA, USA) was prepared in ethanol at the concentration of 0.1% wt/wt. Then, genipin was added to each PVA-A solution and homogenized. Aliquots of homogenized solution (500 µL) were placed in silicon molds and subsequently cross-linked in incubator at 37 °C for 3 days [34].

3D Bioprinting of PVA-A/SDCM Biomaterial Inks: To evaluate printability of the PVA-A/SDCM, different biomaterial inks were formulated from 10 wt% (PVA-A + SDCM) and 0.1% wt/wt genipin (Figure 1). Three different compositions were prepared consisting: (A) PVA-A, (B) PVA-A/SDCM30 (70% PVA-A:30% SDCM), and (C) PVA-A/SDCM50 (50% PVA-A:50% SDCM). The biomaterial inks were loaded into syringes, assembled into a custom holder, designed to hold a cartridge. Bioprinting was carried out with a G15 needle (1.4 mm ID) on a BioScaffolder (GeSiM—Gesellschaft fuür Silizium-Mikrosysteme mbH, Radeberg, Germany) controlled through proprietary software. The hydrogels were printed in the geometry of a cube (10 mm × 10 mm × 2 mm), with a designed side length of 10 mm (3 strands, 3.50 mm between strands), and a total height of 2 mm (80 µm/layer, 25 layers).

PVA-Nb/SDCM Hydrogel Fabrication and Characterization: Due to the step growth of thiol-ene reaction, the selection of cross-linker has a key outcome on the resulted network, such as mechanical properties and mesh size [13]. We chose to use a small bifunctional thiol 2,2′-(Ethylenedioxy)diethanethiol (DT) to cross-link the hydrogel system. Lithium phenyl-2,4,6-trimethylbenzoylphosphinate (LAP) was employed as the photoinitiator since it is efficient at 365 nm, is water-soluble, and has shown low toxicity when used to cross-link cell-laden hydrogels [58]. Stock solutions of the PVA-Nb, SDCM, DT, and the photoinitiator LAP were prepared in phosphate-buffered saline (PBS) and combined to obtain the desired hydrogel formulations (Figure 1). Three formulation consisting of PVA-Nb/SDCM (100:0), PVA-Nb/SDCM30, and PVA-Nb/SDCM50 were prepared. All formulations contained 10 wt% macromere (PVA + SDCM), 0.5 equivalent DT (with respect to Nb subunits), and 2 mM LAP. Adequate volumes of each formulation were pipetted into a PDMS cylindrical molds (well diameter of 8 mm and thickness of 3 mm). The molds were exposed to UV light (50 mW/cm^2 at 10 cm distance) for 60 s to fabricate bulk hydrogel samples.

4.6. Swelling Degree and Volume Changes

The swelling of the hydrogels was measured via the mass swelling ratio to evaluate the density changes over time of the network structure. Bulk hydrogel samples were equilibrated for 48 h and the swollen weight was recorded at different time points. The swelling degree was calculated as the ratio of swollen weight to initial weight [35]. The geometric mean diameter of each gel was measured directly using a caliper to calculate gel dimension on day 1, 2, and 28.

4.7. Mechanical Testing of the Bulk Hydrogels

The gels were allowed to swell to equilibrium in PBS and mechanical testing was performed at day 2 and 28. The gels (n = 3) were compressed with load cell of 1000 g in an unconfined setup until mechanical failure at a rate of 0.01 mm/s, and the compressive modulus was calculated as the slope of the linear portion of the stress–strain curve (i.e., 5–15% strain) [35].

4.8. Cell Culture

A teratocarcinoma-derived chondrogenic cell line (ATDC5) was used for cell study. The cells were cultured at a density of 1×10^5 cells.cm^2 in DMEM (F-12) media supplemented with 1% penicillin streptavidin and 5% fetal bovine serum. Cells were subculture at 80% confluence [59].

4.9. Live/Dead Assay of Hydrogels

The viability of the ATDC5 exposed to the hydrogel environments was evaluated using a LIVE/DEAD viability/cytotoxicity kit (Thermo Fisher, Waltham, MA, USA). Before cell seeding, the materials were sterilized using filtration, freeze drying, and UV irradiation for 2 h. After a week culture, 1 mL of calcein-AM stock solution (1 µM) was added to each scaffold and incubated for 20 min at 37 °C. After that, 1 mL of the ethidium homodimer-1 stock solution (0.036 µM) was added to the wells and incubated for an additional 10 min at 37 °C. The dye solutions were then aspirated from the wells, and 1 mL of phenol red-free medium (DMEM, Sigma-Aldrich) was added to wells before imaging. Calcein–AM, a cell-permeant dye is converted to a green-fluorescent calcein by viable cells, and ethidium homodimer-1 binds to nucleic acids of cells with damaged membranes to produce red fluorescence. For cell imaging, a live cell imaging Nikon TI-E with environmental control with a 20× objective (WD = 15 mm, NA = 0.3) was used. Images were typically acquired via 1024 µm × 1024 µm scans with Z stacks of 5–10 µm on three different hydrogels. Cell viability was estimated through quantification of the number of live cells over total number of cells using ImageJ software [13].

4.10. Histological Analyses

Alcian blue staining is usually used at pH = 2.5, in order to be fixed with acidic groups of carboxylic muco-polysaccharides by electrostatic binding. Acidic polysaccharides such as glycosaminoglycans in cartilages will be stained blue, cytoplasm will be stained pale pink, and nuclei will be stained pink to red. Alcian blue staining was performed in cultured constructs to investigate cell remodeling after encapsulation in the hydrogels. Briefly, after a week culture, ATDC5-containing constructs were fixed during 48 h at 4 °C in 10% buffered formalin in PBS solution. Constructs were processed by dehydration, clearing, paraffin embedding, and sectioning. Xylene-cleared sections were treated with Alcian blue 1% for 45 min and then with 0.1% nuclear Fast red for 3 min; rinsed with distilled water; and dehydrated, cleared, and mounted on microscope slides [47].

4.11. 3D Bioprinting of the PVA-Nb/SDCM Biomaterial Inks and Bioinks

To evaluate printability of the PVA-Nb/SDCM, biomaterial inks were formulated from 10 wt% (PVA-Nb + SDCM), 2 mM LAP, and 0.5 equivalent DT. Three different compositions were prepared consisting: (A) PVA-Nb, (B) PVA-Nb/SDCM30, and (C) PVA-Nb/SDCM50. To evaluate cell compatibility of the bioprinting approach and synthesized PVA-Nb/SDCM biomaterials inks, PVA-Nb with specific formulations of photoinitiator and cross-linker used as bioinks were prepared with the addition of ATDC5 cells (10 million cells/mL). The biomaterial inks and bioinks were loaded into black syringes, assembled into a custom holder, designed to hold a cartridge and LED light source (Thorlabs, 365 nm, 10 mW/cm^2). Bioprinting was carried out with a G15 needle (1.4 mm ID) on a BioScaffolder (GeSiM—Gesellschaft fuür Silizium-Mikrosysteme mbH, Germany) controlled through proprietary software. In general, scaffold geometries and settings were set to a cube (10 mm × 10 mm × 2 mm), comprising 5 meandered strands placed at a distance of 2 mm apart. The deposition angle was turned 90° after each layer. Height of each layer was set to 0.2 mm, and the number of layers was varied according to experiment requirements [13].

4.12. Statistical Analysis

For the statistical analysis, data were evaluated using the Student's *t*-test.

5. Conclusions

The aim of this study was to develop cell-compatible PVA/SDCM biomaterial inks that can be used for cartilage bioprinting purposes. PVA-based synthetic hydrogels allow for good control of mechanical properties, but suffer from a lack of biological properties. In contrast, ECM components contain structures that are ideal for cell viability. Hence, hybrid hydrogels consisting of PVA and SDCM, combining the advantages of both PVA and ECM derivatives, are suitable candidates for TE applications such as bioprinted constructs for cartilage regeneration. The initial PVA-A system developed in this study will require further cross-linking optimization to allow suitable bioprintability. Conversely, the thiol-ene reaction enabled by the PVA-Nb system allowed a suitable bioprintability that was further enhanced with the addition of SDCM. Different composition of the PVA and SDCM supported a high cell viability after bioprinting in the PVA/SDCM50 formulation. The structural integrity of PVA-Nb/SDCM50 hydrogels was well-preserved during culture. Our results indicated that the mixing of well-defined synthetic polymers like PVA-Nb, can be combined with SDCM in order to create promising bioink formulations.

Supplementary Materials: The following are available online at https://www.mdpi.com/article/10.3390/ijms22083901/s1, Figures S1–S8.

Author Contributions: M.S.: Investigation; Methodology; Validation, Writing original draft; Funding acquisition. S.H.: Investigation; Methodology; review and editing. C.v.B.: Funding acquisition, review and editing. L.M.: Conceptualization; Funding acquisition; Data curation; review and editing. C.M.: Visualization; Data curation; Form analysis; Writing—review and editing; M.B.B.:

Conceptualization; Visualization; Data curation; Form analysis; Software; Writing—review and editing. All authors have read and agreed to the published version of the manuscript.

Funding: This research was funded by the Isfahan University of Medical Science under grant number 198274, the Province of Limburg, and the European Research Council (ERC) under the Horizons 2020 research and innovation programme grant number 694,801.

Institutional Review Board Statement: Not applicable for this study.

Informed Consent Statement: Not applicable for this study.

Data Availability Statement: The data presented in this study are available in this article Int. J. Mol. Sci. and its Supplementary Material.

Conflicts of Interest: The authors declare no conflict of interest.

References

1. Mota, C.; Puppi, D.; Chiellini, F.; Chiellini, E. Additive manufacturing techniques for the production of tissue engineering constructs. *J. Tissue Eng. Regen. Med.* **2015**, *9*, 174–190. [CrossRef] [PubMed]
2. Ouyang, L.; Highley, C.B.; Sun, W.; Burdick, J.A. A generalizable strategy for the 3d bioprinting of hydrogels from nonviscous photo-crosslinkable inks. *Adv. Mater.* **2017**, *29*, 1604983. [CrossRef] [PubMed]
3. Moroni, L.; Burdick, J.A.; Highley, C.; Lee, S.J.; Morimoto, Y.; Takeuchi, S.; Yoo, J.J. Biofabrication strategies for 3D in vitro models and regenerative medicine. *Nat. Rev. Mater.* **2018**, *3*, 21–37. [CrossRef] [PubMed]
4. Moroni, L.; Boland, T.; Burdick, J.A.; De Maria, C.; Derby, B.; Forgacs, G.; Groll, J.; Li, Q.; Malda, J.; Mironov, V.A. Biofabrication: A guide to technology and terminology. *Trends Biotechnol.* **2018**, *36*, 384–402. [CrossRef] [PubMed]
5. Murphy, S.V.; Atala, A. 3D bioprinting of tissues and organs. *Nat. Biotechnol.* **2014**, *32*, 773. [CrossRef]
6. You, F.; Eames, B.F.; Chen, X. Application of extrusion-based hydrogel bioprinting for cartilage tissue engineering. *Int. J. Mol. Sci.* **2017**, *18*, 1597. [CrossRef]
7. Malda, J.; Visser, J.; Melchels, F.P.; Jüngst, T.; Hennink, W.E.; Dhert, W.J.; Groll, J.; Hutmacher, D.W. 25th anniversary article: Engineering hydrogels for biofabrication. *Adv. Mater.* **2013**, *25*, 5011–5028. [CrossRef]
8. Highley, C.B.; Rodell, C.B.; Burdick, J.A. Direct 3D printing of shear-thinning hydrogels into self-healing hydrogels. *Adv. Mater.* **2015**, *27*, 5075–5079. [CrossRef]
9. Donderwinkel, I.; van Hest, J.C.; Cameron, N.R. Bio-inks for 3D bioprinting: Recent advances and future prospects. *Polym. Chem.* **2017**, *8*, 4451–4471. [CrossRef]
10. Bertlein, S.; Brown, G.; Lim, K.S.; Jungst, T.; Boeck, T.; Blunk, T.; Tessmar, J.; Hooper, G.J.; Woodfield, T.B.; Groll, J. thiol-ene clickable gelatin: A platform bioink for multiple 3D biofabrication technologies. *Adv. Mater.* **2017**, *29*, 1703404. [CrossRef]
11. Ying, G.; Jiang, N.; Yu, C.; Zhang, Y.S. Three-dimensional bioprinting of gelatin methacryloyl (GelMA). *Bio-Des. Manuf.* **2018**, *1*, 215–224. [CrossRef]
12. Rhee, S.; Puetzer, J.L.; Mason, B.N.; Reinhart-King, C.A.; Bonassar, L.J. 3D bioprinting of spatially heterogeneous collagen constructs for cartilage tissue engineering. *ACS Biomater. Sci. Eng.* **2016**, *2*, 1800–1805. [CrossRef]
13. Ooi, H.W.; Mota, C.; Ten Cate, A.T.; Calore, A.; Moroni, L.; Baker, M.B. thiol-ene Alginate Hydrogels as Versatile Bioinks for Bioprinting. *Biomacromolecules* **2018**, *19*, 3390–3400. [CrossRef]
14. Beck, E.C.; Barragan, M.; Tadros, M.H.; Gehrke, S.H.; Detamore, M.S. Approaching the compressive modulus of articular cartilage with a decellularized cartilage-based hydrogel. *Acta Biomater.* **2016**, *38*, 94–105. [CrossRef]
15. Gao, G.; Yonezawa, T.; Hubbell, K.; Dai, G.; Cui, X. Inkjet-bioprinted acrylated peptides and PEG hydrogel with human mesenchymal stem cells promote robust bone and cartilage formation with minimal printhead clogging. *Biotechnol. J.* **2015**, *10*, 1568–1577. [CrossRef]
16. Müller, M.; Becher, J.; Schnabelrauch, M.; Zenobi-Wong, M. Nanostructured Pluronic hydrogels as bioinks for 3D bioprinting. *Biofabrication* **2015**, *7*, 035006. [CrossRef]
17. Ng, W.; Yeong, W.; Naing, M. Polyvinylpyrrolidone-based bio-ink improves cell viability and homogeneity during drop-on-demand printing. *Materials* **2017**, *10*, 190. [CrossRef]
18. Baudis, S.; Bomze, D.; Markovic, M.; Gruber, P.; Ovsianikov, A.; Liska, R. Modular material system for the microfabrication of biocompatible hydrogels based on thiol-ene-modified poly (vinyl alcohol). *J. Polym. Sci. Part A Polym. Chem.* **2016**, *54*, 2060–2070. [CrossRef]
19. Sakai, S.; Kamei, H.; Mori, T.; Hotta, T.; Ohi, H.; Nakahata, M.; Taya, M. Visible Light-Induced Hydrogelation of an Alginate Derivative and Application to Stereolithographic Bioprinting Using a Visible Light Projector and Acid Red. *Biomacromolecules* **2018**, *19*, 672–679. [CrossRef]
20. Stichler, S.; Jungst, T.; Schamel, M.; Zilkowski, I.; Kuhlmann, M.; Böck, T.; Blunk, T.; Teßmar, J.; Groll, J. Thiol-ene clickable poly (glycidol) hydrogels for biofabrication. *Ann. Biomed. Eng.* **2017**, *45*, 273–285. [CrossRef]
21. Vega, S.L.; Kwon, M.Y.; Song, K.H.; Wang, C.; Mauck, R.L.; Han, L.; Burdick, J.A. Combinatorial hydrogels with biochemical gradients for screening 3D cellular microenvironments. *Nat. Commun.* **2018**, *9*, 1–10. [CrossRef] [PubMed]

22. Senyurt, A.F.; Wei, H.; Hoyle, C.E.; Piland, S.G.; Gould, T.E. Ternary thiol– ene/acrylate photopolymers: Effect of acrylate structure on mechanical properties. *Macromolecules* **2007**, *40*, 4901–4909. [CrossRef]
23. Lee, H.J.; Fernandes-Cunha, G.M.; Myung, D. In situ-forming hyaluronic acid hydrogel through visible light-induced thiol-ene reaction. *React. Funct. Polym.* **2018**, *131*, 29–35. [CrossRef]
24. Nair, B. Final report on the safety assessment of polyvinyl alcohol. *Int. J. Toxicol.* **1998**, *17*, 67–92. [CrossRef]
25. Muppalaneni, S.; Omidian, H. Polyvinyl alcohol in medicine and pharmacy: A perspective. *J. Dev. Drugs* **2013**, *2*, 1–5. [CrossRef]
26. Marin, E.; Rojas, J.; Ciro, Y. A review of polyvinyl alcohol derivatives: Promising materials for pharmaceutical and biomedical applications. *Afr. J. Pharm. Pharmacol.* **2014**, *8*, 674–684.
27. Schmedlen, R.H.; Masters, K.S.; West, J.L. Photocrosslinkable polyvinyl alcohol hydrogels that can be modified with cell adhesion peptides for use in tissue engineering. *Biomaterials* **2002**, *23*, 4325–4332. [CrossRef]
28. Ossipov, D.A.; Brännvall, K.; Forsberg-Nilsson, K.; Hilborn, J. Formation of the first injectable poly (vinyl alcohol) hydrogel by mixing of functional PVA precursors. *J. Appl. Polym. Sci.* **2007**, *106*, 60–70. [CrossRef]
29. Ossipov, D.A.; Piskounova, S.; Hilborn, J.N. Poly (vinyl alcohol) cross-linkers for in vivo injectable hydrogels. *Macromolecules* **2008**, *41*, 3971–3982. [CrossRef]
30. Binder, W.H.; Sachsenhofer, R. 'Click'chemistry in polymer and material science: An update. *Macromol. Rapid Commun.* **2008**, *29*, 952–981. [CrossRef]
31. Smetana, K., Jr. Cell biology of hydrogels. *Biomaterials* **1993**, *14*, 1046–1050. [CrossRef]
32. Garakani, S.S.; Khanmohammadi, M.; Atoufi, Z.; Kamrava, S.K.; Setayeshmehr, M.; Alizadeh, R.; Faghihi, F.; Bagher, Z.; Davachi, S.M.; Abbaspourrad, A. Fabrication of chitosan/agarose scaffolds containing extracellular matrix for tissue engineering applications. *Int. J. Biol. Macromol.* **2020**, *143*, 533–545. [CrossRef]
33. Setayeshmehr, M.; Esfandiari, E.; Rafieinia, M.; Hashemibeni, B.; Taheri-Kafrani, A.; Samadikuchaksaraei, A.; Kaplan, D.L.; Moroni, L.; Joghataei, M.T. Hybrid and composite scaffolds based on extracellular matrices for cartilage tissue engineering. *Tissue Eng. Part B Rev.* **2019**, *25*, 202–224. [CrossRef]
34. Cheng, N.-C.; Estes, B.T.; Young, T.-H.; Guilak, F. Genipin-crosslinked cartilage-derived matrix as a scaffold for human adipose-derived stem cell chondrogenesis. *Tissue Eng. Part A* **2012**, *19*, 484–496. [CrossRef]
35. Setayeshmehr, M.; Esfandiari, E.; Hashemibeni, B.; Tavakoli, A.H.; Rafienia, M.; Samadikuchaksaraei, A.; Moroni, L.; Joghataei, M.T. Chondrogenesis of human adipose-derived mesenchymal stromal cells on the [devitalized costal cartilage matrix/poly (vinyl alcohol)/fibrin] hybrid scaffolds. *Eur. Polym. J.* **2019**, *118*, 528–541. [CrossRef]
36. Skardal, A.; Devarasetty, M.; Kang, H.-W.; Mead, I.; Bishop, C.; Shupe, T.; Lee, S.J.; Jackson, J.; Yoo, J.; Soker, S. A hydrogel bioink toolkit for mimicking native tissue biochemical and mechanical properties in bioprinted tissue constructs. *Acta Biomater.* **2015**, *25*, 24–34. [CrossRef]
37. Honarvar, A.; Karbasi, S.; Hashemibeni, B.; Setayeshmehr, M.; Kazemi, M.; Valiani, A. Effects of cartilage acellular solubilised ECM on physicomechanical and biological properties of polycaprolactone/fibrin hybrid scaffold fabricated by 3D-printing and salt-leaching methods. *Mater. Technol.* **2020**, 1–9. [CrossRef]
38. Song, B.R.; Yang, S.S.; Jin, H.; Lee, S.H.; Lee, J.H.; Park, S.R.; Park, S.-H.; Min, B.-H. Three dimensional plotted extracellular matrix scaffolds using a rapid prototyping for tissue engineering application. *Tissue Eng. Regen. Med.* **2015**, *12*, 172–180. [CrossRef]
39. Dubbin, K.; Hori, Y.; Lewis, K.K.; Heilshorn, S.C. Dual-Stage Crosslinking of a Gel-Phase Bioink Improves Cell Viability and Homogeneity for 3D Bioprinting. *Adv. Healthc. Mater.* **2016**, *5*, 2488–2492. [CrossRef]
40. Rathan, S.; Dejob, L.; Schipani, R.; Haffner, B.; Möbius, M.E.; Kelly, D.J. Fiber reinforced cartilage ECM functionalized bioinks for functional cartilage tissue engineering. *Adv. Healthc. Mater.* **2019**, *8*, 1801501. [CrossRef]
41. Crapo, P.M.; Gilbert, T.W.; Badylak, S.F. An overview of tissue and whole organ decellularization processes. *Biomaterials* **2011**, *32*, 3233–3243. [CrossRef] [PubMed]
42. Vernengo, A.J.; Grad, S.; Eglin, D.; Alini, M.; Li, Z. Bioprinting Tissue Analogues with Decellularized Extracellular Matrix Bioink for Regeneration and Tissue Models of Cartilage and Intervertebral Discs. *Adv. Funct. Mater.* **2020**, *30*, 1909044. [CrossRef]
43. Kheir, E.; Stapleton, T.; Shaw, D.; Jin, Z.; Fisher, J.; Ingham, E. Development and characterization of an acellular porcine cartilage bone matrix for use in tissue engineering. *J. Biomed. Mater. Res. Part A* **2011**, *99*, 283–294. [CrossRef] [PubMed]
44. Ingavle, G.C.; Frei, A.W.; Gehrke, S.H.; Detamore, M.S. Incorporation of aggrecan in interpenetrating network hydrogels to improve cellular performance for cartilage tissue engineering. *Tissue Eng. Part A* **2013**, *19*, 1349–1359. [CrossRef] [PubMed]
45. Illien-Jünger, S.; Sedaghatpour, D.D.; Laudier, D.M.; Hecht, A.C.; Qureshi, S.A.; Iatridis, J.C. Development of a bovine decellularized extracellular matrix-biomaterial for nucleus pulposus regeneration. *J. Orthop. Res.* **2016**, *34*, 876–888. [CrossRef] [PubMed]
46. Wachs, R.A.; Hoogenboezem, E.N.; Huda, H.I.; Xin, S.; Porvasnik, S.L.; Schmidt, C.E. Creation of an injectable in situ gelling native extracellular matrix for nucleus pulposus tissue engineering. *Spine J.* **2017**, *17*, 435–444. [CrossRef]
47. Noor, N.; Shapira, A.; Edri, R.; Gal, I.; Wertheim, L.; Dvir, T. 3D printing of personalized thick and perfusable cardiac patches and hearts. *Adv. Sci.* **2019**, *6*, 1900344. [CrossRef] [PubMed]
48. Rutz, A.L.; Hyland, K.E.; Jakus, A.E.; Burghardt, W.R.; Shah, R.N. A multimaterial bioink method for 3D printing tunable, cell-compatible hydrogels. *Adv. Mater.* **2015**, *27*, 1607–1614. [CrossRef]
49. Zhang, J.; Yang, Z.; Li, C.; Dou, Y.; Li, Y.; Thote, T.; Wang, D.-A.; Ge, Z. Cells behave distinctly within sponges and hydro-gels due to differences of internal structure. *Tissue Eng. Part A* **2013**, *19*, 2166–2175. [CrossRef]

50. Guilak, F.; Jones, W.R.; Ting-Beall, H.P.; Lee, G.M. The deformation behavior and mechanical properties of chondrocytes in articular cartilage. *Osteoarthr. Cartil.* **1999**, *7*, 59–70. [CrossRef]
51. Vidal-Lesso, A.; Ledesma-Orozco, E.; Daza-Benítez, L.; Lesso-Arroyo, R. Mechanical characterization of femoral cartilage under unicompartimental osteoarthritis. *Ing. MecánicaTecnol. Desarro.* **2014**, *4*, 239–246.
52. Rowland, C.R.; Lennon, D.P.; Caplan, A.I.; Guilak, F. The effects of crosslinking of scaffolds engineered from cartilage ECM on the chondrogenic differentiation of MSCs. *Biomaterials* **2013**, *34*, 5802–5812. [CrossRef] [PubMed]
53. Damanik, F.F.; Spadolini, G.; Rotmans, J.; Farè, S.; Moroni, L. Biological activity of human mesenchymal stromal cells on polymeric electrospun scaffolds. *Biomater. Sci.* **2019**, *7*, 1088–1100. [CrossRef] [PubMed]
54. Fermor, H.L.; Russell, S.L.; Williams, S.; Fisher, J.; Ingham, E. Development and characterisation of a decellularised bovine osteochondral biomaterial for cartilage repair. *J. Mater. Sci. Mater. Med.* **2015**, *26*, 1–11. [CrossRef] [PubMed]
55. Farndale, R.W.; Sayers, C.A.; Barrett, A.J. A direct spectrophotometric microassay for sulfated glycosaminoglycans in cartilage cultures. *Connect. Tissue Res.* **1982**, *9*, 247–248. [CrossRef] [PubMed]
56. Edwards, C.; O'Brien, W.D. Modified assay for determination of hydroxyproline in a tissue hydrolyzate. *Clin. Chim. Acta* **1980**, *104*, 161–167. [CrossRef]
57. Rafat, M.; Rotenstein, L.S.; You, J.-O.; Auguste, D.T. Dual functionalized PVA hydrogels that adhere endothelial cells synergistically. *Biomaterials* **2012**, *33*, 3880–3886. [CrossRef]
58. Fairbanks, B.D.; Schwartz, M.P.; Bowman, C.N.; Anseth, K.S. Photoinitiated polymerization of PEG-diacrylate with lithium phenyl-2, 4, 6-trimethylbenzoylphosphinate: Polymerization rate and cytocompatibility. *Biomaterials* **2009**, *30*, 6702–6707. [CrossRef]
59. Camarero-Espinosa, S.; Calore, A.; Wilbers, A.; Harings, J.; Moroni, L. Additive manufacturing of an elastic poly (ester) urethane for cartilage tissue engineering. *Acta Biomater.* **2020**, *102*, 192–204. [CrossRef]

International Journal of
Molecular Sciences

MDPI

Article

Alginate-*g*-PNIPAM-Based Thermo/Shear-Responsive Injectable Hydrogels: Tailoring the Rheological Properties by Adjusting the LCST of the Grafting Chains

Konstantinos Safakas, Sofia-Falia Saravanou, Zacharoula Iatridi and Constantinos Tsitsilianis *

Department of Chemical Engineering, University of Patras, 26500 Patras, Greece; kostassaf@hotmail.com (K.S.); faliasaravanou@hotmail.com (S.-F.S.); iatridi@upatras.gr (Z.I.)
* Correspondence: ct@chemeng.upatras.gr; Tel.: +30-2610-969531

Abstract: Graft copolymers of alginate backbone and *N*-isopropylacrylamide/*N*-tert-butylacrylamide random copolymer, P(NIPAM$_X$-*co*-NtBAM$_Y$), side chains (stickers) with various NtBAM content were designed and explored in aqueous media. Self-assembling thermoresponsive hydrogels are formed upon heating, in all cases, through the hydrophobic association of the P(NIPAM$_X$-*co*-NtBAM$_Y$) sticky pendant chains. The rheological properties of the formulations depend remarkably on the NtBAM hydrophobic content, which regulates the lower critical solution temperature (LCST) and, in turn, the stickers' thermo-responsiveness. The gelation point, T_{gel}, was shifted to lower temperatures from 38 to 20 °C by enriching the PNIPAM chains with 20 mol % NtBAM, shifting accordingly to the gelation temperature window. The consequences of the T_{gel} shift to the hydrogels' rheological properties are significant at room and body temperature. For instance, at 37 °C, the storage modulus increases about two orders of magnitude and the terminal relaxation time increase about 10 orders of magnitude by enriching the stickers with 20 mol % hydrophobic moieties. Two main thermo-induced behaviors were revealed, characterized by a sol–gel and a weak gel–stiff gel transition for the copolymer with stickers of low (0.6 mol %) and high (14, 20 mol %) NtBAM content, respectively. The first type of hydrogels is easily injectable, while for the second one, the injectability is provided by shear-thinning effects. The influence of the type of media (phosphate buffer (PB), phosphate-buffered saline (PBS), Dulbecco's modified Eagle's medium (DMEM)) on the hydrogel properties was also explored and discussed. The 4 wt % NaALG-g-P(NIPAM$_{80}$-*co*-NtBAM$_{20}$)/DMEM formulation showed excellent shear-induced injectability at room temperature and instantaneous thermo-induced gel stiffening at body temperature, rendering it a good candidate for cell transplantation potential applications.

Keywords: alginate; PNIPAM-based graft copolymers; hydrogel; thermo-responsive; shear-responsive; LCST adjustment; sol–gel transition; rheological properties; injectability

Citation: Safakas, K.; Saravanou, S.-F.; Iatridi, Z.; Tsitsilianis, C. Alginate-*g*-PNIPAM-Based Thermo/Shear-Responsive Injectable Hydrogels: Tailoring the Rheological Properties by Adjusting the LCST of the Grafting Chains. *Int. J. Mol. Sci.* **2021**, *22*, 3824. https://doi.org/10.3390/ijms22083824

Academic Editor: Eric Guibal

Received: 21 March 2021
Accepted: 5 April 2021
Published: 7 April 2021

1. Introduction

In the last decades, hydrogels based on polysaccharide-biosourced natural macromolecules have attracted much interest due to their inherent biocompatibility, biodegradability, and nontoxicity, making them suitable candidates for applications in biomedicine and other healthcare applications [1–5]. Thanks to the pendant functional groups that they bear in their monomer units, polysaccharides can be easily modified by effortless reactions that enable them to form reversible three-dimensional (3D) networks through various non-covalent interactions. A commonly used modification is the grafting-to reaction with end-functional short polymeric chains that transforms them into associative graft copolymers [6].

Among others, alginate-based graft copolymers [7] have been designed to form 3D networks through the association of the pendant grafting chains, responding to a trigger like a temperature. In this case, poly(*N*-isopropylacrylamide), PNIPAM, was mainly used as the type of grafting chain due to its appropriate lower critical solution temperature

(LCST), appearing at about 32 °C, below the physiological temperature [8–16]. Above a percolation concentration of alginate-g-PNIPAM (gelator) in water, the graft copolymer forms a 3D network upon heating, above a critical temperature, T_{gel}, exhibiting a sol-to-gel transition. The rheological properties of the as-formed hydrogel and, in turn, its suitability for specific applications, depending on a number of factors, namely gelator concentration, grafting density, nature and length of grafts, and on the environmental conditions like temperature, pH, salinity and other solutes. To fulfill the demands for specific applications, all these factors must be regulated towards targeting tailor-made hydrogel properties, including injectability.

The regulation of T_{gel} is crucial, provided that it determines the rheological properties of the hydrogel at room and physiological temperature. If it is very close to 37 °C, then the elastic modulus might be low, and the strength of the physical crosslinks might be weak, resulting in a viscoelastic response characterized by low relaxation times and relatively low viscosities, which is not desired for several applications, e.g., immobilization of stem cells. A favorable strategy to control T_{gel}, keeping all the other factors constant, suggests the regulation of the LCST and/or the cloud point of the grafting chains [17–20]. It has been reported that the incorporation of hydrophobic or hydrophilic monomers in the LCST polymers in a random copolymer topology decreases or increases the LCST, respectively [18,21].

In a recent paper, we reported the properties of a hydrogel, based on a sodium alginate backbone, highly grafted by a thermo-responsive random copolymer of *N*-isopropylacrylamide, incorporating 10% mol of the hydrophobic *N-tert*-butylacrylamide, NaALG-g-P(NIPAM$_{90}$-*co*-NtBAM$_{10}$) [22]. This hydrogel system exhibited a T_{gel} (determined at tanδ = 1) at about 32 °C, for C_p = 13 wt % (at a fixed pH 7.4 and 0.2 M salinity). At room temperature, it behaves as a viscous liquid with low shear viscosity (three orders of magnitude higher than that of water), while above T_{gel}, it forms a viscoelastic liquid, evolving to a gel at 37 °C, with an elastic modulus of the order of 100 Pa. More importantly, the system exhibits a shear/thermo-induced injectability and self-healing ability, as it responds instantly to sudden changes of shear and temperature. Nevertheless, all these properties may further be tuned, depending on the targeting application. Another issue is the gelator concentration, which is correlated with the mesh size of the network and, in turn, with its capacity to incorporate payloads, especially for cell transplantation. Thus, the gelator concentration should be as low as possible.

As demonstrated, the rheological properties of the system depend strongly on the hydrophobic strength of the P(NIPAM$_{90}$-*co*-NtBAM$_{10}$) grafting chains, which were reinforced by the presence of the NtBAM monomer [22]. It is known that the relative hydrophobicity of PNIPAM is low above its LCST, as PNIPAM never becomes totally hydrophobic, even well above its cloud point [23]. Thus, by using random copolymers enriched with hydrophobic moieties, we can regulate not only the LCST but also the hydrophobic association in terms of exchange dynamics of the stickers of the network that control the rheological properties [22,24–27]. The objective of the present work is to address the above issue by studying the effect of the hydrophobic content of the associative grafting P(NIPAM$_x$-*co*-NtBAM$_y$) chains of alginate-based graft copolymers, attempting to tailor the T_{gel}, and in turns all the properties of the hydrogel formulations, including injectability at relatively low gelator concentration. This study ends up with a NaALG-g-P(NIPAM$_{80}$-*co*-NtBAM$_{20}$)/DMEM formulation, exhibiting excellent shear-induced injectability and instantaneous thermo-induced gel stiffening at physiological temperature, which can be a promising candidate for cell transplantation potential applications.

2. Results and Discussion

2.1. Synthesis and Characterization of Graft Copolymers

In this research study, a series of amino-functionalized P(NIPAM$_x$-*co*-NtBAM$_y$)-NH_2 random copolymers with comparable molecular weight and different NtBAM content as well as a PNIPAM-NH_2 homopolymer, were synthesized through conventional free radical polymerization (FRP) (synthesis details are presented in the Supplementary Materials). These polymers were characterized by potentiometric titration and proton nuclear magnetic resonance (^{1}H NMR). From the acid–base titration of the polymers aqueous solutions, the number average molecular weight (M_n) was determined, while from ^{1}H NMR, the monomer molar composition, NIPAM/NtBAM, of the copolymers was calculated (more details in Figures S1–S4 in the Supplementary Materials). Table 1 summarizes the molecular characterization of the grafting chains. From the results on the molecular composition of the P(NIPAM$_x$-*co*-NtBAM$_y$)-NH$_2$ copolymers, it can be seen that the NIPAM/NtBAM monomers molar ratio is in good accordance with the feed monomer composition throughout the synthesis of the copolymers.

Table 1. Molecular characteristics of the grafting chains.

Polymer	M_n [a] (g/mol)	Theoretical NIPAM/NtBAM Molar Ratio	NIPAM/NtBAM Molar Ratio [b]	T_{cp} [c] (°C)
PNIPAM-NH$_2$	22,700	100/0	100/0	32
P(NIPAM$_{94}$-*co*-NtBAM$_6$)-NH$_2$	14,800	95/5	94/6	30
P(NIPAM$_{86}$-*co*-NtBAM$_{14}$)-NH$_2$	17,000	85/15	86/14	22
P(NIPAM$_{80}$-*co*-NtBAM$_{20}$)-NH$_2$	16,900	80/20	80/20	20

[a] from acid–base titration; [b] from proton nuclear magnetic resonance (^{1}H NMR); [c] from turbidimetry, defined at the onset of the optical density abrupt increase.

At a second step, the aforementioned polymers were grafted onto an alginate backbone in aqueous media by forming an amide bond between the -NH$_2$ groups of the amine-functionalized P(NIPAM$_x$-*co*-NtBAM$_y$) copolymers and the carboxylate groups of alginate, with the aid of EDC [11,22]. The alginate-based graft copolymers were characterized by ^{1}H NMR (see details in Figures S5–S6 in the Supplementary Materials). The weight composition (NaALG/side chains, wt/wt) and the grafting density (grafting chains/NaALG, mol/mol) of the graft copolymers are displayed in Table 2.

Table 2. Molecular characteristics of the graft copolymers from ^{1}H NMR.

Polymer	Mw ($\times 10^3$ g/mol) [a]	% Weight Composition Alg/Grafting Chains (wt/wt)	Grafting [b] Density
NaALG-*g*-PNIPAM	203	69/31	3
NaALG-*g*-P(NIPAM$_{94}$-*co*-NtBAM$_6$)	280	50/50	8
NaALG-*g*-P(NIPAM$_{86}$-*co*-NtBAM$_{14}$)	222	63/37	5
NaALG-*g*-P(NIPAM$_{80}$-*co*-NtBAM$_{20}$)	179	78/22	3

[a] Calculated from the alginate Mw = 140,000 g/mol and its% weight composition from ^{1}H NMR. [b] number of grafting chains per alginate backbone by ^{1}H NMR.

A UV-vis spectrophotometer was utilized to investigate the thermo-responsiveness of the grafting chains since PNIPAM is a well-known thermo-sensitive homopolymer with characteristic LCST behavior (at about 32 °C in water) [28,29]. The turbidimetry method was used to explore the thermo-responsiveness of all the synthesized grafting chains. In Figure 1a, the variation of the optical density at 500 nm at the temperature range 15–45 °C of aqueous solutions (0.5% *w/v*) of the P(NIPAM$_x$-*co*-NtBAM$_y$) random copolymers, including the PNIPAM homopolymer, is presented.

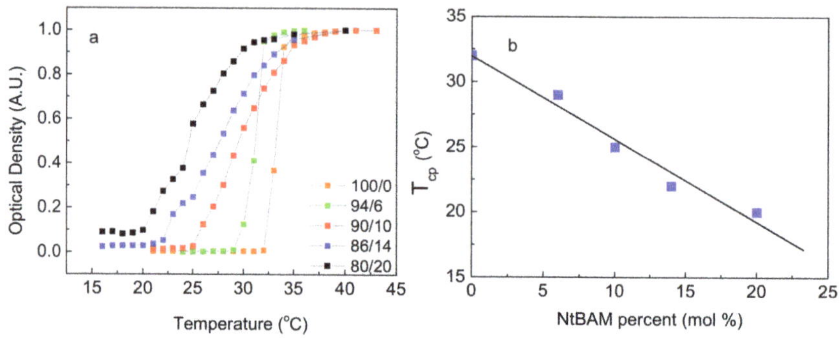

Figure 1. (**a**) Temperature dependence of the normalized optical density at 500 nm of 0.5% *w/v* aqueous solutions of PNIPAM-NH_2 (orange), P($NIPAM_{94}$-*co*-$NtBAM_6$)-NH_2 (green), P($NIPAM_{90}$-*co*-$NtBAM_{10}$)-NH_2 (red) [22], P($NIPAM_{86}$-*co*-$NtBAM_{14}$)-NH_2 (blue), P($NIPAM_{80}$-*co*-$NtBAM_{20}$)-NH_2 (black). (**b**) T_{cp} versus NtBAM content (mol %), the line is the linear fitting of the data (R^2 = 0.962).

As can be seen, the optical density increases abruptly above a certain temperature for the PNIPAM-NH_2 homopolymer and the P($NIPAM_{94}$-*co*-$NtBAM_6$)-NH_2 copolymer. However, for the three copolymers with the higher NtBAM content, P($NIPAM_{90}$-*co*-$NtBAM_{10}$)-NH_2, P($NIPAM_{86}$-*co*-$NtBAM_{14}$)-NH_2 and P($NIPAM_{80}$-*co*-$NtBAM_{20}$)-NH_2, the optical density increases gradually. This is likely due to the heterogeneity of the monomer composition in the copolymers, arisen from the conventional free radical polymerization method. The NtBAM content is an average value, and the sample is constituted of a mixture of statistical copolymers with NtBAM-richer and NIPAM-richer rather than the average value compositions [22,30]. The cloud point temperature (T_{cp}) was defined as the temperature where the optical density of the studied polymer solution increases promptly. The T_{cp} values of the copolymers are included in Table 1. Figure 1b depicts a decreasing linear dependence of T_{cp} as a function of the NtBAM hydrophobic monomer content, as expected [31]. This behavior allows fine T_{cp}-tuning by adjusting the molar content of the hydrophobic comonomer during copolymerization with NIPAM.

2.2. Thermo-Induced Gelation of Graft Copolymers

As we have already reported, graft copolymers of NaALG, grafted with PNIPAM-based thermoresponsive chains, form 3D percolated networks upon heating, exhibiting a sol-to-gel transition. To explore the influence of the hydrophobic NtBAM content of the PNIPAM-rich grafting chains on the thermo-induced rheological properties of the alginate-based graft copolymer hydrogels, oscillatory shear measurements were conducted. Keeping constant the polymer concentration, C_p, at 5 wt % (above the percolation threshold), temperature sweeps were performed in the linear viscoelastic regime (constant strain amplitude of 0.1%) and at a fixed frequency of 1 Hz. First, a cooling procedure was conducted from high to low temperatures, followed by a heating procedure with the same rate of 1 °C/min in all cases. Figure 2a demonstrates an example of this procedure concerning the NaALG-*g*-P($NIPAM_{80}$-*co*-$NtBAM_{20}$) graft copolymer. As seen, at high temperatures, G' dominates to G'', manifesting a gel-like behavior. Upon decreasing temperature, both moduli start to decrease, crossing each other at a certain temperature, denoted as T_{gel}. Below this temperature, the system behaves as a flowing liquid (sol) since the loss modulus prevails the storage one ($G'' > G'$). Upon heating, the system follows a different pathway exhibiting hysteresis, which is more pronounced in the vicinity of the G''/G' crossover point. The T_{gel} (heating) appears about 2–3 °C lower than the T_{gel} (cooling). This hysteresis tends to vanish as the NtBAM content decreases (Figure S7 in the Supplementary Materials). In the copolymer bearing pure PNIPAM side chains, the hysteresis seems to be negligible. The observed hysteresis depends on the ramp rate. As observed for the NaALG-*g*-PNIPAM, the difference between T_{gel} (heating) and T_{gel} (cooling) is practically zero at 0.5 °C/min (slow ramp) and reaches ~4 °C when a 10 °C /min, fast ramp, is applied (Figure S8 in the Supplementary Materials), revealing that it is a matter of equilibration, correlated with the dynamics of the network. This is corroborated by the fact that in the sol state, the system is in an equilibrium state (no network exists). This also appears at high temperatures (gel state), where sufficient time during heating allows equilibration again. In the following, we discuss only the heating procedure data with a heating rate of 1 °C/min. In Figure 2b, the results of all the polymer systems are displayed. The data are shifted to lower temperatures as the hydrophobic content (NtBAM moieties) of the thermo-responsive associative side chains increases.

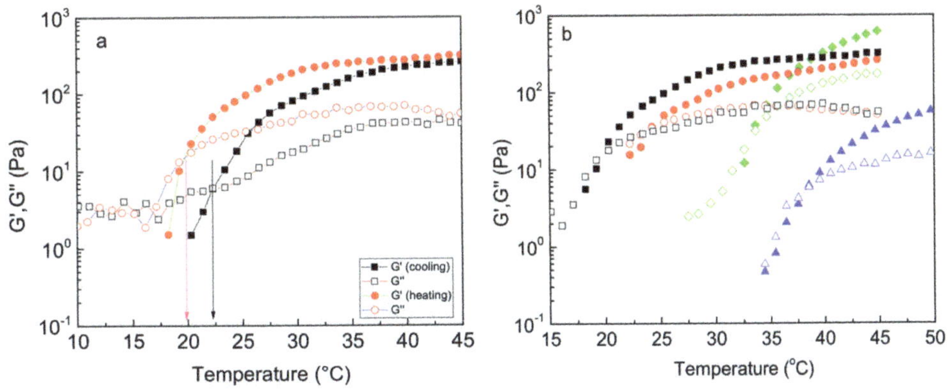

Figure 2. (**a**) Temperature dependence of G' (closed), G'' (open) at 1 Hz and strain amplitude of 0.1% for the NaALG-*g*-P(NIPAM$_{80}$-*co*-NtBAM$_{20}$) graft copolymer in water (C_p = 5 wt %). The different symbols indicate heating (red) and cooling (black) ramp both at a rate of 1 °C/min. The arrows indicate the temperature at $G' = G''$. (**b**) G' (closed), G'' (open) versus temperature (heating ramp) at the same conditions and C_p for different graft copolymers: NaALG-*g*-PNIPAM (triangles, blue), NaALG-*g*-P(NIPAM$_{94}$-*co*-NtBAM$_6$) (diamonds, green), NaALG-*g*-P(NIPAM$_{86}$-*co*-NtBAM$_{14}$) (circles, red), NaALG-*g*-P(NIPAM$_{80}$-*co*-NtBAM$_{20}$) (squares, black).

In Figure 3a, T_{gel} is plotted as a function of NtBAM content (mol %), exhibiting a linear decay with the hydrophobic enrichment of the grafts. It should be emphasized that although the grafting density and, in turn, the total NIPAM/NtBAM percentage of the copolymers differ, and despite their molecular polydispersity, T_{gel} is very well correlated with the NtBAM content of the grafting chain (R^2 = 0.993), showing that this is the determining factor. Any other correlation, e.g., T_{gel} vs. NtBAM$_{total}$ (mol %), fails. However, when the grafting density is remarkably high (see our previous work [22]), there is a deviation from this relationship. The above finding implies that T_{gel} should also be correlated with T_{cp}, which is controlled by the NtBAM content, as shown in Figure 1. Hence, the lower the LCST of the grafting chains, the lower the T_{gel} should be. Indeed, by plotting T_{gel} versus T_{cp} (Figure 3b), a linear function can be observed, clearly showing their relationship. Another interesting observation arises from the comparison of the data with the line $T_{gel} = T_{cp}$. At high NtBAM content (20% mol), the two temperatures coincide, while an increasing deviation $T_{gel} > T_{cp}$ appears as NtBAM content decreases to pure PNIPAM. This should be attributed to the strength of the side chains' hydrophobic association, which depends both on temperature and NtBAM content. Note that PNIPAM is never entirely hydrophobic, even at sufficiently higher temperatures than its cloud point [23]. Thus, by enriching PNIPAM with NtBAM hydrophobic moieties, the side chains exhibit increasing stickiness, and the gel point is manifested closer to the phase transition (LCST type) of the grafting chains.

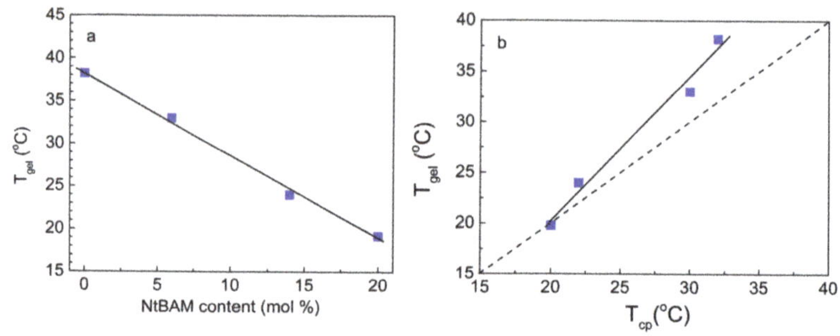

Figure 3. (**a**) T_{gel} as a function of NtBAM content (mol %) of the grafting chains; the straight line is a linear fit of the data (R^2 = 0.993). (**b**) T_{gel} of the graft copolymer versus T_{cp} of the grafting chains. The dashed line denotes $T_{gel} = T_{cp}$, and the solid line is a linear fit of the data (R^2 = 0.973).

Here we should discuss T_{gel}, as it is defined at the crossover temperature where G′ = G″, as usually reported in the literature. This is not a real first-order transition as at $T > T_{gel}$ close to T_{gel}, the solution still flows, behaving as a viscoelastic liquid, and it becomes an opaque free-standing gel at temperatures well above the sol-to-gel transition. Thus, T_{gel} denotes, in fact, the onset of thermo-induced gelation either by heating (LCST-type) or by cooling (UCST-type) [32]. Provided that T_{gel}, determined by the temperature sweep experiments, depends on the applied frequency (verified with the NaALG-*g*-PNIPAM sample), it is referred to as apparent gelation temperature [5,32,33]. This is attributed to the dynamic character of association and deals with the lifetime of the physically formed reversible crosslinks concerning the experimental time. Thus, to evaluate whether real gelation occurs, frequency sweep measurements at various temperatures above T_{gel} should be examined.

The data of Figure 2b can also be demonstrated using a single parameter, namely loss tangent (tanδ = G''/G'), which is a measure of the viscoelasticity of the material (Figure 4). The temperature at tanδ = 1 denotes T_{gel}, and as tanδ decreases below unity upon heating, the material's elasticity increases, in expense to its viscous response. As observed, tanδ decreases steadily with temperature (Figure 4a) in all systems, tending to converge at high temperature, well above T_{gel}. By plotting tanδ as a function of temperature normalized by T_{gel} (Figure 4b), the data almost superimpose to a master curve (with small deviation for the 86/14 polymer), showing that the enrichment of the sticky side chains of the graft copolymer, simply shifts the gel window at lower temperatures and that the gelation mechanism, i.e., heat-induced hydrophobic association, is identical. Evidently, gelation is a gradual process and not a sharp sol-to-gel transition.

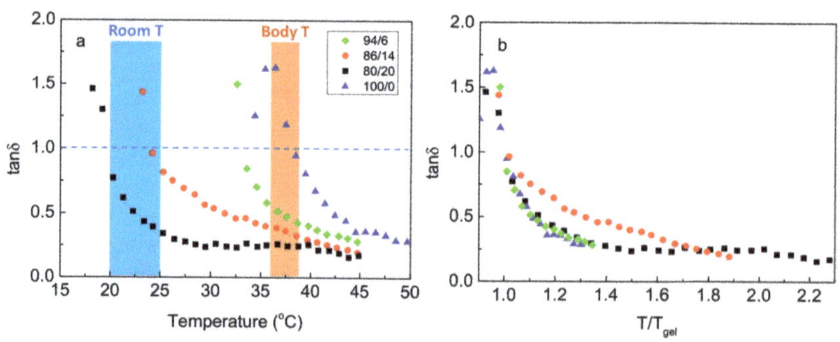

Figure 4. (**a**) Temperature dependence of the loss tangent (tanδ) at 1 Hz for the systems indicated. The two-colored zones mark the room and body temperature. (**b**) Loss tangent (tanδ) as a function of the normalized temperature T/T_{gel} (symbols as in (**a**)).

2.3. Rheological Properties at a Given Temperature and Responsiveness

The comparison of the viscoelastic response of the different systems is useful to also be examined at specific temperature values. For instance, in biomedical applications, the interest focuses on the state of the formulations at room and body temperature (marked zones in Figure 4a). In Figure 5a, the frequency (ω) dependence of the moduli for all the systems at 37 °C is displayed. As can be observed, the shift of T_{gel} at lower temperatures obviously affects the gelation state at 37 °C. The copolymer solution with pure PNIPAM side chains exhibits nearly liquid-like behavior as G' and G'' almost coincide (tanδ close to unity), and both are dependent on frequency. The exponent of the power-law dependence of both moduli is slightly higher and close to 0.5, showing that the system at 37 °C is in the vicinity of the transition point between liquid-like and solid-like behavior [34]. As the side chains are enriched with the hydrophobic NtBAM moieties, the storage modulus dominates the loss modulus in the entire frequency range investigated, exhibiting solid-like behavior. The G' (Figure 5b) increases and the tanδ (Figure S9 in the Supplementary Materials) decreases with increasing the hydrophobic content of the side chains of the graft copolymers, clearly indicating the reinforcement of the polymer networks. Considering that the rubber elasticity theory is valid for the present network systems, the elastic modulus, G_N, is proportional to the number density of the elastically active chains, n (intermolecular bridging), $G_N = nK_BT$, where K_B, T are the Boltzmann's constant and the absolute temperature, respectively. Therefore, the increase of the storage modulus (obtained at 1 rad/s) at 37 °C, demonstrated in Figure 5b, suggests that the hydrophobic enrichment of the sticky side chains imposes a higher number of the network strands at a given temperature. Particularly, from the pure PNIPAM to 80/20 PNIPAM/NtBAM grafting chains, the storage modulus increases two orders of magnitude, while the loss tangent decreases from 0.97 to 0.22 (Figure 5b), suggesting remarkable improvement of the elasticity of the hydrogel at the physiological temperature.

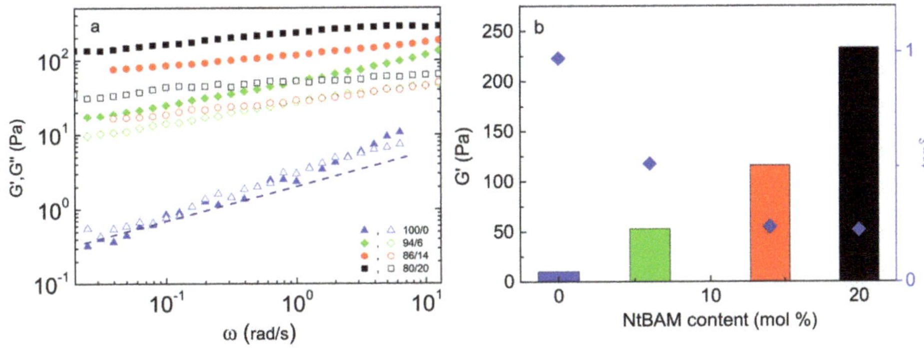

Figure 5. Frequency dependence of G′ (closed), G″ (open) (**a**) for the NaALG-*g*-P(NIPAM$_x$-*co*-NtBAM$_y$)/water systems of different monomer compositions in the grafting chains, as indicated, at 37 °C. The slope of the dashed line is 0.5. (**b**) Storage modulus, G' (at 1 rad/s) (bars) and tanδ (diamonds, right axis) versus NtBAM content (mol %) of the grafting chains.

Furthermore, as seen in Figure 5a, the terminal relaxation zone is not visible for the NaALG-*g*-P(NIPAM-*co*-NtBAM) systems, implying long relaxation times. To evaluate them, frequency sweep experiments were accomplished in various temperatures from 22–37 °C for the 80/20 copolymer (Figure S10 in the Supplementary Materials) to apply the time–temperature superposition principle. Figure 6 demonstrates b_TG' and b_TG'' versus the reduced frequency ωa_T, where a_T, b_T is the horizontal and vertical shift factors, respectively, using 22 °C as the reference temperature. A satisfactory master curve was achieved for the NaALG-*g*-P(NIPAM$_{80}$-*co*-NtBAM$_{20}$) graft copolymer/water system, which, however, exhibits a peculiarity. The loss modulus, G'', continuous to increase above the crossover point. Such behavior has not been reported by well-defined ABA triblock copolymer hydrogelators with sticky ends, while similar behavior has been reported for entangled star-shaped branched polymers [35] or to our previous report that dealt with the highly branched NaALG-*g*-P(NIPAM$_{90}$-*co*-NtBAM$_{10}$) (C_p = 10 wt %) copolymer [22]. This trend may be attributed to the exponential retarded relaxation modes of the grafting chains [36].

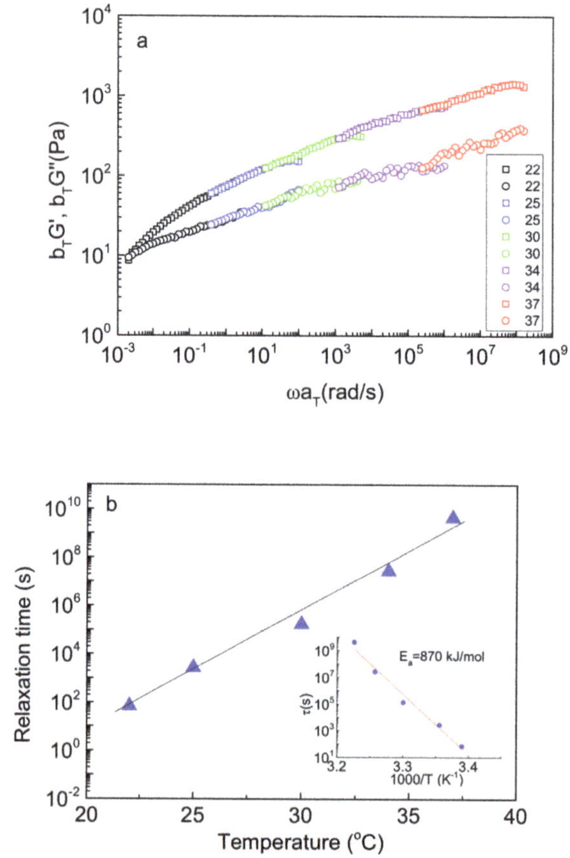

Figure 6. (**a**) Time–temperature superposition master curve of G′ (squares) and G″ (circles) at the reference temperature of 22 °C and (**b**) temperature dependence of relaxation time, τ and the corresponding Arrhenius plot (inset) for the NaALG-*g*-P(NIPAM$_{80}$-*co*-NtBAM$_{20}$) graft copolymer/water system (C_p = 5 wt %). The solid lines are linear fits of the data.

The terminal relaxation time, τ, at the reference temperature, T = 22 °C, was determined, by the crossover frequency, $\tau = 1/\omega_c$, at 70 s. By using the a_T shift factors, τ can be evaluated at various temperatures through the equation 1 $\tau = \tau_{ref}\, a_T$. Figure 6b demonstrates an exponential increase of τ with temperature. The relaxation time increases about eight orders of magnitude from 22 to 37 °C. More important, from pure PNIPAM to 80/20 PNIPAM/NtBAM grafting chains, τ increases from 0.1 to 4.3 × 10^9 s, hence more than ten orders of magnitude, revealing a strong effect for just 20 mol % hydrophobic enrichment of the sticky grafting chains. Thus, at 37 °C, the NaALG-*g*-P(NIPAM$_{80}$-*co*-NtBAM$_{20}$)/water system behaves as a "frozen" hydrogel, resembling a permanent, chemically crosslinked gel.

In the inset of Figure 6b, the data were plotted to apply the Arrhenius Equation (1):

$$\tau = \tau_o\, exp[-E_a/RT] \tag{1}$$

where E_a is the apparent activation energy, which represents the energy barrier the network stickers need to overcome to escape from their crosslinking nanodomains (micellar cores), allowing relaxation [24,32]. E_a was estimated 870 kJ/mol, which is comparable with the 619 kJ/mol reported for a triblock copolymer constituted of poly(*N*,*N*-dimethylacrylamide), end-capped with PNIPAM-based random copolymers of poly($NIPAM_{82}$-*co*-butyl $acrylate_{18}$) and poly($NIPAM_{95}$-*co*-butyl $acrylate_5$) [37]. In this copolymer, the PNIPAM sticky end-blocks have been enriched with the hydrophobic butyl acrylate moieties in 82/18 (mol %) content. As it is known, E_a is proportional to $N^{2/3}\gamma$, where N is the degree of polymerization of the associative chains, while γ is their surface tension with the solvent, which, in turn, is related to the Flory–Huggins polymer/solvent interaction parameter χ [38]. Thus, the enrichment of PNIPAM with hydrophobic monomers increases γ, analogously affecting E_a. Provided that in our case, N is higher (146 versus 89) and γ slightly higher (hydrophobic content: 20 mol % versus 18 mol %), the higher E_a estimated herein seems fairly reasonable (see Supplementary Materials). Note that the different copolymer architecture (graft versus triblock) may also affect E_a when the number of stickers per chain (grafting chains or end blocks) is considerably different since the network structure and connectivity affect the dynamics of the network as well [22,32].

The rheological properties of the hydrogel formulations at room temperature (20–25 °C) are also very important since they decisively influence their injectability. For the gelators with low hydrophobic content (e.g., 100/0, 94/6) in their stickers, the T_{gel} is well above the room temperature (Figure 4a), and hence, they exhibit a sol behavior at room temperature being easily injectable. For the gelators with higher hydrophobic content, exhibiting T_{gel} in the room temperature regime, the hydrogels show significant viscoelasticity, as can be seen in Figure S11 (see in the Supplementary Materials). Especially for the NaALG-*g*-P($NIPAM_{80}$-*co*-$NtBAM_{20}$) graft copolymer, with the characteristics of $G' = 81.7$ Pa, $\tan\delta = 0.366$ (at 10 rad/s) and $\tau = 79$ s, signifying a weak "gel" (viscoelastic fluid), injectability is not obvious.

As reported recently, the injectability is correlated with the thermal and shear rate responsiveness of the formulation, as arisen from their thermo-sensitivity (Figure 2) and shear thinning behavior (Figure S12 in the Supplementary Materials), respectively [22]. Thus, consecutive shear viscosity time-sweep experiments were conducted at specific temperatures (20, 25, 37 °C) and under shear rates of 0.01 and 17.25 s^{-1}. The first low value approaches the zero-shear viscosity (rest) and the second one exemplifies the shear rate applied through a 28-gauge syringe needle [39]. First, the hydrogel was subjected to consecutive stepwise temperature switch, under constant shear rates and at time intervals of 60 s. Figure 7a,b shows the shear viscosity changes upon temperature switch, responding instantaneously to the stepwise temperature variation. More important, the viscosity profiles are reproducible, either on increasing or decreasing temperature, demonstrating excellent temperature responsiveness.

Comparing the viscosity profiles at the same temperature but under different shear rates (0.01 s^{-1}/Figure 7a versus 17.25 s^{-1}/Figure 7b), the following effects can be observed.

First, the lower the shear rate, the higher the viscosity, at the same temperature, in agreement with the shear-thinning behavior of the formulation (Figure S12 in the Supplementary Materials). Second, a steady-state cannot be established under the low shear rate of 0.01 s^{-1} within the time interval investigated, while at the high shear rate of 17.25 s^{-1}, a steady-state can be achieved. These phenomena should be attributed to the long relaxation times of the network that are higher than the experimental time, especially under low shear. At a high shear rate, the network is disrupted, and the increasing mobility of the macromolecules allows steady-state establishment. Finally, the viscosity profiles at the same temperature differ notably when the temperature changes from lower (increasing T) versus higher temperatures (decreasing T). This should be correlated with longer relaxation times at higher temperatures (Figure 6b).

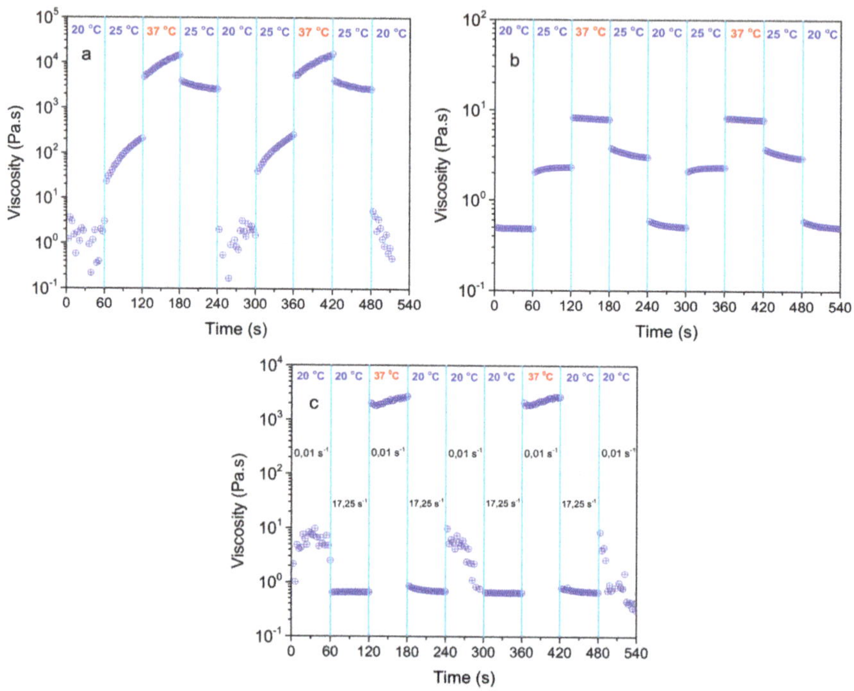

Figure 7. Shear viscosity versus time of a 5 wt% NaALG-g-P(NIPAM$_{80}$-co-NtBAM$_{20}$) aqueous solution subjected to stepwise temperature switch at a constant shear rate of 0.01 s^{-1} (**a**) and of 17.25 s^{-1} (**b**). In (**c**), the sample was subjected to consecutive conditions by simultaneously varying shear rate and temperature, e.g., from 0.01 s^{-1}/ 20 °C to 17.25 s^{-1}/ 20 °C to 0.01 s^{-1}/ 37 °C in three cycles.

In Figure 7c the experiment was designed to simulate conditions similar to those of injection through a 28 gauge needle syringe. The injection temperature was set at 20 °C since, at this temperature, the shear viscosity under the shear rate of 17.25 s^{-1} (Figure 7b) is lower than 1 Pa, which is an acceptable value for injection. As observed, the shear viscosity rises instantly more than three orders of magnitude when the conditions are switched from 17.25 s^{-1}/ 20 °C to 0.01 s^{-1}/ 37 °C. The viscosity profiles are fairly reproducible, irrespectively of the direction of temperature change (heating or cooling). This experiment confirms the excellent injectability of the system.

2.4. Rheological Properties in Various Media

In the next step, we examined the rheological behavior of the NaALG-g-P(NIPAM$_{80}$-co-NtBAM$_{20}$) graft copolymer in various media used for biological research and/or biomedicine applications, namely phosphate buffer (PB, 1 mM), phosphate-buffered saline (PBS, 1 mM, 0.135 M NaCl) and Dulbecco's modified Eagle's medium (DMEM). The same temperature sweep experiments were conducted, as described in Figure 2a, applying cooling/heating of the formulations from 45 to 10 to 45 °C with a rate of 1 °C/min (Figure S13 in the Supporting Information). In Figure 8, the temperature dependence of G', G'' moduli and tanδ are demonstrated for the heating sweep procedure. The general thermal behavior in all media was similar; that is, a remarkable increase of the network elasticity above a certain temperature, T_{gel}, as manifested by the storage moduli rise and the loss tangent decrease upon heating. However, in PBS media, although the G', G'' augment promptly above 15 °C (Figure 8a), they never cross each other, regardless of cooling or heating procedure. G′ always prevails G″ within the temperature range investigated, implying that the network is not entirely disrupted at low temperatures, even well below the T_{cp}

of the sticky grafting chains. More important, tanδ decreases again at low temperatures below 15 °C (Figure 8b), showing an increased elasticity and revealing the so-called cold gelling [14]. This effect should be attributed to the presence of relatively high salt content, and especially potassium cations, which promote the cold gelling through the association of the mannuronic sequences of the alginate backbone [14].

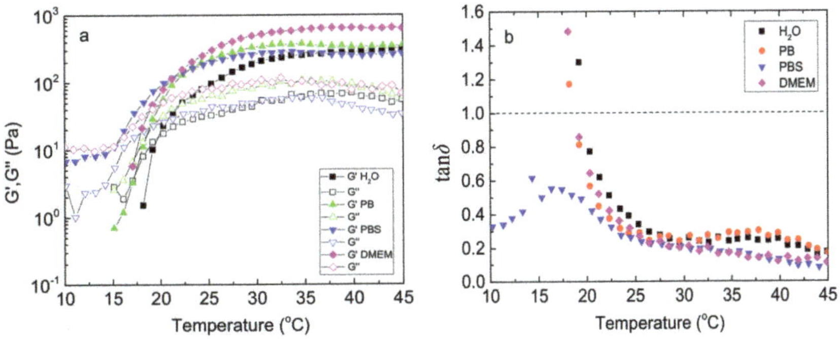

Figure 8. Temperature dependence of G′ (closed), G″ (open) (**a**) and tanδ (**b**) (at 1 Hz, γ = 0.1%, heating rate 1 °C/min) for 5 wt % NaALG-*g*-P(NIPAM$_{80}$-*co*-NtBAM$_{20}$) graft copolymer in various media: water, phosphate buffer (PB), phosphate-buffered saline (PBS), Dulbecco's modified Eagle's medium (DMEM) (indicated in the inset).

Furthermore, the presence of salts slightly affects T_{gel} (Table S2 in the Supplementary Materials). This should be attributed to the shift of the T_{cp} of the thermosensitive PNIPAM-rich pendant sticky grafting chains due to the well-known salting-out effect [20,40]. More pronounced effects were observed to the G′, G″ moduli. Especially in DMEM media, the storage modulus is higher than the one in pure water, more than 240% at the physiological temperature, although T_{gel} shifted to a lower value just 1 °C (Table S2 in the Supporting Information), meaning that the gelation window practically remains in the same temperature range. DMEM is a multi-constituent medium comprising salts, amino acids, vitamins, and glucose and is used for cell culture and transplantation. All these ingredients may affect the network structure and connectivity since they interact with the gelator through non-covalent interactions, either with the alginate backbone (polyelectrolyte screening) and/or PNIPAM-based grafting chains (salting-out effect). Provided that in salts, divalent cations like Ca^{2+} are also included, it is very likely that additional crosslinking occurs due to ionic interactions with the -COO^- anions of the alginate backbone, which could justify the G′ augmentation [39].

2.5. Injectable Hydrogel in Cell Culture Fluid Media (DMEM)

Cell transplantation through injection strategies into host tissues is one of the potential biomedical applications that have attracted much attention. It is known that the injection of cells within simple formulations through a needle significantly reduces their viability due to exposing the cells to substantial extensional flows that can damage cell membranes [41]. A plausible strategy to overcome this problem is to use as the cell carrier a two-step gelling system, exhibiting shear-thinning and self-healing properties [42,43]. Shear- and thermo-responsive hydrogels seem to be good candidates for this purpose since they form a weak gel at room temperature, protecting the cells, and a stiff gel after injection at a physiological temperature that can retain the cell at the target location. Here, we explore the response of the DMEM formulation to the injection requirements.

Figure 9 shows the heating sweep data for the NaALG-*g*-P(NIPAM$_{80}$-*co*-NtBAM$_{20}$)/DMEM formulations at 4 and 5 wt % polymer concentrations. The gelator concentration is another factor that can tune the hydrogel properties. From the data of Figure 9, it seems that concentration affects slightly T_{gel}, about 1 °C (Table S2 in the Supplementary Materials), while it significantly affects the network elasticity well above T_{gel}. For instance, at room temperature, G' is about similar for both 4 and 5 wt % concentrations. On the contrary, at body temperature, when C_p decreases from 5 to 4 wt %, G' also decreases (by a factor of 2.4) (inset of Figure 9). The formulation exhibits sol state at low temperature (e.g., at 5 °C), viscoelastic liquid (soft gel) at 20 °C and free-standing gel (stiff gel) at 37 °C (Figure 9, digital photos).

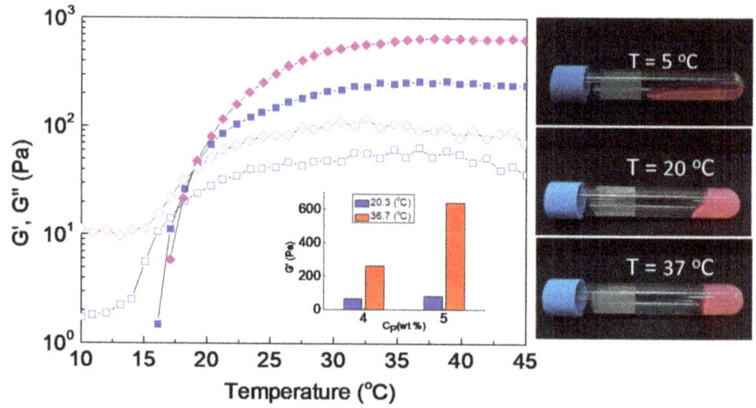

Figure 9. Temperature dependence of G' (closed), G'' (open) (at 1 Hz, $\gamma = 0.1\%$, heating rate 1 °C/min) for NaALG-*g*-P(NIPAM$_{80}$-*co*-NtBAM$_{20}$) graft copolymer in Dulbecco's modified Eagle's medium (DMEM) at 5 wt % (diamonds) and 4 wt % (squares) polymer concentration. The inset demonstrates the storage modulus, G', at room and body temperatures. The photos (**right**) illustrate the solutions of 4 wt %, equilibrated at the indicated temperatures.

The enhanced values of moduli in DMEM media (C_p = 5 wt %) allow lowering of C_p, which is beneficial, as we have already mentioned in the Introduction. Thus, focusing on the 4 wt % NaALG-*g*-P(NIPAM$_{80}$-*co*-NtBAM$_{20}$)/DMEM formulation, we examined its injectability by consecutive shear and oscillatory time sweep experiments. Figure 10 illustrates the shear viscosity experiment that comprises sudden changes of the shear rate/temperature conditions, simulating an injection with a 28 g needle. The shear viscosity decreases under high shear at 20 °C, obtaining values lower than 1 Pa.s, while it rises instantaneously more than three orders of magnitude (6 orders of magnitude higher than the viscosity of the medium) at physiological temperature, forming a stiff gel. The results are quite reproducible, as can be observed by the second cycle of changes in the same experiment. These data imply excellent injectability, which was also verified by optical observation when injecting the formulation at room temperature in a water medium, regulated at 37 °C (see the digital photo in Figure 10).

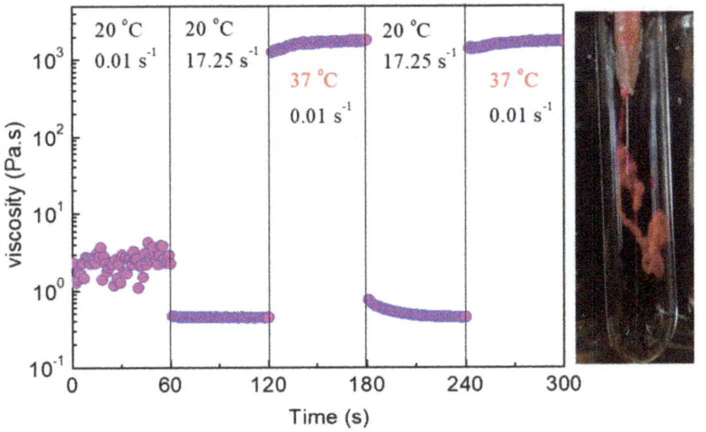

Figure 10. Shear viscosity versus time of a 4 wt % NaALG-*g*-P(NIPAM$_{80}$-*co*-NtBAM$_{20}$)/DMEM formulation subjected to consecutive variations of shear rate and/or temperature, as indicated, in time intervals of 60 s. The digital photo (**right**) shows injection of the formulation through a 28 g needle in water, regulated at 37 °C.

An oscillatory shear experiment was designed to explore the formulation' response to high shear deformation and gel recovery upon cessation of strain. The formulation was subjected to stepwise time sweep under different conditions: at 20 °C, applying a shear amplitude, γ, of 0.1% (linear viscoelastic regime); after 60 s, γ was switched to 300% (well beyond linear viscoelastic regime); after further 60 s, the temperature was increased at 37 °C and simultaneously γ was lowered back to 0.1%. The results are presented in Figure 11. As observed, at 20 °C and low strain, the formulation behaves as a weak gel ($G' > G''$), with a storage modulus of 35 Pa and tanδ = 0.47. Upon applying high strain in the nonlinear regime, the network is disrupted instantly, and the formulation flows ($G' < G''$ and tanδ = 3.3) with low G' ~2 Pa. Finally, in the third step, a stiff gel is recovered at 37 °C ($G' > G''$), as suggested by the high value of G' (260 Pa) and the low value of tanδ = 0.13.

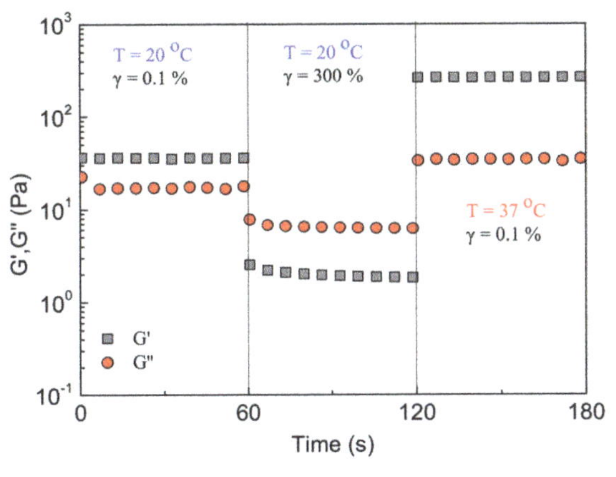

Figure 11. Time dependence of G′ (squares) and G″ (circles) at 1 Hz, under different conditions of temperature and strain amplitude, as indicated, for the 4 wt % NaALG-*g*-P(NIPAM$_{80}$-*co*-NtBAM$_{20}$)/DMEM formulation.

Again, the system responds instantly to any variation of temperature and strain. Notably, the storage modulus instantaneously increases more than two orders of magnitude after network disruption, showing excellent self-healing ability as well (see also Figure S14 in the Supplementary Materials).

It is interesting to compare the thermomechanical response of the present system with the two-component material, namely SHIELD (shear-thinning hydrogel for injectable encapsulation and long-term delivery), constituted of star-shaped polyethylene glycol (8-armed PEG tethered with proline-rich peptides (P) and a PNIPAM chain) copolymer and an engineered recombinant protein (C7). This sophisticated system forms a weak physical network ex situ through P/C7 peptide binding and in situ network reinforcement through the thermo-induced hydrophobic association of PNIPAM, forming additional junctions [43,44]. As has been shown, soft, medium, and stiff SHIELD variants all resulted in statistically higher levels of cell protection compared to cell delivery in saline. Importantly, the presence or absence of cell-adhesive domain within the SHIELD material did not affect the cell membrane protection. Thus, the shear-thinning/self-healing properties of the gel (and not the gel cell-adhesive properties) are responsible for the cell membrane protection [43]. It seems that the highly tunable one-component simple system (graft copolymer) presented herein can meet the required thermomechanical properties of SHIELD, i.e., two-step gelling (ex situ soft gel, in situ network reinforcement), shear-thinning injectability and instantaneous self-healing, combined with thermo-induced gel stiffening, used for the successful delivery of transplanted stem cells [45].

Shear-thinning alginate hydrogels, crosslinked through ionic interactions in the presence of Ca^{2+} divalent cations, were first identified as material carriers to protect cells from mechanical damage during injection [39]. Provided that the gelator suggested herein is an alginate-based graft copolymer, it is worthy to compare them. As reported, the alginate formulation that produced a hydrogel with $G' \sim 30$ Pa provided the most effective cell protection for all the cell types investigated. Either increasing or decreasing, the hydrogel storage modulus reduced this protective effect [39]. In the present system, $G' = 35$ Pa for the 4 wt % P(NIPAM$_{80}$-co-NtBAM$_{20}$) in DMEM at 20 °C (Figure 11), which is very close to the value reported for the best cell viability after injection.

As far as the biocompatibility of the NaALG-g-P(NIPAM-co-NtBAM) is concerned, there are some previous reports dealing with analogous grafted polysaccharide-based materials. Particularly, no cytotoxicity or acute systemic toxicity was associated with hyaluronan derivatives, namely HA-g-P(NIPAAm-co-NtBAAm). This injectable and self-assembling scaffold was also compatible with the production of mesenchymal stem/stromal cells (MSC)-derived chondrocytes [46]. Moreover, viability studies with human corneal epithelial (HCE) cells, performed on nanogels of methylcellulose hydrophobized with N-tert-butylacrylamide, demonstrated in vitro biocompatibility [47]. We should finally mention the in vitro and in vivo biocompatibility studies of alginate/Polyacrylamide IPN gels that showed minimal effects on cells, then pure alginate hydrogels [48]. Although these reports are promising, biocompatibility studies are necessary for the polymers under investigation, and this will be an objective of our ongoing research.

3. Materials and Methods

3.1. Materials

The monomers *N*-isopropylacrylamide (NIPAM) and *N*-tert-butylacrylamide (NtBAM) were used as acquired by Fluorochem (Derbyshire, UK) and Alfa Aesar (Ward Hill, MA, USA), respectively. Potassium peroxodisulfate (KPS, Fluorochem, Derbyshire, UK), 2-aminoethanethiol hydrochloride (ET HCl, Alfa Aesar, Ward Hill, MA, USA), *N*-(3-Dimethylaminopropyl)-*N*′-ethylcarbodiimide hydrochloride (EDC, Alfa Aesar, Ward Hill, MA, USA) and 1-Hydroxybenzotriazole hydrate (HOBt, Fluka, NC, USA) were used as received. Dimethylformamide (DMF, Aldrich, St. Louis, MO, USA), hydrochloric acid (HCl, Panreac, IL, USA), sodium hydroxide (NaOH, Panreac, IL, USA), deuterated water (D_2O, Sigma-Aldrich, St. Louis, MO, USA) and Dulbecco's modified Eagle's medium (DMEM, Sigma-Aldrich, St. Louis, MO, USA) were used as obtained by the provider without purification. Ultrapure water was received by an ELGA Medica-R7/15 device (ELGA Labwater, IL, USA).

Sodium alginate (NaALG, Sigma-Aldrich, St. Louis, MO, USA, no. 180947, molecular weight range: 120,000–190,000 g/mol, the ratio of mannuronic and guluronic units (M/G): 1.53) was further purified (a solution of 7% *w/v* NaALG in NaOH (0.005 M) was repeatedly purified by dialysis against ultrapure water (membrane MWCO: 12,000–14,000 Da), and the final NaALG product was obtained in its solid-state through lyophilization.

3.2. Synthesis of the Graft Copolymers

Carbodiimide chemistry was applied for the synthesis of the NaALG based graft copolymers using EDC as condensation agent and HOBt as coupling agent [22,49–51]. The experimental procedure concerning the synthesis of the grafting chains is presented in the Supplementary Materials. A typical synthesis of a NaALG-g-PNIPAM graft copolymer was conducted as follows: Primarily, 4 g (0.02 mol of repeat units) of NaALG and 4 g (0.18 mmol) PNIPAM-NH_2 were separately dissolved in 80 mL ultrapure water each and left under stirring at 23 °C for 24 h. After 24 h, the two aqueous solutions were mixed and left to stir at 23 °C overnight. After full homogeneity, the pH of the solution was adjusted at pH$\tilde{6}$ by NaOH (1 M). Next, 0.12 g (0.9 mmol) HOBt was added to the mixture, followed by the addition of 0.67 g (0.0035 mol) EDC. The solution was left under stirring at 23 °C for three days. The graft copolymer was first precipitated in acetone and then redissolved in ultrapure water, while the pH of the final aqueous solution was elevated at pH > 12, adding NaOH (1 M). Finally, the reaction product was received after dialysis against ultrapure water (membrane MWCO: 25,000 Da) and lyophilization. Four different graft copolymers were received through this method, enduing the same procedure by altering the feed composition.

3.3. Polymer Characterization

3.3.1. Proton Nuclear Magnetic Resonance, ^{1}H NMR

^{1}H NMR spectra of NaALG, grafting chains, and NaALG-graft copolymers in D_2O were obtained using a BRUKER AVANCE III HD PRODIGY ASCEND TM 600 MHz spectrometers (Billerica, MA, USA). NaALG and PNIPAM-NH_2 were studied at room temperature. The amino-terminated grafting chains P($NIPAM_{94}$-co-$NtBAM_6$)-NH_2 and the graft copolymer NaALG-g-P($NIPAM_{94}$-*co*-$NtBAM_6$) were studied at 20 °C. The side chains P($NIPAM_{86}$-*co*-$NtBAM_{14}$)-NH_2 and P($NIPAM_{80}$-*co*-$NtBAM_{20}$)-NH_2 along with the graft copolymers NaALG-g-P($NIPAM_{86}$-*co*-$NtBAM_{14}$) and NaALG-g-P($NIPAM_{80}$-*co*-NtBAM)$_{20}$ were studied at 15 °C (details are presented in the Supplementary Materials).

3.3.2. Potentiometric Titration

Acid–base titration of the -NH_2 end groups was performed to estimate the number average molar mass of the thermo-responsive amino-functionalized grafting chains. Briefly, 0.25 g of the polymer was dissolved in 10 mL ultrapure water. Aiming to fully deprotonated -NH_2 end-groups, an appropriate amount of NaOH (0.1 M) was added in the polymer solution (pH > 11 was achieved). HCl (0.01 M) was used as the titrant.

3.3.3. Cloud Point Measurements

The phase transition of 0.5% *w/v* aqueous solution of the side chains was observed by turbidimetry at 500 nm using a HITACHI U2001 spectrometer (Illinois/USA). The cloud point, T_{cp}, was defined as the temperature above, which the optical density starts to increase promptly.

3.4. Hydrogels Preparation

Aqueous solutions of NaALG graft copolymers at a concentration of 5 wt % were prepared and left under stirring at 20 °C on a lab shaker (200 rpm) equipped with a refrigerated bath circulator. After homogeneity, the pH of the solutions was tuned at pH 7.4, using NaOH (1 M).

3.5. Rheological Studies

A stress-controlled AR-2000ex (TA Instruments, New Castle, DE, USA) rheometer with a cone and plate geometry (diameter 20 mm, angle 3°, truncation 111 μm) was used for the rheological study of the NaALG graft copolymer aqueous solutions. The experiments were performed in the linear viscoelastic regime, which was determined by strain sweep tests at a frequency of 1 Hz. The temperature was controlled by a Peltier system with an accuracy of ± 0.1 °C. The rheometer was equipped with a solvent trap to avoid concentration changes due to water evaporation.

4. Conclusions

A series of graft copolymers (with relatively low grafting density) consisted of alginate backbone and P(NIPAM$_x$-*co*-NtBAM$_y$) grafting chains were synthesized and explored as gelators in various media. All the NaALG-*g*-P(NIPAM$_x$-*co*-NtBAM$_y$) copolymers, at a concentration of 5 wt %, exhibit thermo-induced gelation in aqueous media upon heating, owing to the hydrophobic intermolecular association of their pendant grafting chains. The enrichment of PNIPAM with the hydrophobic NtBAM monomer units resulted in a linear decline of their cloud points with NtBAM content. This induced a considerable shift of the gelation point, T_{gel}, to lower temperatures, shifting the gelation window accordingly. T_{gel} was lowered from 38 to 20 °C by varying the NtBAM content from 0 to 20 mol %. The linear dependence of T_{gel} with the NtBAM content of the grafting chains seems to be the determining factor of T_{gel}, at least at low grafting density, which allows fine-tuning of the gelation temperature window. The impact of the T_{gel} shift on the rheological properties of the formed hydrogels is significant at room and body temperature. For instance, at 37 °C, the storage modulus increases about two orders of magnitude and the terminal relaxation time increase about 10 orders of magnitude by enriching the sticky grafting chains with 20% mol hydrophobic moieties.

The overall temperature-dependent rheological properties of the graft copolymers reveal two main behaviors, a sol–gel and a weak gel-stiff gel transition for the copolymer with stickers of low (0, 6% mol) and high (14, 20% mol) NtBAM content, respectively. The first type of hydrogels is evidently injectable since they flow easily at room temperature. For the second type of hydrogels, their injectability is based on their additional shear thinning behavior. To prove this, stepwise shear-dependent experiments, designed to simulate injection through a 28-gauge needle syringe, were conducted for the 5 wt % NaALG-*g*-P(NIPAM$_{80}$-*co*-NtBAM$_{20}$)/water formulation. The results showed excellent shear-induced responsiveness. More importantly, the combination of thermo- and shear-responsiveness provides excellent injectability.

To examine its potential use in bioapplications, the NaALG-*g*-P(NIPAM$_{80}$-*co*-NtBAM$_{20}$) copolymer was explored as a gelator in various environments of biological interest. The various constituents of the media, especially salts, affect the rheological properties of the formulations due to their interaction with both alginate backbone and PNIPAM-based stickers. These effects must be taken into account to tailor the rheological behavior of the hydrogels. When a cell cultivating media (e.g., DMEM) was used, the rheological properties differ significantly concerning those in the other media. The 4 wt % NaALG-*g*-P(NIPAM$_{80}$-*co*-NtBAM$_{20}$)/DMEM formulation exhibits excellent shear-induced injectability and instantaneous thermo-induced gel stiffening at physiological temperature. The rheological properties (e.g., storage modulus at room and body temperature, shear-thinning, self-healing) of this formulation resemble those explored for cell transplantation using shear-thinning/self-healing hydrogels (e.g., SHIELD) as carriers. Considering the biocompatibility of PNIPAM-based grafted polysaccharides [46,47,52–54], this formulation seems a good candidate for further investigation, targeting cell transplantation potential applications.

Supplementary Materials: The following are available online at https://www.mdpi.com/article/10.3390/ijms22083824/s1, Table S1, Synthesis conditions of the grafting chains; Figure S1, ^{1}H-NMR spectrum of PNIPAM-NH2; Figure S2, ^{1}H-NMR spectrum of P(NIPAM94-co-NtBAM6)-NH2; Figure S3, ^{1}H-NMR spectrum of P(NIPAM86-co-NtBAM14)-NH2; Figure S4, ^{1}H-NMR spectrum of P(NIPAM80-co-NtBAM20)-NH2; Figure S5, ^{1}H-NMR spectra of left: NaALG-g-PNIPAM; right: NaALG; Figure S6, ^{1}H-NMR spectra of: NaALG-g-P(NIPAM80-co-NtBAM20), NaALG-g-P(NIPAM86-co-NtBAM14), NaALG-g-P(NIPAM94-co-NtBAM6); Figure S7, G′ (closed), G″ (open) versus temperature (cooling/heating ramp) at 1 Hz, strain amplitude of 0.1 % and Cp=5 wt % for different graft copolymers: NaALG-g-PNIPAM, NaALG-P(NIPAM94-co-NtBAM6), NaALG-P(NIPAM86-co-NtBAM14); Figure S8, ΔT = Tgel, heating -Tgel, cooling versus ramp rate (°C/min) for the NaALG-g-PNIPAM (Cp =5 wt %, strain 0.5 %, at 1 Hz); Figure S9, Frequency dependence of tanδ for the NaALG-g-P(NIPAMx-co-NtBAMy)/water systems of different monomer composition in the grafting chains, as indicated (inset), at 37 °C; Figure S10, Frequency dependence of G′, G″ for NaALG-g-P(NIPAM80-co-NtBAM20) Cp = 5 wt%) at γ=0.1% and at various temperatures; Figure S11, Frequency dependence of G′ (closed), G″ (open) for NaALG-g-P(NIPAM86-co-NtBAM14) (left) and NaALG-g-P(NIPAM80-co-NtBAM20) (right) at γ=0.1% and Cp = 5 wt%; Figure S12, Shear viscosity as a function of shear rate (decreasing from 100 s-1) at different temperatures, as indicated; Figure S13, Temperature dependence of G′ (closed), G″ (open) (cooling/heating ramp, 1 °C/min) at 1 Hz and strain amplitude of 0.1 % for the NaALG-g-P(NIPAM80-co-NtBAM20) graft copolymer (Cp=5 wt %) in various media; Table S2, Tgel and G′ (37 °C) of 5 wt% NaAlg-g-P(NIPAM80-co-NtBAM20) in various media; Figure S14, Storage modulus G′ (squares, black) and loss modulus G″ (circles, red) versus strain at 20 °C followed by time sweep, immediately after switching temperature and strain at 37 °C and 0.1%.

Author Contributions: Investigation; K.S. and S.-F.S.; writing—original draft preparation, Z.I.; writing—review and editing, C.T.; supervision, C.T. All authors have read and agreed to the published version of the manuscript.

Funding: This research received no external funding.

Institutional Review Board Statement: Not applicable.

Informed Consent Statement: Not applicable.

Data Availability Statement: Not applicable.

Acknowledgments: We would like to thank Fotoula Kounelaki and Nicoletta-Paraskevi Kouli for their contributions with some additional rheological data.

Conflicts of Interest: The authors declare no conflict of interest.

References

1. Graham, S.; Marina, P.F.; Blencowe, A. Thermoresponsive polysaccharides and their thermoreversible physical hydrogel networks. *Carbohydr. Polym.* **2019**, *207*, 143–159. [CrossRef] [PubMed]
2. Klouda, L.; Mikos, A.G. Thermoresponsive hydrogels in biomedical applications. *Eur. J. Pharm. Biopharm.* **2008**, *68*, 34–45. [CrossRef]
3. Tiwari, S.; Patil, R.; Bahadur, P. Polysaccharide Based Scaffolds for Soft Tissue Engineering Applications. *Polymers* **2019**, *11*, 1. [CrossRef]
4. Hogan, K.J.; Mikos, A.G. Biodegradable thermoresponsive polymers: Applications in drug delivery and tissue engineering. *Polymer* **2020**, *211*, 123063. [CrossRef]
5. Cook, M.T.; Haddow, P.; Kirton, S.B.; McAuley, W.J. Polymers Exhibiting Lower Critical Solution Temperatures as a Route to Thermoreversible Gelators for Healthcare. *Adv. Funct. Mater.* **2020**, *2008123*. [CrossRef]
6. Tsitsilianis, C. Multisegmental block/graft copolymers. In *Macromolecular Engineering*; Matyjaszewski, K., Gnanou, Y., Leibler, L., Eds.; Wiley-Vch: Weinheim, Germany, 2007; Volume 2, pp. 839–873.
7. Sun, J.; Tan, H. Alginate-Based Biomaterials for Regenerative Medicine Applications. *Materials* **2013**, *6*, 1285–1309. [CrossRef] [PubMed]
8. Kim, J.H.; Lee, S.B.; Kim, S.J.; Lee, Y.M. Rapid temperature/pH response of porous alginate-g-poly(N-isopropylacrylamide) hydrogels. *Polymer* **2002**, *43*, 7549–7558. [CrossRef]
9. Karakasyan, C.; Legros, M.; Lack, S.; Brunel, F.; Maingault, P.; Ducouret, G.; Hourdet, D. Cold Gelation of Alginates Induced by Monovalent Cations. *Biomacromolecules* **2010**, *11*, 2966–2975. [CrossRef] [PubMed]
10. Leal, D.; De Borggraeve, W.; Encinas, M.V.; Matsuhiro, B.; Muller, R. Preparation and characterization of hydrogels based on homopolymeric fractions of sodium alginate and PNIPAAm. *Carbohyd. Polym.* **2013**, *92*, 157–166. [CrossRef] [PubMed]

11. Lencina, S.M.M.; Iatridi, Z.; Villar, M.A.; Tsitsilianis, C. Thermoresponsive hydrogels from alginate-based graft copolymers. *Eur. Polym. J.* **2014**, *61*, 33–44. [CrossRef]
12. Martinez-Gomez, F.; Encinas, M.V.; Matsuhiro, B.; Pavez, J. Preparation and swelling properties of homopolymeric alginic acid fractions/poly(N-isopropyl acrylamide) graft copolymers. *J. Appl. Polym. Sci.* **2015**, *132*, 42398. [CrossRef]
13. Lencina, S.M.M.; Ciolino, A.E.; Andreucetti, N.A.; Villar, M.A. Thermoresponsive hydrogels based on alginate-g-poly(N-isopropylacrylamide) copolymers obtained by low doses of gamma radiation. *Eur. Polym. J.* **2015**, *68*, 641–649. [CrossRef]
14. Liu, M.; Song, X.; Wen, Y.; Zhu, J.-L.; Li, J. Injectable Thermoresponsive Hydrogel Formed by Alginate-g-Poly(Nisopropylacrylamide) Releasing Doxorubicin-Encapsulated Micelles as Smart Drug Delivery System. *ACS Appl. Mater. Interfaces* **2017**, *9*, 35673–35682. [CrossRef] [PubMed]
15. Guo, H.; de Magalhaes Goncalves, M.; Ducouret, G.; Hourdet, D. Cold and Hot Gelling of Alginate-graft-PNIPAM: A Schizophrenic Behavior Induced by Potassium Salts. *Biomacromolecules* **2018**, *12*, 576–587. [CrossRef]
16. Chalanqui, M.J.; Pentlavalli, S.; McCrudden, C.; Chambers, P.; Ziminska, M.; Dunne, N.; McCarthy, H.O. Influence of alginate backbone on efficacy of thermo-responsive alginate-g-P(NIPAAm) hydrogel as a vehicle for sustained and controlled gene delivery. *Mater. Sci. Eng. C* **2019**, *95*, 409–421. [CrossRef] [PubMed]
17. Chen, G.; Hoffman, A.S. Graft Copolymers That Exhibit Temperature-Induced Phase Transitions over a Wide Range of pH. *Nature* **1995**, *373*, 49–52. [CrossRef] [PubMed]
18. Tsitsilianis, C.; Gotzamanis, G.; Iatridi, Z. Design of "smart" segmented polymers by incorporating random copolymers as building blocks. *Eur. Polym. J.* **2011**, *47*, 497–510. [CrossRef]
19. Lauber, L.; Chassenieux, C.; Nicolai, T.; Colombani, O. Highlighting the Role of the Random Associating Block in the Self-Assembly of Amphiphilic Block–Random Copolymers. *Macromolecules* **2015**, *48*, 7613–7619. [CrossRef]
20. Pasparakis, G.; Tsitsilianis, C. LCST polymers: Thermoresponsive nanostructured assemblies towards bioapplications. *Polymer* **2020**, *211*, 123146. [CrossRef]
21. Yin, X.; Hoffman, A.S.; Stayton, P.S. Poly(N-isopropylacrylamide-copropylacrylic acid) copolymers that respond sharply to temperature and pH. *Biomacromolecules* **2006**, *7*, 1381–1385. [CrossRef]
22. Iatridi, Z.; Saravanou, S.-F.; Tsitsilianis, C. Injectable self-assembling hydrogel from alginate grafted by P(N-isopropylacrylamide-co-N-tert-butylacrylamide) random copolymers. *Carbohydr. Polym.* **2019**, *219*, 344–352. [CrossRef] [PubMed]
23. Pelton, R. Poly(N-isopropylacrylamide) (PNIPAM) is never hydrophobic. *J. Colloid Interface Sci.* **2010**, *348*, 673–674. [CrossRef] [PubMed]
24. Jung, H.; Gang, S.-E.; Kim, J.-M.; Heo, T.-Y.; Lee, S.; Shin, E.; Kim, B.-S.; Choi, S.-H. Regulating Dynamics of Polyether-Based Triblock Copolymer Hydrogels by End-Block Hydrophobicity. *Macromolecules* **2020**, *53*, 10339–10348. [CrossRef]
25. Tsitsilianis, C.; Serras, G.; Ko, C.-H.; Jung, F.; Papadakis, C.-M.; Rikkou-Kalourkoti, M.; Patrickios, C.S.; Schweins, R.; Chassenieux, C. Thermoresponsive Hydrogels Based on Telechelic Polyelectrolytes: From Dynamic to "Frozen" Networks. *Macromolecules* **2018**, *51*, 2169–2179. [CrossRef]
26. Charbonneau, C.; Chassenieux, C.; Colombani, O.; Nicolai, T. Controlling the Dynamics of Self-Assembled Triblock Copolymer Networks via the pH. *Macromolecules* **2011**, *44*, 4487–4495. [CrossRef]
27. Chassenieux, C.; Tsitsilianis, C. Recent trends on pH/thermo-responsive self-assembling hydrogels: From polyions to peptide-based polymeric gelators. *Soft Matter* **2016**, *12*, 1344–1359. [CrossRef]
28. Schild, H.G. Poly(*N*-isopropylacrylamide): Experiment, theory and application. *Prog. Polym. Sci.* **1992**, *17*, 163–249. [CrossRef]
29. Halperin, A.; Kröger, M.; Winnik, F.M. Poly(N-isopropylacrylamide) Phase Diagrams: Fifty Years of Research. *Angew. Chem. Int. Ed.* **2015**, *54*, 15342–15367. [CrossRef]
30. Kametani, Y.; Tournilhac, F.; Sawamoto, M.; Ouchi, M. Unprecedented Sequence Control and Sequence-Driven Properties in a Series of AB-Alternating Copolymers Consisting Solely of Acrylamide Units. *Angew. Chem. Int. Ed.* **2020**, *59*, 5193–5201. [CrossRef]
31. Rwei, S.-P.; Chuang, Y.-Y.; Way, T.-F.; Chiang, W.-Y. Thermosensitive copolymer synthesized by controlled living radical polymerization: Phase behavior of diblock copolymers of poly(N-isopropyl acrylamide) families. *J. Appl. Polym. Sci.* **2016**, *133*, 43224. [CrossRef]
32. Ye, Y.N.; Cui, K.; Indei, T.; Nakajima, T.; Hourdet, D.; Kurokawa, T.; Gong, J.P. Relaxation Dynamics and Underlying Mechanism of a Thermally Reversible Gel from Symmetric Triblock Copolymer. *Macromolecules* **2019**, *52*, 8651–8661. [CrossRef]
33. Henderson, K.J.; Shull, K.R. Effects of Solvent Composition on the Assembly and Relaxation of Triblock Copolymer-Based Polyelectrolyte Gels. *Macromolecules* **2012**, *45*, 1631–1635. [CrossRef]
34. He, Y.; Boswell, P.G.; Buhlmann, P.; Lodge, T.P. Ion Gels by Self-Assembly of a Triblock Copolymer in an Ionic Liquid. *J. Phys. Chem. B* **2007**, *111*, 4645–4652. [CrossRef]
35. Milner, S.T.; McLeish, T.C.B. Parameter-Free Theory for Stress Relaxation in Star Polymer Melts. *Macromolecules* **1997**, *30*, 2159–2166. [CrossRef]
36. He, Y.; Lodge, T.P. Thermoreversible Ion Gels with Tunable Melting Temperatures from Triblock and Pentablock Copolymers. *Macromolecules* **2008**, *41*, 167–174. [CrossRef]
37. Onoda, M.; Ueki, T.; Tamate, R.; Akimoto, A.M.; Hall, C.C.; Lodge, T.P.; Yoshida, R. Precisely Tunable Sol-Gel Transition Temperature by Blending Thermoresponsive ABC Triblock Terpolymers. *ACS Macro Lett.* **2018**, *7*, 950–955. [CrossRef]

38. Ma, Y.; Lodge, T.P. Chain exchange kinetics in diblock copolymer micelles in ionic liquids: The role of χ. *Macromolecules* **2016**, *49*, 9542–9552. [CrossRef]
39. Aguado, B.A.; Mulyasasmita, W.; Su, J.; Lampe, K.J.; Heilshorn, S.C. Improving viability of stem cells during syringe needle flow through the design of hydrogel cell carriers. *Tissue Eng. Part A* **2012**, *18*, 806–815. [CrossRef]
40. Zhang, Y.; Furyk, S.; Bergbreiter, D.E.; Cremer, P.S. Specific Ion Effects on the Water Solubility of Macromolecules: PNIPAM and the Hofmeister Series. *J. Am. Chem. Soc.* **2005**, *127*, 14505–14510. [CrossRef] [PubMed]
41. Yan, C.; Mackay, M.E.; Czymmek, K.; Nagarkar, R.P.; Schneider, J.P.; Pochan, D.J. Injectable Solid Peptide Hydrogel as a Cell Carrier: Effects of Shear Flow on Hydrogels and Cell Payload. *Langmuir* **2012**, *28*, 6076–6087. [CrossRef]
42. Madl, C.M.; Heilshorn, S.C.; Blau, H.M. Bioengineering strategies to accelerate stem cell therapeutics. *Nature* **2018**, *557*, 335–342. [CrossRef] [PubMed]
43. Marquardt, L.M.; Doulames, V.M.; Wang, A.T.; Dubbin, K.; Suhar, R.A.; Kratochvil, M.J.; Medress, Z.A.; Plant, G.W.; Heilshorn, S.C. Designer, injectable gels to prevent transplanted Schwann cell loss during spinal cord injury therapy. *Sci. Adv.* **2020**, *6*, eaaz1039. [CrossRef]
44. Cai, L.; Dewi, R.E.; Heilshorn, S.C. Injectable hydrogels with in situ double network formation enhance retention of transplanted stem cells. *Adv. Funct. Mater.* **2015**, *25*, 1344–1351. [CrossRef]
45. Cai, L.; Dewi, R.E.; Goldstone, A.B.; Cohen, J.E.; Steele, A.N.; Woo, Y.J.; Heilshorn, S.C. Regulating stem cell secretome using injectable hydrogels with in situ network formation. *Adv. Healthc. Mater.* **2016**, *5*, 2758–2764. [CrossRef] [PubMed]
46. Muramatsu, K.; Saito, Y.; Wada, T.; Hirai, H.; Miyawaki, F. Poly(N-isopropylacrylamide-co-N-tert-butylacrylamide)- grafted hyaluronan as an injectable and self-assembling scaffold for cartilage tissue engineering. *J. Biomed. Sci. Eng.* **2012**, *5*, 639–646. [CrossRef]
47. Jamard, M.; Hoare, T.; Sheardown, H. Nanogels of methylcellulose hydrophobized with N-tert-butylacrylamide for ocular drug delivery. *Drug Del. Transl. Res.* **2016**, *6*, 648–659. [CrossRef]
48. Darnell, M.C.; Sun, J.-Y.; Mehta, M.; Johnson, C.; Arany, P.R.; Suo, Z.; Mooney, D.J. Performance and biocompatibility of extremely tough alginate/polyacrylamide hydrogels. *Biomaterials* **2013**, *34*, 8042–8048. [CrossRef] [PubMed]
49. Durand, A.; Hourdet, D. Synthesis and thermoassociative properties in aqueous solution of graft copolymers containing poly(N-isopropylacrylamide) side chains. *Polymer* **1999**, *40*, 4941–4951. [CrossRef]
50. Iatridi, Z.; Bokias, G. Temperature-Sensitive Water-Soluble Hybrid Organic/Inorganic Nanoparticles Formed through Complexation of Cu^{2+} Ions with Poly(sodium acrylate)-g-poly(N-isopropylacrylamide) Comb-Type Copolymers in Aqueous Solution. *Langmuir* **2009**, *25*, 7695–7703. [CrossRef]
51. Cheaburu, C.N.; Ciocoiu, O.-N.; Staikos, G.; Vasile, C. Thermoresponsive Sodium Alginate-g-Poly(N-Isopropylacrylamide) Copolymers III. Solution Properties. *J. Appl. Polym. Sci.* **2013**, *127*, 3340–3348. [CrossRef]
52. Pentlavalli, S.; Chambers, P.; Sathy, B.N.; O'Doherty, M.; Chalanqui, M.; Kelly, D.J.; Haut-Donahue, T.; McCarthy, H.O.; Dunne, N.J. Simple Radical Polymerization of Poly(Alginate-Graft-N-Isopropylacrylamide) Injectable Thermoresponsive Hydrogel with the Potential for Localized and Sustained Delivery of Stem Cells and Bioactive Molecules. *Macromol. Biosci.* **2017**, *17*, 1700118. [CrossRef] [PubMed]
53. Diekjürgen, D.; Grainger, D.W. Polysaccharide Matrices Used in 3D in Vitro Cell Culture Systems. *Biomaterials* **2017**, *141*, 96–115. [CrossRef] [PubMed]
54. Puscaselu, R.G.; Lobiuc, A.; Dimian, M.; Covasa, M. Alginate: From Food Industry to Biomedical Applications and Management of Metabolic Disorders. *Polymers* **2020**, *12*, 2417. [CrossRef] [PubMed]

International Journal of
Molecular Sciences

MDPI

Article

Pathogenic Hydrogel? A Novel-Entrapment Neuropathy Model Induced by Ultrasound-Guided Perineural Injections

Ming-Yen Hsiao [1,2], Ya-Wen Wu [2], Wen-Shiang Chen [1,2], Yu-Ling Lin [1,2], Po-Ling Kuo [2,3] and Chueh-Hung Wu [1,2,4,*]

1 Department of Physical Medicine and Rehabilitation, College of Medicine, National Taiwan University, Taipei 10048, Taiwan; myhsiao@ntu.edu.tw (M.-Y.H.); wenshiang@gmail.com (W.-S.C.); cj801107@gmail.com (Y.-L.L.)
2 Department of Physical Medicine and Rehabilitation, National Taiwan University Hospital, Taipei 10048, Taiwan; 0607091@gmail.com (Y.-W.W.); poling@ntu.edu.tw (P.L.-K.)
3 Department of Electrical Engineering, National Taiwan University, Taipei 10617, Taiwan
4 Department of Physical Medicine & Rehabilitation, National Taiwan University Hospital Hsin-Chu Branch, Hsinchu 302058, Taiwan
* Correspondence: nojred@gmail.com

Abstract: Entrapment neuropathy (EN) is a prevalent and debilitative condition caused by a complex pathogenesis that involves a chronic compression–edema–ischemia cascade and perineural adhesion that results in excessive shear stress during motion. Despite decades of research, an easily accessible and surgery-free animal model mimicking the mixed etiology is currently lacking, thus limiting our understanding of the disease and the development of effective therapies. In this proof-of-concept study, we used ultrasound-guided perineural injection of a methoxy poly(ethylene glycol)-b-Poly(lactide-co-glycoilide) carboxylic acid (mPEG-PLGA-BOX) hydrogel near the rat's sciatic nerve to induce EN, as confirmed sonographically, electrophysiologically, and histologically. The nerve that was injected with hydrogel appeared unevenly contoured and swollen proximally with slowed nerve conduction velocities across the injected segments, thus showing the compressive features of EN. Histology showed perineural cellular infiltration, deposition of irregular collagen fibers, and a possible early demyelination process, thus indicating the existence of adhesions. The novel method provides a surgery-free and cost-effective way to establish a small-animal model of EN that has mixed compression and adhesion features, thus facilitating the additional elucidation of the pathophysiology of EN and the search for promising treatments.

Keywords: entrapment neuropathy; animal model; hydrogel; ultrasound; carpal tunnel syndrome

Citation: Hsiao, M.-Y.; Wu, Y.-W.; Chen, W.-S.; Lin, Y.-L.; Kuo, P.-L.; Wu, C.-H. Pathogenic Hydrogel? A Novel-Entrapment Neuropathy Model Induced by Ultrasound-Guided Perineural Injections. *Int. J. Mol. Sci.* **2021**, 22, 3494. https://doi.org/10.3390/ijms22073494

Academic Editor: Peter John Jervis

Received: 16 March 2021
Accepted: 25 March 2021
Published: 28 March 2021

Publisher's Note: MDPI stays neutral with regard to jurisdictional claims in published maps and institutional affiliations.

1. Introduction

Entrapment neuropathy (EN) is a prevalent and debilitative condition that causes pain, sensory impairment, and muscle atrophy in severe cases [1]. The etiology of EN is multifactorial and is hypothesized to be attributed to repetitive compression and shear stress injuries during motion [2–4]. Despite decades of research, the exact pathophysiology and the most effective treatment of EN are still debated. One of the reasons for this uncertainty is the lack of an easily accessible animal model that presents the complex, mixed pathological features of the disease.

Previously-used animal models can be classified into chemical and mechanical methods. In chemical methods, hypertonic dextrose or a proinflammatory agent is injected perineurally to induce inflammation and subsequent fibrosis [5,6]. Injection of hypertonic dextrose in the forepaws of rabbits has been used as a research model of carpal tunnel syndrome, the most common type of EN [5,6]. Being highly similar to the human carpal tunnel structurally, the rabbit model has potential for clinical translation. However, inflammation is not a key feature in EN, in which degeneration of myelin and perineural fibrosis mark the change [3,7]. Furthermore, multiple injections are required to induce obvious histology changes, partly attributed to brief injectate retention after injection. An injectate that can stay in situ after injection may be more suitable for the development of this model.

Mechanical methods include direct neural ligation with sutures [8], or perineural compression by angioplasty balloon or silicone cuff [9–11]. While offering consistently repeatable effects, these methods induce neuropathy predominantly owing to mechanical compression. The presence of balloon, cuff, or suture prevents direct interaction of the epineurium and surrounding connective tissue, a condition that is gaining attention as a significant contributing factor of EN [3,7]. Furthermore, the methods require surgery, which is time consuming and technically challenging. The tissue injury induced during surgical procedures can also complicate the pathogenesis.

Thermosensitive hydrogels, which are liquids in ambient temperature and convert to a gel state at physiological temperatures, could be injected in vivo and be retained in situ for up to several weeks [12]. We hypothesized that perineurally-injected hydrogel could encompass and compress the nerve by gelation, and that tissue reaction during resorption of the hydrogel could cause adhesion of the nerve and surrounding connective tissues, thus creating a mixed etiology model of EN. Therefore, the present study aimed to establish an easily accessible, surgery-free, and cost-effective small-animal model that presents a mixed etiology of EN by using ultrasound (US)-guided perineural injection of thermosensitive hydrogels, which form a gel state at physiological temperatures.

2. Results

2.1. US-Guided Perineural Injection

In total, 11 rats were used in the experiment. US imaging demonstrated perineural delivery of hydrogel, which appeared as homogeneous hypoechoic immediately postinjection, distributed both superficially and at deep sciatic nerve locations (Figure 1). The middle part of the sciatic nerve was completely surrounded by the hydrogel. Retention of perineural hydrogel at days 7 and 14 was confirmed sonographically. The hydrogel appeared to be slightly heterogeneous echogenic, possibly owing to the partial degradation and tissue reaction. Note that the surrounding hydrogel displaced the sciatic nerve, which became tortuous 7 and 14 days postinjection. In addition, the hydrogel-injected nerves appeared swollen proximal to the injection site, and thus demonstrated the pathognomonic feature of EN (Figure 2).

2.2. Changes of Sciatic Nerve Diameter

The segment of the sciatic nerve proximal to the hydrogel injection site increased its diameter considerably at day 14 compared with the pre-injection status ($p < 0.001$). The diameter of the proximal segment was also larger than that of the saline group at days 7 and 14 (both $p < 0.001$). The diameter of the middle segment of the sciatic nerve (injection site) in the hydrogel group exhibited a tendency to become smaller at day 14 compared with that of the saline group ($p = 0.06$). The diameter of the distal segment of the hydrogel group remained unchanged when it was compared with the saline group or the pre-injection status (Figure 3).

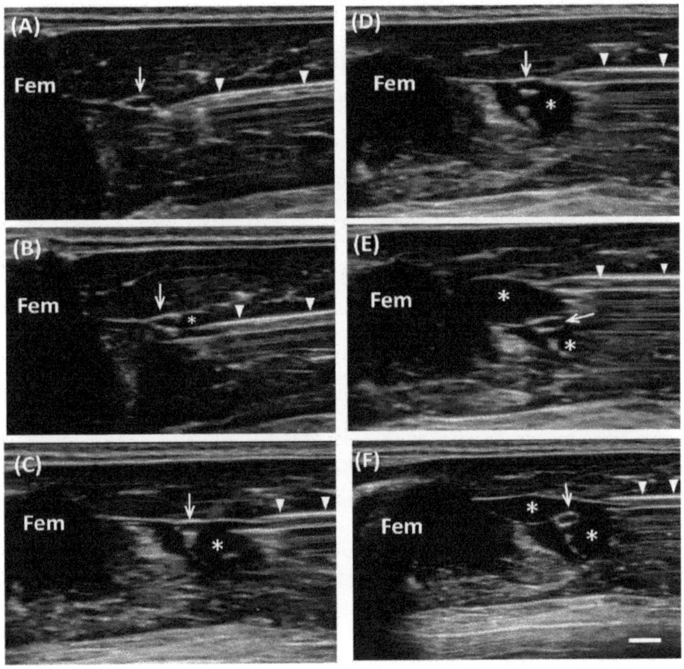

Figure 1. Ultrasound (US)-guided perineural injection of hydrogel. Short-axis view, in-plane approach of US-guided perineural injection of hydrogel (asterisks) showing the insertion of the needle (arrowheads) in the deep regions around the sciatic nerve (arrow) initially (**A–C**), and then superficially (**D–F**) (Fem, femur) (Scale bar: 1 mm).

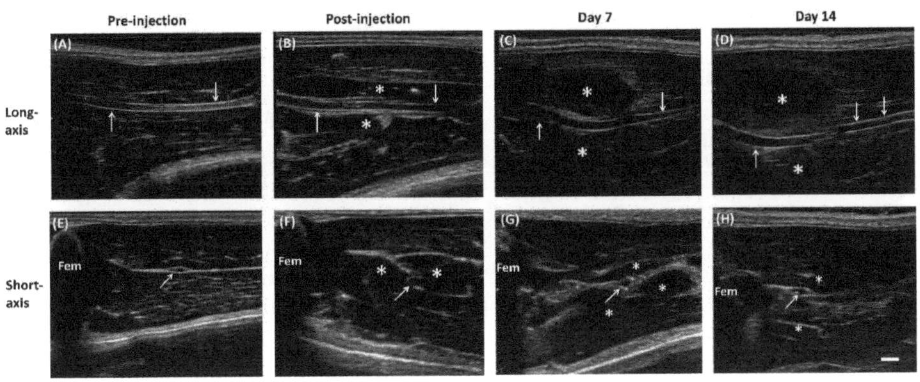

Figure 2. Sonography of sciatic nerve before and after injection. Long-axis (**A–D**) and short-axis (**E–H**) views of the sciatic nerve (arrows) before, immediately after, and 7 days and 14 days after hydrogel (asterisk) injection (Scale bar: 1 mm).

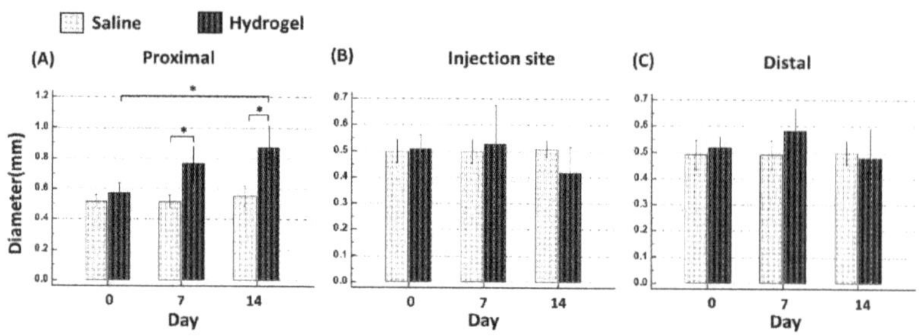

Figure 3. Changes of sciatic nerve diameter at the proximal (**A**), injection site (**B**), and at a distal (**C**) segment at days 7 and 14 (asterisk, $p < 0.05$).

2.3. NCV

Nerve conduction velocity (NCV) of the injected segment of the sciatic nerve was tested at day 14 after US evaluation. Considerable reduction of NCV by more than 50% was noted compared with that of the saline group (hydrogel group: 74.1 ± 12.3 m/s, saline group: 36.6 ± 7.8 m/s, $p = 0.002$). The results confirm that hydrogel injection resulted in impaired nerve conduction, a key feature of electrophysiological diagnosis of EN, in addition to morphological changes. Additionally, the encasement of the sciatic nerve by the hydrogel with the adhesion of the surrounding connective tissue was observed during the exposure of the nerve for the NCV procedure.

2.4. Histology

Histologically, hydrogel-injected nerves had perineural infiltration of nuclear cells and deposition of irregular collagen fibers with disrupted fibrillary structure of axons at day 14. The saline-injected group showed well-organized axons and scarce nuclear cells perineurally (Figure 4). Triple immunohistochemical staining of sciatic nerves 14 days after hydrogel injection showed patchy breakdown of myelin basic protein (MBP) and increased expression of glial fibrillary acidic protein (GFAP) compared with those of the saline-injection group (Figure 5), indicating a possible demyelinating process.

3. Discussion

In this proof-of concept study, US-guided perineural injection of mPEG-PLGA-BOX hydrogel successfully induced sciatic nerve neuropathy, as confirmed sonographically, electrophysiologically, and histologically. The hydrogel-injected nerve appeared irregular and swollen proximally with slowed NCV across the injected segment, demonstrating the compressive features of EN. Histology demonstrated perineural cellular infiltration, deposition of irregular collagen fibers, and a possible demyelination process, indicative of the existence of adhesions. The novel method provides a surgical-free, minimally invasive way to establish small animal models of EN with mixed compression and adhesion features. The versatile model could be customized to create the EN of different mechanisms, representing different etiologies or stages of the disease, thus facilitating further understanding of the disease and the development of effective treatment.

The pathophysiology of EN is complex and multifactorial. Following chronic nerve compression, the nerves show disrupted and thinning of the myelin [2,13,14]. Decreased number and deformed collagen fibrils of perineural tissues was noted in patients with carpal tunnel syndrome [7]. Cross-sectional enlargement of the nerves and increased content of surrounding connective tissue is a common finding, even in subclinical individuals [15]. Of note, there is prominent thickening of the subsynovial connective tissue, without obvious inflammatory phenomenon [3,7,16], indicating a chronic degenerative feature of the disease.

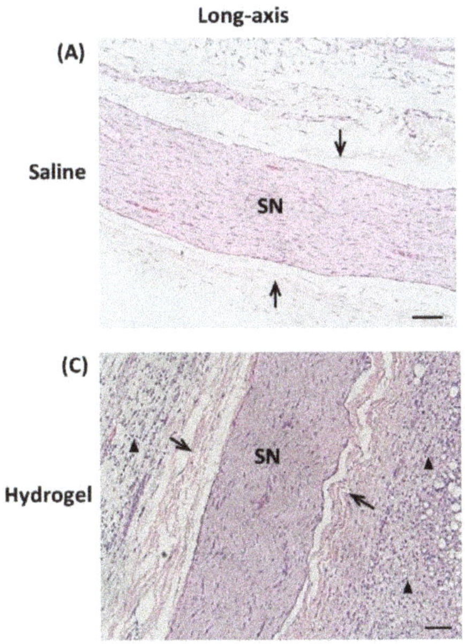

Figure 4. Representative micrographs of hematoxylin and eosin (H&E) stained sciatic nerve (SN) 14 days after injection of normal saline (**A**,**B**) and hydrogel (**C**,**D**). There is prominent cellular infiltration (arrowheads) surrounding epineurium (arrows) in the hydrogel-injected group with an irregular alignment of neural fascicles (scale bar: 100 μm).

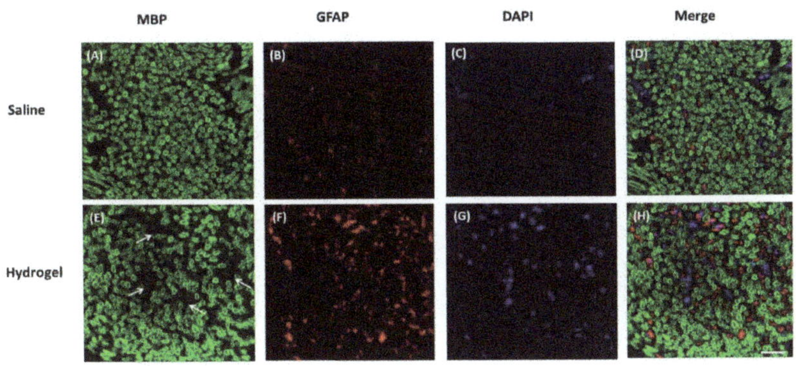

Figure 5. Representative micrographs of triple immunohistochemical staining of sciatic nerve, short-axis view 14 days after the injection of normal saline (**A–D**) and hydrogel (**E–H**). In the hydrogel group, the myelin basic protein (MBP) showed patchy areas (arrows) of breakdown (**E**), and glial fibrillary acidic protein (GFAP) showed increased expression (**F**). 4′,6-Diamidino-2-Phenylindole, Dihydrochloride (DAPI) was used for nucleic acid staining (scale bar: 40 μm).

While repetitive impingement of the nerve due to the increased interstitial pressure or external compression has been considered as the most important etiology, accumulating evidence has indicated the contributing role of shear stress caused by perineural adhesion and fibrosis [2–4,7,17]. Chronic impingement of the nerve results in perineural edema that impairs local circulation [2] that causes ischemia and reperfusion injury [18]. Surrounding connective tissue fibrosis and adhesion of the nerve ensue as a result of accumulated injury [3,7], which increase the shear stress of the nerve during motion and lead to additional damage [4].

In this regard, a model that presents both the compression and adhesion features of EN can describe the disease entity and may expedite further translational research. The proposed model presents a versatile choice of inducing EN with different characteristics, potentially by adjusting the composition of hydrogel or by combining different chemicals. For example, with quick gelation and long degradation times, the mPEG-PLGA-BOX hydrogel, as used in the present study, can induce compression-predominant EN by the mass effect (although tissue reaction may still cause perineural adhesion). By contrast, hydrogel with a lower degree of cross-linking has higher fluidity and longer gelation time, and tends to spread in a wider area rather than be retained focally after injection. By incorporating sclerosing or adhesive agents, adhesion-predominant EN could be simulated.

Sonographically, the injected hydrogel was retained perineurally on day 14. This is consistent with previous studies that reported the degradation rate of mPEG-PLGA-BOX. Peng et al. demonstrated that mPEG-PLGA-BOX was degraded by only 30% at 2 weeks in cultures in vitro [19]. In-vivo tests by Wu et al. showed that as a drug carrier, the hydrogel was able to provide steady drug release for at least 1 week, and the residual volume was still observable at day 14, and was measured to be approximately equal to 1/3 of the initial injection volume [20]. Longer degradation times were also possible by the modification of the degree of cross-linking of mPEG-PLGA. While the duration of the time needed to induce EN was highly variable, in most animal models there was an observable reduction of NCV and thinning of the myelin sheaths by week 2 based on histological evaluations [21]. Two in vitro studies showed that shear stress induced downregulation of myelin-associated glycoprotein and MBP, which are related to the demyelination process, and this occurred as early as a few hours after stimulation [22,23]. Our results indicate that the mPEG-PLGA-BOX hydrogel was retained long enough in vivo to produce morphological features of EN, which appeared as early as 7 days postinjection.

Recently, perineural injection of 5% dextrose (hydrodissection) has gained increased popularity as a treatment option in an attempt to release the adhesions between nerves and surrounding connective tissue [24–27]. However, controversy exists regarding the underlying mechanism and the optimal regiment of injection. As mentioned above, one of the reasons for the uncertainty is the lack of suitable animal models for testing the effects of injection therapy. Mechanical compression models by neural ligation [8], balloon, or silicone cuff [9–11] have generated tremendous insights into the pathophysiology of EN, but the presence of a balloon or cuff hinders the investigation of the perineural adhesion process and the injection effect. In this regard, our proposed method could serve as an optimal model to evaluate the efficacy of hydrodissection therapy. The hydrogel degraded gradually with controllable retention time. The increased perineural cellularity suggested the reaction of the tissue with the hydrogel, which resulted in possible adhesion. Based on the current results, our future study will optimize the hydrogel regimen with respect to the adhesion–induction properties and investigate the effect of hydrodissection on EN.

The study had several limitations. First, the time course of pathological changes of EN was not determined. Histology and NCV were evaluated at day 14 but were not sequentially examined. Needle stimulation electrodes had been evaluated in an attempt to measure NCV in vivo at day 7, but no consistent signals were obtained, possibly owing to the difficulty associated with the correct placement of the electrodes perineurally after hydrogel injection. Thus, NCV was only measured after the euthanasia of rats and dissection of the sciatic nerve at day 14. Second, the presence of compression and adhesions was demonstrated based on indirect morphological evidence, NCV, and histology. In future studies, the gliding resistance of the nerve could be measured before and after injection to demonstrate perineural adhesion. Furthermore, interstitial pressure could be measured to confirm the compression of the nerve owing to the hydrogel. Serial evaluation and longer follow-ups are also warranted to establish a comprehensive understanding of the disease process.

4. Materials and Methods

All experiments were performed in accordance with the guidelines established by the Institutional Animal Care and Use Committee at the National Taiwan University College of Medicine and the ARRIVE guidelines, and they were approved by the ethics committee of the Laboratory Animal Center at the National Taiwan University College of Medicine (Approval No. 20170470, 18 October 2018). All rats (Sprague–Dawley (SD), 6–8 weeks, weighing approximately 250–300 g) were procured from the National Laboratory Animal Center (Taipei City, Taiwan).

4.1. Experimental Protocols

US-guided hydrogel injection targeted the right sciatic nerves of male Sprague–Dawley rats at the mid-thigh level. The left sciatic nerves were injected with normal saline as the control group.

US imaging of bilateral sciatic nerves was conducted, and the diameter of the nerves was measured before injection and at 7 and 14 days after injection. The retention of the injected hydrogel was also confirmed by US scanning at day 7 and day 14.

The nerve conduction studies of the sciatic nerve were performed at day 14 after the rats were euthanized, by completely exposing the nerve proximal and distal to the injection site. The nerves were then harvested for histology (hematoxylin and eosin (H&E) staining) and immunohistochemical staining of the myelin basic protein (MBP), glial fibrillary acidic protein (GFAP), and nuclear staining with 4′,6-Diamidino-2-Phenylindole, Dihydrochloride (DAPI) (Figure 6).

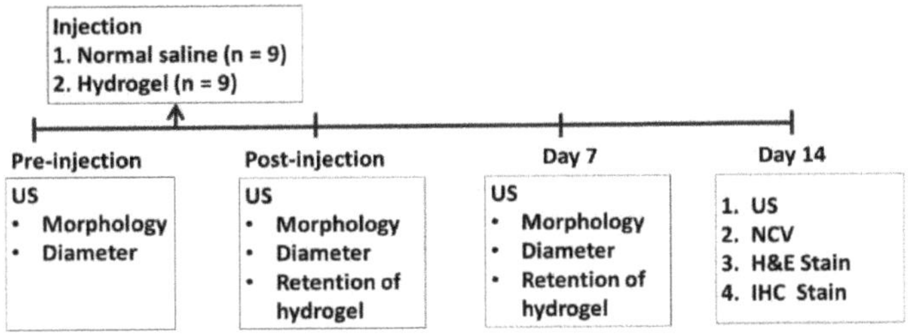

Figure 6. Schematic summary of the experimental procedure and outcome measures (US, ultrasound imaging; NCV, nerve conduction velocities; H&E, hematoxylin amd eosin; IHC, immunohistochemical).

4.2. US Evaluation and US-Guided Perineural Injection of Hydrogel

US-guided perineural injection of 1 mL of hydrogel was performed by a physician who had 10 years of experience with US-guided intervention procedures. US-guided injection was performed with the use of a 15 MHz linear array transducer (L18-4, SONIMAGE® HS1, Konica Minolta Healthcare Americas, Inc., Wayne, NJ, USA). The SD rat was anesthetized with isofluorane (2–3%, 2.1–3.3 L/min) and placed in a prone position with the hind legs fixed (Figure 7). The greater trochanter, knee joint, and the mid-point that connected these two sites were marked. Short-axis views were acquired first to locate the nerve, by placing the transducer perpendicular to the femur, with the center of the transducer aligned with the mid-point. The sciatic nerve appears as a hollow, small ovoid structure at the interfascial plane medial to the femur (Figures 1 and 2).

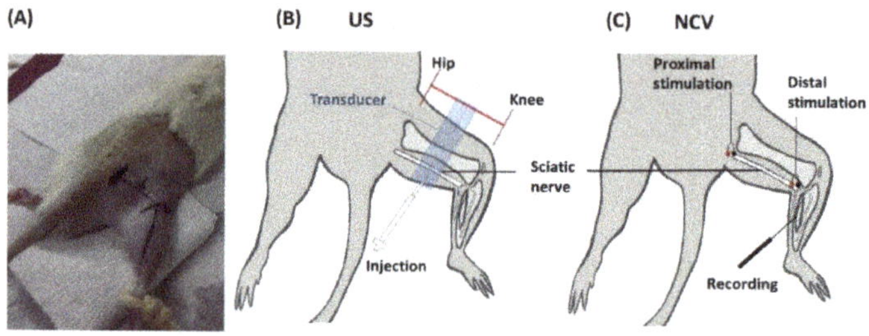

Figure 7. Positioning (**A**) and localization of sciatic nerve in US-guided injection (**B**) and NCV exam (**C**). Blue rectangle: placement of US probe.

The transducer was rotated 90° degrees to align with the femur to obtain the long-axis view of the sciatic nerve. The nerve appeared as a long, double contour structure. The mid-point of the transducer was aligned with the mid-point that connected the greater trochanter and knee joint. The diameter of the nerve was measured at proximal, middle, and distal sites.

US-guided injection was performed with short-axis acquisitions. The needle was inserted in plane with the transducer in a posterior–anterior direction and slowly advanced toward the nerve. Hydrogel or saline was injected after placing the needle tip between the nerve and fascia. The injectate (1 mL total) was dispersed in superficial and deep regions with respect to the nerve equally by repositioning the needle so that the injectate completely surrounded the nerve (Figure 1).

4.3. Preparation of mPEG-PLGA-BOX Hydrogel

The preparation of the mPEG-PLGA-BOX hydrogel has been described previously [19,28]. Briefly, the mPEG-PLGA copolymer was synthesized by mixing D, L-lactide (20 g) and glycolide (5.64 g) (both from Corbion PURAC Co., Provincie Zuid-Holland, Netherlands) with a mPEG (10.04 g) (Polysciences, Inc., Warrington, US) solution in a reactor. In total, 0.5 mL of tin (II) 2-ethylhexanoate or tin (II) octoate ($Sn(Oct)_2$) solution (0.1 M concentration) was added to dried toluene under a nitrogen atmosphere at room temperature. The temperature of the reactor was then maintained at 160 °C for 9 h to generate the mPEG-PLGA copolymer.

The copolymer was dissolved in Dimethyl Sulfoxide (DMSO) followed by dialysis at 4 °C for 5 d, lyophilization for 2 d, and then it was mixed with 200 mL of DMSO in a three-neck reactor with a mechanical stirrer. The final volume of DMSO was adjusted to 90 mL. Succinic anhydride (SA) (1.84 g) was added under a nitrogen atmosphere to the mPEG-PLGA-DMSO solution at room temperature. The temperature of the reactor was maintained at 180 °C for 3 h. Finally, the linker 2,2′-Bis (2-oxazoline) (BOX) (1.28 g) (Tokyo Chemical Industry Co., Ltd. Toyko, Japan) was added to the reactor. The mPEG-PLGA-BOX copolymer solution was dissolved at 4 °C for 4 d, followed by lyophilization for 5 d. The rheological analysis and cytotoxicity were elaborated in our previous study [20].

4.4. Nerve Conduction Studies

The nerve conduction velocities (NCV) of the sciatic nerves were measured by dissecting the nerves to expose the segments of the injected sites, after the rats were euthanized with isofluorane (3.5–5%, 3.9–5.9 L/min). A pair of hooked-wire electrodes (micrograbbers #085-416800, Natus Neurology, Inc., Middleton, WI, USA) was placed distally and then proximally at the injection site to deliver electric stimulation. Needle electrodes (27 G, #019-476900, Natus Neurology, Inc., Middleton, WI, USA) were placed in the gastrocnemius muscle to record compound motor action potentials (CMAP) (Figure 2). An average of 10 CMAP was obtained for each stimulation site. Motor latency was measured as the mean duration from stimulation to the onset of CMAP. Thereafter, NCV was calculated by dividing the distance between proximal and distal stimulation sites by the difference of motor latency. All the recordings used electrodiagnostic equipment (Nicolet VIkingQuest, Natus Neurology, Inc., Middleton, WI, USA).

4.5. Histology and Immunohistochemistry Staining

The rats were euthanized with isofluorane (3.5–5%, 3.9–5.9 L/min) at the end of the experiment, and the sciatic nerve was harvested and fixed with 10% formalin at room temperature overnight. The sample was embedded in paraffin and pasted on a glass slide (thickness = 7 μm), followed by the removal of paraffin, rehydration, and antigen retrieval (120 °C, 10 min). The sections were washed with tris buffered saline (TBS) that contained 0.025% Triton X-100 for 10 min and were blocked with 10% newborn calf serum and 1% bovine serum albumin in TBS for 2 h at room temperature. Primary antibodies were added to the sections to react overnight at 4 °C followed by the addition of fluorescence-labeled secondary antibodies. After 2 h at room temperature, the samples were washed with TBS and mounted with the EverBrite™ hardset mounting medium that contained DAPI to label the nuclei (1:5000, 40043, Biotium, Fremont, CA). The following primary antibodies were used: anti-MBP (1:1000, ab40390, Abcam, Cambridge, UK) and anti-GFAP (1:1000, Thermo Scientific, Waltham, MA, USA). The secondary antibodies used were Alexa Fluor 555-conjugated goat antimouse IgG (1:100; Thermo Scientific, Waltham, MA, USA) and Alexa Fluor 488-conjugated goat antirabbit IgG (1:100; Thermo Scientific, Waltham, MA, USA). Slides were viewed and images were captured with an LSM780 confocal microscope (Zeiss, Jena, Germany).

4.6. Statistical Analysis

All results are expressed as mean ± standard deviation. Normality was assessed using the Shapiro–Wilk test. The comparison between the hydrogel and saline groups was performed based on the Mann–Whitney U test for nonparametric nerve diameter data, and by an independent t-test for parametric NCV data. Wilcoxon two-sample tests were used to analyze the nerve diameter data collected at the follow-up compared with those at baseline. A test result was considered statistically significant at $p < 0.05$. All data obtained were analyzed with the MedCalc statistical software (version 15.11.3, MedCalc Software, Ostend, Belgium).

5. Conclusions

Ultrasound-guided perineural injection of hydrogel provides a surgery-free, minimally invasive method for the establishment of an animal model of EN of mixed etiology that could serve as a versatile tool for research on the pathophysiology and development of effective therapy of the disease.

Author Contributions: M.-Y.H.: research design; acquisition, analysis, and interpretation of data, drafting of the paper. Y.-W.W.: data acquisition, analysis, and interpretation of data. W.-S.C.: interpretation of data, revision of the paper, supervision of the work. Y.-L.L.: data acquisition, analysis, and interpretation of data. P.-L.K.: interpretation of data, revision of the paper. C.-H.W.: research design, interpretation of data, revision of the paper, supervision of the work. All authors have read and agreed to the published version of the manuscript.

Funding: This research was funded by the National Taiwan University Hospital (NTUH 108-M4366, 108-M4277, 109-M4670), the Ministry of Science and Technology, Taiwan (MOST 107-2314-B-002-045-MY3, MOST 107-2314-B-002-232-MY3). The APC was funded by the National Taiwan University Hospital Hsin-Chu Branch (110-BIH009).

Institutional Review Board Statement: The study was conducted according to the guidelines of the Declaration of Helsinki, and approved by the ethics committee of the Laboratory Animal Center at the National Taiwan University College of Medicine (Approval No. 20170470, 18 October 2018).

Informed Consent Statement: Not applicable.

Data Availability Statement: The datasets generated during and/or analyzed during the current study are available from the corresponding author upon reasonable request.

Acknowledgments: The authors thank Wei Chun Chien for assistance in the experiments.

Conflicts of Interest: The authors declare no conflict of interest.

References

1. Padua, L.; Coraci, D.; Erra, C.; Pazzaglia, C.; Paolasso, I.; Loreti, C.; Caliandro, P.; Hobson-Webb, L.D. Carpal tunnel syndrome: Clinical features, diagnosis, and management. *Lancet Neurol.* **2016**, *15*, 1273–1284. [CrossRef]
2. Aboonq, M.S. Pathophysiology of carpal tunnel syndrome. *Neurosciences* **2015**, *20*, 4–9. [PubMed]
3. Ettema, A.M.; Amadio, P.C.; Zhao, C.; Wold, E.L.; O'Byrne, M.M.; Moran, S.L.; An, K.-N. Changes in the Functional Structure of the Tenosynovium in Idiopathic Carpal Tunnel Syndrome: A Scanning Electron Microscope Study. *Plast. Reconstr. Surg.* **2006**, *118*, 1413–1422. [CrossRef]
4. Festen-Schrier, V.; Amadio, P. The biomechanics of subsynovial connective tissue in health and its role in carpal tunnel syndrome. *J. Electromyogr. Kinesiol.* **2018**, *38*, 232–239. [CrossRef] [PubMed]
5. Yoshii, Y.; Zhao, C.; Schmelzer, J.D.; Low, P.A.; An, K.-N.; Amadio, P.C. The Effects of Hypertonic Dextrose Injection on Connective Tissue and Nerve Conduction Through the Rabbit Carpal Tunnel. *Arch. Phys. Med. Rehabil.* **2009**, *90*, 333–339. [CrossRef]
6. Yoshii, Y.; Zhao, C.; Schmelzer, J.D.; Low, P.A.; An, K.-N.; Amadio, P.C. Effects of Multiple Injections of Hypertonic Dextrose in the Rabbit Carpal Tunnel: A Potential Model of Carpal Tunnel Syndrome Development. *Hand* **2013**, *9*, 52–57. [CrossRef]
7. Oh, J.; Zhao, C.; Zobitz, M.E.; Wold, L.E.; An, K.N.; Amadio, P.C. Morphological changes of collagen fibrils in the subsynovial connective tissue in carpal tunnel syndrome. *J. Bone Jt. Surg. Am.* **2006**, *88*, 824–831.
8. Chen, N.F.; Chen, W.F.; Sung, C.S.; Lu, C.H.; Chen, C.L.; Hung, H.C.; Feng, C.W.; Chen, C.H.; Tsui, K.H.; Kuo, H.M.; et al. Contributions of p38 and ERK to the antinociceptive effects of TGF-beta1 in chronic constriction injury-induced neuropathic rats. *J. Headache Pain* **2016**, *17*, 72. [CrossRef]
9. MacKinnon, S.E.; Dellon, A.L.; Hudson, A.R.; Hunter, D.A. Chronic Nerve Compression—An Experimental Model in the Rat. *Ann. Plast. Surg.* **1984**, *13*, 112–120. [CrossRef]
10. O'brien, J.P.; MacKinnon, S.E.; MacLean, A.R.; Hudson, A.R.; Dellon, A.L.; Hunter, D.A. A Model of Chronic Nerve Compression in the Rat. *Ann. Plast. Surg.* **1987**, *19*, 430–435. [CrossRef]
11. Phame, K.; Nassiri, N.; Gupta, R. c-Jun, krox-20, and integrin beta4 expression following chronic nerve compression injury. *Neurosci. Lett.* **2009**, *465*, 194–198. [CrossRef] [PubMed]
12. Qu, Y.; Wang, B.; Chu, B.; Liu, C.; Rong, X.; Chen, H.; Peng, J.; Qian, Z. Injectable and Thermosensitive Hydrogel and PDLLA Electrospun Nanofiber Membrane Composites for Guided Spinal Fusion. *ACS Appl. Mater. Interfaces* **2018**, *10*, 4462–4470. [CrossRef] [PubMed]
13. Rempel, D.M.; Diao, E. Entrapment neuropathies: Pathophysiology and pathogenesis. *J. Electromyogr. Kinesiol.* **2004**, *14*, 71–75. [CrossRef]
14. Kerwin, G.; Williams, C.S.; Seiler, J.G. The pathophysiology of carpal tunnel syndrome. *Hand Clin.* **1996**, *12*, 243–251. [CrossRef]
15. Neary, D.; Ochoa, J.; Gilliatt, R. Sub-clinical entrapment neuropathy in man. *J. Neurol. Sci.* **1975**, *24*, 283–298. [CrossRef]
16. Werthel, J.-D.R.; Zhao, C.; An, K.-N.; Amadio, P.C. Carpal Tunnel Syndrome Pathophysiology: Role of Subsynovial Connective Tissue. *J. Wrist Surg.* **2014**, *3*, 220–226. [CrossRef] [PubMed]
17. Uchiyama, S.; Itsubo, T.; Nakamura, K.; Kato, H.; Yasutomi, T.; Momose, T. Current concepts of carpal tunnel syndrome: Pathophysiology, treatment, and evaluation. *J. Orthop. Sci.* **2010**, *15*, 1–13. [CrossRef] [PubMed]
18. Sud, V.; Freeland, A.E. Biochemistry of carpal tunnel syndrome. *Microsurgery* **2005**, *25*, 44–46. [CrossRef]
19. Peng, K.-T.; Hsieh, M.Y.; Lin, C.T.; Chen, C.F.; Lee, M.S.; Huang, Y.Y.; Chang, P.J. Treatment of critically sized femoral defects with recombinant BMP-2 delivered by a modified mPEG-PLGA biodegradable thermosensitive hydrogel. *BMC Musculoskelet. Disord.* **2016**, *17*, 286. [CrossRef] [PubMed]
20. Wu, C.-H.; Sun, M.-K.; Kung, Y.; Wang, Y.-C.; Chen, S.-L.; Shen, H.-H.; Chen, W.-S.; Young, T.-H. One injection for one-week controlled release: In vitro and in vivo assessment of ultrasound-triggered drug release from injectable thermoresponsive biocompatible hydrogels. *Ultrason. Sonochem.* **2020**, *62*, 104875. [CrossRef]
21. Pham, K.; Gupta, R. Understanding the mechanisms of entrapment neuropathies. *Neurosurg. Focus* **2009**, *26*, E7. [CrossRef] [PubMed]
22. Gupta, R.; Truong, L.; Bear, D.; Chafik, D.; Modafferi, E.; Hung, C.T. Shear stress alters the expression of myelin-associated glycoprotein (MAG) and myelin basic protein (MBP) in Schwann cells. *J. Orthop. Res.* **2005**, *23*, 1232–1239. [CrossRef] [PubMed]
23. Zhang, L.; Yang, X.; Yue, Y.; Ye, J.; Yao, Y.; Fu, Y.; Li, G.; Yao, Q.; Lin, Y.; Gong, P. Cyclic mechanical stress modulates neurotrophic and myelinating gene expression of Schwann cells. *Cell Prolif.* **2014**, *48*, 59–66. [CrossRef] [PubMed]
24. Wu, Y.-T.; Chen, S.-R.; Li, T.-Y.; Ho, T.-Y.; Shen, Y.-P.; Tsai, C.-K.; Chen, L.-C. Nerve hydrodissection for carpal tunnel syndrome: A prospective, randomized, double-blind, controlled trial. *Muscle Nerve* **2019**, *59*, 174–180. [CrossRef] [PubMed]
25. Fried, S.M.; Nazarian, L.N. Ultrasound-Guided Hydroneurolysis of the Median Nerve for Recurrent Carpal Tunnel Syndrome. *Hand* **2019**, *14*, 413–421. [CrossRef]

26. Wu, Y.T.; Ho, T.Y.; Chou, Y.C.; Ke, M.J.; Li, T.Y.; Tsai, C.K.; Chen, L.C. Six-month Efficacy of Perineural Dextrose for Carpal Tunnel Syndrome: A Prospective, Randomized, Double-Blind, Controlled Trial. *Mayo Clin. Proc.* **2017**, *92*, 1179–1189. [CrossRef] [PubMed]
27. Lin, M.-T.; Liao, C.-L.; Hsiao, M.-Y.; Hsueh, H.-W.; Chao, C.-C.; Wu, C.-H. Volume Matters in Ultrasound-Guided Perineural Dextrose Injection for Carpal Tunnel Syndrome: A Randomized, Double-Blinded, Three-Arm Trial. *Front. Pharmacol.* **2020**, *11*, 625830. [CrossRef]
28. Hu, C.-C.; Chaw, J.-R.; Chen, C.-F.; Liu, H.-W. Controlled release bevacizumab in thermoresponsive hydrogel found to inhibit angiogenesis. *Bio-Med. Mater. Eng.* **2014**, *24*, 1941–1950. [CrossRef]

International Journal of
Molecular Sciences

MDPI

Article

Enhanced Regeneration of Vascularized Adipose Tissue with Dual 3D-Printed Elastic Polymer/dECM Hydrogel Complex

Soojin Lee [1,2], Hyun Su Lee [2], Justin J. Chung [1], Soo Hyun Kim [1,3], Jong Woong Park [4], Kangwon Lee [5,*] and Youngmee Jung [1,6,*]

1 Center for Biomaterials, Biomedical Research Institute, Korea Institute of Science and Technology, Seoul 02792, Korea; dltnwls830@snu.ac.kr (S.L.); chungjj@kist.re.kr (J.J.C.); soohkim@kist.re.kr (S.H.K.)
2 Program in Nanoscience and Technology, Graduate School of Convergence Science and Technology, Seoul National University, Seoul 08826, Korea; hyun118soo@snu.ac.kr
3 NBIT, KU-KIST Graduate School of Converging Science and Technology, Korea University, Seoul 02841, Korea
4 Department of Orthopedic Surgery, Korea University Anam Hospital, Seoul 02841, Korea; ospark@korea.ac.kr
5 Department of Applied Bioengineering, Graduate School of Convergence Science and Technology, Seoul National University, Seoul 08826, Korea
6 School of Electrical and Electronic Engineering, YU-KIST Institute, Yonsei University, Seoul 03722, Korea
* Correspondence: kangwonlee@snu.ac.kr (K.L.); winnie97@kist.re.kr (Y.J.)

Abstract: A flexible and bioactive scaffold for adipose tissue engineering was fabricated and evaluated by dual nozzle three-dimensional printing. A highly elastic poly (L-lactide-co-ε-caprolactone) (PLCL) copolymer, which acted as the main scaffolding, and human adipose tissue derived decellularized extracellular matrix (dECM) hydrogels were used as the printing inks to form the scaffolds. To prepare the three-dimensional (3D) scaffolds, the PLCL co-polymer was printed with a hot melting extruder system while retaining its physical character, similar to adipose tissue, which is beneficial for regeneration. Moreover, to promote adipogenic differentiation and angiogenesis, adipose tissue-derived dECM was used. To optimize the printability of the hydrogel inks, a mixture of collagen type I and dECM hydrogels was used. Furthermore, we examined the adipose tissue formation and angiogenesis of the PLCL/dECM complex scaffold. From in vivo experiments, it was observed that the matured adipose-like tissue structures were abundant, and the number of matured capillaries was remarkably higher in the hydrogel–PLCL group than in the PLCL-only group. Moreover, a higher expression of M2 macrophages, which are known to be involved in the remodeling and regeneration of tissues, was detected in the hydrogel–PLCL group by immunofluorescence analysis. Based on these results, we suggest that our PLCL/dECM fabricated by a dual 3D printing system will be useful for the treatment of large volume fat tissue regeneration.

Keywords: 3d printing; PLCL; decellularization; angiogenesis; dECM hydrogel; adipose tissue regeneration

Citation: Lee, S.; Lee, H.S.; Chung, J.J.; Kim, S.H.; Park, J.W.; Lee, K.; Jung, Y. Enhanced Regeneration of Vascularized Adipose Tissue with Dual 3D-Printed Elastic Polymer/dECM Hydrogel Complex. *Int. J. Mol. Sci.* **2021**, *22*, 2886. https://doi.org/10.3390/ijms22062886

Academic Editor: Peter John Jervis

Received: 5 February 2021
Accepted: 9 March 2021
Published: 12 March 2021

1. Introduction

Development of enhanced vascularized adipose tissue is a prospective issue for not only adipose tissue but also soft tissue engineering in modern healthcare [1–4]. Adipose tissue engineering has been studied and is required in soft tissue deficiency, such as traumatic injuries, extensive burn defects, innate defects, and defects after cancer treatments. Most of all, as the number of patients diagnosed with breast cancer increases, mammotomies or lumpectomies are performed more often. Therefore, interest has increased in breast regeneration to improve the quality of life of cancer patients. Adipose tissue is the largest and ubiquitous tissue in the body; thus, unlike other organs or tissues, it can easily be acquired [5–7]. Although autologous fat grafts have been regarded as an ideal way to augment soft tissue lost because of fewer immunological rejections and complications, their problem is that the engrafted fat has a poor volume maintenance rate after surgery, with a 40–60% volume loss because of ineffective integration with the mature adipose tissue [8–11]. In the case of small defects, it may be efficient; however, not for large defects.

Various trials for adipose tissue regeneration have been conducted [10]. Many studies with diverse materials have been performed to replace autografting, but the results were not satisfactory [12,13]. Tissue engineering with scaffolds has potential for developing and reconstructing adipose tissue. Scaffolds are one of fundamental components [11,14], which contribute significantly to tissue engineering. A scaffold primarily supports a three-dimensional (3D) structure implanted into host tissue [8]. To fabricate a 3D-constructed scaffold, three-dimensional (3D) printing technology has been shown to be a promising method due to its various abilities, including the creation of complex constructs and their precise placement as well as a wide permissible range for materials. For effective tissue engineering, a scaffold must be easy to handle, biocompatible, and biodegradable to help cells to adhere, function, migrate, and proliferate well and to prevent inflammation [15]. In addition, the scaffold should have similar mechanical properties, especially the Young's modulus, as the original region into which it is to be implanted to avoid stress shielding and implantation failure [16,17]. Synthetic and natural polymers such as PLA, PCL, collagen, and fibrin have been used in adipose tissue engineering; however, mismatched mechanical properties or unmatched degradation rates have been thresholds to overcome [18,19]. Poly (L-lactide-co-ε-caprolactone) (PLCL) is a co-polymer synthesized with L-lactide and ε-caprolactone through ring-opening polymerization. The mechanical properties of the PLCL co-polymer can be controlled by changing the monomer component [20]. It has flexible, stretchable, and elastic properties and degrades slowly, which means that it not only has more suitable properties than other synthetic polymers for soft tissue engineering, but also overcomes the limitations of native hydrogels and provides a three-dimensional spatial structure [21].

In regeneration for large defects, an efficient blood supply is absolutely necessary to deliver oxygen and nutrients and to remove waste products. Fat injection can be a good solution for a small defect; however, 3D-vascularized engineered tissue is a necessary challenge for the regeneration of large-sized adipose tissue with implantable constructs [22]. Thick implants, especially the central part, require vascularization to be used in vivo [1]; moreover, O_2 is diffused less than 100–200 μm through tissue [23]. There are various strategies to enhance vascularization in tissue development, such as the pre-vascularization technique, growth factor-based technique, cell-based technique, and scaffold- and material-based techniques. Medical cost, initial angiogenesis, and functionality of the newly formed blood vessels have been considered as important conditions for the recovery of a critical size defect. Various strategies have been tried to develop vascularized tissue; however, they still have problems in forming functional vascularized tissues or organs [24]. Most cells are generally found within 100–200 μm of the nearest capillary for survival, and the maximum distance between capillaries is 200 μm. For these reasons, a highly developed neo-vascularized system of large- and micro-size blood vessels is required for tissue engineering [25–29].

The extracellular matrix (ECM) is a widely used material for scaffolds because it is composed of functional proteins, collagen, fibronectin, glycosaminoglycans (GAGs) and proteoglycans, which means that it is a favorable condition, providing a suitable microenvironment for cells and tissues to maintain their morphology and to differentiate [30]. The decellularization procedure is the protocol that removes the cellular components from the organs or tissues while maintaining the ECM components [31]. A complex mixture of decellularized ECM (dECM) is usually made into mixed hydrogels with collagen, hyaluronic acid, and fibroin for enhanced tissue regeneration [32]. In this regard, adipose tissue-derived dECM (adECM) is a potential material for adipose tissue regeneration because it has been reported that the adECM promotes regeneration and repair by providing a supportive microenvironment for adipogenic differentiation of induced cells [6,33] and by enhancing angiogenesis because it has growth factors related to neovascularization, such as VEGF [7,34].

The aim of this study was to develop a vascularized adipose tissue with a PLCL/dECM scaffold with the capability of cell recruitment and growth by providing the proper mechanical properties and microenvironment. The main focus of this study was to prepare a construct using the dual nozzle 3D printing technique with an elastic synthetic polymer and dECM hydrogel that induces angiogenesis and forms adipose tissue.

2. Results and Discussion

2.1. PLCL Characterization

For adipose tissue engineering, a flexible and bioactive scaffold was fabricated, as shown in Figure 1. The PLCL co-polymer was synthesized by the ring-opening polymerization of L-lactide and ε-caprolactone in the presence of Sn(Oct)2 with continuous stirring under bulk conditions. A high conversion rate was expected under these conditions after a predetermined optimum reaction time. The polymerizations were carried out at 150 °C with a constant molar feed ratio (fL/fC) of 5:5. The distribution ratio of the PLCL was calculated by taking the ratio between the integration area of the peaks of the lactyl (Quartet, 5.160 ppm) and caproyl (Triplet, 4.051 ppm) units by 600 Hz ^{1}H-NMR (Figure 2b). The molar ratio of L-lactide and ε-caprolactone in the co-polymer was 4.8:5.2. The overall yield of the PLCL co-polymer was 70%. Mn (number-average), Mw (weight-average), and molecular weight distributions (polydispersity, PDI = Mw/Mn) were determined by gel permeation chromatography (GPC) (Figure 2c). It was revealed that the Mn, Mw, and PDI of the PLCL co-polymer were 1.50×10^5, 1.70×10^5, and 1.14, respectively [20].

2.2. Characterization of dECM Hydrogel for 3D Priting Ink

Various scaffolds, including the extracellular matrix (ECM), have been used for diverse tissue or organ regeneration applications and medical strategies. The ECM produced by the resident cells of each tissue and organ can serve as an ideal biomimetic microenvironment for target cells which can accelerate tissue remodeling and healing in many organs [35–39]. Adipose tissue-derived ECM induces the migration and adipogenic differentiation of cells to form fat [40].

To promote adipogenic differentiation and angiogenesis, adipose tissue-derived decellularized extracellular matrix-based hydrogels were used to make a biocompatible ink. To preserve the residual ECM and maximize the cellular material loss [41], we used a modified method for adipose tissue with the physical, chemical, and enzymatic processes described elsewhere resulting in the successful decellularization of the tissues [42]. The successfully decellularized adECM powder was made into an ink with collagen type I for 3D printing (Figure 3). The adECM powder was dissolved with 10 mg of pepsin in 3% (v/v) acetic acid solution and neutralized with 10× Minimum Essential Media (MEM) and 5 N NaOH solution to adjust pH. The collagen was diluted to 3 mg/mL and neutralized with 10× MEM media, distilled water, and 1 N NaOH solution. Finally, the 0.3% collagen and 3% adECM neutralized solutions were mixed well and incubated at 37 °C for 30 min for gelation. Different adECM-to-collagen mass ratios were used, ranging from 1:3, 1:1, and 3:1 to determine the suitability of the ink in terms of printability and cell viability. Cell viability results (Figure 4) showed that all composition ratios of the mixed hydrogel had no cell toxicity and provided a proper environment for both adipose-derived stem cells (ADSCs) and human umbilical vein endothelial cells (HUVECs), which was evident by the more than 90% cell viability in each hydrogel. The collagen only, 1:1-adECM:collagen, and 1:3-adECM:collagen had better cell viability results for ADSCs in that order, although there was some difference, whereas the 1:3, 1:1, and 3:1-adECM:collagen had better cell viability results for the HUVECs (Figure S1). The compressive moduli of the 1:3-adECM:collagen, 1:1-adECM:collagen, and 3:1-adECM:collagen were 16.95, 12.80, and 8.20 kPa, respectively. In the printability test, the 1:3-adECM:collagen had the clearest printability compared to the others; thus, the 1:3-adECM:collagen hydrogel was selected as the ink for fabrication of the 3D scaffold. The 3:1-adECM:collagen hydrogel tended to be too runny and would spread during printing, so it was excluded from the printing test.

2.3. Characterization of the PLCL Scaffolds and the Hydrogel–PLCL Constructs

Porosity and pore size are significant factors for tissue regeneration. Proper scaffolds should have an excess of 50% porosity with at least a pore size of 100 μM [43–45]. Porosity and pore size are directly related to their functionality for tissue engineering applications. Moreover, openly porous and interconnected pores are essential conditions for providing oxygen, nutrients, and cell migration, as well as proliferation for vascularization and tissue formation. Generally, macropores are necessary to have enough space for vascularization, cellular infiltration, tissue ingrowth, and waste product removal [46–51].

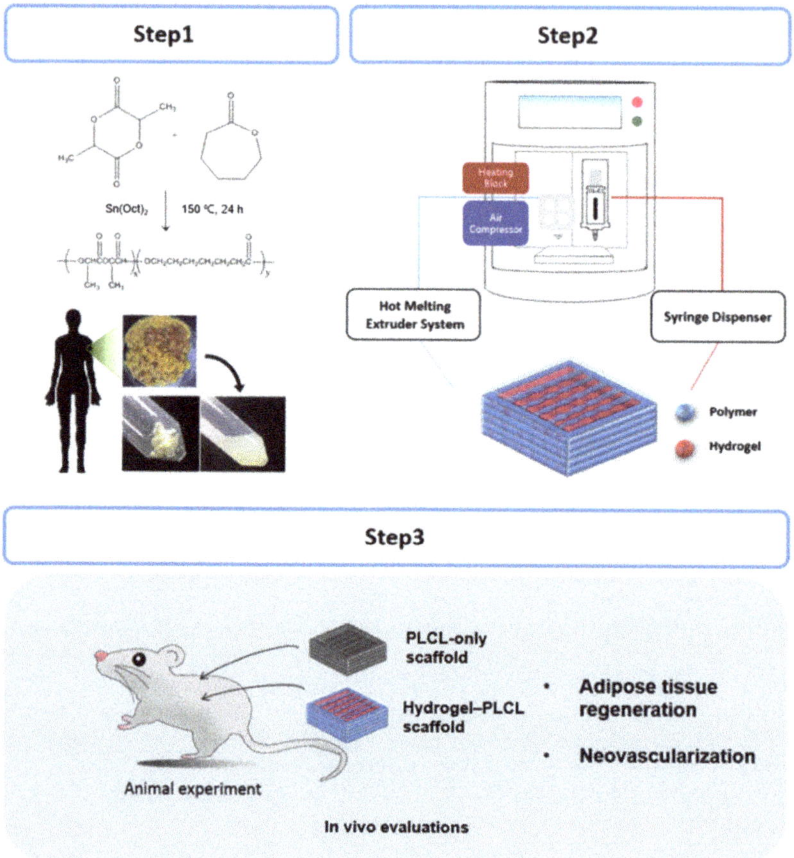

Figure 1. Schematic explanation of the research strategy. An elastic copolymer, PLCL, was synthesized by ring-opening polymerization, and the hydrogel was made with decellularized adipose tissue extracellular matrix (ECM) and collagen. Three-dimensional (3D) printing was performed with the PLCL and hydrogel inks with a hot melting extruder and syringe, respectively. The 3D-printed scaffolds were subcutaneously implanted in rats for the in vivo experiments.

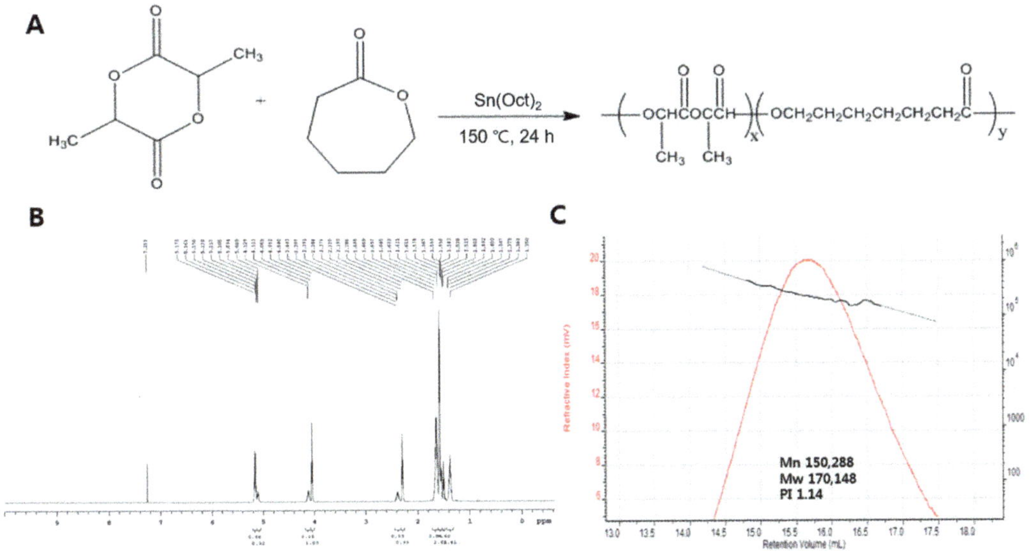

Figure 2. PLCL synthesis and characterization. (**A**) Scheme for the PLCL copolymer synthesis. (**B**) 600 MHz ^{1}H-NMR spectrum of the PLCL copolymer (50:50). (**C**) Gel permeation chromatography (GPC) curve of the PLCL copolymer.

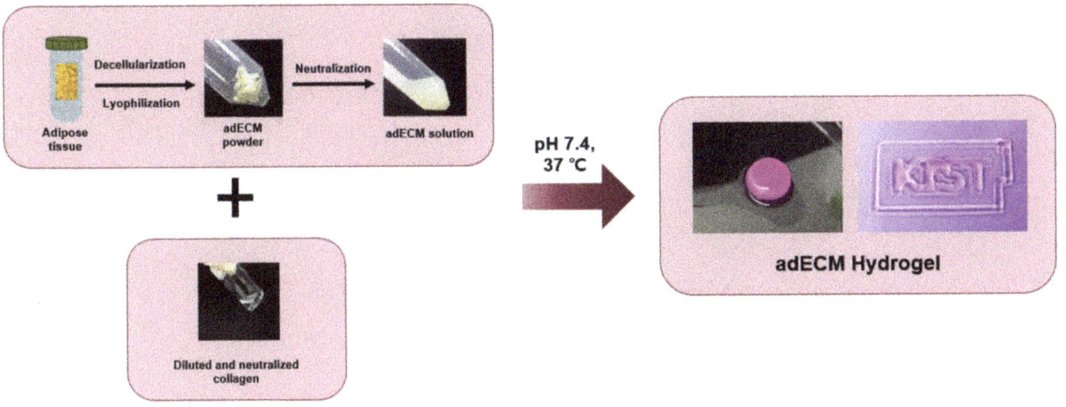

Figure 3. Preparation scheme for the mixed hydrogel using adipose tissue-derived decellularized ECM (adECM) and collagen. Human adipose tissue was decellularized by various processes and lyophilized. The resulting adECM was crushed into a powder and made into a hydrogel ink with collagen type I for 3D printing. Gelation and 3D printing of the hydrogel with various compositions of adECM and collagen was performed to optimize the printing condition. Bottom picture shows the optimized hydrogel (adECM:Col = 1:3).

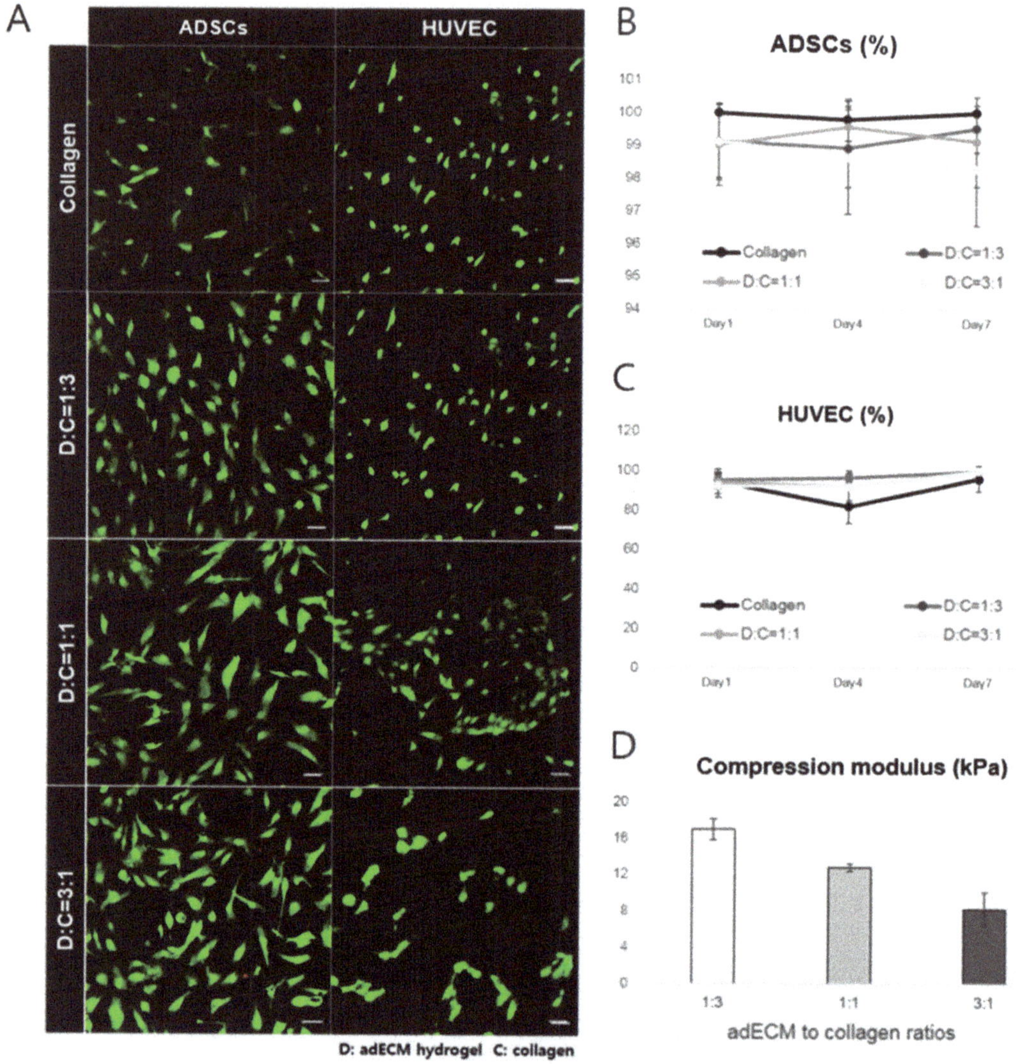

Figure 4. (**A**) LIVE/DEAD assay. Cell viability test was performed to optimize the hydrogel composition. Adipose-derived stem cells (ADSCs) and human umbilical vein endothelial cells (HUVECs) were seeded onto each of the hydrogels. The confocal images above were obtained after a 7-day incubation of cells. Green and red fluorescence represents live and dead cells, respectively. Scale bar = 50 μM. (**B**) Cell viability results of ADSCs. (**C**) Cell viability results of HUVEC. (**D**) Compression modulus of each hydrogel. All samples were evaluated in triplicate, and the error bar indicates the SD.

Thus, the cellular behavior for new tissue reconstruction can be affected by the material properties of the scaffold such as the porosity, pore size, and interconnectivity [52]. Angiogenesis means the growth of new blood vessels from vasculature existing originally; thus, it is very important to supply oxygen and nutrients for tissue accumulation and wound healing during long-term tissue regeneration [53]. Larger pore sizes of approximately 160 to 270 μM facilitate angiogenesis throughout a scaffold, and many studies have shown that average pore sizes larger than 300 μM provide a better condition for vascularization [54–56].

Figure 5 shows the Standard Triangle Language (STL) image files and printed scaffolds by the 3D printer (In vivo premium, Rokit, Republic of Korea) for the PLCL-only scaffold and hydrogel–PLCL construct, respectively. Each scaffold was printed quite accurately when compared to the bird's-eye view images of the STL files as intended. The PLCL scaffolds were very flexible and stretchable, as shown in Figure S2. The morphology of the PLCL-only scaffolds and the hydrogel–PLCL constructs were observed by scanning electron microscopy (SEM). The open pores were highly interconnected and created surfaces on both the outside and inside of the scaffolds. The interconnected pore channels in the PLCL-only or hydrogel–PLCL scaffolds had a generally uniformed pore size of 920.75 ± 6.25 and 997.5 ± 122.1, while the mean rod thicknesses of the scaffolds were 404.47 ± 39.33 and 403.17 ± 41.83, respectively.

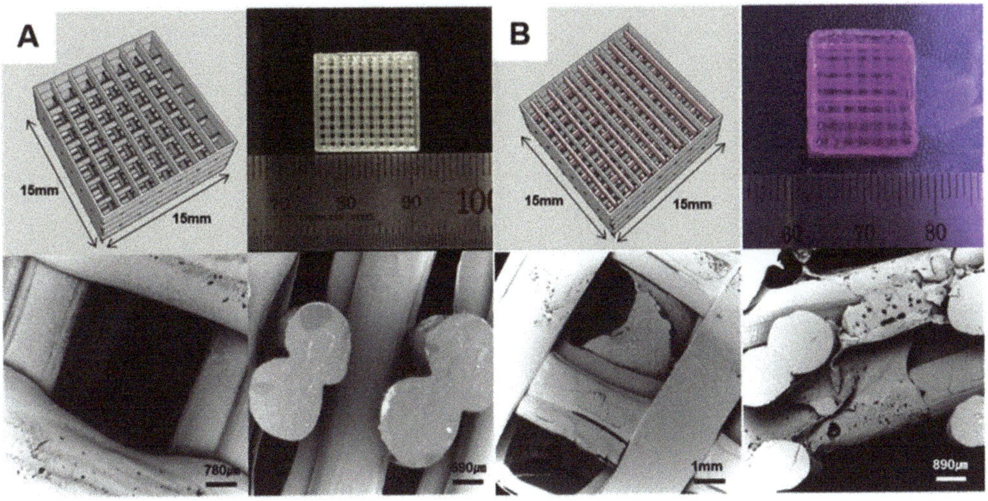

Scaffold	Pore size (μm)	Rod thickness (μm)
PLCL-only	920.75 ± 6.25	404.67 ± 39.33
Hydrogel–PLCL	997.9 ± 122.1	403.17 ± 41.83

Figure 5. 3D-printed scaffold. Standard Triangle Language (STL) file image for 3D printing, macroscopic image, and SEM images of the (**A**) PLCL-only scaffold and (**B**) Hydrogel–PLCL complex construct. The pore size and rod thickness were measured.

To evaluate the ultimate tensile strength change of the PLCL polymer before and after 3D printing, the ultimate strength and its Young's modulus were determined and plotted as a graph (Figure 6). The PLCL copolymer is degraded by heat [57,58], which could be confirmed by molecular weight changes and Young's modulus loss after 3D printing. The number (Mn) and average molecular weight (Mw) decreased from 150 kD to 114 kD and from 170 kD to 139 kD. The ultimate strength and Young's modulus decreased from 17.50 ± 5.43 MPa to 11.99 ± 2.52 MPa, and from 2.65 ± 0.39 MPa to 1.85 ± 0.64 MPa. The compressive moduli of the PCL and PLCL were measured to confirm that their physical properties matched those of human adipose tissue, to avoid stress shielding and implantation failure. The compressive moduli of the scaffolds were approximately 3.6 MPa for the PCL, 122 kPa for the PLCL, and 40 kPa for adipose tissue. Tissue engineering scaffolds are required to have both synthetic materials with adjustable mechanical properties and natural materials with biomimetic properties. Synthetic polymers with a well-defined structure and without immunological problems such as PLA, PCL, and PLGA are widely used as a biomedical scaffold which has a crucial role in tissue engineering [59]. Among them, PCL is a representative material of tissue engineering [59–61]. PCL scaffolds used to be fabricated for soft tissue engineering [62,63]; however, as shown in this result, PCL has a very high modulus compared to native tissue. Meanwhile, in the case of the PLCL scaffold, it has a modulus value much closer to adipose tissue than that of the PCL scaffold. This result demonstrates that PLCL has a physical property that is more similar to adipose tissue, which is beneficial for regeneration, compared to the PCL.

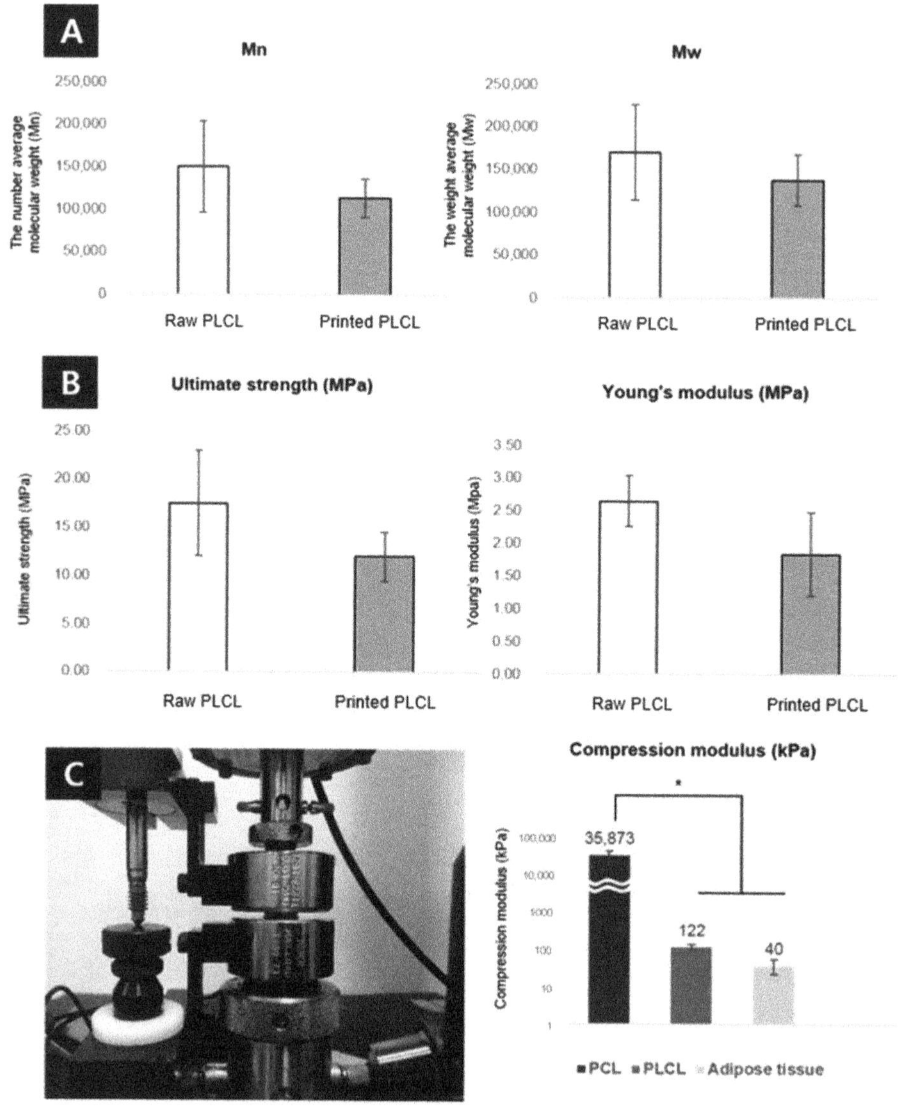

Figure 6. Evaluation of the molecular weight changes and mechanical properties. (**A**) The number (Mn, left) and weight (Mw, right) average molecular weight changes of the PLCL copolymer after 3D printing. Analysis of the mechanical properties of the PLCL copolymer before and after 3D printing, including the (**B**) ultimate strength (left) and Young's modulus (right). (**C**) Compressive modulus value of the PCL, PLCL scaffolds and adipose tissue was evaluate. All samples were evaluated in triplicate, and the error bar indicates SD (* $p < 0.01$).

2.4. In Vivo Studies

All animals for the in vivo experiments were treated in accordance with the standard operating protocols of the Institutional Animal Care and Use Committee at the Korea Institute of Science and Technology (KIST). All protocols were approved by the Institutional Review Board of Animal Experiments at KIST (Approval Number Kist-2019-009). Four and eight weeks after subcutaneous implantation, the scaffolds were harvested from the rats. Figure 7 shows the macroscopic view of the implanted scaffolds by each time point and group. Plenty more blood vessels were observed in the hydrogel–PLCL scaffold group than in the PLCL-only group.

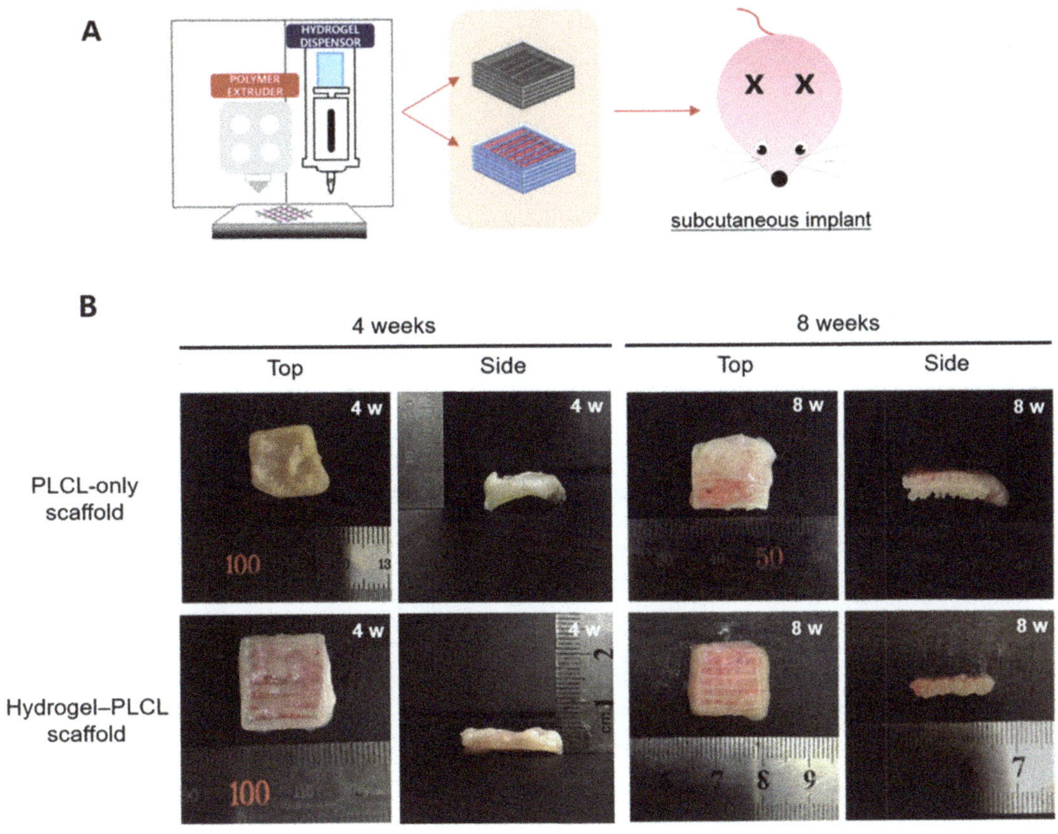

Figure 7. In vivo experiment. (**A**) Schematic image of the in vivo experiments. (**B**) Macroscopic images of the PLCL-only scaffolds and hydrogel–PLCL complex constructs after 4 and 8 weeks, respectively.

Evaluation for the molecular weight change of the PLCL polymer in vivo was performed for eight weeks. The results show that the tendency for the changes in the number (Mn) and average molecular weights (Mw) gradually decreased or increased over the entire experiment period. The PLCL-only scaffolds had Mn decreases of 57.58% and 58.12%, and Mw increases of 200.67% and 219.03% at four and eight weeks after the subcutaneous implants, respectively. The hydrogel–PLCL constructs had larger Mn decreases of 60.37% and 83.41% and Mw increases of 153.48% and 94.46% at four and eight weeks, respectively (Figure 8). The result of increasing Mw can be interpreted as the short chains were unstable and degraded rapidly compared to long chains. The polydispersity index (Mw/Mn) increased gradually up to eight weeks post-implantation, indicating cleaved chains from the polymer. The PLCL has both crystalline and amorphous phases; the amorphous phase is more easily attacked by water compared to the crystalline regions [64]. Therefore, the changes in the average molecular weight of both the scaffolds could be explained by the following reason: the amorphous phases composed of CL, which is degraded faster than the crystalline regions composed of LA, were degraded in an earlier stage in the body.

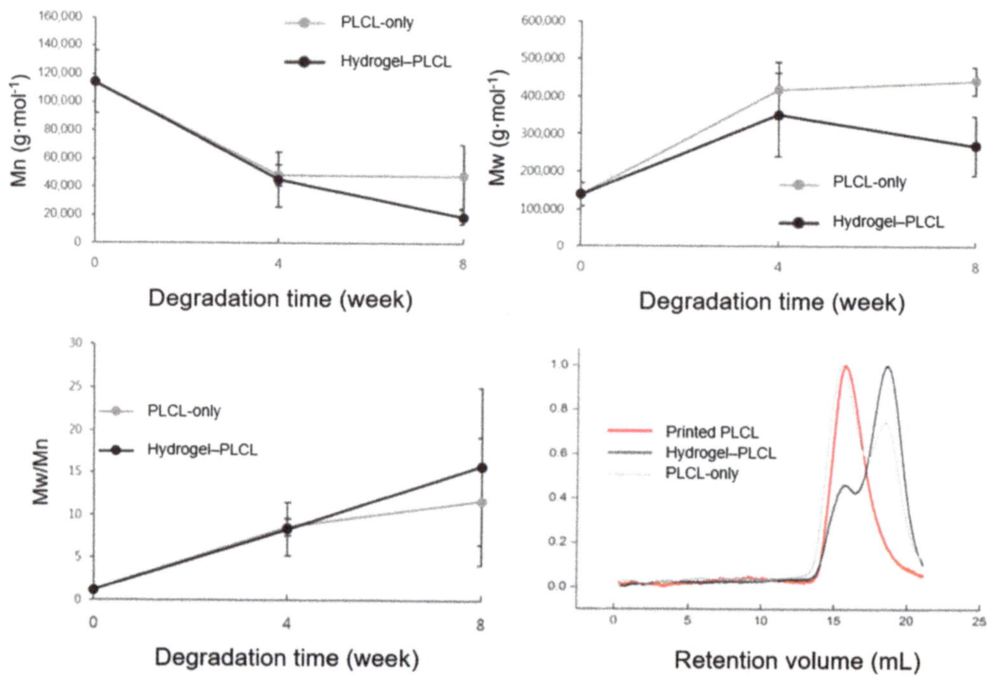

Figure 8. The number (Mn), average molecular weight (Mw), and polydispersity (PDI) value changes of the scaffolds from the in vivo degradation study.

2.5. Histological Analysis with Hematoxylin and Eosin (H&E) and Masson's Trichrome (MT) Staining

For qualitative analysis, explanted scaffolds were made into 5 μM slides and stained with hematoxylin and eosin (H&E) and Masson's Trichrome (MT) for histological analysis. Figure 9 shows the results of the staining of each group for each time point. As shown in the figure, in the fourth week, both groups had little fat accumulation and tissue formation. However, the difference is that tissue formation only occurred on the outside edge in the PLCL-only group, whereas the tissue was coming into and moving out of the scaffold in the hydrogel–PLCL group. In addition, fat accumulation and tissue formation were observed more in the center of the hydrogel–PLCL scaffold compared to the PLCL-only group, which means the hydrogel degradation behavior enhanced tissue formation [65,66]. Red blood cells and small blood vessels were observed in both groups. More significant differences between the two groups were observed at week 8. In the case of the PLCL-only group, a large amount of fat accumulation was observed on the outside of the scaffolds. Red blood cells, veins, and arteries were also found, but not in the center of the scaffold. There was no significant tissue or vessel formation in the center at eight weeks compared to four weeks. On the other hand, in the hydrogel–PLCL group, there were obvious differences such as the accumulation of fat, tissue formation, and vascularization. There was much tissue and fat coming into the scaffold as well as moving out; especially, there was plenty of adipose tissue with blood vessels in the center of the scaffold. A number of mature veins and arteries were found outside and inside of the scaffold, as well as lymphatic vessels. Lymphatic vessels are crucial for the recirculation of tissue fluid balance and cells that enter tissues from blood vessels [67]. An amount of sulfated GAG deposition was also observed in the hydrogel–PLCL group by the blue color of Masson's trichrome, and it was found that the blood vessels, GAG, and tissue were formed between the scaffold both outside and inside of the scaffold. These results show that the hydrogel–PLCL constructs enhanced the tissue formation and promoted angiogenesis and vascularization.

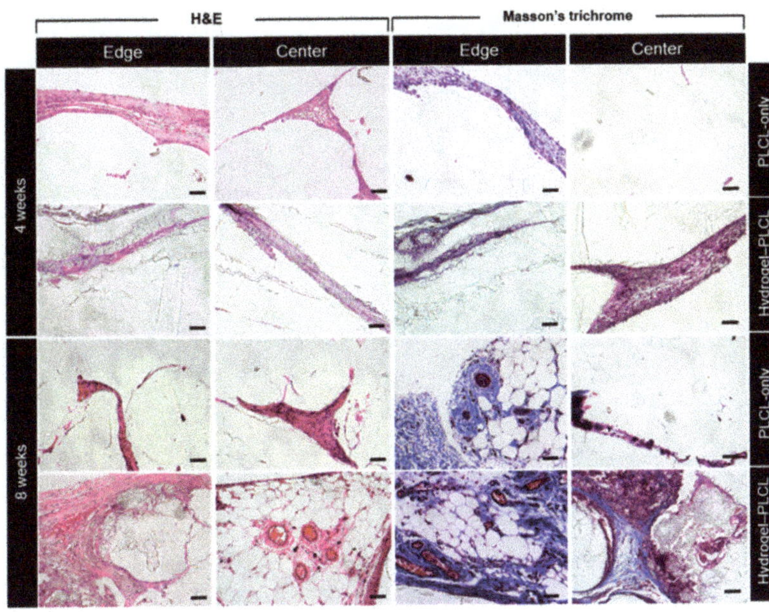

Figure 9. Histological analysis of the scaffolds after specific time points. The specimens were stained with hematoxylin and eosin (H&E) and Masson's trichrome (MT). Scale bar = 50 μm.

2.6. Macrophage Infiltration Analysis

Immunofluorescence staining with CD68 and CD206 was performed to investigate macrophage infiltration, as shown in Figure 10a. CD68 and CD206 were stained with a red and green color, respectively. In addition, DAPI was stained as a blue color. CD68 is a pan-monocyte/macrophage marker, and CD206 is an M2 macrophage marker. Macrophages are an important factor to detect early inflammatory responses in tissue engineering [68,69]. Macrophage differentiation is affected by the cytokine environment. M1 macrophages produce the pro-inflammatory cytokines, phagocytize microbes, and induce an immune response. These factors predominantly have an effect in inflammation, tissue injuries, and apoptosis [70–72]. Activated M2 macrophages are alternatively associated with tissue remodeling and angiogenesis by downregulating inflammation and eliminating tissue wastes and apoptotic bodies. Additionally, they also produce polyamines to induce collagen accumulation [73–76]. Significant amounts of CD68+ and CD206+ cells were found in the hydrogel–PLCL constructs. For quantitative analysis, the macrophage infiltration area (μm^2) for the CD68+ and CD206+ cells were calculated to identify the M1 and M2 macrophages, and the area ratio of CD206+/CD68+ cells ($\mu m^2/\mu m^2$) was also calculated. The calculated values were plotted as a graph (Figure 10b). The area ratio of the CD206+/68+ cells ($\mu m^2/\mu m^2$) for the PLCL-only group was 0.61 ± 0.29 and 0.52 ± 0.16 for four weeks and eight weeks, respectively, and that of the hydrogel–PLCL group was 0.82 ± 0.21 and 0.80 ± 0.16, respectively. Intriguingly, the ratio of the CD206+/CD68+

cells was significantly higher in the hydrogel–PLCL group compared with that in the PLCL-only group, which might reflect the macrophages in the former moving toward the tissue regeneration state [76,77].

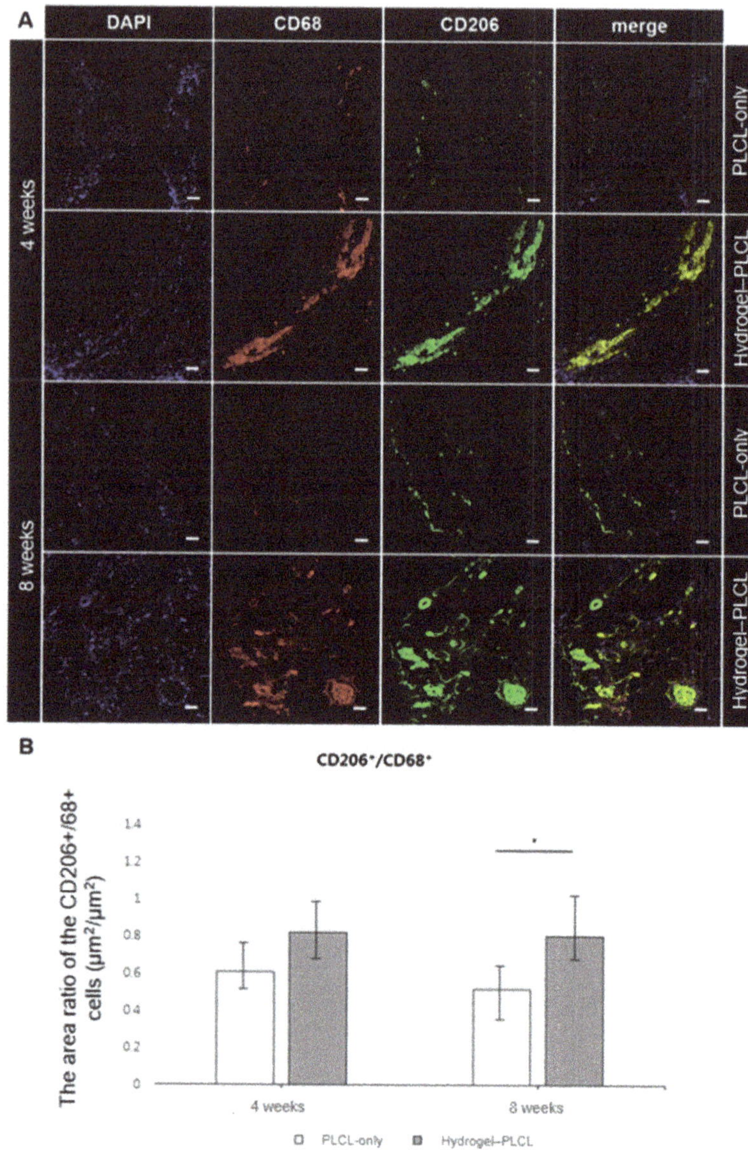

Figure 10. (**A**) Immunofluorescence images stained with CD68 and CD206 for investigation of macrophage infiltration. The images were obtained by confocal microscopy after 4 and 8 weeks in vivo (* $p < 0.01$). Scale bar = 50 μm. (**B**) Quantitative analysis of the macrophage infiltration area for identification of the M1 and M2 macrophages.

2.7. Analysis of Angiogenesis and Vascularization in the Scaffolds

To confirm the endothelial lineage differentiation in the hydrogel–PLCL construct, the expression of von Willebrand factor (vWF) and α-smooth muscle actin (α-SMA) was studied by immunofluorescent staining at four and eight weeks after implantation. vWF and α-SMA are well-known specific endothelial markers. vWF is used to detect panendothelial cells, and α-SMA targets pericytes and mature vessels surrounded with actin-positive vascular smooth cells. vWF staining is observed for both immature and mature vessels; however, α-SMA is observed for mature vessels [78–81]. The confocal image in Figure 11 shows that the formation of matured blood vessels gradually increased in the hydrogel–PLCL group from the fourth to the eighth week. Early angiogenesis at four weeks occurred in both the PLCL-only scaffolds and hydrogel–PLCL constructs. vWF and α-SMA densities of each scaffold were analyzed by immunofluorescence staining. A few small, immature vessels were detected. However, there was no significant formation of mature capillaries in the PLCL-only group at eight weeks, while matured capillaries did form in the hydrogel–PLCL constructs (Figure S3). The density of the capillaries increased from 4218.7 ± 1787.1 μm^2 to 4400.1 ± 696.8 μm^2, and 6156.6 ± 2357.4 μm^2 to 14736.3 ± 1955.0 μm^2 for the PLCL-only scaffolds and the hydrogel–PLCL constructs, respectively. Moreover, the arterial density detected by α-SMA staining was also higher in the hydrogel–PLCL constructs at 7603.2 ± 2109.4 μm^2 compared to the PLCL-only scaffolds at 2913.7 ± 1620.6 μm^2. The immunofluorescent staining showed that the adECM hydrogel in the hydrogel–PLCL constructs promoted the recruitment of endothelial cells and their differentiation.

2.8. Quantitative Analysis by Real-Time Polymerase Chain Reaction

Figure 12 shows the gene expression data at 8 weeks for both scaffolds. The PLCL-only scaffold had a higher gene expression for peroxisome proliferator-activated receptor γ (PPARγ); however, there was no significant statistical difference between the groups. In contrast, C/EBPα was expressed much higher in the hydrogel–PLCL construct group compared to the PLCL-only group. The relative gene expression analysis of lipoprotein lipase (LPL), PPARγ, glyceraldehyde-3-phosphate dehydrogenase (GAPDH), and CCAAT/enhancer binding protein β (C/EBPα) was performed at eight weeks after implantation. Markers related to adipogenic differentiation were identified at the mRNA level, and the expression levels of all genes were normalized relative to that of the GAPDH level. Lipoprotein lipase (LPL) is a very early marker of adipogenic differentiation induced by cell–cell contact. Peroxisome proliferator-activated receptor γ (PPARγ) is mediated by CCAAT/enhancer binding protein β (C/EBPβ), and C/EBPα is also mediated by C/EBPβ. PPARγ is cross-regulated with C/EBPβ, but C/EBPα cooperates with PPARγ, inducing the transcription of adipocyte genes which leads to the creation and maintenance of the adipocyte phenotype. C/EBPα is known as a terminal differentiation marker and works in a synergistic manner with PPARγ. Therefore, LPL is expressed at a very early stage of adipogenic differentiation, and PPARγ is gradually expressed in the intermediate stage; then, C/EBPα has a critical role in the late stage of differentiation [82]. As shown by the results in this study, the PLCL-only group was in the intermediate stage of adipogenic differentiation. Furthermore, it means that the hydrogel–PLCL construct group was favorable for the promotion of adipose tissue regeneration by leading to full adipogenic lineage differentiation.

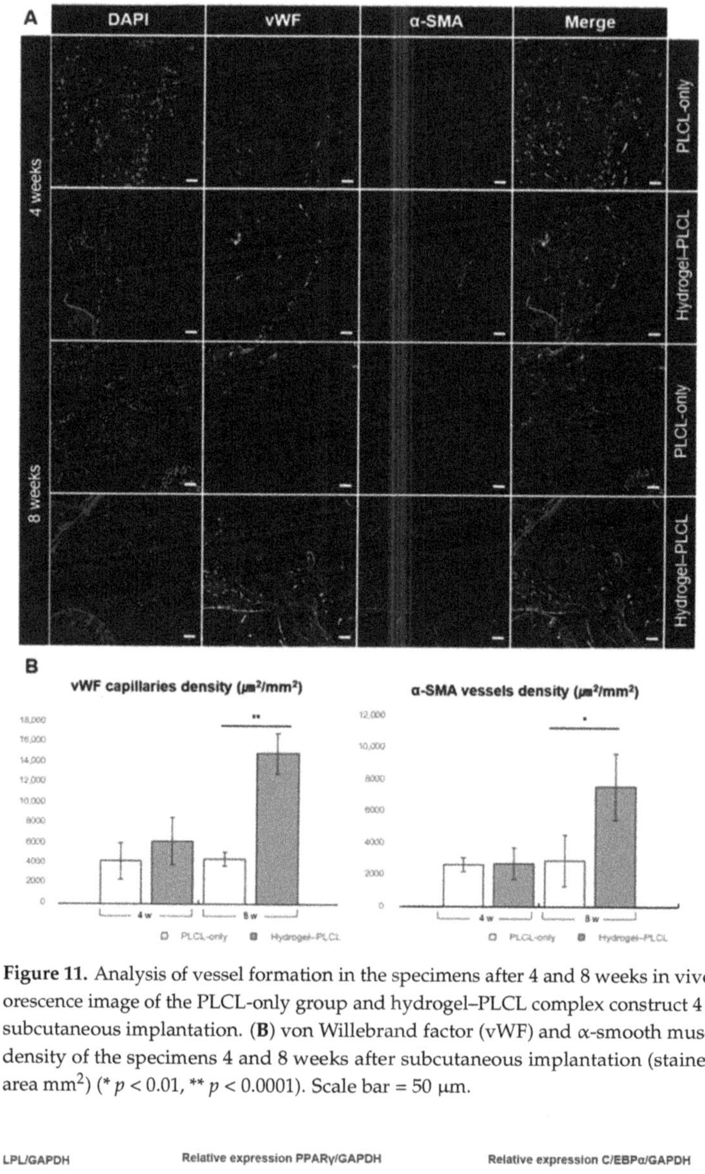

Figure 11. Analysis of vessel formation in the specimens after 4 and 8 weeks in vivo. (**A**) Immunofluorescence image of the PLCL-only group and hydrogel–PLCL complex construct 4 and 8 weeks after subcutaneous implantation. (**B**) von Willebrand factor (vWF) and α-smooth muscle actin (α-SMA) density of the specimens 4 and 8 weeks after subcutaneous implantation (stained area μm^2/total area mm^2) (* $p < 0.01$, ** $p < 0.0001$). Scale bar = 50 μm.

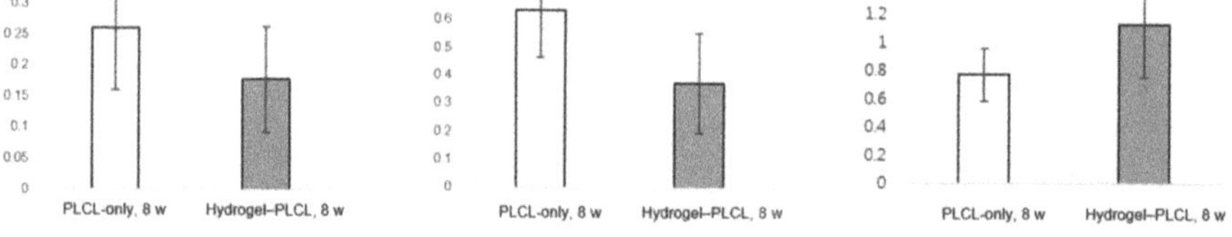

Figure 12. Analysis of quantitative real-time polymerase chain reaction results. Relative mRNA expression of LPL, PPARγ and C/EBPα in the PLCL-only scaffolds and the hydrogel–PLCL complex constructs (* $p < 0.01$).

3. Materials and Methods

3.1. Materials

L-Lactide (Purac Biochem, Gorinchem, Netherlands) and ε-Caprolactone (Sigma Aldrich, St. Louis, MO, USA) were used as received. As an initiator, 1-dodecanol (Sigma Aldrich, St. Louis, MO, USA) was used, and tin (II) 2-ethylhexanoate (Sigma Aldrich, St. Louis, MO, USA) was used as a catalyst. All chemicals and solvents were of analytical grade and were used as received.

3.2. Synthesis and Characterization of PLCL

PLCL co-polymer was synthesized by the ring-opening polymerization of L-lactide (100 mmol) and ε-caprolactone (100 mmol), which was catalyzed by Sn(Oct)2, as described elsewhere [20,83]. Briefly, the reaction mixture containing the monomers, catalyst and initiator diluted in dried toluene was poured into a 50 mL glass ampoule. The ampoule was completely sealed under a vacuum after purging three times with nitrogen and then polymerized at 150 °C for 24 h (Figure 2a). After the reaction, the product was dissolved in chloroform and filtered through an 8 μm pore filter paper. Then, the polymer was precipitated into an excess of methanol, filtered, and dried under a vacuum over 3 days.

To identify the distribution ratio of the PLCL copolymer, nuclear magnetic resonance (NMR) spectroscopy was performed on a 600 MHz ^{1}H-NMR spectrometer (Agilent, Santa Clara, CA, USA) with deuterated chloroform ($CDCl_3$, Aldrich, St. Louis, MO, USA) solutions. Spectra were obtained with 1% (*w*/*v*) solutions in chloroform with 0.03% (*v*/*v*) tetramethylsilane as the standard. The distribution ratio was calculated from the NMR spectra. The number and weight of the molecular weights were determined using gel permeation chromatography (GPC, Viscotek TDA 302, Malvern Instruments Ltd, Worcestershire, United Kingdom) calibrated with polystyrene. Chloroform was used as a mobile phase at a flow rate of 1.0 mL/min.

3.3. Preparation of the Decellularized Extracellular Matrix (dECM)-Based Ink

The decellularization process for adipose tissue was performed by following the method published elsewhere with a small modification [42]. Adipose tissue was obtained from Seoul National University. The adipose tissue was chopped into fine pieces first and decellularized in 0.5% sodium dodecyl sulfate (SDS) solution for 48 h. The 0.5% SDS solution was replaced every 12 h. After that, the tissue was removed from the 0.5% SDS solution and treated with isopropyl alcohol to eliminate the lipids for 48 h, changing the solution every 12 h. The decellularized and dilapidated tissues were then washed with PBS and treated with a solution of 0.1% peracetic acid in 4% ethanol for 4 h. Finally, the tissues were washed with PBS for 72 h and lyophilized. The lyophilized tissues were crushed into a fine powder with a cryomiller (FreezerMill, SPEX Sample Prep., Metuchen, NJ, USA). The adECM powder was stored in a −80 °C freezer until the gelation protocol.

The adECM powder was sterilized with a supercritical carbon dioxide system at 200 bar for 1 h [84]. The sterilized adECM powder was taken as required and dissolved with 10 mg of pepsin (P7125, Sigma, St. Louis, MO, USA) in 3% (*v*/*v*) acetic acid (Sigma, St. Louis, MO, USA) solution using a mixing rotator at 4 °C for 48 h. The adECM solution had an acidic pH, so it was neutralized with sterile 10× MEM (11430-030, Gibco, Waltham, MA, USA) and sterile 5 N NaOH solution to adjust the pH.

Rat tail collagen I (354236, Corning, St. Louis, MO, USA) was diluted to 3 mg/mL and neutralized with sterile 10× MEM media, sterile distilled water, and sterile 1 N NaOH solution, while the temperature was kept below 10 °C to avoid gelation. After adding the calculated volume of all components which were neutralized separately, the 0.3% collagen and 3% adECM solutions were mixed well. To optimize the biological and mechanical conditions, adECM and collagen solutions were mixed at 3:1, 1:1, and 1:3 (*v*/*v*) ratios.

The mechanical properties of the hydrogels with different adECM-to-collagen ratios were measured by unconfined compression tests (n = 3). A cylinder mold with a 6 mm diameter was prepared to make regular-shaped hydrogels. The mixtures of the hydrogels were poured into the mold. After allowing 30 min for gelation, uniformed cylindrical hydrogels were finally prepared and compressed using a 10 kN load cell at the rate of mm/min (Instron 5900, Norwood, MA, USA). The compressive modulus was determined as the force per unit area applied to the hydrogels during compression and calculated as the slope of the initial straight line in the stress–strain curve.

The cell viability measurements for the various adECM and collagen ratios were performed with live/dead cell viability assay kits (L3224, Thermo Fisher scientific, Waltham, MA, USA) (n = 3). First, 200 µL of the mixed hydrogels were poured into a 96-well plate and incubated at 37 °C in a humidified atmosphere containing 5% (v/v) CO_2 for 30 min for gelation. ADSCs and HUVECs were then seeded onto each of the hydrogels (1.5×10^4 cells/well). Later, on days 1, 4 and 7, Calcein AM 5 µL and Ethidium homodimer (EthD-1, 20 µL) were mixed in 10 mL of PBS and spread onto the mixed hydrogel. Calcein AM (green)- and EthD-1 (red)-stained live and dead cells, respectively. The stained cells were observed and counted manually using confocal microscopy to calculate the percentage of live and dead cells (Zeiss 520, Jena, Germany).

3.4. Fabrication of the PLCL Scaffolds and the Hydrogel–PLCL Constructs

Figure 1 shows a schematic illustration of all steps for our experiments. The PLCL-only scaffolds and hydrogel–PLCL scaffolds were fabricated with a dual-nozzle 3D printer (In vivo premium, Rokit, Seoul, Republic of Korea). The PLCL was extruded from a hot-melting extruder nozzle using a pneumatic air compressor with high pressure for the PLCL, which is very elastic [85].

The hydrogel–PLCL constructs were fabricated with the dual-nozzle system. The hydrogel solution containing the collagen and adECM was printed with a 10 mL syringe while maintaining temperatures below 10 °C to avoid gelation before printing. The PLCL was printed with the same method that was used for the PLCL-only scaffold. The hydrogel solution and PLCL were printed alternately at 37 °C in a humidified atmosphere condition to prevent the hydrogel solution from drying out before gelation. Both the PLCL-only and hydrogel–PLCL complex scaffolds were placed under a UV light for 15 min for sterilization before the in vivo experiments.

3.5. Scanning Electron Microscopy (SEM) Analysis

The surface morphology of the constructs was examined by scanning electron microscopy (SEM) (Phenom-world, Eindhoven, Netherlands). The top surface and cross-section of the samples were observed. The samples were coated for 60 s with gold using a sputter-coater (Bal-tec SCD 005, Baltac prä parathion, Niesgrau, Germany).

3.6. Mechanical Tests

The mechanical properties of the PLCL scaffold were measured by unconfined compression tests (n = 3). Uniformed scaffolds with a constant shape and size were fabricated with the PLCL co-polymer and PCL polymer. Human adipose tissue was cut into a cylindrical shape with a stamping mold to compare the compression modulus with the PCL and PLCL scaffolds. Adipose tissue and the PCL and PLCL scaffolds were prepared and compressed using a 10 kN load cell at a rate of 1 mm/min (Instron 5900, Instron®, Norwood, MA, USA). The compressive modulus was determined as the force per unit area applied to the hydrogels during compression and calculated as the slope of the initial straight line in the stress–strain curve.

To evaluate the ultimate tensile strength, raw PLCL (before printing) and printed PLCL were made into thin films. The films were mounted on a universal testing machine and subjected to tensile testing at a crosshead speed of 1 mm/min. Means and standard deviations were expressed in MPa, and the Young's modulus was calculated as the slope of the initial straight line in the stress–strain curve.

3.7. In Vivo Experiments

SD rats (7-week-old, male; Dae-han biolink co., Ltd, the Republic of Korea) were randomly allocated into two groups: PLCL-only group (n = 4) and PLCL with hydrogel (adECM:collagen = 1:3) group (n = 4). All equipment for the surgery was sterilized with an autoclave in advance. Sterilized constructs were individually implanted into the bilateral dorsal subcutis pouches of the rats to evaluate the ability of adipose tissue regeneration and vascularization. The rats were temporarily anesthetized with ethyl ether. Gentamicin (Gibco, Waltham, MA, USA) was used as antibiotics (dose, 80 mg kg^{-1}) after surgery. The constructs were harvested at 4 and 8 weeks after implantation.

The molecular weight changes during the in vivo study were analyzed by collecting samples after specific time points. The samples were chopped and dissolved in $CHCl_3$ and then purified in methanol. The resulting samples were dried for 3 days at 60 °C, after which the samples were dissolved in $CHCl_3$ to evaluate their molecular weight.

3.8. Evaluation of Adipose Tissue Regeneration with Histological Analysis

The constructs were fixed in 10% (v/v) buffered formalin (HT110116, Sigma, St. Louis, MO, USA) for 24 h, dehydrated in a graded ethanol series, embedded, and then made into paraffin blocks for histological analysis. Each sample was sectioned by a microtome at a 5 μm thickness. The sectioned samples (n = 4) were stained with hematoxylin and eosin (H&E) and Masson's trichrome (MT) to examine the tissue morphology and collagen composition. The samples were examined under light microscopy (Nikon, Tokyo, Japan) [86].

3.9. Macrophage and Angiogenesis Assessment of the Constructs

To evaluate the immune response by macrophage infiltration 4 and 8 weeks after implantation, the sample section was blocked and incubated with mouse monoclonal anti-cluster of differentiation 68 (CD 68, 1:100, ab955, Abcam, Cambridge, United Kingdom) for both M1 and M2 macrophages, and CD 206 (1:100, SC-34577, Santa Cruz Biotechnology, Inc., Dallas, TX, USA) for M2 macrophages, overnight.

To examine angiogenesis, rabbit anti-human von Willebrand factor antibody (vWF, 1:400, ab6994, Abcam, Cambridge, United Kingdom) and monoclonal mouse anti-human α-smooth muscle actin antibody (α-SMA, 1:200, ab7817, Abcam, Cambridge, United Kingdom) were used as primary antibodies to stain endothelial cells (ECs) and vascular smooth muscle cells (SMCs). After an overnight incubation at 4 °C, the samples were incubated with Alexa Fluor 488 goat anti-rabbit IgG (1:1000, A11034, Life Technologies, Carlsbad, CA, USA) and Alexa Fluor 594 goat Anti-mouse IgG (1:1000, A11062, Life Technologies, Carlsbad, CA, USA) for 1 h at RT. Cell nuclei were observed by staining with 4′, 6-diamidino-2-phenylindole (DAPI, Molecular Probes) for up to 5 min. Slides were observed under a confocal microscope (Zeiss 520, Jena, Germany) [83,84,86,87]. The area (μm^2) of the DAPI+, CD68+, and CD206 staining was measured with the Image J program to estimate the immune response in the scaffolds by the ratio of macrophages per whole cell. The value of the CD206+/CD68+ indicates the ratio of the M2 macrophages/all macrophages in the scaffolds which enhance angiogenesis and tissue regeneration.

3.10. Adipose Tissue mRNA Expression Analysis by RT-PCR

The RNA of the cells from each of the samples was isolated and purified using the RNeasy®mini kit (Qiagen, Hilden, Germany), following the manufacturer's instructions, to quantify the gene expression. RNA extraction was measured using Nanodrop (Nanodrop®ND-1000, Thermo Fisher Scientific, Waltham, MA, USA), after which reverse transcription for RNA was performed with the Omniscript®System (Qiagen, Hilden, Germany). RNA samples were incubated in a dry bath (WiseTherm®HB-R, DAIHAN Scientific, Seoul, Republic of Korea) at 37 °C for 60 min. The real-time PCR was performed with the Power SYBR®Green PCR Master Mix (Applied Biosystems, Foster City, CA, USA) using the 7500 Real-Time PCR system (ABI prism 7500, Applied Biosystems, Foster City, CA, USA). The oligonucleotide primers for lipoprotein lipase (LPL), peroxisome proliferator-activated receptor gamma (PPARγ), glyceraldehyde-3-phosphate dehydrogenase (GAPDH) and CCAAT/enhancer binding protein alpha (C/EBPα) were designed based on references or published gene sequences (NCBI) [88,89]. GAPDH primers were used for the control genes, and the other genes were normalized by GAPDH. Relative gene expression levels were analyzed using the $2^{-\Delta\Delta CT}$ method.

3.11. Statistical Analysis

The samples were assessed at least in triplicate, and all results are expressed as the mean ± standard deviation values. All statistical analyses were performed by Student's *t*-tests, and statistical significance was set at * $p < 0.01$ or ** $p < 0.0001$ for all results.

4. Conclusions

We have developed adipose tissue-derived dECM hydrogel based PLCL constructs to develop vascularized adipose tissue. The 3D printing technique was used for scaffold fabrication, which could fabricate patient-specific scaffold for later clinical application. It was confirmed that the highly elastic PLCL co-polymer used in this study has more approximate physical properties as native adipose tissue compared to the other polymers. Furthermore, we revealed that adECM hydrogel promotes adipose tissue reconstruction by encouraging neovascularization and tissue formation. A high level of PPARγ gene expression was observed in the PLCL-only group; however, histological and immunological staining analysis revealed that the tissue formation and vessel formation were markedly poor compared to that of the hydrogel–PLCL construct group. Indeed, C/EBPα, the late-stage marker, was expressed much more in the hydrogel–PLCL construct group, with significantly different values. Our results demonstrated that the hydrogel–PLCL constructs promoted efficient adipogenic differentiation with a developed vascularized structure. Consequently, we have demonstrated the feasibility of a therapeutic method for large-sized adipose tissue regeneration by developing vascularized tissue with the 3D-printed hydrogel–PLCL scaffold. Therefore, it could be a good alternative for adipose tissue engineering, and we expect that it is applicable to not only adipose tissue but also to other soft tissue regeneration.

Supplementary Materials: The following are available online at https://www.mdpi.com/1422-0067/22/6/2886/s1, Figure S1: Cell viability test was performed to optimize the hydrogel composition. The confocal images above were obtained after a 7-day incubation of ADSCs and HUVECs. All samples were evaluated in triplicate, and the error bar indicates the SD. Scale bar = 50 μm; Figure S2: Flexible and stretchable test for the 3D-printed PLCL scaffold. The scaffold bent easily and recovered to its original shape after printing; Figure S3: Confocal images of the in vivo experiments for immune response: DNA, vWF, a-SMA, and merged images at 4 and 8 weeks after subcutaneous implantation; 8w, 20× confocal. Scale bar = 50 μm; Video S1: Flexibility test for the 3D-printed PLCL scaffold; Video S2: Stretchability test for the 3D-printed PLCL scaffold.

Author Contributions: Conceptualization, K.L., and Y.J.; methodology, S.L., J.J.C., S.H.K., and Y.J.; formal analysis; S.L. and H.S.L.; resources, K.L. and Y.J; data curation, S.L., H.S.L. and J.J.C.; writing—original draft preparation, S.L., H.S.L. and Y.J.; writing—review and editing, S.L., H.S.L., J.J.C., S.H.K., J.W.P., K.L., and Y.J.; visualization, S.L. and H.S.L.; funding acquisition, Y.J. All authors have read and agreed to the published version of the manuscript.

Funding: This research was supported by the KIST Institutional Program (2V08550), a grant of the Basic Science Research Program (2021R1A2C2004634) through the National Research Foundation of Korea funded by the Ministry of Science and ICT, and the Technology Innovation Program (Project No. 20008686) funded By the Ministry of Trade, Industry & Energy (MOTIE, Korea).

Institutional Review Board Statement: All animals for the in vivo experiments were treated in accordance with the standard operating protocols of the Institutional Animal Care and Use Committee at the Korea Institute of Science and Technology (KIST). All protocols were approved by the Institutional Review Board of Animal Experiments at KIST (Approval Number KIST-2019-009, Approval Date 2019.02.26).

Informed Consent Statement: Informed consent was obtained from all subjects involved in the study.

Data Availability Statement: All data are reported in the manuscript and in the supplementary information.

Conflicts of Interest: The authors declare no conflict of interest.

References

1. Benavides, O.M. Amniotic Fluid-Derived Stem Cells as a Source of In Situ Vascularization within Fibrin/Poly (Ethylene Glycol) Hydrogels. Ph.D. Thesis, Rice University, Houston, TX, USA, 2014.
2. Lequeux, C.; Rodriguez, J.; Boucher, F.; Rouyer, O.; Damour, O.; Mojallal, A.; Auxenfans, C. In vitro and in vivo biocompatibility, bioavailability and tolerance of an injectable vehicle for adipose-derived stem/stromal cells for plastic surgery indications. *J. Plast. Reconstr. Aesth. Surg.* **2015**, *68*, 1491–1497. [CrossRef] [PubMed]
3. Auclair, E.; Blondeel, P.; Del Vecchio, D.A. Composite breast augmentation: Soft-tissue planning using implants and fat. *Plast. Reconstr. Surg.* **2013**, *132*, 558–568. [CrossRef]
4. Sterodimas, A.; de Faria, J.; Nicaretta, B.; Pitanguy, I. Tissue engineering with adipose-derived stem cells (ADSCs): Current and future applications. *J. Plast. Reconstr. Aesth. Surg.* **2010**, *63*, 1886–1892. [CrossRef]
5. Choi, J.S.; Yang, H.-J.; Kim, B.S.; Kim, J.D.; Lee, S.H.; Lee, E.K.; Park, K.; Cho, Y.W.; Lee, H.Y. Fabrication of porous extracellular matrix scaffolds from human adipose tissue. *Tissue Eng. Part C Methods* **2010**, *16*, 387–396. [CrossRef] [PubMed]
6. Flynn, L. The use of decellularized adipose tissue to provide an inductive microenvironment for the adipogenic differentiation of human adipose-derived stem cells. *Biomaterials* **2010**, *31*, 4715–4724. [CrossRef]
7. Han, T.T.Y.; Toutounji, S.; Amsden, B.G.; Flynn, L.E. Adipose-derived stromal cells mediate in vivo adipogenesis, angiogenesis and inflammation in decellularized adipose tissue bioscaffolds. *Biomaterials* **2015**, *72*, 125–137. [CrossRef]
8. Chang, K.-H.; Liao, H.-T.; Chen, J.-P. Preparation and characterization of gelatin/hyaluronic acid cryogels for adipose tissue engineering: In vitro and in vivo studies. *Acta Biomater.* **2013**, *9*, 9012–9026. [CrossRef] [PubMed]
9. Kaufman, M.R.; Bradley, J.P.; Dickinson, B.; Heller, J.B.; Wasson, K.; O'Hara, C.; Huang, C.; Gabbay, J.; Ghadjar, K.; Miller, T.A. Autologous fat transfer national consensus survey: Trends in techniques for harvest, preparation, and application, and perception of short-and long-term results. *Plast. Reconstr. Surg.* **2007**, *119*, 323–331. [CrossRef]
10. Patrick, C.W., Jr. Adipose tissue engineering: The future of breast and soft tissue reconstruction following tumor resection. *Semin. Surg. Oncol.* **2000**, *19*, 302–311. [CrossRef]
11. Lanza, R.; Langer, R.; Vacanti, J.P.; Atala, A. *Principles of Tissue Engineering*; Academic Press: Cambridge, MA, USA, 2020.
12. Klein, A.W.; Rish, D.C. Substances for soft tissue augmentation: Collagen and silicone. *J. Dermatol. Surg. Oncol.* **1985**, *11*, 337–339. [CrossRef]
13. Siggelkow, W.; Klosterhalfen, B.; Klinge, U.; Rath, W.; Faridi, A. Analysis of local complications following explantation of silicone breast implants. *Breast* **2004**, *13*, 122–128. [CrossRef]
14. Patrick, C.W.; Mikos, A.G.; McIntire, L.V. *Frontiers in Tissue Engineering*; Elsevier: Amsterdam, The Netherlands, 1998.
15. O'brien, F.J. Biomaterials & scaffolds for tissue engineering. *Mater. Today* **2011**, *14*, 88–95.
16. Khor, H.L.; Ng, K.W.; Schantz, J.-T.; Phan, T.-T.; Lim, T.C.; Teoh, S.-H.; Hutmacher, D. Poly (ε-caprolactone) films as a potential substrate for tissue engineering an epidermal equivalent. *Mater. Sci. Eng. C* **2002**, *20*, 71–75. [CrossRef]
17. Suzuki, Y.; Nomura, N.; Hanada, S.; Kamakura, S.; Anada, T.; Fuji, T.; Honda, Y.; Masuda, T.; Sasaki, K.; Kokubun, S.; et al. Osteoconductivity of porous titanium having young's modulus similar to bone and surface modification by OCP. *Key Eng. Mater.* **2007**, *330–332*, 951–954. [CrossRef]
18. Cho, S.-W.; Kim, S.-S.; Rhie, J.W.; Cho, H.M.; Choi, C.Y.; Kim, B.-S. Engineering of volume-stable adipose tissues. *Biomaterials* **2005**, *26*, 3577–3585. [CrossRef] [PubMed]
19. Liu, H.; Slamovich, E.B.; Webster, T.J. Less harmful acidic degradation of poly (lactic-co-glycolic acid) bone tissue engineering scaffolds through titania nanoparticle addition. *Int. J. Nanomed.* **2006**, *1*, 541. [CrossRef] [PubMed]
20. Jung, Y.; Kim, S.H.; You, H.J.; Kim, S.-H.; Ha Kim, Y.; Min, B.G. Application of an elastic biodegradable poly (L-lactide-co-ε-caprolactone) scaffold for cartilage tissue regeneration. *J. Biomater. Sci. Polym. Ed.* **2008**, *19*, 1073–1085. [CrossRef] [PubMed]
21. Jung, Y.; Chung, Y.-I.; Kim, S.H.; Tae, G.; Kim, Y.H.; Rhie, J.W.; Kim, S.-H.; Kim, S.H. In situ chondrogenic differentiation of human adipose tissue-derived stem cells in a TGF-β1 loaded fibrin–poly (lactide-caprolactone) nanoparticulate complex. *Biomaterials* **2009**, *30*, 4657–4664. [CrossRef] [PubMed]
22. Kayabolen, A.; Keskin, D.; Aykan, A.; Karslıoglu, Y.; Zor, F.; Tezcaner, A. Native extracellular matrix/fibroin hydrogels for adipose tissue engineering with enhanced vascularization. *Biomed. Mater.* **2017**, *12*, 035007. [CrossRef]
23. Novosel, E.C.; Kleinhans, C.; Kluger, P.J. Vascularization is the key challenge in tissue engineering. *Adv. Drug Deliv. Rev.* **2011**, *63*, 300–311. [CrossRef]
24. Rouwkema, J.; Khademhosseini, A. Vascularization and angiogenesis in tissue engineering: Beyond creating static networks. *Trends Biotechnol.* **2016**, *34*, 733–745. [CrossRef]
25. Lovett, M.; Lee, K.; Edwards, A.; Kaplan, D.L. Vascularization strategies for tissue engineering. *Tissue Eng. Part B Rev.* **2009**, *15*, 353–370. [CrossRef]

26. Phelps, E.A.; García, A.J. Engineering more than a cell: Vascularization strategies in tissue engineering. *Curr. Opin. Biotechnol.* **2010**, *21*, 704–709. [CrossRef] [PubMed]
27. Rouwkema, J.; Rivron, N.C.; van Blitterswijk, C.A. Vascularization in tissue engineering. *Trends Biotechnol.* **2008**, *26*, 434–441. [CrossRef]
28. Carmeliet, P.; Jain, R.K. Angiogenesis in cancer and other diseases. *Nature* **2000**, *407*, 249–257. [CrossRef]
29. Kannan, R.Y.; Salacinski, H.J.; Sales, K.; Butler, P.; Seifalian, A.M. The roles of tissue engineering and vascularisation in the development of micro-vascular networks: A review. *Biomaterials* **2005**, *26*, 1857–1875. [CrossRef] [PubMed]
30. Badylak, S.F. The extracellular matrix as a scaffold for tissue reconstruction. *Semin. Cell Dev. Biol.* **2002**, *13*, 377–383. [CrossRef]
31. Wong, M.L.; Griffiths, L.G. Immunogenicity in xenogeneic scaffold generation: Antigen removal vs. decellularization. *Acta Biomater.* **2014**, *10*, 1806–1816. [CrossRef] [PubMed]
32. Petrovic, L.; Schiegel, A.K.; Schultze-Mosgau, S.; Wiltfang, J. Different substitute biomaterials as potential scaffolds in tissue engineering. *Int. J. Oral Maxillofac. Implants* **2006**, *21*, 225–231. [PubMed]
33. Cheung, H.K.; Han, T.T.Y.; Marecak, D.M.; Watkins, J.F.; Amsden, B.G.; Flynn, L.E. Composite hydrogel scaffolds incorporating decellularized adipose tissue for soft tissue engineering with adipose-derived stem cells. *Biomaterials* **2014**, *35*, 1914–1923. [CrossRef]
34. Wang, L.; Johnson, J.A.; Zhang, Q.; Beahm, E.K. Combining decellularized human adipose tissue extracellular matrix and adipose-derived stem cells for adipose tissue engineering. *Acta Biomater.* **2013**, *9*, 8921–8931. [CrossRef] [PubMed]
35. Badylak, S.F.; Tullius, R.; Kokini, K.; Shelbourne, K.D.; Klootwyk, T.; Voytik, S.L.; Kraine, M.R.; Simmons, C. The use of xenogeneic small intestinal submucosa as a biomaterial for Achille's tendon repair in a dog model. *J. Biomed. Mater. Res.* **1995**, *29*, 977–985. [CrossRef]
36. Brown, B.; Lindberg, K.; Reing, J.; Stolz, D.B.; Badylak, S.F. The basement membrane component of biologic scaffolds derived from extracellular matrix. *Tissue Eng.* **2006**, *12*, 519–526. [CrossRef]
37. Badylak, S.F. The extracellular matrix as a biologic scaffold material. *Biomaterials* **2007**, *28*, 3587–3593. [CrossRef]
38. Gilbert, T.W.; Stewart-Akers, A.M.; Simmons-Byrd, A.; Badylak, S.F. Degradation and remodeling of small intestinal submucosa in canine Achilles tendon repair. *J. Bone Jt. Surg.* **2007**, *89*, 621–630. [CrossRef]
39. Lutolf, M.P.; Hubbell, J.A. Synthetic biomaterials as instructive extracellular microenvironments for morphogenesis in tissue engineering. *Nat. Biotechnol.* **2005**, *23*, 47–55. [CrossRef]
40. Poon, C.J.; Cotta, M.V.P.E.; Sinha, S.; Palmer, J.A.; Woods, A.A.; Morrison, W.A.; Abberton, K.M. Preparation of an adipogenic hydrogel from subcutaneous adipose tissue. *Acta Biomater.* **2013**, *9*, 5609–5620. [CrossRef]
41. Crapo, P.M.; Gilbert, T.W.; Badylak, S.F. An overview of tissue and whole organ decellularization processes. *Biomaterials* **2011**, *32*, 3233–3243. [CrossRef] [PubMed]
42. Pati, F.; Jang, J.; Ha, D.-H.; Kim, S.W.; Rhie, J.-W.; Shim, J.-H.; Kim, D.-H.; Cho, D.-W. Printing three-dimensional tissue analogues with decellularized extracellular matrix bioink. *Nat. Commun.* **2014**, *5*, 3935. [CrossRef] [PubMed]
43. Hutmacher, D.W. Scaffolds in tissue engineering bone and cartilage. *Biomaterials* **2000**, *21*, 2529–2543. [CrossRef]
44. Karageorgiou, V.; Kaplan, D. Porosity of 3D biomaterial scaffolds and osteogenesis. *Biomaterials* **2005**, *26*, 5474–5491. [CrossRef]
45. Yook, S.-W.; Jung, H.-D.; Park, C.-H.; Shin, K.-H.; Koh, Y.-H.; Estrin, Y.; Kim, H.-E. Reverse freeze casting: A new method for fabricating highly porous titanium scaffolds with aligned large pores. *Acta Biomater.* **2012**, *8*, 2401–2410. [CrossRef] [PubMed]
46. Sung, H.-J.; Meredith, C.; Johnson, C.; Galis, Z.S. The effect of scaffold degradation rate on three-dimensional cell growth and angiogenesis. *Biomaterials* **2004**, *25*, 5735–5742. [CrossRef] [PubMed]
47. Hollister, S.J. Porous scaffold design for tissue engineering. *Nat. Mater.* **2005**, *4*, 518–524. [CrossRef] [PubMed]
48. Causa, F.; Netti, P.A.; Ambrosio, L. A multi-functional scaffold for tissue regeneration: The need to engineer a tissue analogue. *Biomaterials* **2007**, *28*, 5093–5099. [CrossRef]
49. Chevalier, E.; Chulia, D.; Pouget, C.; Viana, M. Fabrication of porous substrates: A review of processes using pore forming agents in the biomaterial field. *J. Pharm. Sci.* **2008**, *97*, 1135–1154. [CrossRef]
50. Salerno, A.; Di Maio, E.; Iannace, S.; Netti, P. Tailoring the pore structure of PCL scaffolds for tissue engineering prepared via gas foaming of multi-phase blends. *J. Porous Mater.* **2012**, *19*, 181–188. [CrossRef]
51. Cheng, A.; Schwartz, Z.; Kahn, A.; Li, X.; Shao, Z.; Sun, M.; Ao, Y.; Boyan, B.D.; Chen, H. Advances in porous scaffold design for bone and cartilage tissue engineering and regeneration. *Tissue Eng. Part B Rev.* **2019**, *25*, 14–29. [CrossRef] [PubMed]
52. Loh, Q.L.; Choong, C. Three-dimensional scaffolds for tissue engineering applications: Role of porosity and pore size. *Tissue Eng. Part B Rev.* **2013**, *19*, 485–502. [CrossRef]
53. Sun, H.; Wang, X.; Hu, X.; Yu, W.; You, C.; Hu, H.; Han, C. Promotion of angiogenesis by sustained release of rhGM-CSF from heparinized collagen/chitosan scaffolds. *J. Biomed. Mater. Res. Part B Appl. Biomater.* **2011**, *100*, 788–798. [CrossRef]
54. Murphy, C.M.; O'Brien, F.J. Understanding the effect of mean pore size on cell activity in collagen-glycosaminoglycan scaffolds. *Cell Adhes. Migr.* **2010**, *4*, 377–381. [CrossRef] [PubMed]
55. Kuboki, Y.; Jin, Q.; Takita, H. Geometry of carriers controlling phenotypic expression in BMP-induced osteogenesis and chondrogenesis. *J. Bone Jt. Surg.* **2001**, *83*, S105–S115. [CrossRef]
56. Götz, H.; Müller, M.; Emmel, A.; Holzwarth, U.; Erben, R.; Stangl, R. Effect of surface finish on the osseointegration of laser-treated titanium alloy implants. *Biomaterials* **2004**, *25*, 4057–4064. [CrossRef]

57. Fernández, J.; Etxeberria, A.; Sarasua, J.-R. Synthesis, structure and properties of poly (L-lactide-co-ε-caprolactone) statistical copolymers. *J. Mech. Behav. Biomed. Mater.* **2012**, *9*, 100–112. [CrossRef]
58. Larrañaga, A.; Sarasua, J.-R. Effect of bioactive glass particles on the thermal degradation behaviour of medical polyesters. *Polym. Degrad. Stab.* **2013**, *98*, 751–758. [CrossRef]
59. Guo, B.; Ma, P.X. Synthetic biodegradable functional polymers for tissue engineering: A brief review. *Sci. China Chem.* **2014**, *57*, 490–500. [CrossRef] [PubMed]
60. Siddiqui, N.; Asawa, S.; Birru, B.; Baadhe, R.; Rao, S. PCL-based composite scaffold matrices for tissue engineering applications. *Mol. Biotechnol.* **2018**, *60*, 506–532. [CrossRef] [PubMed]
61. Izquierdo, R.; Garcia-Giralt, N.; Rodriguez, M.; Caceres, E.; Garcia, S.; Gómez Ribelles, J.; Monleon, M.; Monllau, J.C.; Suay, J. Biodegradable PCL scaffolds with an interconnected spherical pore network for tissue engineering. *J. Biomed. Mater. Res. Part A* **2008**, *85*, 25–35. [CrossRef] [PubMed]
62. Pati, F.; Ha, D.-H.; Jang, J.; Han, H.H.; Rhie, J.-W.; Cho, D.-W. Biomimetic 3D tissue printing for soft tissue regeneration. *Biomaterials* **2015**, *62*, 164–175. [CrossRef]
63. Dai, R.; Wang, Z.; Samanipour, R.; Koo, K.-I.; Kim, K. Adipose-derived stem cells for tissue engineering and regenerative medicine applications. *Stem Cells Int.* **2016**, *2016*, 6737345. [CrossRef] [PubMed]
64. Jeong, S.I.; Kim, B.-S.; Kang, S.W.; Kwon, J.H.; Lee, Y.M.; Kim, S.H.; Kim, Y.H. In vivo biocompatibilty and degradation behavior of elastic poly (l-lactide-co-ε-caprolactone) scaffolds. *Biomaterials* **2004**, *25*, 5939–5946. [CrossRef] [PubMed]
65. Montalbano, G.; Toumpaniari, S.; Popov, A.; Duan, P.; Chen, J.; Dalgarno, K.; Scott, W., III; Ferreira, A. Synthesis of bioinspired collagen/alginate/fibrin based hydrogels for soft tissue engineering. *Mater. Sci. Eng. C* **2018**, *91*, 236–246. [CrossRef] [PubMed]
66. Benoit, D.S.; Durney, A.R.; Anseth, K.S. Manipulations in hydrogel degradation behavior enhance osteoblast function and mineralized tissue formation. *Tissue Eng.* **2006**, *12*, 1663–1673. [CrossRef] [PubMed]
67. Baluk, P.; Fuxe, J.; Hashizume, H.; Romano, T.; Lashnits, E.; Butz, S.; Vestweber, D.; Corada, M.; Molendini, C.; Dejana, E.; et al. Functionally specialized junctions between endothelial cells of lymphatic vessels. *J. Exp. Med.* **2007**, *204*, 2349–2362. [CrossRef] [PubMed]
68. Papadimitriou, J.C.; Drachenberg, C.B.; Munivenkatappa, R.; Ramos, E.; Nogueira, J.; Sailey, C.; Klassen, D.K.; Haririan, A. Glomerular inflammation in renal allografts biopsies after the first year: Cell types and relationship with antibody-mediated rejection and graft outcome. *Transplantation* **2010**, *90*, 1478–1485. [CrossRef]
69. Tinckam, K.J.; Djurdjev, O.; Magil, A.B. Glomerular monocytes predict worse outcomes after acute renal allograft rejection independent of C4d status. *Kidney Int.* **2005**, *68*, 1866–1874. [CrossRef] [PubMed]
70. Brown, B.N.; Badylak, S.F. Expanded applications, shifting paradigms and an improved understanding of host–biomaterial interactions. *Acta Biomater.* **2013**, *9*, 4948–4955. [CrossRef] [PubMed]
71. Wentworth, J.M.; Naselli, G.; Brown, W.A.; Doyle, L.; Phipson, B.; Smyth, G.K.; Wabitsch, M.; O'Brien, P.E.; Harrison, L.C. Pro-inflammatory CD11c+ CD206+ adipose tissue macrophages are associated with insulin resistance in human obesity. *Diabetes* **2010**, *59*, 1648–1656. [CrossRef]
72. Hu, W.; Jiang, Z.; Zhang, Y.; Liu, Q.; Fan, J.; Luo, N.; Dong, X.; Yu, X. Characterization of infiltrating macrophages in high glucose-induced peritoneal fibrosis in rats. *Mol. Med. Rep.* **2012**, *6*, S83–S99.
73. Ng, Y.-Y.; Hou, C.-C.; Wang, W.; Huang, X.R.; Lan, H.Y. Blockade of NFκB activation and renal inflammation by ultrasound-mediated gene transfer of Smad7 in rat remnant kidney. *Kidney Int.* **2005**, *67*, 83–91. [CrossRef]
74. Nie, J.; Hao, W.; Dou, X.; Wang, X.; Luo, N.; Lan, H.Y.; Yu, X. Effects of Smad7 overexpression on peritoneal inflammation in a rat peritoneal dialysis model. *Perit. Dial. Int.* **2007**, *27*, 580–588. [CrossRef]
75. Wang, W.; Huang, X.R.; Li, A.G.; Liu, F.; Li, J.-H.; Truong, L.D.; Wang, X.J.; Lan, H.Y. Signaling mechanism of TGF-β1 in prevention of renal inflammation: Role of Smad7. *J. Am. Soc. Nephrol.* **2005**, *16*, 1371–1383. [CrossRef]
76. Hirata, Y.; Tabata, M.; Kurobe, H.; Motoki, T.; Akaike, M.; Nishio, C.; Higashida, M.; Mikasa, H.; Nakaya, Y.; Takanashi, S.; et al. Coronary atherosclerosis is associated with macrophage polarization in epicardial adipose tissue. *J. Am. Coll. Cardiol.* **2011**, *58*, 248–255. [CrossRef]
77. Mokarram, N.; Merchant, A.; Mukhatyar, V.; Patel, G.; Bellamkonda, R.V. Effect of modulating macrophage phenotype on peripheral nerve repair. *Biomaterials* **2012**, *33*, 8793–8801. [CrossRef]
78. Acker, T.; Beck, H.; Plate, K.H. Cell type specific expression of vascular endothelial growth factor and angiopoietin-1 and-2 suggests an important role of astrocytes in cerebellar vascularization. *Mech. Dev.* **2001**, *108*, 45–57. [CrossRef]
79. Cremona, O.; Savoia, P.; Marchisio, P.C.; Gabbiani, G.; Chaponnier, C. The alpha 6 and beta 4 integrin subunits are expressed by smooth muscle cells of human small vessels: A new localization in mesenchymal cells. *J. Histochem. Cytochem.* **1994**, *42*, 1221–1228. [CrossRef] [PubMed]
80. Mirza, A.; Hyvelin, J.-M.; Rochefort, G.Y.; Lermusiaux, P.; Antier, D.; Awede, B.; Bonnet, P.; Domenech, J.; Eder, V. Undifferentiated mesenchymal stem cells seeded on a vascular prosthesis contribute to the restoration of a physiologic vascular wall. *J. Vasc. Surg.* **2008**, *47*, 1313–1321. [CrossRef] [PubMed]
81. Zhang, Y.; Zhang, R.; Li, Y.; He, G.; Zhang, D.; Zhang, F. Simvastatin augments the efficacy of therapeutic angiogenesis induced by bone marrow-derived mesenchymal stem cells in a murine model of hindlimb ischemia. *Mol. Biol. Rep.* **2012**, *39*, 285–293. [CrossRef]

82. Ntambi, J.M.; Young-Cheul, K. Adipocyte differentiation and gene expression. *J. Nutr.* **2000**, *130*, 3122S–3126S. [CrossRef] [PubMed]
83. Lee, S.; Lee, K.; Kim, S.H.; Jung, Y. Enhanced cartilaginous tissue formation with a cell aggregate-fibrin-polymer scaffold complex. *Polymers* **2017**, *9*, 348. [CrossRef]
84. Seo, Y.; Jung, Y.; Kim, S.H. Decellularized heart ECM hydrogel using supercritical carbon dioxide for improved angiogenesis. *Acta Biomater.* **2018**, *67*, 270–281. [CrossRef] [PubMed]
85. Jung, Y.; Kim, S.-H.; Kim, Y.H.; Kim, S.H. The effects of dynamic and three-dimensional environments on chondrogenic differentiation of bone marrow stromal cells. *Biomed. Mater.* **2009**, *4*, 055009. [CrossRef] [PubMed]
86. Kim, S.H.; Kim, J.E.; Kim, S.H.; Jung, Y. Substance P/dexamethasone-encapsulated PLGA scaffold fabricated using supercritical fluid process for calvarial bone regeneration. *J. Tissue Eng. Regen. Med.* **2017**, *11*, 3469–3480. [CrossRef] [PubMed]
87. Kim, T.H.; Jung, Y.; Kim, S.H. Nanofibrous electrospun heart decellularized extracellular matrix-based hybrid scaffold as wound dressing for reducing scarring in wound healing. *Tissue Eng. Part A* **2018**, *24*, 830–848. [CrossRef] [PubMed]
88. Neville, M.J.; Collins, J.M.; Gloyn, A.L.; McCarthy, M.I.; Karpe, F. Comprehensive human adipose tissue mRNA and microRNA endogenous control selection for quantitative real-time-PCR normalization. *Obesity* **2011**, *19*, 888–892. [CrossRef]
89. Tan, Q.-W.; Zhang, Y.; Luo, J.-C.; Zhang, D.; Xiong, B.-J.; Yang, J.-Q.; Xie, H.-Q.; Lv, Q. Hydrogel derived from decellularized porcine adipose tissue as a promising biomaterial for soft tissue augmentation. *J. Biomed. Mater. Res. Part A* **2017**, *105*, 1756–1764. [CrossRef]

International Journal of *Molecular Sciences*

MDPI

Review

Dehydropeptide Supramolecular Hydrogels and Nanostructures as Potential Peptidomimetic Biomedical Materials

Peter J. Jervis *, Carolina Amorim, Teresa Pereira, José A. Martins and Paula M. T. Ferreira**

Centre of Chemistry, University of Minho, Campus de Gualtar, 4710-057 Braga, Portugal; carolinaamorim753@gmail.com (C.A.); pg37045@alunos.uminho.pt (T.P.); jmartins@quimica.uminho.pt (J.A.M.); pmf@quimica.uminho.pt (P.M.T.F.)

* Correspondence: peterjervis@quimica.uminho.pt

Abstract: Supramolecular peptide hydrogels are gaining increased attention, owing to their potential in a variety of biomedical applications. Their physical properties are similar to those of the extracellular matrix (ECM), which is key to their applications in the cell culture of specialized cells, tissue engineering, skin regeneration, and wound healing. The structure of these hydrogels usually consists of a di- or tripeptide capped on the *N*-terminus with a hydrophobic aromatic group, such as Fmoc or naphthalene. Although these peptide conjugates can offer advantages over other types of gelators such as cross-linked polymers, they usually possess the limitation of being particularly sensitive to proteolysis by endogenous proteases. One of the strategies reported that can overcome this barrier is to use a peptidomimetic strategy, in which natural amino acids are switched for non-proteinogenic analogues, such as D-amino acids, β-amino acids, or dehydroamino acids. Such peptides usually possess much greater resistance to enzymatic hydrolysis. Peptides containing dehydroamino acids, i.e., dehydropeptides, are particularly interesting, as the presence of the double bond also introduces a conformational restraint to the peptide backbone, resulting in (often predictable) changes to the secondary structure of the peptide. This review focuses on peptide hydrogels and related nanostructures, where α,β-didehydro-α-amino acids have been successfully incorporated into the structure of peptide hydrogelators, and the resulting properties are discussed in terms of their potential biomedical applications. Where appropriate, their properties are compared with those of the corresponding peptide hydrogelator composed of canonical amino acids. In a wider context, we consider the presence of dehydroamino acids in natural compounds and medicinally important compounds as well as their limitations, and we consider some of the synthetic strategies for obtaining dehydropeptides. Finally, we consider the future direction for this research area.

Keywords: hydrogel; supramolecular; dehydrodipeptide; drug delivery; wound healing; cancer; smart materials; peptidomimetic

Citation: Jervis, P.J.; Amorim, C.; Pereira, T.; Martins, J.A.; Ferreira, P.M.T. Dehydropeptide Supramolecular Hydrogels and Nanostructures as Potential Peptidomimetic Biomedical Materials. *Int. J. Mol. Sci.* **2021**, *22*, 2528. https://doi.org/10.3390/ijms22052528

Academic Editor: Bice Conti

Received: 2 February 2021
Accepted: 25 February 2021
Published: 3 March 2021

1. Introduction

Dehydroamino acid residues are commonly employed in peptidomimetic medicinal chemistry strategies [1]. Furthermore, they feature in their own right in many naturally occurring peptides and several medicinally important molecules [2–5]. This review summarizes the reports of where dehydroamino acid residues have been successfully employed in the development of supramolecular molecular assemblies, including hydrogels, nanoparticles, nanotubes, nanospheres, and other nanostructures. We begin with a general overview of supramolecular hydrogels and related nanostructures; then, we consider the structure and properties of dehydroamino acids and how these may influence the assembly of supramolecular structures.

1.1. Supramolecular Peptide Hydrogels

Supramolecular hydrogels consist of non-covalently cross-linked polymers and peptides [6]. Peptide-based hydrogels are gaining popularity owing to their intrinsic biocompatibility. Short peptides attached to an aromatic capping group are particularly attractive alternatives as minimalist low-molecular weight gelators [7]. In some specific cases, short peptides without an aromatic capping group can form hydrogels [8,9]. The combination of intermolecular interactions, such as hydrogen bonds, π–π stacking, and van der Waals forces allows molecular self-assembly to take place, with the formation of nanofibrils. When these nanofibrils are able to trap water molecules, self-supporting soft materials, such as hydrogels, may form [6]. The gelation process is often initiated by means of an external trigger, which lowers the solubility of a hydrogelator in solution [10]. The most commonly employed triggers for hydrogelation are the use of a heating–cooling cycle, a pH change, the addition of chelating metal ions, a solvent swap, or an enzymatic, chemical, or photolytic cleavage of a solubilizing group (Figure 1A) [11–15]. A balance of hydrophilic and hydrophobic properties is crucial to the gelation process. If a putative gelator is too hydrophilic, it may stay in aqueous solution, if it is too hydrophobic, then precipitation may occur before the onset of the gelation process [16]. Generally, a clogP value of between 3.4 and 5.5 is considered ideal [7]. In addition to a favorable polarity, two or three aromatic groups are usually required to promote the intramolecular stacking [17]. Hydrogels can be characterized using a variety of techniques, including tube inversion tests, rheology, fluorescence spectroscopy, circular dichroism (CD) spectroscopy, and various microscopic techniques such as scanning electron microscopy (SEM), tunneling electron microscopy (TEM), and atomic force microscopy (AFM) [18].

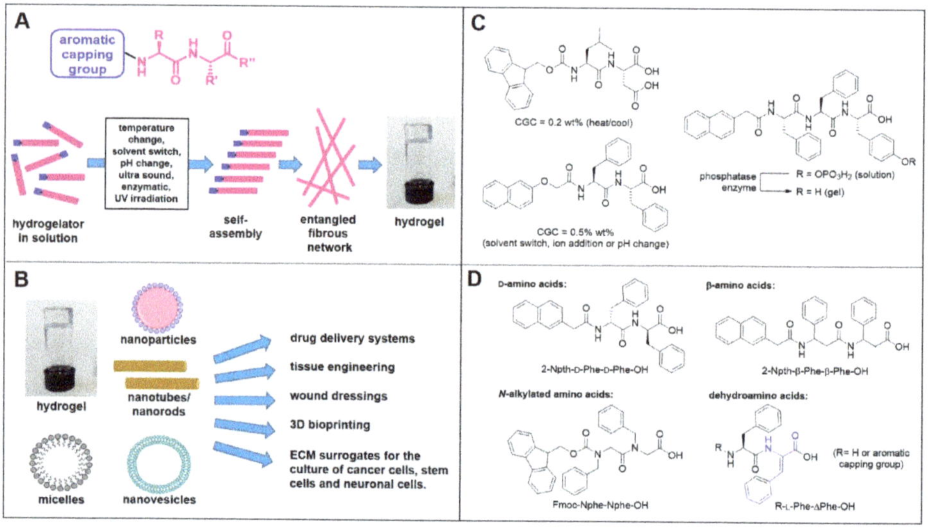

Figure 1. (**A**) Schematic of the formation of peptide hydrogels. (**B**) Biological applications of supramolecular peptide hydrogels and related nanostructures. (**C**) Examples of well-studied supramolecular peptide hydrogelators. (**D**) Some peptidomimetic strategies in the design of supramolecular hydrogels.

The properties of hydrogels often resemble those of the extracellular matrix, and as such, they have many biomedical applications in tissue engineering, skin regeneration and wound healing, 3D bioprinting, and biosensors [19]. Alongside other related nanostructures, they are also promising materials for delivery vehicles in sustained and/or targeted drug release (Figure 1B) [20]. The most commonly studied hydrogelators consist of di- or tripeptides attached at the *N*-terminus to an aromatic group such as Fmoc or naphthalene

(Figure 1C) [21]. The free carboxylic acid group at the C-terminus provides a handle for dissolving a suspension of hydrogelator, as the addition of a base, such as a dilute solution of sodium hydroxide, provides the more aqueous-soluble carboxylate salt. A subsequent reduction of pH using dilute acid protonates the carboxylate back to the neutral carboxylic acid, reducing the solubility and initiating the gelation process. As an alternative to direct mineral acid addition, the group of Adams pioneered the use of glucono-δ-lactone (GdL) as a method for controlled pH reduction, *via* its slow hydrolysis to gluconic acid in the presence of water [22]. More uniform gel structures are obtained in this way, as the rate of pH decrease is now slower than the rate of diffusion. The group of Webber recently demonstrated that the rate of GdL-induced pH change can be used to tune the morphology of supramolecular nanostructures of discotic trimeric peptides [23].

Some important examples of hydrogelators are shown in Figure 1C, but countless more have been described. These are presented within many detailed reviews published in this area [24,25]. Despite the progress being made in this field, there are still areas for improvement. Much of the work in this area has been carried out on fluorenylmethoxycarbonyl(Fmoc)-capped hydrogels, which imparts favorable gelation properties. However, with regard to biological applications, there are concerns over the stability of these hydrogelators owing to the base-lability of the Fmoc group and the toxicity associated with the degradation products of the Fmoc group [26,27]. For example, the group of Thordarson studied the leaching of monomers from a Fmoc-Phe-Phe-OH hydrogel into the surrounding medium, and they found that the degraded hydrogel was cytotoxic to various cell lines [27]. Furthermore, the peptide chain itself, consisting of canonical amino acids, is susceptible to proteolytic degradation by endogeneous enzymes in vivo [28]. This has prompted research groups to search for alternative structures. Peptidomimetic strategies, where the standard canonical amino acids have been replaced by non-coding amino acids, are gaining interest in this area, as these are not recognized by naturally occurring enzymes in biological systems. The most commonly employed strategies involve the use of D-amino acid, β-amino acid, *N*-alkylated amino acid, α-aminobutyric acid, and dehydroamino acid residues [29,30]. The group of Xu investigated hydrogelators containing D-amino acids, and they found increased resistance to protease enzymes [31]. The group of Xu also investigated hydrogelators constructed from β-amino acids, again with favorable results [32]. The group of Nilsson investigated *N*-benzylated glycine residues as replacements for phenylalanine residues (Figure 1D) [33]. This review focuses on hydrogelators and nanostructures containing a dehydroamino acid residue, the presence of which not only increases enzymatic stability but also significantly alters the structural conformation and overall properties of the peptide.

1.2. Structure of Dehydroamino Acids and Dehydropeptides

Dehydroamino acid residues differ from their corresponding canonical proteinogenic amino acids by the presence of unsaturation (i.e., a double bond), which is usually between the carbons C-α and C-β. When the double bond is in this position, they are known as α,β-didehydro-α-amino acids. Although alternative positional isomers can be invoked, for the purpose of this review, α,β-didehydro-α-amino acids and α,β-didehydropeptides shall be simply referred to as dehydroamino acids and dehydropeptides, respectively. The presence of the double bond has a number of effects. Structurally, the planar geometry around the double bond means that the stereogenic center of the corresponding canonical amino acid is no longer present. Molecular flexibility is restricted, with fixed bond angles around the C-α and C-β carbon atoms. The overlapping p-orbitals ensure that bond rotation around Cα=Cβ is completely suppressed, and therefore, if R^1 and R^2 are different, then two possible geometric isomers are possible, namely *E* and *Z*. The *E* isomer features the substituent *cis* to the carbonyl, whereas in the Z isomer, the substituent is *cis* to the nitrogen atom (Figure 2A). In dehydroamino acids, the *Z* form is the more thermodynamically stable on steric grounds [34]. Synthetic methods exist for accessing both possible isomers (*vide infra*), although methods producing the Z-isomer are more commonly reported. The most important dehydroamino acid residues involved in supramolecular structures are dehydrophenylalanine, dehydroalanine, and dehydro-2-aminobutyric acid (Figure 2B).

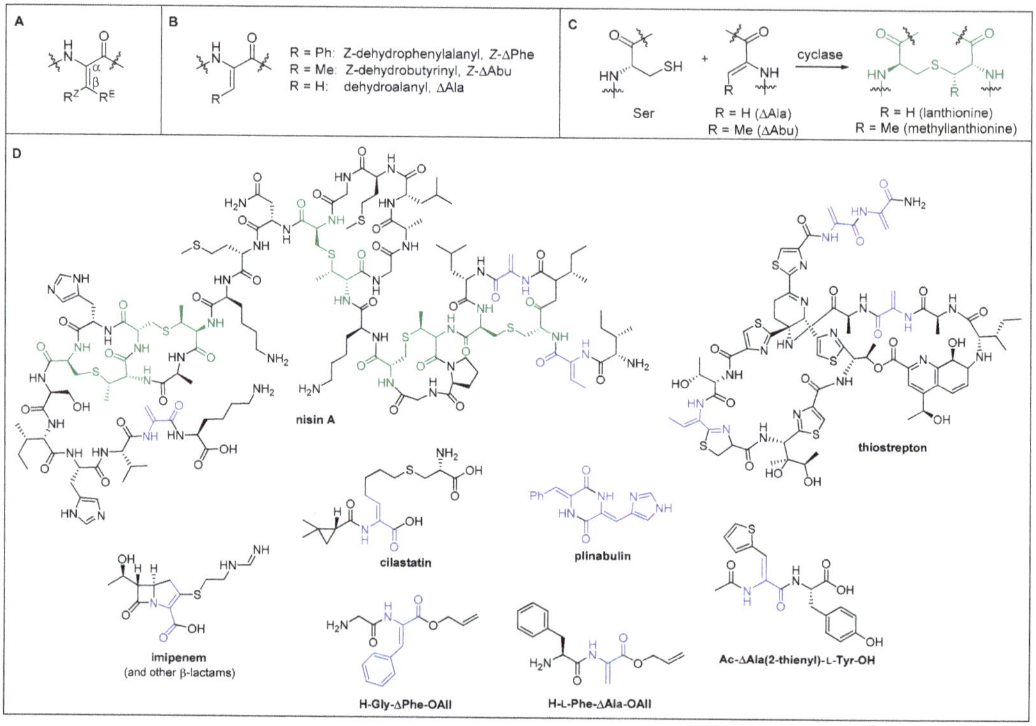

Figure 2. (**A**) General structure of dehydroamino acid residue. (**B**) Important dehydroamino acid residues relevant to supramolecular hydrogels and nanostructures. (**C**) Lanthionine residues are formed through an intramolecular reaction involving serine and dehydroalanine, in the biosynthesis of lantibiotic peptides. (**D**) Examples of biologically active dehydropeptides.

The presence of the double bond within a peptide chain can have a profound effect on the secondary structure, often predictably, compared with the corresponding peptides comprised of solely canonical amino acids. The group of Broda studied the effects of the dehydroamino acid residue double bond geometry on the secondary structure of short peptides by comparing the conformations of Boc-Gly–*E*-ΔPhe-NHMe and Boc-Gly–*Z*-ΔPhe-NHMe. They found that the geometry has a profound effect on the secondary structure, with the *E*-isomer far less likely to adopt β-turn conformations [35]. The effect of dehydroamino acid residues on the structure and stability of helical tetrapeptides has been investigated by Joaquin et al., who found that the position of a dehydroamino acid residue within a tetrapeptide has a substantial effect on the secondary structure [36]. The conformations of dehydropeptides have also been extensively studied by the group of Chauhan [37–39]. Perhaps most significantly, they found that switching a parent phenylalanine residue for a *Z*-dehydrophenylalanine (*Z*-ΔPhe) prevents β-sheet formation (i.e., "β-breakers"), suggesting a role in inhibiting the amyloid formation and fibrillization implicated in Alzheimer's disease [40].

1.3. Occurrence in Nature and Pharmaceutics

The purpose of this section is to demonstrate that dehydroamino acid residues can serve as biocompatible and non-cytotoxic structural motifs, which as well as providing increased enzymatic stability also offer their own unique biological activities. We will also address some of their limitations. In addition to conventional dehydroamino acid residues, other relevant motifs include lanthionine and methyllanthionine residues, which are special cases where the thiol group of a serine residue reacts intramolecularly (via a dedicated cyclase enzyme) with ΔAla or ΔAbu, to form a thioether bridge (Figure 2C) [41].

The dehydroamino acid motif is frequently encountered in nature, usually as constituents of large peptides [2,42]. In these cases, the main amino sequences are ribosomally assembled, and then the unsaturation is introduced post-translation. They are most commonly produced by bacteria, followed by fungi. For example, dehydroalanine, dehydro-2-aminobutyric acid, lanthionine, and methyllanthionine residues are constituents of nisin A, which is a polycyclic antibacterial and food preservative produced by *Lactococcus lactis* [43]. Thiostrepton is an oligopeptide antibiotic used in veterinary medicine that is produced by various *Streptomyces* bacteria, which contains ΔAla and ΔAbu [4]. The dehydroamino acid motif is also present in medicinally important small molecules, such as several classes of β-lactam antibiotics, for example cephalosporins, oxacephems, cephamycins, carbapenems, and penems [44]. The carbapenem class includes thienamycin and its more stable analogue, imipenum. Imipenum is hydrolyzed by renal dehydrogenase I enzyme in the liver, and therefore, it is co-administered with another dehydropeptide, cilastatin, which acts as a dehydrogenase inhibitor [5,45]. Plinabulin, a cyclic dipeptide containing a dehydrophenylalanine residue and a dehydrohistidine residue, is currently involved in world-wide phase 3 clinical trials for non-small cell lung cancer (Figure 2D) [3].

There are some examples in the literature of biologically active synthetic dehydrodipeptides. Schorlemmer et al. reported that acetyldehydro-3-Z-(2-thienyl)alanyltyrosine (Ac-ΔAla(2-thienyl))-L-Tyr-OH) and acetyldehydro-3-(2-furyl)alanyltyrosine (Ac-ΔAla(2-furyl))-L-Tyr-OH) were able to activate macrophages in mice. Following in vivo activation, the macrophages were able to kill several tumor cell-lines in vitro [46]. Latajka et al., and then later Makowski et al., reported that various short dehydropeptide esters and amides are capable of inhibiting cathepsin B enzymes, which are implicated in several inflammatory diseases (Figure 2D) [47–49].

1.4. Pharmacological Considerations

α,β-Unsaturated carbonyl groups are potentially electrophilic in reactivity at C-β by means of conjugate addition. Oxy-Michael, aza-Michael, and thia-Michael reactions can occur by reaction with alcohols, amines, and thiols, respectively. In biological settings, the reaction is particularly selective for "soft" nucleophiles, such as thiols, which in nature are provided by the side chain of cysteine. Michael acceptors can be a challenge for medicinal chemists to work with if this reactivity is undesired. Sometimes, they can provide false hits in screening assays due to off-target binding, be deactivated in vivo through their reaction with cellular thiols such as glutathione (GSH), or exhibit increased cytotoxic effects [50]. On the other hand, medicinal chemists are often able to exploit this reactivity if a cysteine (thiol-containing) residue is present within the target enzyme active site in a strategy known as "targeted covalent inhibition", which most notably is often able to overcome the problem of drug resistance.

In the context of dehydropeptides, we have already seen that ΔAla and ΔAbu can undergo intramolecular thio-Michael reactions in the formation of lantibiotics, by the action of specific cyclases (Figure 2C). Generally, Michael reactivity decreases with the presence of electron-donating β-substituents, such as alkyl groups or electron-rich aromatics. Electron-withdrawing substituents can increase the reactivity, but the rate of the reverse reaction (retro-Michael) is also increased [50]. With this in mind, the group of Joaquin et al. examined the reactivity of dehydroamino acid residues present in various tetrapeptides [36]. They found that peptides containing ΔAla were reactive to nucleophiles, e.g., the thiol group of cystamine. The residues Z-ΔAbu and ΔVal were resistant to nucleophilic attack, presumably on steric grounds, suggesting that β-substituted dehydroamino acid residues are more suited to in vivo applications.

2. Dehydropeptide Hydrogels and Other Nanostructures

In the structures of reported dehydropeptide hydrogels and nanostructures, the geometry around the Cα=Cβ double usually exists as the *Z* geometric isomer, which is the most thermodynamically stable isomer and the one for which the most synthetic methods of preparation exist [34]. Often, reports of dehydropeptide hydrogels do not explicitly state the *Z*/*E* stereochemistry present or provide proof of the stereochemistry. In these cases, the structures are shown, inferring the stereochemistry from the known outcome of the synthetic method employed (if possible) whilst omitting the stereo-descriptor from the text of the short-hand name.

2.1. *Dehydropeptides without Capping Groups (Free Amine and Acid Groups at N-Terminus and C-Terminus)*

2.1.1. Nanostructures from Uncapped Dehydropeptides

A short time before the first dehydropeptide hydrogelators were reported, the group of Chauhan reported two important discoveries. In 2007, the group reported on the ability of H-L-Phe-ΔPhe-OH to form ordered nanotubes [51]. The structures were longer and thinner than those obtained from the corresponding dipeptide composed of canonical amino acids (H-L-Phe-L-Phe-OH). They were stable to high temperatures, had a large pH range, and non-specific proteases (proteinase K). Around the same time, the same group reported two more dehydrodipeptides, H-L-Lys-ΔPhe-OH and H-L-Glu-ΔPhe-OH, as nanovesicles capable of encapsulating various model drug compounds, including vitamin B12, hemin, insulin, riboflavin, Pf MSP-I19, Pf MSP-3N, Pf HRP-II, BSA, lysozyme, and anti-mouse immunoglobin G (IgG) [52]. These nanovesicles were again stable to various proteases and proved non-toxic to Vero and HeLa cells.

Later studies showed that nanoparticles of H-L-Met-ΔPhe-OH were effective delivery vehicles for curcumin that were superior to H-L-Leu-ΔPhe-OH and H-L-Ile-ΔPhe-OH [53]. Curcumin possesses anti-cancer and anti-inflammatory properties, but as in many hydrophobic drugs, poor aqueous solubility is responsible for low bioavailability. Encapsulation within amphiphilic nanostructures offers a potential solution, and the authors found that when curcumin-loaded nanoparticles of H-L-Met-ΔPhe-OH were dispersed in aqueous solution, bioavailability was improved. This provided an increased cytotoxicity to the cancer cells lines HeLa, MCF-7, and HUH-7, whilst being non-cytotoxic to human fibroblasts (L-929) in concentration up to 50 μM. In animal studies, the loaded nanoparticles delayed tumor growth and enhanced survival rates in B6F10 melanoma mice.

Panda et al. reported that H-L-Arg-ΔPhe-OH and H-L-Lys-ΔPhe-OH were able to bind plasmid DNA and be taken up by cells. The plasmid DNA was resistant to the action of DNase enzymes [54]. A similar result was observed by Khatri et al., who from a panel of dehydrodipeptides found that H-L-Arg-ΔPhe-OH was the best suited for forming spherical nanoparticles capable of delivering plasmid DNA to HEK293T cells [55]. The release was pH dependent, with the plasmid DNA retaining their native conformation at endosomal pH.

Basker et al. functionalized iron oxide nanoparticles with H-L-Arg-ΔPhe-OH for targeting cancer cells in the presence of a pulsed magnetic field, providing heat-induced cell death by altering the membrane permeability of lung cancer cells [56]. This strategy provides a targeted delivery at low cost whilst avoiding problems associated with toxicity and drug resistance.

Varshney et al. attached lactabionic acid (LA) to a dehydrodipeptide to form a H-L-Arg-ΔPhe-LA conjugate [57]. In turn, this was used to create LA-L-Arg-ΔPhe-OH/miR-199a-3p nanoparticles, for delivering microRNA to hepatocellular carcinoma. The conjugated LA is a ligand for asialoglycoprotein receptors, which are over expressed on hepatocytes.

Folic acid receptors are known to be overexpressed on cancer cells, which lead Panda et al. to derivatize nanoparticles of H-L-Arg-ΔPhe-OH with folic acid (FA) in order to provide a tumor-targeted drug delivery of the cancer drug, doxorubicin. Doxorubicin entrapped within FA-L-Arg-ΔPhe-OH nanoparticles was able to provide an enhanced target specificity and anti-tumor effect compared with free doxorubicin or non-folic acid conjugated nanoparticles. The stability and low toxicity of these peptide nanoparticles provides advantages over the alternative types of nanoparticles previously studied, such as polymeric or metallic nanoparticles [58].

The group of Chauhan combined *Z*-dehydrophenylalanine with β-phenylalanine (another non-coding amino acid). The dipeptide H-β-Phe-ΔPhe-OH formed nanotubes with different properties to the related H-β-Phe-L-Phe-OH and H-L-Phe-L-Phe-OH dipeptides. Nanotubes of H-β-Phe-ΔPhe-OH were stable over a range of temperatures, stable to non-specific proteases, and demonstrated non-cytotoxicity to HeLa and L929 cells up to concentrations of 250 μM. The authors found that H-β-Phe-ΔPhe-OH-encapsulated mitoxanthrone was more efficient in killing HeLa and B6F10 cancer cells than free mitoxanthrone [59].

The uncapped dehydrodipeptides able to form nanostructures are summarized in Figure 3.

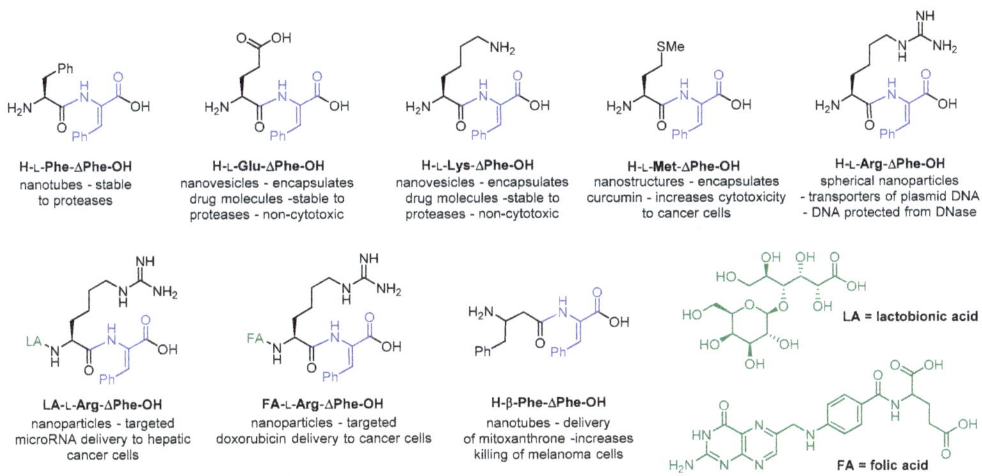

Figure 3. Summary of uncapped dehydrodipeptides capable of forming nanostructures.

2.1.2. Hydrogels from Uncapped Dehydropeptides

In 2008, Chauhan reported the first dehydrodipeptide hydrogel [60]. Dehydrodipeptide, H-L-Phe-ΔPhe-OH, was able to form a stable supramolecular hydrogel by adding a 50 mg/mL 1,1,1,3,3,3-hexafluoroisopropanol (HFIP) solution of hydrogelator to 0.8 M sodium acetate buffer, to achieve a final concentration as low as 0.2 wt % at neutral pH. Notably, unlike other known small dipeptide hydrogelators, this compound had no aromatic capping group attached to the *N*-terminus and therefore features a free amine group and carboxylic acid group at the *N*-terminus and *C*-terminus, respectively. With a molecular weight of 310 g/mol, it was one of the smallest known hydrogelators. The conformational constraint conferred by the double bond proved to be crucial to gelation, as the corresponding saturated dipeptide, H-L-Phe-L-Phe-OH, failed to provide a gel under the same conditions. In addition, the double bond imparted proteolytic stability, with the hydrogelator being inert to the action of chemotrypsin and other enzymes present in cell culture supernatant. The hydrogel possessed high mechanical strength whilst being non-toxic to HeLa and L929 mammalian cells in cell viability assays. The hydrogel was able to encapsulate and release model drug compounds, and therefore, it possessed potential for sustained drug delivery. The gel strength increased with increasing salt concentration, and it decreased with increasing temperature above a threshold temperature of 50 °C. The gels were stable at neutral or basic pH, but reducing the pH to below 7 °C caused gel disassembly. Therefore, the hydrogels are responsive to a range of physical conditions (temperature, pH, and salt concentration) and provide an opportunity for tunable drug delivery [60].

In 2010, the same team extended the peptide sequence to form the heptadipeptide sequence H-L-Phe-Δ-Phe-L-Arg-Gly-L-Asp-Gly-Gly-OH, which combined the gelation capacity of H-L-Phe-ΔPhe-OH with the cell-adhesion properties of "RGD" (L-Arg-Gly-L-Asp) [61]. The resulting gel was able to support 3D growth and proliferation of mammalian cells (HeLa and L929) for two weeks. The low-toxicity combined with cell growth-promoting properties provides a promising candidate tissue engineering and cell biology applications.

In 2016, the group of Chauhan extended their search for hydrogels based on the H-L-Phe-ΔPhe-OH structure by synthesizing a panel of 16 variants of H-L-AA-ΔPhe-OH (where AA = a canonical amino acid residue) and testing for gelation ability at 0.4 wt % [62]. They found that only H-L-Leu-ΔPhe-OH provided a strong hydrogel. The hydrogel was injectable (i.e., converts between gel and liquid state in response to applied strain), stable to trypsin, non-toxic to HEK293T cells, and was able to encapsulate a range of hydrophilic and hydrophobic drug molecules. The release rate was found to increase with increasing hydrophilicity of the drug molecule. An in vivo study showed that the sustained release of the cancer drug, mitoxanthrone, from the hydrogel was able to reduce tumor growth.

In 2020, Kumar Thota et al. reported hybrid gels of H-L-Leu-ΔPhe-OH and Fo-L-Met-L-Leu-L-Phe (fMLF, a macrophage attractant) as possible wound-healing materials [63]. Co-assembly afforded injectable gels, which were non-toxic to a variety of human cells including HEK293T cells, macrophages, and fibroblasts in viability assays. The gels were also able to act as 3D platforms for the culture of macrophages and fibroblasts. The presence of Fo-L-Met-L-Leu-L-Phe did not alter the secondary structure of gels of the parent gel, and Fo-L-Met-L-Leu-L-Phe retained its ability to attract macrophages, which is beneficial for the wound-healing process. In drug delivery studies, the hydrogel provided a slow release of various antibiotic molecules, including ciprofloxacin, suggesting an ability to prevent infections at the wound site over a prolonged time period.

The uncapped dehydrodipeptides able to form hydrogels are summarized in Figure 4.

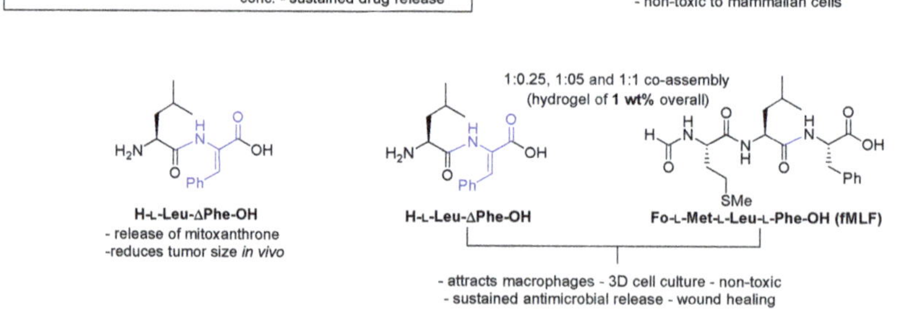

Figure 4. Summary of uncapped short dehydropeptides capable of forming hydrogels.

2.2. Dehydropeptides Modified at the C-Terminus

In 2012, the group of Kumar Sharma reported glucosamine-conjugated dehydropeptide hydrogelators, H-L-Phe-ΔPhe-εAhx-GA and Boc-L-Phe-ΔPhe-εAhx-GA, as potential drug delivery vehicles (Figure 5) [64]. Conjugation to glucosamine increased the aqueous solubility. At low concentrations (0.1 wt %), the conjugates provide spherical nanostructures of approximately 55 and 175 nm for H-L-Phe-ΔPhe-εAhx-GA and Boc-L-Phe-ΔPhe-εAhx-GA, respectively, determining by techniques such as dynamic light scattering (DLS), AFM, and TEM. At higher concentrations (1.0–2.0 wt %), the conjugates formed hydrogels. Furthermore, the conjugates possessed intrinsic antimicrobial activity, as evidenced by disk diffusion assays. In particular, the conjugates were active against *Micrococcus flavus*, *Bacillus subtilis*, and *Pseudomonas aeruginosa*. Successful drug encapsulation by nanostructures of the peptides (0.5 wt %) was achieved with model hydrophobic dye compounds, namely eosin (0.1 wt %) and *N*-fluoresceinyl-2-aminoethanol (FAE) (0.1 wt %). The peptide containing a free amine group was able to reduce auric chloride to form gold nanoparticles conjugated to the peptide. No cytotoxicity to HeLa cells was observed at concentrations of up to 25 μM.

Then, in 2017, the same team conjugated dehydropeptides to the natural polysaccharide, inulin [65]. By using carbodiimide (CDI) coupling chemistry, the team polyesterified the primary alcohols along the inulin chain to form a dehydropeptide array. The resulting conjugate, (Boc-L-Phe-ΔPhe-εAhx)$_n$-inulin, was able to self-assemble into nanostructures of 146–486 nm dimensions, which were able to encapsulate the antibiotic, ornidazole, and then release the drug in a controlled manner. These structures have great potential for targeted drug delivery to treat colonic diseases, as the nanostructures can be disassembled by enzymatic degradation using inulinase, which is produced by colonic bacteria. This degradation of the nanostructures is accompanied by the accelerated release of the drug cargo (Figure 5). Changes to the pH had little effect on the rate of release. In cell viability assays, the nanostructures were non-toxic to HEK293 cells at concentrations of up to 60 μg/mL.

Rekha Deka et al. discovered that tripeptides Boc-L-Pro-L-Phe-Gly-OMe and Boc-L-Pro-Z-ΔPhe-Gly-OMe were able to form nanostructures in aqueous solution (Figure 5) [66]. The presence of the dehydrophenylalanine residue was found to have a significant effect on the drug loading ability, as Boc-L-Pro-Z-ΔPhe-Gly-OMe was able to encapsulate the antibiotic compound ornidazole and the anti-cancer compound curcumin at higher loading levels than Boc-L-Pro-L-Phe-Gly-OMe. The nanostructures of the dehydropeptide were able to provide a sustained drug release of ornidazole and curcumin of 34% and 23%, respectively, after 6 days. These nanostructures could be further stabilized with D-α-tocopheryl polyethylene glycol (TPGS or vitamin E TPGS), which reduced the overall release of ornidazole and curcumin to 30% and 19%, respectively, over the same time period. In MTT viability assays, the dehydropeptide showed little toxicity to MCF-7 cells up to a concentration of 48 μg/mL.

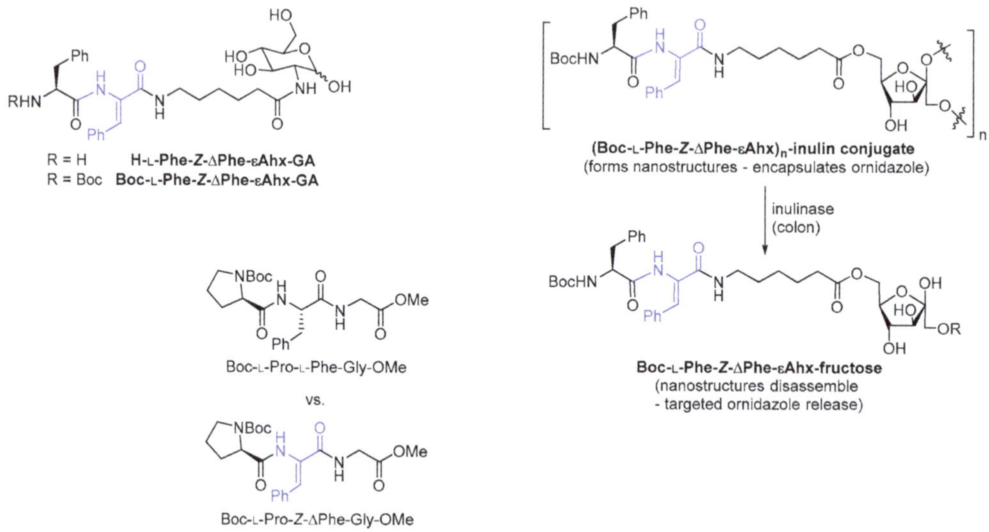

Figure 5. Summary of dehydrodipeptides modified at the *C*-terminus.

2.3. Dehydropeptides Modified at the N-Terminus

2.3.1. *N*-Conjugated with Non-Steroidal Anti-Inflammatory Drugs (NSAIDs)

Non-steroidal anti-inflammatory drugs function by inhibiting cyclooxygenase (COX) enzymes, which are responsible for the biosynthesis of prostaglandins and thromboxanes responsible for pain and inflammation in response to injury [67]. The unwanted side effects of many NSAIDs means that their administration as prodrugs or as nanoformulations are attractive strategies. COX-mediated inflammation plays a role in tumor development. COX enzymes are also over-expressed on the surface of tumor cells and therefore hold potential for targeted cancer therapy [68]. Important examples of NSAIDs include aspirin, ibuprofen naproxen, and ketoprofen [36]. First, we will briefly mention some of the other examples of peptidomimetic supramolecular hydrogels conjugated to NSAIDs [69]. The group of Xu reported that conjugates of naproxen with peptides of D-amino acids were able to form hydrogels. They were not only stable to proteases (relative to the equivalent conjugates containing L-amino acids) but retained the anti-inflammatory activity of the parent NSAID. More significantly, they provided increased COX-2 vs. COX-1 selectivity, potentially minimizing the gastric side effects of COX inhibitors [70]. The group of Cui reported similar observations for ketoprofen-capped D-amino acid-containing peptides [71]. Majumder et al. investigated conjugates of naproxen with β-amino acids, which also provided increased enzymatic stability whilst retaining the anti-inflammatory properties of naproxen [72]. A naproxen-dipeptide conjugate was very recently identified as a folate receptor ligand from a DNA-encoded chemical library [73].

Our research group synthesized the first dehydropeptides to be conjugated with NSAIDs. The focused library of Npx-L-Phe-ΔPhe-OH, Npx-L-Phe-ΔAbu-OH, Npx-L-Val-ΔPhe-OH, and Npx-L-Ala-ΔPhe-OH were able to form hydrogels, with critical gelation concentrations (CGCs) of 0.4–0.8 wt %. The properties of Npx-L-Phe-ΔPhe-OH were compared with the parent Npx-L-Phe-L-Phe-OH, and it was found that Npx-L-Phe-ΔPhe-OH offers increased resistance to chemotrypsin, a lower CGC (0.4 compared with 0.8 wt %), and a higher pH tolerance [74].

This work was followed up with a more in-depth study focused on Npx-L-Trp-Z-ΔPhe-OH and Npx-L-Trp-Z-ΔAla-OH as protease-resistant gelators. Hydrogels of Npx-L-Trp-Z-ΔPhe-OH possessed higher elasticity and lower CGCs, whilst Npx-L-Trp-Z-ΔAla-OH possessed self-healing and injectable properties. Both hydrogels showed promise as delivery systems for hydrophobic cancer drugs [75].

In a further study, the peptide structure of a minimum gelator module, Npx-L-Ala-ΔPhe-OH, was extended so as to incorporate L-Arg-Gly-Asp ("RGD"), a cell-binding motif that is overexpressed on cancer cells. Assembled using a combination of solution phase and solid phase peptide synthesis, the resulting construct, Npx-L-Ala-Z-ΔPhe-Gly-L-Arg-Gly-L-Asp-Gly-OH, was an effective nanocarrier [76].

The gelation of the dehydrodipeptides Npx-L-Tyr-Z-ΔPhe-OH and Npx-L-Asp(OH)-Z-ΔPhe-OMe was performed in the presence of superparamagnetic iron oxide nanoparticles (SPIONs) to incorporate the SPIONs within the hydrogel network. The magnetic behavior of the SPIONs was retained within the gel network. These hydrogel composites are stimuli-responsive, as magnetic excitation with AMF generates heat accompanied by a gel-to-sol transition, which may allow targeted drug delivery [77].

The hydrogel of Npx-L-Met-Z-ΔPhe-OH containing core/shell manganese ferrite/gold nanoparticles or gold-decorated manganese ferrite nanoparticles were created, exploiting the affinity of sulfur for gold, and these hydrogel composites showed promising results toward targeted cancer therapy. The dehydroamino acid provided proteolytic stability, while the naproxen capping group is a ligand for COX enzymes, which are overexpressed on cancer cells. The release of the anti-cancer drug, curcumin, was investigated, which was released at a faster rate from irradiated gels compared with non-irradiated gels, and therefore potentially had a dual action in cancer therapy [78].

The increased proteolytic resistance of these naproxen-dehydropeptides, compared with naproxen-dipeptides consisting of canonical amino acids, could potentially affect their ability to release naproxen and act as anti-inflammatory prodrugs. Moreira et al. considered that these conjugates may themselves retain the anti-inflammatory properties of the conjugated NSAID and therefore be suitable as hydrogels for "self-delivery" [79]. With this in mind, the biological activities of a panel of hydrogelator naproxen conjugates was assessed. Many of the examples were found to retain anti-inflammatory activity. The most interesting compound was Npx-L-Ala-Z-ΔPhe-OH, which could inhibit enzymes involved in the inflammation process, COX-2 and lipoxygenase (LOX), to a similar extent as the parent naproxen, indicating a dual action. Interestingly, COX-1 inhibition was significantly reduced, meaning that this conjugate is, in fact, a COX-2-selective inhibitor, which is a desired profile for minimizing the side effects of NSAIDs. The compound library was also assessed for alternative biological activities, and Npx-L-Tyr-ΔPhe-OH was found to be an effective proteasome inhibitor. The inhibition of proteasome enzymes is reported to have potential in cancer therapy. A conjugate that can both bind COX enzymes and inhibit proteasome enzymes may be a powerful combination for targeting cancer cells [79].

Naproxen-capped dehydropeptide hydrogelators are summarized in Figure 6.

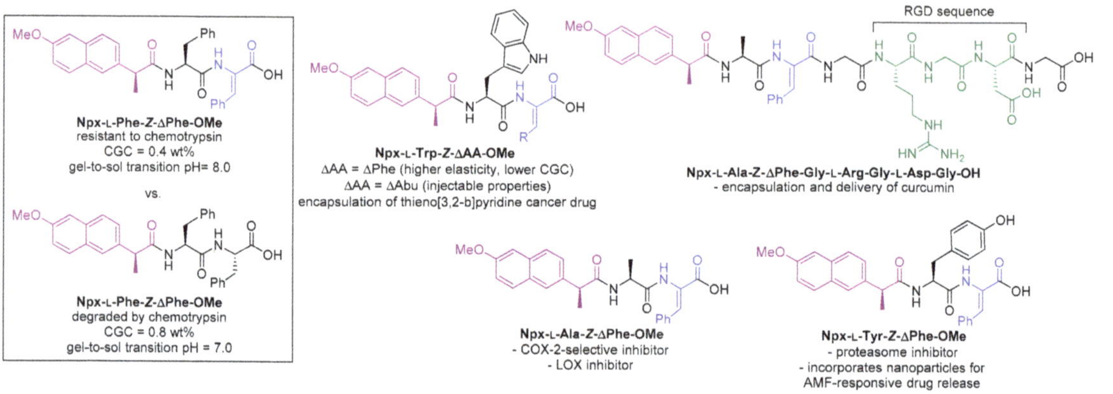

Figure 6. Summary of naproxen-capped peptides capable of forming hydrogels.

2.3.2. *N*-Conjugated with Carboxybenzyl (Benzyloxycarbonyl) Groups

Veloso et al. reported the synthesis of a panel of carboxybenzyl-protected dehydrodipeptides and investigated their hydrogelation properties [80]. Cbz-L-Phe-ΔPhe-OH, Cbz-L-Tyr-ΔPhe-OH, and Cbz-L-Met-ΔPhe were all able to form hydrogels, at 0.1, 0.2, and 0.2 wt %, respectively (Figure 7). Cbz-L-Ala-ΔPhe-OH and Cbz-L-Gly-ΔPhe-OH both failed to provide hydrogels. The successful hydrogelators showed promising drug delivery properties using membrane models. The anti-cancer compounds curcumin and doxorubicin could be encapsulated and then sustainably released. Furthermore, the hydrogelators were found to be non-toxic to the keratynocyte cell line, HaCat, at concentrations of up to 100 μM.

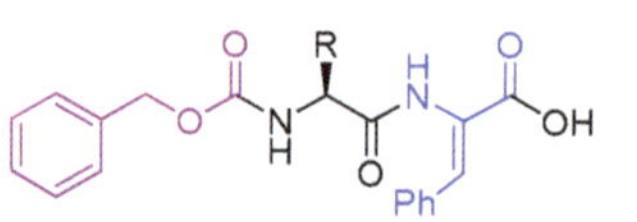

Cbz-L-AA-ΔPhe-OH

peptide	CGC (wt%)
Cbz-L-Phe-ΔPhe-OH	0.1
Cbz-L-Tyr-ΔPhe-OH	0.2
Cbz-L-Met-ΔPhe-OH	0.2
Cbz-L-Ala-ΔPhe-OH	no gelation
Cbz-L-Gly-ΔPhe-OH	no gelation

Figure 7. Summary of carboxybenzyl-capped peptides capable of forming hydrogels.

3. Synthesis of Dehydropeptides

The synthesis of short peptides containing canonical amino acids is usually a routine procedure for organic chemists, whether achieved by solution-phase or solid-phase synthesis. However, the presence of a dehydroamino acid residue complicates the synthetic process because the nucleophilicity of the amine is reduced by the conjugation of its lone pair of electrons with the double bond. Indeed, this enamine moiety is in equilibrium with the corresponding imine tautomer, which is subject to hydrolysis to the corresponding α-ketocarbonyl. A number of strategies exist for introducing the unsaturation, which have been reviewed by Humphrey [81] and later Bonauer [82]. We shall only briefly consider synthetic methods, focusing on those that are commonly used in the synthesis of dehydrodipeptide hydrogelators or those that appear especially useful for the synthesis of such dehydrodipeptides. The examples discussed here are not intended to be exhaustive but rather serve as an overview of the various reaction classes available.

Many groups, including our own [74], have accessed dehydrodipeptides via a β-elimination strategy, with the initial step involving the conversion of a β-hydroxy group to a better leaving group, which is then followed by a base-induced elimination, often *via* a alkylidene-substituted azlactone intermediate if the C-terminus is present as the free acid (Scheme 1A) [60]. In complex syntheses of large peptides, a dehydroamino acid residue is often introduced to solid-phase synthesis as part of a pre-formed dehydrodipeptide, where the dehydroamino acid residue is present at the C-terminus. Therefore, the solution phase strategies employed to synthesize these smaller units can be exploited to make the short peptides present in supramolecular dehydropeptide hydrogelators. With this in mind, the Staudinger ligation routes [83] and Cu(II)-mediated coupling routes [84] developed by Inoue, in a route to yaku'amide, may well prove useful in the synthesis of molecules able to form supramolecular structures (Scheme 1B). These methods have the advantage that the E/Z stereochemistry is pre-installed, albeit slightly eroded during the coupling reaction.

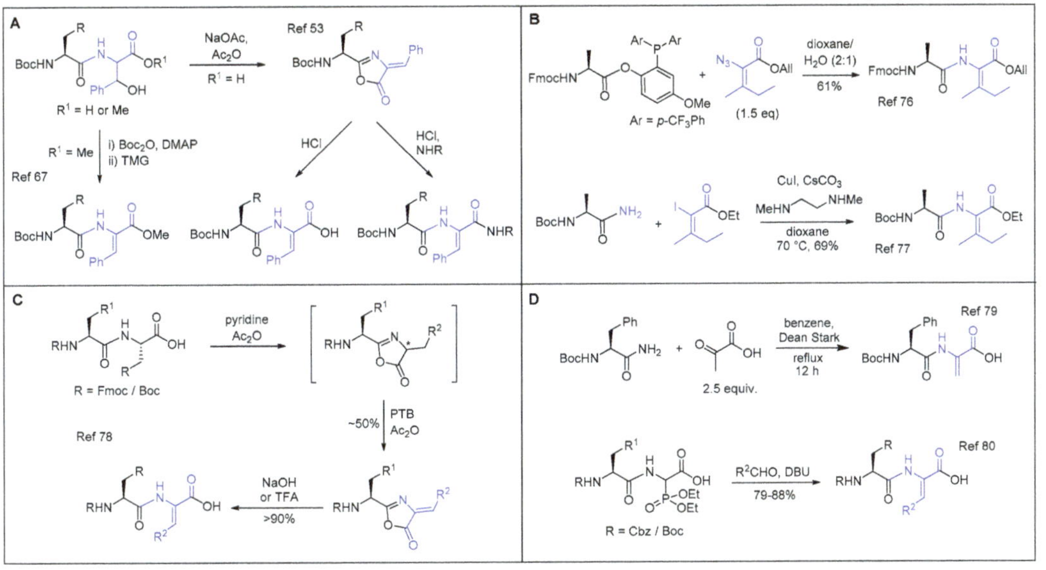

Scheme 1. (**A**) Examples of synthesis *via* β-elimination. (**B**) Examples of synthesis via Staudinger ligation and Cu(II)-mediated coupling. (**C**) Examples of synthesis via azlactone oxidation. (**D**) Examples of synthesis via condensation with pyruvic acid and Horner–Wadsworth–Emmons (HWE) olefination.

Wołczański and Lisowski recently reported a useful method where a dipeptide azlactone intermediate is oxidized by pyridinium tribromide, which effectively allows canonical amino acids to be converted to their dehydro- versions (Scheme 1C) [85]. More classical methods involve condensation reactions, such as the reaction of Boc-L-Phe-NH_2 with pyruvic acid [86] or Horner–Wadsworth–Emmons reactions of α-phosphonates with aldehydes (Scheme 1D) [87].

4. Conclusions

In this review, we have demonstrated how dehydropeptide supramolecular hydrogels and nanostructures have great potential for many biological applications, particularly as targeted and sustained drug delivery vehicles, as wound treatment materials, and as platforms for 3D cell culture with tissue engineering potential. The presence of a non-proteinogenic dehydroamino acid residue greatly increases resistance to endogenous proteases. Unlike other commonly employed peptidomimetic residues, such as D-amino acids and β-amino acids, dehydroamino acids also decrease the molecular flexibility, which often has a significant effect on the overall secondary structure.

In a general sense, we have seen how dehydropeptides often possess their own intrinsic biological properties, such as anti-bacterial, anti-cancer, anti-inflammatory, and β-breaker activities. Therefore, future work in this area may look to adapt the structures of these biologically active peptides towards hybrid materials with hydrogelation ability. These chimeric materials would then be suitable for "self-delivery" of their own biological properties, perhaps in addition to the encapsulation and then targeted release of other drug molecules. This may be further combined with conjugated drug pharmacophores and/or specific receptor ligands in multi-modal therapies.

Dehydroamino acids, particularly those containing dehydroalanine, can be reactive at C-β to various thiol nucleophiles, through conjugate addition (thia-Michael) reactions. This can be both an advantage and a disadvantage, depending on the system under study. The putative Michael addition will covalently attach the peptide to other biological molecules containing amines and thiols. If this Michael addition is unwanted, then increasing the steric bulk around the double bond reduces the reactivity. In contrast, sometimes, an in vivo conjugation can be specifically targeted. Therefore, short dehydropeptide mimetics of known enzyme inhibitors, capable of hydrogelation, should be investigated as targeted alkylating inhibitors of specific enzymes.

The future direction of dehydropeptide hydrogels should also look to incorporate the aforementioned properties into stimuli-responsive dehydropeptides to form so-called "smart materials". There are many known methods for either forming or disassembling a hydrogel "on demand", such as through the application of light, magnetic field, enzymatic action, or pH change. Such strategies can allow the release of drug cargo with spatial and temporal control, can "uncage" an intrinsic biological function, or can allow the facile removal of hydrogel-based wound dressings [88–90]. As well as the a

Considering the low toxicity, ease of synthesis, and favorable mechanical and biological properties, we expect to see more of these types of nanostructures developed in the clinic. The required properties of a particular hydrogel/nanostructure may vary, in terms of proteolytic resistance and structural flexibility, depending on the intended biological application. This may dictate if a dehydropeptide is the right choice of building block for a particular purpose. Having discussed their advantages and limitations, we consider the dehydropeptides discussed in this review as complementary to other specific peptide/amino acid motifs, as part of the structural toolbox within the wider area of supramolecular nanostructures. The area of low molecular weight peptide hydrogels has enjoyed incredible growth over the last two decades, from a curiosity of materials science to successful biomedical tools. Research in the areas of drug delivery, tissue engineering, wound dressings, 3D printing, and cell culture will no doubt continue for the foreseeable future, perhaps alongside new applications still to be discovered (Figure 8). New applications may well expand into the food, cosmetics, and arts industries. Possible applications as synthetic nanofactories or even as model biological cells can also be envisaged. The realization of these possibilities is clearly going to require the joint effort of materials chemists, nanotechnologists, biologists, and many others in this interdisciplinary field.

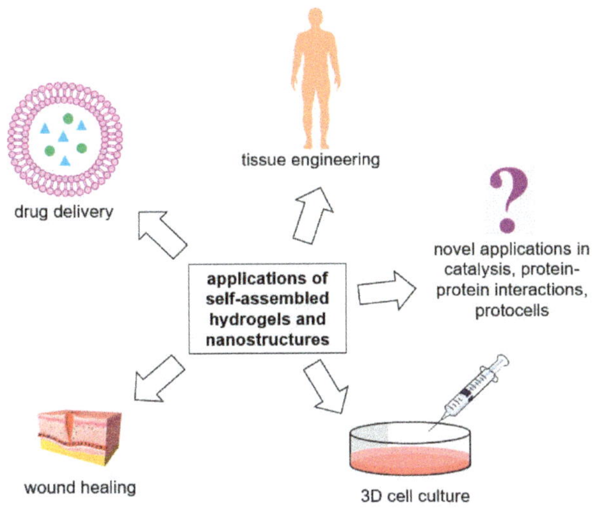

Figure 8. Summary of current and possible future direction of self-assembled hydrogels and nanostructures.

Author Contributions: Writing—original draft preparation, P.J.J.; writing—review and editing, P.J.J., C.A., T.P., J.A.M., P.M.T.F. All authors have read and agreed to the published version of the manuscript.

Funding: This work was supported by the Portuguese Foundation for Science and Technology (FCT) in the framework of the Strategic Funding of CQUM (UID/QUI/00686/2019). FCT, FEDER, PORTUGAL2020 and COMPETE2020 are also acknowledged for funding under research project PTDC/QUI-QOR/29015/2017 (POCI-01-0145-FEDER-029015).

Conflicts of Interest: The authors declare no conflict of interest.

Abbreviations

Ac: acetyl; AFM, atomic force microscopy; εAhx, 6-aminohexanoic aicd; All, allyl; AMF, alternating magnetic field; Boc, *tert*-butyloxycarbonyl; BSA, bovine serum albumin; Cbz, carboxybenzyl or benzyloxycarbonyl; CGC, critical gelation concentration; COX, cyclooxygenase; DBU, 1,8-diazabicyclo[5.4.0]undec-7-ene; DLS, dynamic light scattering; DMAP, 4-dimethylaminopyridine; ECM, extracellular matrix; FA, folic acid; FAE, *N*-fluoresceinyl-2-aminoethanol; Fmoc, fluorenylmethyloxycarbonyl; Fo, formyl; GdL, glucono-delta-lactone; GSH, glutathione; HFIP, 1,1,1,3,3,3-hexafluoroisopropanol; HWE, Horner–Wadsworth–Emmons; LA, lactobionic acid; LOX, lipoxygenase; Me, methyl; MTT, 3-(4,5-dimethylthiazol-2-yl)-2,5-diphenyl tetrazolium bromide; PTB, pyridinium tribromide; TPGS, D-α-tocopherol polyethylene glycol succinate; SEM, scanning electron microscopy; SPION, superparamagnetic iron oxide nanoparticle; TEM, tunneling electron microscopy; TFA, trifluoroacetic acid; TMG, 1,1,3,3-tetramethylguanidine.

References

1. Lenci, E.; Trabocchi, A. Peptidomimetic toolbox for drug discovery. *Chem. Soc. Rev.* **2020**, *49*, 3262–3277. [CrossRef] [PubMed]
2. Dong, R.; Pang, Y.; Su, Y.; Zhu, X. Supramolecular hydrogels: Synthesis, properties and their biomedical applications. *Biomater. Sci.* **2015**, *3*, 937–954. [CrossRef]
3. Johnson, E.K.; Adams, D.J.; Cameron, P.J. Peptide based low molecular weight gelators. *J. Mater. Chem.* **2011**, *21*, 2024–2027. [CrossRef]

4. Frederix, P.W.J.M.; Scott, G.G.; Abul-Haija, Y.M.; Kalafatovic, D.; Pappas, C.G.; Javid, N.; Hunt, N.T.; Ulijn, R.V.; Tuttle, T. Exploring the sequence space for (tri-)peptide self-assembly to design and discover new hydrogels. *Nat. Chem.* **2015**, *7*, 30–37. [CrossRef] [PubMed]
5. Sahoo, J.K.; Nazareth, C.; VandenBerg, M.A.; Webber, M.J. Self-assembly of amphiphilic tripeptides with sequence-dependent nanostructure. *Biomater. Sci.* **2017**, *5*, 1526–1530. [CrossRef] [PubMed]
6. Seow, W.Y.; Hauser, C.A.E. Short to ultrashort peptide hydrogels for biomedical uses. *Mater. Today* **2014**, *17*, 381–388. [CrossRef]
7. Chen, J.; Zou, X. Self-assemble peptide biomaterials and their biomedical applications. *Bioact. Mater.* **2019**, *4*, 120–131. [CrossRef] [PubMed]
8. Adams, D.J. Dipeptide and Tripeptide Conjugates as Low-Molecular-Weight Hydrogelators. *Macromol. Biosci.* **2011**, *11*, 160–173. [CrossRef]
9. Draper, E.R.; Adams, D.J. Low-Molecular-Weight Gels: The State of the Art. *Chem* **2017**, *3*, 390–410. [CrossRef]
10. Yang, Z.; Liang, G.; Xu, B. Enzymatic Hydrogelation of Small Molecules. *Acc. Chem. Res.* **2008**, *41*, 315–326. [CrossRef]
11. Shi, J.; Gao, Y.; Zhang, Y.; Pan, Y.; Xu, B. Calcium Ions to Cross-Link Supramolecular Nanofibers to Tune the Elasticity of Hydrogels over Orders of Magnitude. *Langmuir* **2011**, *27*, 14425–14431. [CrossRef] [PubMed]
12. Draper, E.R.; Adams, D.J. Controlling the Assembly and Properties of Low-Molecular-Weight Hydrogelators. *Langmuir* **2019**, *35*, 6506–6521. [CrossRef]
13. Liu, C.; Zhang, Q.; Zhu, S.; Liu, H.; Chen, J. Preparation and applications of peptide-based injectable hydrogels. *RSC Adv.* **2019**, *9*, 28299–28311. [CrossRef]
14. Fichman, G.; Gazit, E. Self-assembly of short peptides to form hydrogels: Design of building blocks, physical properties and technological applications. *Acta Biomater.* **2014**, *10*, 1671–1682. [CrossRef] [PubMed]
15. Li, J.; Xing, R.; Bai, S.; Yan, X. Recent advances of self-assembling peptide-based hydrogels for biomedical applications. *Soft Matter* **2019**, *15*, 1704–1715. [CrossRef]
16. Shah, A.; Malik, M.S.; Khan, G.S.; Nosheen, E.; Iftikhar, F.J.; Khan, F.A.; Shukla, S.S.; Akhter, M.S.; Kraatz, H.-B.; Aminabhavi, T.M. Stimuli-responsive peptide-based biomaterials as drug delivery systems. *Chem. Eng. J.* **2018**, *353*, 559–583. [CrossRef]
17. Martin, A.D.; Thordarson, P. Beyond Fmoc: A review of aromatic peptide capping groups. *J. Mater. Chem. B* **2020**, *8*, 863–877. [CrossRef]
18. Adams, D.J.; Butler, M.F.; Frith, W.J.; Kirkland, M.; Mullen, L.; Sanderson, P. A new method for maintaining homogeneity during liquid–hydrogel transitions using low molecular weight hydrogelators. *Soft Matter* **2009**, *5*, 1856–1862. [CrossRef]
19. VandenBerg, M.A.; Sahoo, J.K.; Zou, L.; McCarthy, W.; Webber, M.J. Divergent Self-Assembly Pathways to Hierarchically Organized Networks of Isopeptide-Modified Discotics under Kinetic Control. *ACS Nano* **2020**, *14*, 5491–5505. [CrossRef] [PubMed]
20. Yadav, N.; Chauhan, M.K.; Chauhan, V.S. Short to ultrashort peptide-based hydrogels as a platform for biomedical applications. *Biomater. Sci.* **2020**, *8*, 84–100. [CrossRef] [PubMed]
21. Du, X.; Zhou, J.; Shi, J.; Xu, B. Supramolecular Hydrogelators and Hydrogels: From Soft Matter to Molecular Biomaterials. *Chem. Rev.* **2015**, *115*, 13165–13307. [CrossRef]
22. Carpino, L.A.; Han, G.Y. 9-Fluorenylmethoxycarbonyl amino-protecting group. *J. Org. Chem.* **1972**, *37*, 3404–3409. [CrossRef]
23. Truong, W.T.; Su, Y.; Gloria, D.; Braet, F.; Thordarson, P. Dissolution and degradation of Fmoc-diphenylalanine self-assembled gels results in necrosis at high concentrations in vitro. *Biomater. Sci.* **2015**, *3*, 298–307. [CrossRef]
24. Swanekamp, R.J.; Welch, J.J.; Nilsson, B.L. Proteolytic stability of amphipathic peptide hydrogels composed of self-assembled pleated β-sheet or coassembled rippled β-sheet fibrils. *Chem. Commun.* **2014**, *50*, 10133–10136. [CrossRef]
25. Michele, M.; Katie, E.S.; Silvia, M. The Unexpected Advantages of Using D-Amino Acids for Peptide Self- Assembly into Nanostructured Hydrogels for Medicine. *Curr. Top. Med. Chem.* **2016**, *16*, 2009–2018. [CrossRef]
26. Luca, G.; Rossella De, M.; Lucia, C. Chemical Modifications Designed to Improve Peptide Stability: Incorporation of Non-Natural Amino Acids, Pseudo-Peptide Bonds, and Cyclization. *Curr. Pharm. Des.* **2010**, *16*, 3185–3203. [CrossRef]
27. Li, X.; Du, X.; Li, J.; Gao, Y.; Pan, Y.; Shi, J.; Zhou, N.; Xu, B. Introducing d-Amino Acid or Simple Glycoside into Small Peptides to Enable Supramolecular Hydrogelators to Resist Proteolysis. *Langmuir* **2012**, *28*, 13512–13517. [CrossRef] [PubMed]
28. Yang, Z.; Liang, G.; Xu, B. Supramolecular hydrogels based on β-amino acid derivatives. *Chem. Commun.* **2006**, 738–740. [CrossRef] [PubMed]
29. Rajbhandary, A.; Nilsson, B.L. Investigating the effects of peptoid substitutions in self-assembly of Fmoc-diphenylalanine derivatives. *Pept. Sci.* **2017**, *108*, e22994. [CrossRef]
30. Bogart, J.W.; Bowers, A.A. Dehydroamino acids: Chemical multi-tools for late-stage diversification. *Org. Biomol. Chem.* **2019**, *17*, 3653–3669. [CrossRef]
31. Buczek, A.; Siodłak, D.; Bujak, M.; Makowski, M.; Kupka, T.; Broda, M.A. Impact of the ΔPhe configuration on the Boc-Gly-ΔPhe-NHMe conformation: Experiment and theory. *Struct. Chem.* **2019**, *30*, 1685–1697. [CrossRef]
32. Joaquin, D.; Lee, M.A.; Kastner, D.W.; Singh, J.; Morrill, S.T.; Damstedt, G.; Castle, S.L. Impact of Dehydroamino Acids on the Structure and Stability of Incipient 310-Helical Peptides. *J. Org. Chem.* **2020**, *85*, 1601–1613. [CrossRef] [PubMed]
33. Jain, R.M.; Rajashankar, K.R.; Ramakumar, S.; Chauhan, V.S. First Observation of Left-Handed Helical Conformation in a Dehydro Peptide Containing Two l-Val Residues. Crystal and Solution Structure of Boc-l-Val-ΔPhe-ΔPhe-ΔPhe-l-Val-OMe. *J. Am. Chem. Soc.* **1997**, *119*, 3205–3211. [CrossRef]

34. Mathur, P.; Ramakumar, S.; Chauhan, V.S. Peptide design using α,β-dehydro amino acids: From β-turns to helical hairpins. *Pept. Sci.* **2004**, *76*, 150–161. [CrossRef] [PubMed]
35. Gupta, M.; Chauhan, V.S. De novo design of α,β-didehydrophenylalanine containing peptides: From models to applications. *Biopolymers* **2011**, *95*, 161–173. [CrossRef]
36. Gupta, M.; Acharya, R.; Mishra, A.; Ramakumar, S.; Ahmed, F.; Chauhan, V.S. Dehydrophenylalanine (ΔPhe) as a β Breaker: Extended Structure Terminated by a ΔPhe-Induced Turn in the Pentapeptide Boc-Phe1-Ala2-Ile3-ΔPhe4-Ala5-OMe. *ChemBioChem* **2008**, *9*, 1375–1378. [CrossRef]
37. Paul, M.; Donk, W.A.v.d. Chemical and Enzymatic Synthesis of Lanthionines. *Mini-Rev. Org. Chem.* **2005**, *2*, 23–37. [CrossRef]
38. Siodłak, D. α,β-Dehydroamino acids in naturally occurring peptides. *Amino Acids* **2015**, *47*, 1–17. [CrossRef]
39. Ma, S.; Zhang, Q. Linaridin natural products. *Nat. Prod. Rep.* **2020**, *37*, 1152–1163. [CrossRef]
40. Gross, E.; Morell, J.L. Structure of nisin. *J. Am. Chem. Soc.* **1971**, *93*, 4634–4635. [CrossRef] [PubMed]
41. Sandu, C.; Chandramouli, N.; Glickman, J.F.; Molina, H.; Kuo, C.-L.; Kukushkin, N.; Goldberg, A.L.; Steller, H. Thiostrepton interacts covalently with Rpt subunits of the 19S proteasome and proteasome substrates. *J. Cell. Mol. Med.* **2015**, *19*, 2181–2192. [CrossRef]
42. Demain, A.L.; Elander, R.P. The β-lactam antibiotics: Past, present, and future. *Antonie Leeuwenhoek* **1999**, *75*, 5–19. [CrossRef]
43. Pastel, D.A. Imipenem-Cilastatin Sodium, A Broad-Spectrum Carbapenem Antibiotic Combination. *Am. J. Hosp. Pharm.* **1986**, *43*, 2630–2644. [CrossRef]
44. Balfour, J.A.; Bryson, H.M.; Brogden, R.N. Imipenem/Cilastatin. *Drugs* **1996**, *51*, 99–136. [CrossRef]
45. Blayney, D.W.; Bazhenova, L.; Lloyd, G.K.; Huang, L.; Mohanlal, R. Plinabulin, a Novel Small Molecule That Ameliorates Chemotherapy-Induced Neutropenia, Is Administered on the Same Day of Chemotherapy and Has Anticancer Efficacy. *Blood* **2016**, *128*, 2508. [CrossRef]
46. Schorlemmer, H.-U.; Opitz, W.; Etschenberg, E.; Bitter-Suermann, D.; Hadding, U. Killing of Tumor Cells in vitro by Macrophages from Mice Treated with Synthetic Dehydrodipeptides. *Cancer Res.* **1979**, *39*, 1847–1853. [PubMed]
47. Latajka, R.; Makowski, M.; Jewgiński, M.; Pawełczak, M.; Koroniak, H.; Kafarski, P. Peptide p-nitrophenylanilides containing (E)-dehydrophenylalanine—synthesis, structural studies and evaluation of their activity towards cathepsin C. *New J. Chem.* **2006**, *30*, 1009–1018. [CrossRef]
48. Latajka, R.; Jewginski, M.; Makowski, M.; Pawełczak, M.; Huber, T.; Sewald, N.; Kafarski, P. Pentapeptides containing two dehydrophenylalanine residues-synthesis, structural studies and evaluation of their activity towards cathepsin C. *J. Pept. Sci.* **2008**, *14*, 1084–1095. [CrossRef]
49. Makowski, M.; Lenartowicz, P.; Oszywa, B.; Jewgiński, M.; Pawełczak, M.; Kafarski, P. Synthesis of dehydrodipeptide esters and their evaluation as inhibitors of cathepsin C. *Med. Chem. Res.* **2015**, *24*, 3157–3165. [CrossRef]
50. Jackson, P.A.; Widen, J.C.; Harki, D.A.; Brummond, K.M. Covalent Modifiers: A Chemical Perspective on the Reactivity of α,β-Unsaturated Carbonyls with Thiols via Hetero-Michael Addition Reactions. *J. Med. Chem.* **2017**, *60*, 839–885. [CrossRef]
51. Gupta, M.; Bagaria, A.; Mishra, A.; Mathur, P.; Basu, A.; Ramakumar, S.; Chauhan, V.S. Self-Assembly of a Dipeptide- Containing Conformationally Restricted Dehydrophenylalanine Residue to Form Ordered Nanotubes. *Adv. Mater.* **2007**, *19*, 858–861. [CrossRef]
52. Mishra, A.; Panda, J.J.; Basu, A.; Chauhan, V.S. Nanovesicles Based on Self-Assembly of Conformationally Constrained Aromatic Residue Containing Amphiphilic Dipeptides. *Langmuir* **2008**, *24*, 4571–4576. [CrossRef]
53. Alam, S.; Panda, J.J.; Chauhan, V.S. Novel dipeptide nanoparticles for effective curcumin delivery. *Int. J. Nanomed.* **2012**, *7*, 4207–4222. [CrossRef]
54. Panda, J.J.; Varshney, A.; Chauhan, V.S. Self-assembled nanoparticles based on modified cationic dipeptides and DNA: Novel systems for gene delivery. *J. Nanobiotechnol.* **2013**, *11*, 18. [CrossRef]
55. Khatri, A.; Mishra, A.; Chauhan, V.S. Characterization of DNA Condensation by Conformationally Restricted Dipeptides and Gene Delivery. *J. Biomed. Nanotechnol.* **2017**, *13*, 35–53. [CrossRef]
56. Baskar, G.; Ravi, M.; Panda, J.J.; Khatri, A.; Dev, B.; Santosham, R.; Sathiya, S.; Babu, C.S.; Chauhan, V.S.; Rayala, S.K.; et al. Efficacy of Dipeptide-Coated Magnetic Nanoparticles in Lung Cancer Models Under Pulsed Electromagnetic Field. *Cancer Investig.* **2017**, *35*, 431–442. [CrossRef] [PubMed]
57. Varshney, A.; Panda, J.J.; Singh, A.K.; Yadav, N.; Bihari, C.; Biswas, S.; Sarin, S.K.; Chauhan, V.S. Targeted delivery of microRNA-199a-3p using self-assembled dipeptide nanoparticles efficiently reduces hepatocellular carcinoma in mice. *Hepatology* **2018**, *67*, 1392–1407. [CrossRef] [PubMed]
58. Panda, J.J.; Kaul, A.; Kumar, S.; Alam, S.; Mishra, A.K.; Kundu, G.C.; Chauhan, V.S. Modified dipeptide-based nanoparticles: Vehicles for targeted tumor drug delivery. *Nanomedicine* **2013**, *8*, 1927–1942. [CrossRef] [PubMed]
59. Parween, S.; Misra, A.; Ramakumar, S.; Chauhan, V.S. Self-assembled dipeptide nanotubes constituted by flexible β-phenylalanine and conformationally constrained α,β-dehydrophenylalanine residues as drug delivery system. *J. Mater. Chem. B* **2014**, *2*, 3096–3106. [CrossRef] [PubMed]
60. Panda, J.J.; Mishra, A.; Basu, A.; Chauhan, V.S. Stimuli Responsive Self-Assembled Hydrogel of a Low Molecular Weight Free Dipeptide with Potential for Tunable Drug Delivery. *Biomacromolecules* **2008**, *9*, 2244–2250. [CrossRef]

61. Panda, J.J.; Dua, R.; Mishra, A.; Mittra, B.; Chauhan, V.S. 3D Cell Growth and Proliferation on a RGD Functionalized Nanofibrillar Hydrogel Based on a Conformationally Restricted Residue Containing Dipeptide. *ACS Appl. Mater. Interfaces* **2010**, *2*, 2839–2848. [CrossRef]
62. Thota, C.K.; Yadav, N.; Chauhan, V.S. A novel highly stable and injectable hydrogel based on a conformationally restricted ultrashort peptide. *Sci. Rep.* **2016**, *6*, 31167. [CrossRef]
63. Thota, C.K.; Berger, A.A.; Elomaa, L.; Nie, C.; Böttcher, C.; Koksch, B. Coassembly Generates Peptide Hydrogel with Wound Dressing Material Properties. *ACS Omega* **2020**, *5*, 8557–8563. [CrossRef]
64. Mahato, M.; Arora, V.; Pathak, R.; Gautam, H.K.; Sharma, A.K. Fabrication of nanostructures through molecular self-assembly of small amphiphilic glyco-dehydropeptides. *Mol. Biosyst.* **2012**, *8*, 1742–1749. [CrossRef]
65. Shivhare, K.; Garg, C.; Priyam, A.; Gupta, A.; Sharma, A.K.; Kumar, P. Enzyme sensitive smart inulin-dehydropeptide conjugate self-assembles into nanostructures useful for targeted delivery of ornidazole. *Int. J. Biol. Macromol.* **2018**, *106*, 775–783. [CrossRef]
66. Deka, S.R.; Yadav, S.; Kumar, D.; Garg, S.; Mahato, M.; Sharma, A.K. Self-assembled dehydropeptide nano carriers for delivery of ornidazole and curcumin. *Colloids Surf. B Biointerfaces* **2017**, *155*, 332–340. [CrossRef] [PubMed]
67. Brune, K.; Patrignani, P. New insights into the use of currently available non-steroidal anti-inflammatory drugs. *J. Pain. Res.* **2015**, *8*, 105–118. [CrossRef] [PubMed]
68. Xu, X.C. COX-2 inhibitors in cancer treatment and prevention, a recent development. *Anti Cancer Drugs* **2002**, *13*, 127–137. [CrossRef] [PubMed]
69. Jervis, P.J.; Amorim, C.; Pereira, T.; Martins, J.A.; Ferreira, P.M.T. Exploring the properties and potential biomedical applications of NSAID-capped peptide hydrogels. *Soft Matter* **2020**, *16*, 10001–10012. [CrossRef] [PubMed]
70. Li, J.; Kuang, Y.; Gao, Y.; Du, X.; Shi, J.; Xu, B. d-Amino Acids Boost the Selectivity and Confer Supramolecular Hydrogels of a Nonsteroidal Anti-Inflammatory Drug (NSAID). *J. Am. Chem. Soc.* **2013**, *135*, 542–545. [CrossRef]
71. Chen, Z.; Xing, L.; Fan, Q.; Cheetham, A.G.; Lin, R.; Holt, B.; Chen, L.; Xiao, Y.; Cui, H. Drug-Bearing Supramolecular Filament Hydrogels as Anti-Inflammatory Agents. *Theranostics* **2017**, *7*, 2003–2014. [CrossRef]
72. Majumder, J.; Das, M.R.; Deb, J.; Jana, S.S.; Dastidar, P. β-Amino Acid and Amino-Alcohol Conjugation of a Nonsteroidal Anti-Inflammatory Drug (NSAID) Imparts Hydrogelation Displaying Remarkable Biostability, Biocompatibility, and Anti-Inflammatory Properties. *Langmuir* **2013**, *29*, 10254–10263. [CrossRef]
73. Huang, Y.; Meng, L.; Nie, Q.; Zhou, Y.; Chen, L.; Yang, S.; Fung, Y.M.E.; Li, X.; Huang, C.; Cao, Y.; et al. Selection of DNA-encoded chemical libraries against endogenous membrane proteins on live cells. *Nat. Chem.* **2021**, *13*, 77–88. [CrossRef]
74. Vilaça, H.; Pereira, G.; Castro, T.G.; Hermenegildo, B.F.; Shi, J.; Faria, T.Q.; Micaêlo, N.; Brito, R.M.M.; Xu, B.; Castanheira, E.M.S.; et al. New self-assembled supramolecular hydrogels based on dehydropeptides. *J. Mater. Chem. B* **2015**, *3*, 6355–6367. [CrossRef]
75. Vilaça, H.; Hortelão, A.C.L.; Castanheira, E.M.S.; Queiroz, M.-J.R.P.; Hilliou, L.; Hamley, I.W.; Martins, J.A.; Ferreira, P.M.T. Dehydrodipeptide Hydrogelators Containing Naproxen N-Capped Tryptophan: Self-Assembly, Hydrogel Characterization, and Evaluation as Potential Drug Nanocarriers. *Biomacromolecules* **2015**, *16*, 3562–3573. [CrossRef]
76. Vilaça, H.; Castro, T.; Costa, F.M.G.; Melle-Franco, M.; Hilliou, L.; Hamley, I.W.; Castanheira, E.M.S.; Martins, J.A.; Ferreira, P.M.T. Self-assembled RGD dehydropeptide hydrogels for drug delivery applications. *J. Mater. Chem. B* **2017**, *5*, 8607–8617. [CrossRef]
77. Carvalho, A.; Gallo, J.; Pereira, D.M.; Valentão, P.; Andrade, P.B.; Hilliou, L.; Ferreira, P.M.T.; Bañobre-López, M.; Martins, J.A. Magnetic Dehydrodipeptide-Based Self-Assembled Hydrogels for Theragnostic Applications. *Nanomaterials* **2019**, *9*, 541. [CrossRef] [PubMed]
78. Veloso, S.R.S.; Martins, J.A.; Hilliou, L.; Amorim, C.O.; Amaral, V.S.; Almeida, B.G.; Jervis, P.J.; Moreira, R.; Pereira, D.M.; Coutinho, P.J.G.; et al. Dehydropeptide-based plasmonic magnetogels: A supramolecular composite nanosystem for multimodal cancer therapy. *J. Mater. Chem. B* **2020**, *8*, 45–64. [CrossRef] [PubMed]
79. Moreira, R.; Jervis, P.J.; Carvalho, A.; Ferreira, P.M.T.; Martins, J.A.; Valentão, P.; Andrade, P.B.; Pereira, D.M. Biological Evaluation of Naproxen–Dehydrodipeptide Conjugates with Self-Hydrogelation Capacity as Dual LOX/COX Inhibitors. *Pharmaceutics* **2020**, *12*, 122. [CrossRef] [PubMed]
80. Veloso, S.R.S.; Jervis, P.J.; Silva, J.F.G.; Hilliou, L.; Moura, C.; Pereira, D.M.; Coutinho, P.J.G.; Martins, J.A.; Castanheira, E.M.S.; Ferreira, P.M.T. Supramolecular ultra-short carboxybenzyl-protected dehydropeptide-based hydrogels for drug delivery. *Mater. Sci. Eng. C* **2021**, *122*, 111869. [CrossRef] [PubMed]
81. Humphrey, J.M.; Chamberlin, A.R. Chemical Synthesis of Natural Product Peptides: Coupling Methods for the Incorporation of Noncoded Amino Acids into Peptides. *Chem. Rev.* **1997**, *97*, 2243–2266. [CrossRef]
82. Bonauer, C.; Walenzyk, T.; König, B. α,β-Dehydroamino Acids. *Synthesis* **2006**, *2006*, 1–20. [CrossRef]
83. Itoh, H.; Miura, K.; Kamiya, K.; Yamashita, T.; Inoue, M. Solid-Phase Total Synthesis of Yaku'amide B Enabled by Traceless Staudinger Ligation. *Angew. Chem. Int. Ed.* **2020**, *59*, 4564–4571. [CrossRef] [PubMed]
84. Kuranaga, T.; Mutoh, H.; Sesoko, Y.; Goto, T.; Matsunaga, S.; Inoue, M. Elucidation and Total Synthesis of the Correct Structures of Tridecapeptides Yaku'amides A and B. Synthesis-Driven Stereochemical Reassignment of Four Amino Acid Residues. *J. Am. Chem. Soc.* **2015**, *137*, 9443–9451. [CrossRef] [PubMed]
85. Wołczański, G.; Lisowski, M. A general method for preparation of N-Boc-protected or N-Fmoc-protected α,β-didehydropeptide building blocks and their use in the solid-phase peptide synthesis. *J. Pept. Sci.* **2018**, *24*, e3091. [CrossRef]

86. Makowski, M.; Rzeszotarska, B.; Kubica, Z.; Pietrzyńnski, G. Synthesis of Peptides with α,β-Dehydroamino Acids, II. Synthesis of tert-Butyloxycarbonyldipeptides of Dehydroalanine and Dehydrophenylalanine. *Liebigs Ann. Chem.* **1985**, *1985*, 893–900. [CrossRef]
87. Buck, R.T.; Clarke, P.A.; Coe, D.M.; Drysdale, M.J.; Ferris, L.; Haigh, D.; Moody, C.J.; Pearson, N.D.; Swann, E. The Carbenoid Approach to Peptide Synthesis. *Chem. A Eur. J.* **2000**, *6*, 2160–2167. [CrossRef]
88. Koetting, M.C.; Peters, J.T.; Steichen, S.D.; Peppas, N.A. Stimulus-responsive hydrogels: Theory, modern advances, and applications. *Mater. Sci. Eng. R Rep.* **2015**, *93*, 1–49. [CrossRef]
89. Kasiński, A.; Zielińska-Pisklak, M. Smart Hydrogels-Synthetic Stimuli-Responsive Antitumor Drug Release Systems. *Int. J. Nanomed.* **2020**, *15*, 4541–4572. [CrossRef]
90. Lu, H.; Yuan, L.; Yu, X.; Wu, C.; He, D.; Deng, J. Recent advances of on-demand dissolution of hydrogel dressings. *Burns Trauma* **2018**, *6*, 35. [CrossRef] [PubMed]

International Journal of *Molecular Sciences*

MDPI

Article

Heparin-Mimicking Polymer-Based In Vitro Platform Recapitulates In Vivo Muscle Atrophy Phenotypes

Hyunbum Kim [1,2,†], Ji Hoon Jeong [1,3,†], Mona Fendereski [1,3,†], Hyo-Shin Lee [1,3], Da Yeon Kang [4], Sung Sik Hur [1], Jhaleh Amirian [5], Yunhye Kim [1,3], Nghia Thi Pham [1,3], Nayoung Suh [4], Nathaniel Suk-Yeon Hwang [2], Seongho Ryu [1,3], Jeong Kyo Yoon [1,3,*] and Yongsung Hwang [1,3,*]

1 Soonchunhyang Institute of Medi-Bio Science (SIMS), Soonchunhyang University, Cheonan-si 31151, Korea; tiggerhy@snu.ac.kr (H.K.); jjh2020@sch.ac.kr (J.H.J.); mona.fendereski@gmail.com (M.F.); hyo7093@sch.ac.kr (H.-S.L.); sstahur@sch.ac.kr (S.S.H.); yhkim@sch.ac.kr (Y.K.); ptnghia@sch.ac.kr (N.T.P.); ryu@sch.ac.kr (S.R.)

2 School of Chemical and Biological Engineering, Institute of Chemical Processes, Seoul National University, Seoul 08826, Korea; nshwang@snu.ac.kr

3 Department of Integrated Biomedical Science, Soonchunhyang University, Asan-si 31538, Korea

4 Department of Pharmaceutical Engineering, Soonchunhyang University, Asan-si 31538, Korea; gubin15@naver.com (D.Y.K.); nysuh@sch.ac.kr (N.S.)

5 Institute of Tissue Regeneration, Soonchunhyang University, Asan-si 31538, Korea; jalehamirian@gmail.com

* Correspondence: jkyoon@sch.ac.kr (J.K.Y.); yshwang0428@sch.ac.kr (Y.H.); Tel.: +82-41-413-5016 (J.K.Y.); +82-41-413-5017 (Y.H.)

† These authors contributed equally to this study.

Abstract: The cell–cell/cell–matrix interactions between myoblasts and their extracellular microenvironment have been shown to play a crucial role in the regulation of in vitro myogenic differentiation and in vivo skeletal muscle regeneration. In this study, by harnessing the heparin-mimicking polymer, poly(sodium-4-styrenesulfonate) (PSS), which has a negatively charged surface, we engineered an in vitro cell culture platform for the purpose of recapitulating in vivo muscle atrophy-like phenotypes. Our initial findings showed that heparin-mimicking moieties inhibited the fusion of mononucleated myoblasts into multinucleated myotubes, as indicated by the decreased gene and protein expression levels of myogenic factors, myotube fusion-related markers, and focal adhesion kinase (FAK). We further elucidated the underlying molecular mechanism via transcriptome analyses, observing that the insulin/PI3K/mTOR and Wnt signaling pathways were significantly downregulated by heparin-mimicking moieties through the inhibition of FAK/Cav3. Taken together, the easy-to-adapt heparin-mimicking polymer-based in vitro cell culture platform could be an attractive platform for potential applications in drug screening, providing clear readouts of changes in insulin/PI3K/mTOR and Wnt signaling pathways.

Keywords: synthetic mimic of heparin; poly(sodium-4-styrenesulfonate); myoblast; myogenic differentiation; fusion; focal adhesion kinase (FAK)

Citation: Kim, H.; Jeong, J.H.; Fendereski, M.; Lee, H.-S.; Kang, D.Y.; Hur, S.S.; Amirian, J.; Kim, Y.; Pham, N.T.; Suh, N.; et al. Heparin-Mimicking Polymer-Based In Vitro Platform Recapitulates In Vivo Muscle Atrophy Phenotypes. *Int. J. Mol. Sci.* **2021**, *22*, 2488. https://doi.org/10.3390/ijms22052488

Academic Editor: Peter John Jervis

Received: 6 February 2021
Accepted: 25 February 2021
Published: 2 March 2021

1. Introduction

Skeletal muscle, the largest organ in the human body, comprises approximately 40% of the total body mass of healthy individuals [1]. Due to the hierarchical structure of skeletal muscles and the intrinsic structure–function relationship between the skeletal muscle extracellular matrix (ECM) and skeletal muscle stem cells (skMSCs), cell–matrix interactions within skeletal muscle have been extensively investigated [2–4]. While bioengineered models for enhancing skeletal muscle function have been extensively developed and explored, there have been fewer advancements in the development of highly functional muscle atrophy-specific organ-on-a-chip models compared to in vivo muscle injury models [5,6].

Cell fusion and cell–matrix interactions within the microenvironment are considered key cellular processes that govern the formation and repair of skeletal muscle, in turn maintaining skeletal muscle homeostasis [7,8]. During this process, activated skMSCs fuse to form multinucleated myotubes, forming parallel bundle structures, which eventually comprise functional skeletal muscle tissue [9,10]. Therefore, it is critical to understand the molecular mechanism underlying myogenic commitment, fusion, and the maturation of activated skMSCs into fully functional myofibers, including the various signaling pathways involved. One way to explore these is through the use of an in vitro skeletal muscle-like drug screening platform [11].

The microenvironment surrounding skMSCs is known to trigger cellular responses, which alter various aspects of cellular function, including the cell adhesion, spreading, proliferation, and differentiation of skMSCs [12–14]. Similarly to the multitude of signaling cues, such as growth factors, small molecules, and hormones, direct regulation of stem cell fate also occurs in response to matrix stiffness, and thus the topographical features of cell culture substrates are considered important factors for the stimulation of signal transduction and subsequent cell lineage determination [15–18]. Molecular dynamics within focal adhesions have been shown to regulate cellular adhesion through integrins and the ECM, with these biomechanical signals being transmitted in a bidirectional manner [19,20]. In addition, physicochemical interactions between cells and their microenvironment consistently provoke intercellular signaling, in turn modulating gene expression and diverse signal transduction [21–24].

Previous studies have demonstrated the important roles of basic fibroblast growth factor (bFGF), a heparin-binding protein, in the regulation of skMSC proliferation and differentiation as well as its contribution to skeletal muscle homeostasis [25]. In addition, cell surface-bound heparin and heparin sulfate proteoglycans (HSPGs) have been proven as necessary for the modulation of EGF signaling and the determination of terminal myogenesis [26,27]. Previously, Sangaj et al. performed molecular docking simulations and reported that the soluble form of synthetic mimics of heparin, poly(sodium-4-styrenesulfonate) (PSS), could be thermodynamically favorable for binding bFGF in a manner equivalent to that of native heparin. Thus, the soluble form of PSS could sequester bFGF to promote myogenic differentiation [28].

In this study, we investigated whether a heparin-mimicking polymer-based cell culture substrate with a negatively charged surface could be further utilized as an in vitro platform for the regulation of cell adhesion-mediated myoblast functions, aiming to recapitulate in vivo muscle atrophy-like phenotypes. Additionally, through transcriptome profiling of cells cultured on various matrices, we further elucidated the underlying molecular mechanisms inducing the inhibition of mononucleated myoblast fusion into multinucleated myotubes.

2. Results

2.1. Characterization of the Physical Properties of GelMA-Based Hydrogels

Previously, numerous studies have demonstrated the potential roles of heparin-mimicking polymers in various biological applications in tissue engineering and regenerative medicine approaches [29,30]. In order to synthesize covalently photo-crosslinkable gelatin-based hydrogels, gelatin was initially methacrylated (GelMA), and GelMA was further co-polymerized with the synthetic heparin mimic, PSS, hereafter termed GelMA-PSS (Figure 1A,B). As shown in Figure 1C, the ^{1}H nuclear magnetic resonance (^{1}H-NMR) spectra revealed the characteristics peaks corresponding to vinyl protons observed at 5.978 and 5.845 ppm, indicative of the successful binding of methacrylate groups into gelatin. Furthermore, the ^{1}H-NMR spectra at 1.783 ppm peak indicated that PSS was successfully co-polymerized with GelMA. To further confirm the synthesis of GelMA-PSS hydrogel, Fourier transform infrared spectroscopy (FTIR) analysis was performed on a pre-polymerized solution containing GelMA and PSS (Figure 1D). The transmittance of the gelatin solution alone showed an NH amine group and a C–O stretch at 1242 and 1148 cm^{-1}, respectively. The percent transmittance of GelMA-specific peaks was reduced when the PSS solution volume was increased. Additionally, the FTIR spectrum of GelMA-PSS revealed an $-SO_3$

stretch at 1186, 1043, and 1014 cm^{-1}, exhibiting the successful co-polymerization of PSS into GelMA.

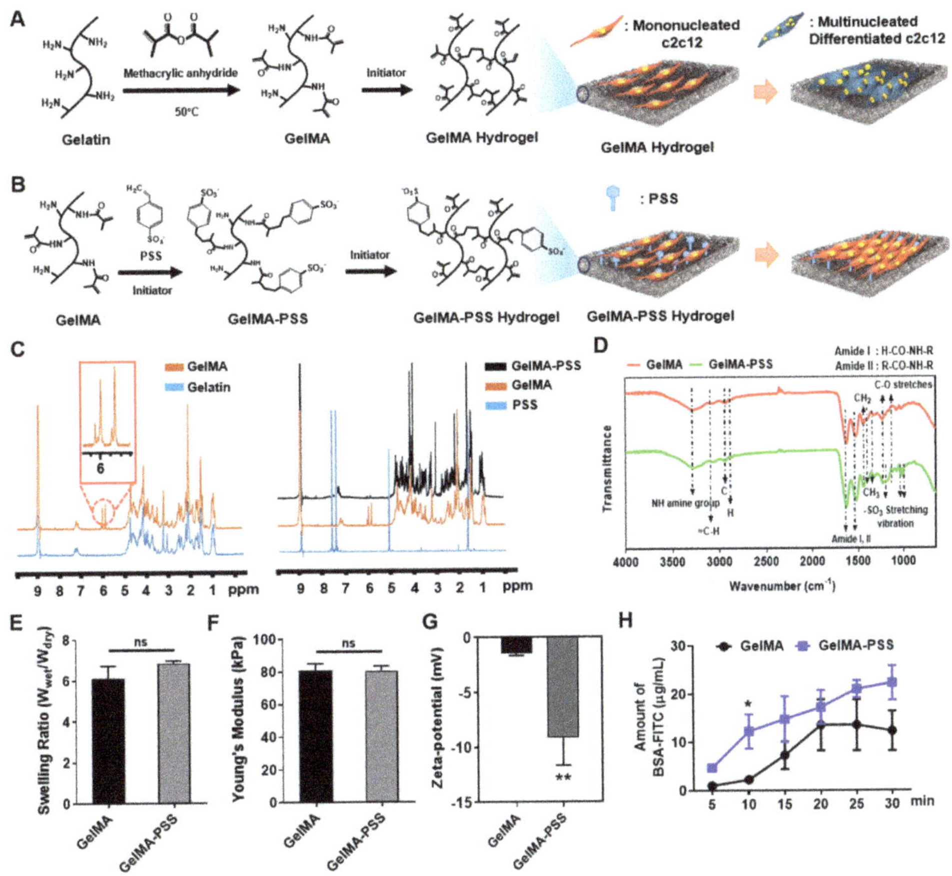

Figure 1. Schematic illustration and characterization of GelMA and GelMA-PSS hydrogels. (**A**,**B**) Synthesis scheme for GelMA and GelMA-PSS hydrogels, respectively. (**C**) ^{1}H-Nuclear magnetic resonance (NMR) spectra of synthetic polymer. (**D**) Fourier transform infrared spectroscopy (FTIR) analysis of GelMA and GelMA-PSS. (**E**–**G**) Characterization of various physical properties including swelling ratio, Young's modulus, and zeta-potential of GelMA and GelMA-PSS. (**H**) Quantification of BSA absorption using FITC-BSA on GelMA and GelMA-PSS hydrogels. Data are presented as the mean ± SD. * $p < 0.05$; ** $p < 0.01$; ns, not significant.

Next, the GelMA and GelMA-PSS hydrogels were polymerized using redox initiators, and various mechanical properties, such as swelling ratios, Young's modulus, and zeta-potential, were characterized. As shown in Figure 1E,F, the swelling ratios and Young's modulus of both GelMA and GelMA-PSS were not significantly different, indicating that the incorporation of PSS into GelMA does not affect the mechanical properties of GelMA hydrogels. We further characterized the PSS-mediated negative charge distribution and its contribution to protein adsorption on the hydrogel surface via zeta-potential analysis and a fluorescein isothiocyanate (FITC)-conjugated bovine serum albumin (FITC-BSA) binding assay (Figure 1G,H). Zeta-potential analysis demonstrated that the anionic charge distribution of the GelMA hydrogel was −1.05 to −1.67 mV, whereas that of the GelMA-PSS hydrogel was −5.34 to −13.8 mV, suggesting that integration of the sulfonate functional groups of PSS into the GelMA hydrogel results in an additive anionic charge on its hydrogel surface. Additionally, the FITC-BSA binding assay revealed that due to the higher degree of negative charges on GelMA-PSS hydrogels, theses could adsorb a higher amount of FITC-BSA when compared to GelMA hydrogels at all time points.

2.2. Effect of Heparin-Mimicking Matrix-Based Cues on the Initial Cell Adhesion and Proliferation of Myoblasts

To assess whether heparin-mimicking matrix-based cues could support cell adhesion and proliferation of murine myoblasts, c2c12 cells were seeded onto tissue culture plates (TCPs), GelMA, and GelMA-PSS hydrogels. As shown in Figure 2, our initial observation found that cells were able to adhere to all cell culture substrates and showed highly similar spindle cell morphology, evident by phase-contrast microscopy. Further, adhered c2c12 cells cultured on all groups reached confluence within 72 h, confirming that the anionic surface introduced by PSS did not have any detrimental effects on cell adhesion, spreading, and growth, which is in accordance with our previous report [30].

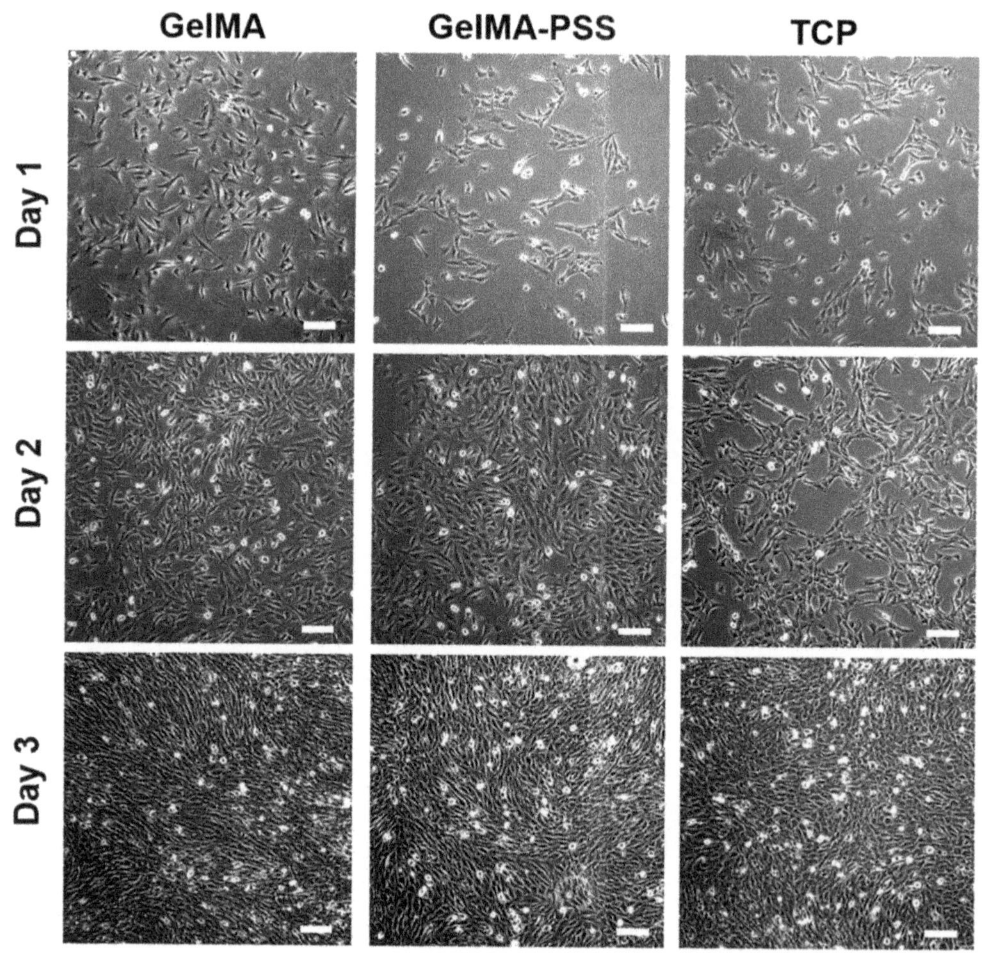

Figure 2. Phase-contrast images of cells adhered onto GelMA (**left column**), GelMA-PSS (**middle column**), and tissue culture plate (TCP) (**right column**) as a function of cell culture time up to day 3. Scale bar = 200 μm.

2.3. Inhibition of Myogenesis by Heparin-Mimicking Matrix-Based Cues

Given the importance of matrix-based cues, including matrix stiffness, protein immobilization, and topographical features of cell culture substrates, with regard to cell adhesion, differentiation, as well as myoblast fusion [17,24,31], we examined whether heparin-mimicking matrix-based cues presented on the surface of GelMA-PSS hydrogels could affect the myogenic differentiation of c2c12 cells. Upon full confluence, myogenic differentiation was induced for up to 3 days using TCP, GelMA, and GelMA-PSS as cell culture substrates. Thereafter, myogenic potential was evaluated via qualitative polymerase chain reaction (qPCR), Western blot, and immunofluorescence staining. As shown in Figure 3A, the gene expression profiles of various myogenic markers revealed that cells cultured on both TCP and GelMA hydrogels could undergo robust myogenic differentiation, confirmed by upregulation of the late muscle-specific gene expression of *MyoG* and *MyHC* in a culture time-dependent manner. In contrast, cells cultured on GelMA-PSS hydrogels showed significantly decreased late myogenic marker expression when compared to the TCP and GelMA hydrogel groups. To further corroborate gene expression profiles, we evaluated the protein expression in cells cultured on TCP, GelMA, and GelMA-PSS hydrogels via Western blot analyses. Similarly to the qPCR results, cells cultured on TCP and GelMA hydrogels exhibited elevated protein levels of myogenic markers, such as MyoD, desmin, and MyHC. In contrast, cells cultured on GelMA-PSS produced significantly lower levels of the muscle-specific proteins desmin and MyHC (Figure 3B).

Next, to determine the degree of myogenic differentiation as a function of culture time, c2c12 cells were cultured on TCP, GelMA, and GelMA-PSS hydrogels and induced to undergo myogenesis for 3 days. Their differentiation potential at each time point was immunohistologically assessed. At the early time point (day 1), cells cultured on both TCP and GelMA hydrogels showed that a few cells were stained positive for desmin, a muscle-specific intermediate filament, and MF20, a sarcomeric myosin heavy chain. Although these cells were initially mononucleated myoblasts on day 1, they were successfully fused to form multinucleated myotubes during days 2–3 (Figure 3C, left and right panels). On the other hand, cells cultured with heparin-mimicking matrix-based GelMA-PSS hydrogels were unable to form MF20-positive myoblasts on day 1 and underwent limited fusion of mononucleated myoblasts into multinucleated myotubes (Figure 3C, middle panel).

To further validate results from the immunofluorescence staining, we quantified the extent of myogenic differentiation by calculating the differentiation index, determined by the ratio of MF20-positive cells to the total number of cells, as well as by the MF20-positive cellular area. Analyses for differentiation index and MF20-positive cellular area indicated that cells cultured on GelMA-PSS hydrogels had a significantly lower number of MF20-positive myoblasts with undoubtedly smaller myotube sizes (Figure 3D,E), when compared to cells cultured on TCP and GelMA hydrogels. In addition to the differentiation index, we also calculated the fusion index, defined as the ratio of myotubes having either 1, 2–6, 7–14, or >15 nuclei to the total number of MF20-positive cells. Fusion index calculation surprisingly revealed that cells cultured with heparin-mimicking matrix-based cues failed to facilitate fusion of mononucleated myoblasts into matured multinucleated myotubes (>15 nuclei, approximately 3%) until day 3 (Figure 3F). In contrast, during the 3 days of myogenic differentiation, cells cultured on TCP and GelMA hydrogels were able to fuse, forming fully matured multinucleated myotubes, and their fusion indices (>15 nuclei) dramatically increased as a function of culture time (~80% at day 3). No significant differences in MF20-positive area, differentiation, and fusion indices were observed between TCP and GelMA hydrogels. Taken together, these results suggest that the incorporation of PSS inhibits both the gene and protein expression of muscle-specific markers. More importantly, PSS culture restricts the fusion of mononucleated myoblasts into multinucleated myotubes through the cell-to-heparin-mimicking matrix interaction.

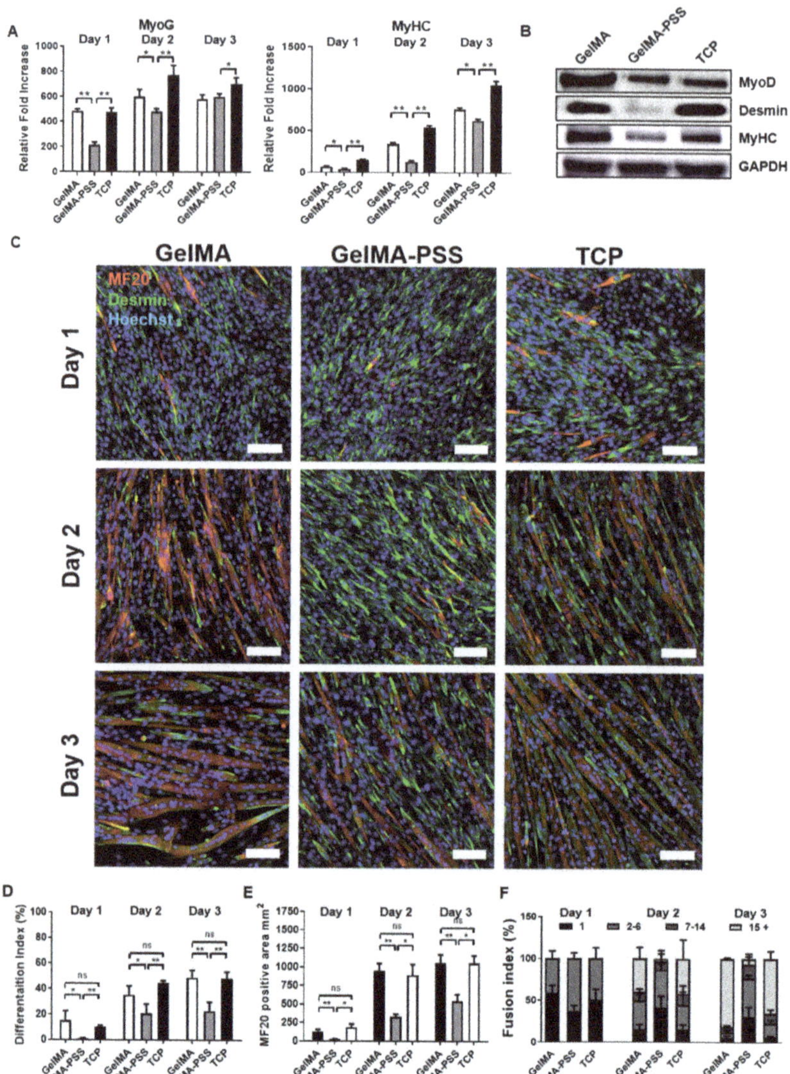

Figure 3. Myogenic potential of cells cultured on GelMA, GelMA-PSS and TCP. (**A**) Gene expression profiles of cells cultured on GelMA, GelMA-PSS, and TCP for 1–3 days in a myogenic induction medium. (**B**) Myogenic protein expression of cells cultured on GelMA, GelMA-PSS, and TCP for 2 days in a myogenic induction medium, as evaluated by Western blot. (**C**) Immunofluorescence staining of MF20 (red) and desmin (green) and nuclei (blue) in cells cultured in a myogenic induction medium on GelMA, GelMA-PSS, and TCP for 1–3 days. Scale bar = 100 μm. (**D–F**) Estimated differentiation and fusion indices of cells cultured in a myogenic induction medium on GelMA, GelMA-PSS, and TCP for 1–3 days (for GelMA, n = 203, 184, 286 at day 1,2,3; for GelMA-PSS, n = 405, 267, 273 at day 1,2,3; for TCP, n = 281, 523, 341, respectively). Data are presented as the mean ± SD, * $p < 0.05$; ** $p < 0.01$; ns, not significant.

2.4. Transcriptional Profiling Reveals the Suppression of Myogenesis-Related Genes by Heparin-Mimicking Polymers

To elucidate the molecular mechanisms underlying the inhibition of myogenesis by heparin-mimicking matrix-based cues, we performed RNA sequencing (RNA-seq) analysis of cells cultured in myogenic induction medium on TCP, GelMA, and GelMA-PSS hydrogels for 2 days in order to identify differentially expressed genes (DEGs) with an expression change greater than 1.5-fold. After obtaining statistically significance DEGs ($^{*}p < 0.05$) with an activation z-score of >2 as a cut-off, we performed pathway analysis of all differentially expressed transcripts using Ingenuity Pathway Analysis (IPA, www.ingenuity.com). As shown in Figure 4A, the canonical mTOR pathway, as an important signaling pathway in skeletal muscle development and myogenic differentiation [32], was downregulated in cells cultured with heparin-mimicking matrix-based cues. Furthermore, we performed comparative function and network analyses, observing a list of diseases and cellular functions related to myogenesis, which indicated significant phenotypic changes in cells cultured on GelMA-PSS hydrogels, when compared to those cultured on both GelMA and TCP (Table 1). The top diseases and biofunctions identified by IPA revealed that cells cultured on GelMA-PSS hydrogels exhibited inferior myogenic phenotypes, including down-regulation in formation, quality, quantity, and contractility of the muscle itself or myoblasts, as well as the inhibition of cell fusion for the formation of multinucleated functional myotubes through cell–cell contact, assembly, and formation of intercellular junctions, and cell aggregation.

Table 1. Top diseases and cellular functions significantly changed in cells cultured on GelMA-PSS hydrogels, when compared to those cultured on TCP and GelMA. Diseases and biofunctions are sorted by the activation of z-score (z-score of > 2).

Diseases or Biofunctions Annotation	*p*-Value	Activation z-Score	# Molecules
Motor dysfunction or movement disorder	0.00191	3.848	91
Apoptosis	0.00686	2.865	255
Formation of muscle cells	0.000459	−3.162	27
Quantity of muscle	0.000396	−3.148	28
Cell–cell contact	1.32 × 10−10	−3.123	105
Quantity of muscle cells	0.00239	−3.119	24
Assembly of intercellular junctions	0.0000109	−3.05	44
Formation of intercellular junctions	0.00000317	−2.936	47
Aggregation of cells	0.00116	−2.84	38
Formation of myofibrils	0.00716	−2.646	11
Quantity of striated muscle	0.00242	−2.578	15
Contractility of muscle	0.000398	−2.324	36
Microtubule dynamics	0.001	−2.098	138

Based on DEG functional annotations, gene ontology (GO) functional enrichment analyses, including biological process (BP), cellular component (CC), and molecular function (MF), were carried out to further understand the signaling pathways modulating myoblast differentiation. The top five upregulated and downregulated BP, CC, and MF terms in cells cultured with heparin-mimicking matrix-based cues when compared to those cultured on TCP and GelMA hydrogels are presented in Figure 4B,C. Upregulated GO terms were associated with the cell cycle and division-associated enrichment. On the other hand, downregulated GO terms of cells cultured with heparin-mimicking matrix-based cues when compared to cells cultured on TCP and GelMA hydrogels revealed a downregulation of insulin, Wnt and mTOR signaling pathways, cell–cell adherens junction, myofibril formation, and cell–matrix interaction (vinculin binding), known to be critical in myogenic differentiation as well as in the maturation of myoblasts and skMSCs [32–34]. Kyoto Encyclopedia of Genes and Genomes (KEGG) pathway analysis revealed upregulated oxidative phosphorylation (OXPHOS) in cells cultured with heparin-mimicking matrix-based cues along with the upregulation of degenerative disease-associated pathways, which is in accordance with a previous study suggesting that OXPHOS is required at early myogenic commitment, rather than the maturation stage [35]. Similarly to the GO functional enrichment analysis, KEGG pathway analysis revealed that the insulin, Wnt, and mTOR signaling pathways as well as focal adhesions were significantly downregulated in cells cultured with heparin-mimicking matrix-based cues as opposed to those cultured on TCP and GelMA hydrogels.

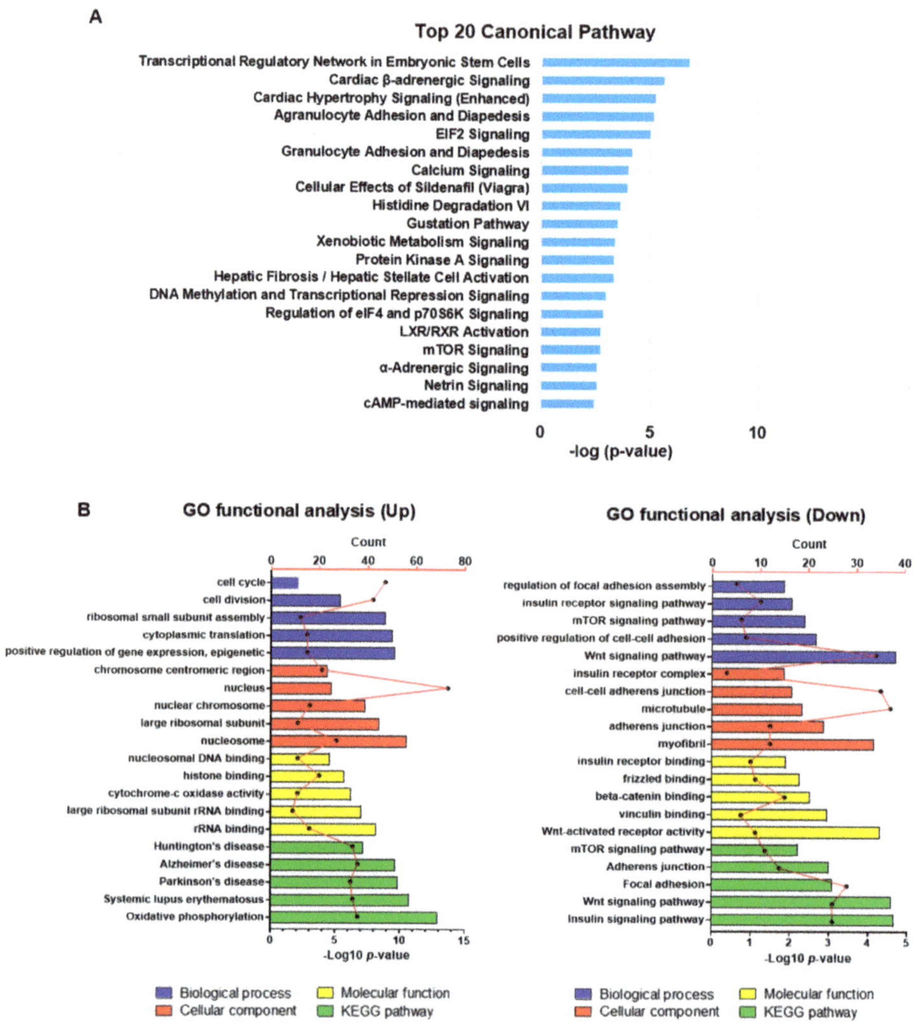

Figure 4. Transcriptome analysis. (**A**) Top 20 canonical pathways commonly enriched in cells cultured on GelMA-PSS. (**B**) Top 5 significantly enriched GO terms functional analysis data based on differentially expressed genes (DEGs) in cells cultured on GelMA-PSS compared to GelMA and TCP. Biological process (Blue), cellular component (red), molecular function (yellow), and KEGG pathways (green).

2.5. Validation of DEGs Using qPCR and Western Blot Analyses

Having demonstrated the inhibition of myogenic differentiation in cells grown on GelMA-PSS hydrogels, we next sought to validate the findings from the RNA-seq analysis, which indicated that myogenesis was inhibited through the suppression of insulin/PI3K/mTOR and Wnt signaling pathways, as well as through the modulation of focal adhesions. As shown in Figure 5, the cells cultured with heparin-mimicking matrix-based cues exhibited significantly lower expression of various genes associated with the insulin/PI3K/mTOR and Wnt signaling pathways, such as *IGF1*, *RIK3R1*, *mTOR*, Cav3, *LRP5*, and *Axin2*. Furthermore, we found that *KLF4* and its transcriptional target, nephronectin (*NPNT*), were downregulated in cell cultured on GelMA-PSS hydrogels when compared to those grown on TCP and GelMA hydrogels. Western blot analyses (Figure 5B,C) revealed that cells cultured with heparin-mimicking matrix-based cues showed significantly lower protein expression of p-mTOR, Cav3, as well as active and total β-catenin. In an accordance with a previous report demonstrating the important role of focal adhesion kinase (FAK) signaling in myoblast fusion, we observed that the levels of total and phosphorylated FAK, associated with the regulation of integrin-mediated cell adhesion and intracellular signal transduction in the PI3K/AKT/mTOR pathway [36], were significantly suppressed in cells cultured with heparin-mimicking matrix-based cues.

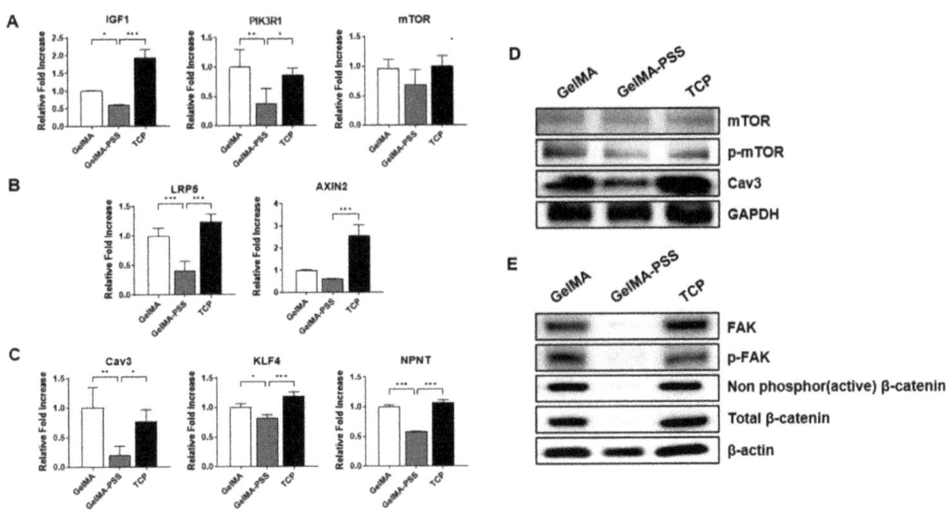

Figure 5. Validation of RNA-seq data via qPCR and Western blot analyses. Gene and protein expression profiles of cells cultured in a myogenic induction medium on GelMA, GelMA-PSS, and TCP for 2 days; (**A**) insulin/PI3K/mTOR signaling pathway, (**B**) Wnt signaling pathway, (**C**) myotube fusion-related markers. Data are presented as the mean ± SD, * $p < 0.05$; ** $p < 0.01$; *** $p < 0.001$; ns, not significant. (**D**) mTOR phosphorylation and Cav protein expression, (**E**) phosphorylated FAK and active β-catenin protein expression.

2.6. Validation of the In Vitro Muscle Injury Screening Platform in Comparison to an In Vivo Muscle Injury Model

To validate the applicability of proof-of-concept that our in vitro heparin-mimicking polymer-based cell culture platform could successfully recapitulate an in vivo muscle injury model, we induced muscle injury in 8-week-old C57BL/6J mice using barium chloride. Tibialis anterior (TA) muscles treated with barium chloride were harvested at different time points up to 14 days post-injury to evaluate whether muscle injury-activated signaling pathways would resemble the in vitro phenotypes of cells cultured with heparin-mimicking matrix-based cues. Non-injured TA muscles served as a control. As shown in Figure 6A, hematoxylin and eosin staining revealed that injection of barium chloride into TA muscles initially caused necrosis of damaged skeletal muscle fibers, as evident by center-located nuclei in degenerating muscle fibers, and a decrease in fiber diameters. Furthermore, at day 5 post-injury, degenerating muscle fibers exhibited a significant muscle atrophy and excessive connective tissue formation within damaged muscle fibers, and finally, damaged muscle fibers were fully regenerated after 14 days.

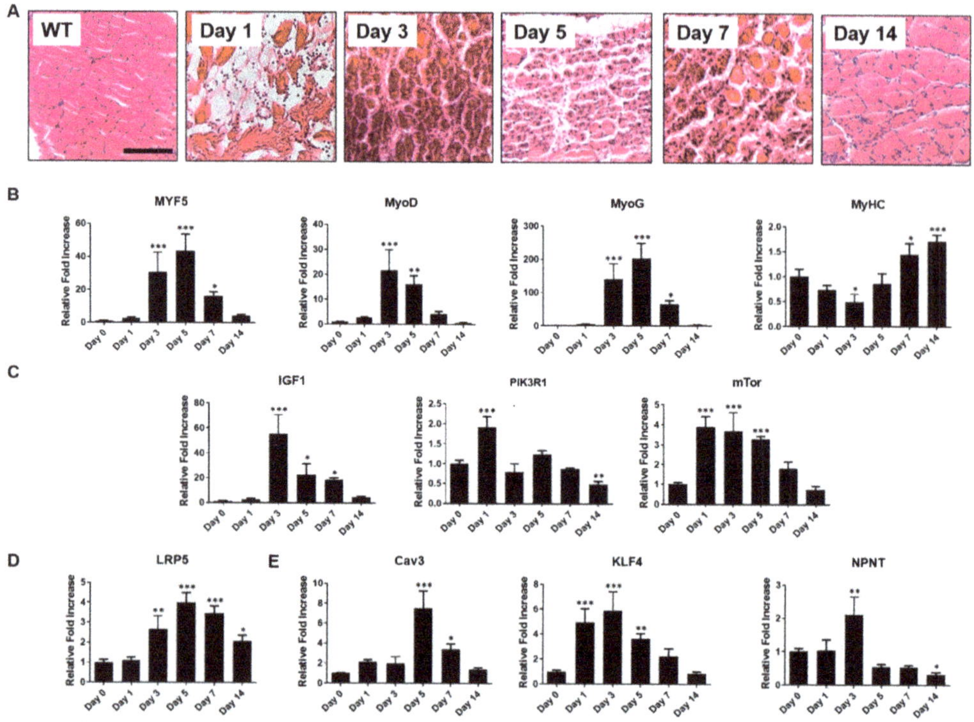

Figure 6. Barium chloride ($BaCl_2$)-induced muscle injury model. (**A**) Histological evaluation of barium chloride ($BaCl_2$)-injured TA muscles using H&E staining. Scale bar = 100 μm. Gene expression profiles of (**B**) myogenic markers during muscle regeneration, (**C**) insulin/PI3K/mTOR signaling pathways, (**D**) Wnt signaling pathway, (**E**) myotube fusion-related markers. Data are presented as the mean ± SD, * $p < 0.05$; ** $p < 0.01$; *** $p < 0.001$ of six biological replicates for each time point.

Since damaged muscle fibers exhibited typical muscle atrophy phenotypes upon barium chloride-induced muscle injury, we determined the expression of various muscle-specific genes, including those encoding early myogenic markers such as Myf5, MyoD, and MyoG, all of which were significantly elevated during the degeneration processes. In contrast, the late myogenic marker MyHC was considerably upregulated at the later stage of regeneration (Figure 6B). For a comprehensive comparison with cells cultured on GelMA-PSS, we also evaluated genes involved in the fusion and maturation of myoblasts into multinucleated myotubes, namely insulin/PI3K/mTOR and Wnt signaling pathway-related genes. As shown in Figure 6C,D, DEGs identified and confirmed by RNA-seq and qPCR analyses, were upregulated during muscle regeneration. Similarly to previous in vitro results, *KLF4* and *NPNT* genes were upregulated during muscle regeneration, whereas cells cultured on our in vitro platform suppressed their expression, thus mirroring muscle atrophy phenotypes rather than muscle regeneration.

Taken together, the comparison to the in vivo barium chloride-induced muscle injury model supported the utility of heparin-mimicking matrix-based cues, suggesting that the in vitro platform could simulate muscle atrophy phenotypes by stimulating various pathological alterations of signaling pathways, characteristic of the muscle degeneration process.

3. Discussion

There has been increasing interest in the development of disease-specific in vitro organ-on-a-chip platforms for the accurate assessment of small molecule drugs, including toxicity tests and pharmacokinetics, by recapitulating cell–cell and cell–matrix interactions within tissue-specific microenvironments [11,37,38]. We previously developed a heparin-mimicking polymer-based cell culture substrate to support the long-term expansion and self-renewal of human pluripotent stem cells (hPSCs) while maintaining their pluripotency [30]. In this study, by harnessing a heparin-mimicking polymer with a negatively charged surface due to the presence of sulfonate groups, we engineered an in vitro cell culture platform to facilitate the cell–matrix interaction-mediated inhibition of myoblast function. Thereafter, we sought to recapitulate in vivo muscle atrophy-like phenotypes in cells cultured on our platform.

Heparin and HSPGs on the cellular surface of myoblasts have previously been shown to bind bFGF with high affinity, promoting the self-renewal of skMSCs [39]. Recently, we harnessed a bFGF-immobilized matrix as a synthetic cell culture substrate to support the cell adhesion and proliferation of skMSCs while maintaining their self-renewal capacity during in vitro expansion [24]. Similarly, we elucidated the roles of various physicochemical cues of a heparin-mimicking anionic polymer PSS, including functional groups, matrix stiffness, surface roughness, and hydrophilicity, with regard to the long-term self-renewal of hPSCs. The presence of negatively charged sulfonate groups in PSS enabled the adsorption of various ECM proteins and bFGF [30]. In our current study, despite their similar physical properties, including swelling ability and matrix stiffness, GelMA-PSS and GelMA hydrogels elicited different cellular responses. These discrepancies in the inhibition of myoblast fusion and terminal differentiation into multinucleated myotubes could be attributed to the presence of negatively charged sulfonate groups in heparin-mimicking polymers (GelMA-PSS hydrogels). Therefore, it is possible that the bioactivity of bFGF bound to sulfonate groups on the GelMA-PSS hydrogel surface through electrostatic interactions can be stabilized without undergoing denaturation and enzymatic degradation, as previously reported [29,40,41], resulting in suppression of the myogenic differentiation of adhered cells. On the other hand, previous studies have demonstrated that the soluble heparin in a culture medium could sequester the bFGF from microenvironment surrounding myoblasts, leading to the induction of terminal differentiation into myocytes [42,43]. Similarly to the pro-myogenic effect of soluble heparin on the myoblast function, exogenous supplement of soluble heparin-mimicking polymers have been shown to downregulate the mitogen activated extracellular regulated signaling kinase (MAPK/ERK) pathway by withdrawing bFGF away from FGFR in a concentration-dependent manner [28,44]. However, it is still unclear whether heparin-mimicking polymers in our study would preferentially regulate the modulation of bFGF stability over sequestering the bFGF.

Another possibility is the presence of negatively charged sulfonate groups in heparin-mimicking polymers, which could sophisticatedly modulate important signaling pathways involved in muscle homeostasis through cell–matrix interactions [36]. Although our initial findings indicated that there were no significant differences in the initial cell adhesion and proliferation of cells cultured on various cell culture substrates such as GelMA, GelMA-PSS, and TCP (Figure 2), only heparin-mimicking matrix-based cues inhibited the fusion of mononucleated myoblasts into multinucleated myotubes, resulting in the suppression of myogenic differentiation (Figure 3). To distinguish the molecular signatures of cells cultured on heparin-mimicking matrix-based cues, RNA-seq analysis and qPCR validation revealed that the insulin/PI3K/mTOR and Wnt signaling pathways were greatly downregulated (Figures 4 and 5A,B,E), both being tightly associated with skeletal muscle development and myogenic differentiation [45,46]. Furthermore, similarities between our in vitro platform and the in vivo barium chloride-induced muscle injury model were validated, including the suppression of insulin/PI3K/mTOR and Wnt signaling (Figure 6).

Insulin/PI3K/mTOR signaling has been extensively investigated as a key pathway in skeletal muscle development and regeneration [45,47]. Risson et al. reported that muscle-specific mTOR knockout mice exhibited significant myopathic phenotypes and decreased dystrophin expression, indicative of the crucial role of mTOR in the maintenance of muscle function and metabolism [48]. Cong et al. recently reported that the IGF–Akt–mTOR signaling pathway activated via insulin receptor substrate 1 (IRS1), known as a critical regulator for IGF signaling, could promote myogenic differentiation of myoblasts as well as skeletal muscle regeneration [49]. Furthermore, the Wnt signaling pathway is known to regulate various cellular functions, including cell proliferation, polarity, and cell fate determination, through membrane proteins LRP5/6 and Frizzled, which orchestrated together as receptors for Wnt ligands in the canonical Wnt pathway, in turn regulating myogenic differentiation and skeletal muscle homeostasis [33,50].

Over the past decades, studies have shown that the formation of multinucleated myotubes through mononucleated myoblast fusion is an event during myogenic differentiation and skeletal muscle formation [51]. Therefore, we assessed myotube fusion-associated markers, including *Cav3, KLF4,* and *NPNT*, as well as the levels of cell adhesion proteins such as focal adhesion kinase (FAK) and its phosphorylated form (p-FAK) (Figures 4 and 5C–E). Fluck et al. previously reported that FAK was highly expressed during the in vitro fusion of myoblasts into myotubes, and its expression was increased in newly formed skeletal muscles [52]. Similarly, Quach et al. and other research groups have confirmed the crucial role of FAK and its subsequent regulation of Cav3 and β1D integrin during the fusion and maturation of myoblasts into functional myotubes [36,53]. In particular, FAK was proposed as a key regulator of myoblast fusion, whose inhibition lead to the suppression of Cav3 and β1D integrin, resulting in impaired in vivo skeletal muscle regeneration. Sunadome et al. demonstrated that overexpression of KLF2/4 in MEK inhibitor-treated myoblasts could restore their ability to form fused myotubes and identified NPNT as a key factor for myoblast fusion [54]. In agreement with these previous studies, we observed that FAK, Cav3, and NPNT were significantly decreased at both gene and protein levels, confirmed by in vitro as well as in vivo results. Cells cultured on our GelMA-PSS platform resembled the myogenically restricted phonotypes of the barium chloride-induced muscle atrophy phenotypes in vivo. Therefore, the current findings suggest that cells on GelMA-PSS hydrogels could interact with heparin-mimicking moieties and thus exhibit decreased expression of pFAK and Cav3, leading to impaired myoblast fusion.

In conclusion, our proof-of-concept study confirmed that the GelMA-PSS hydrogel-based in vitro platform recapitulated in vivo muscle atrophy-like phenotypes, as confirmed by the downregulation of key signaling pathways in cells cultured on heparin-mimicking matrix. Our results suggest that heparin-mimicking matrix-based cues determine focal adhesions and cell–cell interactions of cells at the cell–matrix interface through subtle changes in modulation of protein adsorption and bFGF stabilization, as previously reported [30,41]. In turn, the fusion of myoblasts into matured myotubes and the formation of cell–cell/intercellular junctions were inhibited, recapitulating the initial muscle atrophy phenotypes observed during in vivo muscle injury-induced degeneration of skeletal muscles (Figure 4 and Table 1). Therefore, our in vitro platform may serve as a promising muscle injury model for drug screening and toxicity tests.

4. Materials and Methods

4.1. Synthesis of Methacrylated Gelatin

Methacrylated gelatin (GelMA) was synthesized according to a previously described method [55]. Briefly, 10 g of gelatin (cat# G1890, Sigma-Aldrich, St. Louis, MO, USA) was dissolved in 100 mL of phosphate-buffered saline (PBS) solution at 60 °C under constant stirring. Then, 8 mL of methacrylic anhydride (cat# 276685, Sigma-Aldrich, St. Louis, MO, USA) was added dropwise to the gelatin solution while stirring vigorously for 2 h. The reaction mixture was mixed with preheated PBS at 60 °C and kept for 15 min. The resulting solution was dialyzed in deionized water at 52 °C for 7 days using a dialysis membrane (MWCO, approximately 12–14 kDa, Spectrum Laboratories,

Gardena, CA, USA). Finally, the solution was filtered, lyophilized, and stored at −20 °C. After synthesis, the products were analyzed via ^{1}H-NMR by dissolving in deuterium oxide (D2O; Sigma-Aldrich, Saint Louis, MO, USA).

4.2. Preparation of GelMA-Based hydrogels

GelMA-based hydrogels were synthesized as follows: GelMA was dissolved in PBS at a final concentration of 15% w/v with or without a heparin-mimicking polymer PSS (cat# 94904, Sigma-Aldrich, St. Louis, MO, USA) at a final concentration of 1% w/v. Ammonium persulfate (APS, cat# 161-0700, Bioshop, Burlington, ON, Canada) and N,N,N′,N′-tetramethylethylenediamine (TEMED, cat# 161-0800, Bioshop, Burlington, ON, Canada) were added to this precursor solution at a final concentration of 0.4% w/v and 0.3% w/v, respectively. The reaction mixtures, including GelMA and GelMA-co-PSS, hereafter referred to as GelMA-PSS, were polymerized in Bio-Rad 1-mm-spacer glass plates at room temperature. Finally, polymerized hydrogels were punched into 24-well plates, sterilized with 70% ethanol, and washed with fresh PBS for 48 h prior to cell culture experiments. The rinsed hydrogels were incubated in growth medium (high glucose DMEM with 2 mM L-glutamine and 50 units/mL penicillin/streptomycin) containing 10% fetal bovine serum (Corning, Oneonta, NY, USA) overnight before seeding the cells.

4.3. Fourier Transform Infrared Spectroscopy (FTIR)

The FTIR spectra of gelatin (powder), GelMA (lyophilized powder), and GelMA-PSS (lyophilized powder) were analyzed using an FTIR spectrometer (Spectrum 100, Perkin Elmer) with a spectral width ranging from 500 to 4000 cm^{-1} at a spectral resolution of 4 cm^{-1}. IR peaks corresponding to the methacrylation of gelatin and copolymerization of PSS into GelMA were analyzed.

4.4. Swelling Ratio Measurement

The swelling ratio of each hydrogel construct was measured using the gravimetric method. Briefly, hydrogels were swollen in PBS for 24 h to reach equilibrium, and their wet weights were determined as an equilibrium-swollen state (W_{wet}). The weighed hydrogels were then lyophilized to measure dry weight (W_{dry}). The swelling ratios of each hydrogel were calculated according to the following equation:

$$\text{swelling ratio } (\%) = \frac{Wwet}{Wdry} \times 100\%$$

4.5. Characterization of Young's Modulus

To measure the Young's modulus of hydrogel constructs, hydrogels were equilibrated in PBS for 24 h, and their dimensions, such as height and diameter, were measured prior to the compression test. Compression tests were performed using a universal testing machine (EZ-Test EZ-SX, Shimadzu, Kyoto, Japan) equipped with a 500 N load cell, and samples were compressed up to 30% strain at a strain rate of 1 mm/min. The Young's modulus was calculated from the linear region of the stress–strain curve (0–10% strain).

4.6. Zeta-Potential Measurement

The synthesized GelMA and GelMA-PSS hydrogel samples were lyophilized, freeze-milled, and dispersed in ultrapure distilled water by sonication. Prepared samples were subjected to ζ-potential measurement at room temperature with a ZetasizerNano ZS (Malvern Instruments Ltd., Malvern, UK). All measurements were performed in triplicate.

4.7. Protein Adsorption Assay Using FITC-BSA

The amount of protein adsorbed onto the GelMA and GelMA-PSS hydrogels was quantified by measuring the amount of fluorescein isothiocyanate-conjugated bovine serum albumin (FITC-BSA, A23015, Thermo Fisher Scientific, Waltham, MA, USA) following a previously reported method [56]. Briefly, hydrogels were incubated with 100 μg/mL of FITC-BSA solution in a dark room for 5, 10, 15, 20, 25, and 30 min. Then, each supernatant solution was transferred to a flat-bottom 96-well plate to measure the fluorescence at 488 nm, and the amount of FITC-BSA in the supernatant was used to calculate the amount of FITC-BSA in hydrogels. The adsorption was calculated from a standard curve generated by a series of known concentrations of FITC-BSA.

4.8. *Cell Culture*

Prior to myogenic differentiation, undifferentiated murine myoblast c2c12 cells were seeded into the TCP, GelMA, and GelMA-PSS hydrogels at an initial seeding density of 5×10^3 cells/cm^2 in growth medium containing high-glucose Dulbecco's modified Eagle's medium (DMEM; Gibco-BRL, Waltham, MA, USA) supplemented with 10% *v/v* fetal bovine serum (FBS; Gibco-BRL, Waltham, MA, USA), 1% L-glutamine (200 mM, Gibco-BRL, Waltham, MA, USA), and 1% *v/v* penicillin–streptomycin (10,000 U/mL of penicillin and 10,000 g/mL of streptomycin, Gibco-BRL, Waltham, MA, USA) under a 5% CO_2 atmosphere at 37 °C. When cells reached 80% confluence, the growth medium was replaced with myogenic induction medium containing DMEM supplemented with 2% *v/v* horse serum (Gibco-BRL, Waltham, MA, USA) and 1% *v/v* penicillin–streptomycin (Gibco-BRL, Waltham, MA, USA) to induce myogenic differentiation.

4.9. *Immunocytochemistry*

Cells cultured on TCP, GelMA, and GelMA-PSS were fixed with 4% paraformaldehyde (PFA) for 10 min at room temperature, washed in PBS, blocked with 3% (*w/v*) bovine serum albumin in PBS, and permeabilized with 0.1% (*v/v*) Triton X-100 for 1 h at room temperature. Samples were incubated with primary antibodies (diluted in 1% BSA in PBS), including rabbit anti-Desmin (1:200, Abcam, Cambridge, UK) and mouse anti-MyHC (MF-20) (1:200, Developmental Studies Hybridoma Bank, Iowa City, IA, USA), overnight at 4 °C, washed 3 times in fresh PBS, and incubated with secondary antibodies, including anti-rabbit Alexa 488 (1:200, Thermo Fisher Scientific, Waltham, MA, USA) (diluted in 1% BSA in PBS), anti-mouse Alexa 546 (1:200, Thermo Fisher Scientific, Waltham, MA, USA), as well as with an anti-phalloidin (1:200, Thermo Fisher Scientific, Waltham, MA, USA) antibody for 1 h at room temperature. The nuclei were stained with Hoechst 33342 (2 mg/mL; Thermo Fisher Scientific, Waltham, MA, USA) for 10 min at room temperature. All fluorescence images were acquired by using an inverted microscope (Eclipse Ti-U, Nikon, Tokyo, Japan) at the Soonchunhyang Biomedical Research Core-Facility of the Korea Basic Science Institute (KBSI).

4.10. *Histochemistry*

Barium chloride ($BaCl_2$)-injured TA muscle samples were fixed with 4% paraformaldehyde (PFA) for 24 h, gradually dehydrated in ethanol, and embedded in paraffin. The paraffin-embedded samples were processed into 10 μm-thick sections to be used for staining. Briefly, paraffin-embedded sections were de-paraffinized with CitriSolv, rehydrated with a graded series of ethanol, and washed with fresh distilled water. For histopathological analysis, sections were stained with hematoxylin (Mayer's modified; Abcam, Cambridge, UK) and eosin (Sigma-Aldrich, St. Louis, MO, USA). All bright-field images were acquired using an inverted microscope (Eclipse Ti-U, Nikon, Tokyo, Japan) at the Soonchunhyang Biomedical Research Core-Facility of the KBSI.

4.11. *Image Analysis*

To quantify the degree of myogenic differentiation of murine myoblast c2c12 cells, myosin heavy chain-positive cellular area, differentiation, and fusion indices were calculated using NIH ImageJ software, as previously described [18,57]. The differentiation index was determined as the ratio of MF20-positive cells to the total number of cells, and the fusion index was calculated as the ratio of myotubes having either 1, 2–6, 7–14, or >15 nuclei to the total number of MF20-positive cells. At least 1000 nuclei per group were analyzed from three random fields of view at each time point.

4.12. *qPCR*

Total RNA was isolated using TRIzol reagent (Invitrogen, Carlsbad, CA, USA) and reverse transcribed using ReverTra Ace qPCR RT Master Mix with gDNA Remover (Toyobo, Osaka, Japan) according to the manufacturer's protocols. qPCR was carried out using the SYBR Green Real-time PCR Master Mix (Toyobo, Osaka, Japan) on a StepOnePlus Real-Time PCR System (Applied Biosystems, Foster City, CA, USA) at the Soonchunhyang Biomedical Research Core-Facility of the KBSI. The expression levels of target genes were normalized against glyceraldehyde-3-phosphate dehydrogenase (GAPDH), the ΔCt values were calculated as $Ct_{target} - Ct_{GAPDH}$, and relative fold inductions were calculated using the $2^{-\Delta\Delta Ct}$ method. The primer sequences used in this study are presented in Table 2.

Table 2. List of primers used for quantitative PCR.

Gene	Primer Sequence (5′ to 3′)
B2M	F-ACCGGCCTGTATGCTATCCAG R-AATGTGAGGCGGGTGGAACTG
Myf5	F-CTGCTGTTCTTTCGGGACCA R-TATTACAGCCTGCCGGGACA
MyoG	F-CCTACAGACGCCCACAATC R-CCCAGGCTGACAGACAATC
MyHC	F-ACGCCATCAGGCTCAAGAAGAAGA R-TGAGTGTCCTTGAGGATGCCTTGT
MyoD	F-CCGCCTGAGCAAAGTGAATG R-GCGGTCCAGGTGCGTAGAA
Klf4	F-GCCAACTACCCTCCTTTCCTG R-TCTTTGGCTTGGGCTCCTC
Npnt	F-GGCCAAACAAGTGCAAATGTC R-GGTGGAAGGACTCATCTTGGTT
Ccl5	F-GCTGCTTTGCCTACCTCTCC R-TCGAGTGACAAACACGACTGC
Cxcl1	F-CTGGGATTCACCTCAAGAACATC R-CAGGGTCAAGGCAAGCCTC
Igf1	F-TCTCACTGAAGCCAGCTCTCT R-CAGGCCCAGAAGCATGACA
Pik3r1	F-ACACCACGGTTTGGACTATGG R-GGCTACAGTAGTGGGCTTGG
mTOR	F-ACCGGCACACATTTGAAGAAG R-CTCGTTGAGGATCAGCAAGG
Cav3	F-GGATCTGGAAGCTCGGATCAT R-TCCGCAATCACGTCTTCAAAAT
Lrp5	F-GAAAGCACAATGGGTCCTCCA R-CTGACGCCTGTTCCACTTCT
Axin2	F-TGACTCTCCTTCCAGATCCCA R-TGCCCACACTAGGCTGACA

4.13. *RNA-Seq and Data Analysis*

Total RNA was isolated using TRIzol reagent (Invitrogen, Carlsbad, CA, USA), and samples were sent to Macrogen (Macrogen Inc., Seoul, Korea) for library preparation and sequencing. Library preparation was performed for transcriptome sequencing using the TruSeq RNA Sample Prep Kit v2 (Illumina, San Diego, CA, USA) following the manufacturer's instructions. Each library was sequenced using the HiSeq 2500 platform (Illumina, San Diego, CA, USA), and over 25 million 101 bp strand-specific paired-end reads were obtained per sample. DEGs were determined using fragments per kilobase of transcript per million mapped reads (FPKM) values with a change greater than 2-fold and $p < 0.05$ (GelMa-PSS vs. TCP/GelMA). GO analyses of biological function, cellular component, molecular function, and KEGG pathway were carried out using the bioinformatics resource DAVID tool, which provides relevant GO terms associated with DEGs (https://david.ncifcrf.gov/ (accessed on 2 March 2021)). Additionally, canonical pathway and functional network analyses were performed using IPA software (version 42012434, Qiagen, Hilden, Germany).

4.14. Western Blot

Cells were lysed in protein extraction solution (RIPA, Elpis Biotech, Daejeon, Korea) with protease and phosphatase inhibitors (Sigma-Aldrich, Saint Louis, MO, USA). Samples were centrifuged at 15,000 rcf and at 4 °C. The protein concentration was determined via Bradford assay. Gel electrophoresis was performed using 6% *w/v* or 12% *w/v* polyacrylamide gels at 100 V, based on the molecular weight of proteins. Proteins were transferred to a polyvinylidene fluoride (PVDF) blotting membrane (GE Healthcare, Chicago, IL, USA) at 350 mA, blocked with 5% skim milk in 1X Tris-buffered saline containing Tween-20 (TBS-T) (pH 7.5), and probed with primary antibodies against GAPDH (2118L, Cell Signaling Technology, Danvers, MA, USA), MyoD (sc-377460, Santa Cruz, Dallas, TX, USA), Desmin (ab8592, Abcam, Cambridge, UK), mTOR (2972S, Cell Signaling Technology, Danvers, MA, USA), p-mTOR (2971S, Cell Signaling Technology, Danvers, MA, USA), Caveolin-3 (ab2912, Abcam, Cambridge, UK), β-catenin (9562, Cell Signaling Technology, Danvers, MA, USA), non-phosphorylated (active) β-catenin (8814, Cell Signaling Technology, Danvers, MA, USA), phospho-FAK (Tyr397) (cat# 700255, Thermo Fisher Scientific, Waltham, MA, USA), FAK (sc-1688, Santa Cruz, Dallas, TX, USA), and β-actin (A5441, Sigma Aldrich, Waltham, MA, USA) diluted at 1:1000, overnight at 4 °C with gentle agitation. Membranes were incubated with horseradish peroxidase-conjugated secondary antibodies (Bio-Rad, Hercules, CA, USA) (1:2000 dilution) for 1 h at room temperature. Finally, immunodetection was performed using the ECL Prime Western Blotting Detection Reagent (GE Healthcare, Chicago, IL, USA) on an Amersham Imager 600 (GE Healthcare, Chicago, IL, USA) at the Soonchunhyang Biomedical Research Core-Facility of the KBSI.

4.15. Barium Chloride ($BaCl_2$)-Induced Muscle Injury Model

All animal experimental procedures were performed in accordance with the protocols for animal handling and ethical standards approved by Institutional Animal Care and Use Committee (IACUC) at the Soonchunhyang University (protocol number: SCH17-0033). To induce muscle injury, 8-week-old C57BL/6J mice were anesthetized with 1% isoflurane inhalation, and the TA muscle was injured via intramuscular injection of 30 μL 1.2% *w/v* barium chloride ($BaCl_2$, cat# 449644, Sigma Aldrich, St. Louis, MO, USA). At 1, 3, 5, 7, and 14 days post-injury, the TA muscles were harvested for further histochemistry and qPCR analyses.

4.16. Statistical Analysis

All values are presented as the mean ± standard deviation of three biological replicates for each group, and statistical significance was assessed via one-way analysis of variance (ANOVA) with Tukey's multiple comparison test using GraphPad Prism 9.0 (San Diego, CA, USA) software. A p value less than 0.05 was considered statistically significant (* $p < 0.05$; ** $p < 0.01$; *** $p < 0.001$).

Author Contributions: Conceptualization: H.K., J.H.J., M.F., J.K.Y., and Y.H.; investigation, formal analysis: H.K., J.H.J., M.F., H.-S.L., D.Y.K., S.S.H., J.A., Y.K., N.T.P., N.S., N.S.-Y.H., S.R., J.K.Y., and Y.H.; methodology: H.K., J.H.J., M.F., N.S., N.S.-Y.H., S.R., J.K.Y., and Y.H.; writing—original draft preparation and writing—review and editing: H.K., J.H.J., M.F., J.K.Y., and Y.H.; funding acquisition: J.K.Y. and Y.H. All authors have read and agreed to the published version of the manuscript.

Funding: This research was supported by a Global Research Development Program grant (2016K1A4-A3914725), a research grant (2020R1A2C1013652), Brain Pool Program through the National Research Foundation of Korea (NRF) funded by the Ministry of Science and ICT (2019H1D3A2A02102074), and a research grant (HI17C1193) from the Korea Health Industry Development Institute (KHIDI).

Data Availability Statement: The RNA sequencing datasets generated for this study can be found in the NCBI Gene Expression Omnibus (GSE165480).

Acknowledgments: The authors would like to thank Jinhui Rhee for the valuable discussions.

Conflicts of Interest: The authors declare no conflict of interest. The funders had no role in the design of the study; in the collection, analyses, or interpretation of data; in the writing of the manuscript; or in the decision to publish the results.

Abbreviations

ANOVA	analysis of variance
APS	ammonium persulfate
bFGF	basic fibroblast growth factor
BP	biological process
BSA	bovine serum albumin
CC	cellular component
DEG	differentially expressed gene
DW	distilled water
FAK	focal adhesion kinase
FITC	fluorescein isothiocyanate
FPKM	fragments per kilobase of transcript per million mapped reads
FTIR	Fourier transform infrared spectroscopy
GelMA	methacrylated gelatin
GO	gene ontology
HPSCs	human pluripotent stem cells
HSPG	heparan sulfate proteoglycans
IPA	Ingenuity Pathway Analysis
IRS1	insulin receptor substrate 1
KEGG	Kyoto Encyclopedia of Genes and Genomes
MAPK	mitogen-activated extracellular regulated signaling kinase
MF	molecular function
mTOR	mechanistic target of rapamycin
Myf5	myogenic factor 5
MyHC	myosin heavy chain
MyoG	myogenin
NMR	nuclear magnetic resonance
OXPHOS	oxidative phosphorylation
PBS	phosphate buffered saline
PFA	paraformaldehyde
PSS	poly(sodium-4-styrenesulfonate)
PVDF	polyvinylidene fluoride
SDS	sodium dodecyl sulfate
skMSC	skeletal muscle stem cell
TA	tibialis anterior
TBS	tris-buffered saline
TCP	tissue culture plate
TEMED	*N,N,N′,N′*-tetramethylethylenediamine

References

1. Janssen, I.; Heymsfield, S.B.; Wang, Z.M.; Ross, R. Skeletal muscle mass and distribution in 468 men and women aged 18–88 yr. *J. Appl. Physiol.* **2000**, *89*, 81–88. [CrossRef] [PubMed]
2. Kuang, S.; Gillespie, M.A.; Rudnicki, M.A. Niche regulation of muscle satellite cell self-renewal and differentiation. *Cell Stem Cell* **2008**, *2*, 22–31. [CrossRef]
3. Kuraitis, D.; Giordano, C.; Ruel, M.; Musaro, A.; Suuronen, E.J. Exploiting extracellular matrix-stem cell interactions: A review of natural materials for therapeutic muscle regeneration. *Biomaterials* **2012**, *33*, 428–443. [CrossRef] [PubMed]
4. Stegemann, J.P.; Hong, H.; Nerem, R.M. Mechanical, biochemical, and extracellular matrix effects on vascular smooth muscle cell phenotype. *J. Appl. Physiol.* **2005**, *98*, 2321–2327. [CrossRef] [PubMed]

5. Gholobova, D.; Gerard, M.; Decroix, L.; Desender, L.; Callewaert, N.; Annaert, P.; Thorrez, L. Human tissue-engineered skeletal muscle: A novel 3D in vitro model for drug disposition and toxicity after intramuscular injection. *Sci. Rep.* **2018**, *8*, 12206. [CrossRef]
6. Khodabukus, A.; Prabhu, N.; Wang, J.; Bursac, N. In Vitro Tissue-Engineered Skeletal Muscle Models for Studying Muscle Physiology and Disease. *Adv. Healthc. Mater.* **2018**, *7*, e1701498. [CrossRef] [PubMed]
7. Le Grand, F.; Rudnicki, M.A. Skeletal muscle satellite cells and adult myogenesis. *Curr. Opin. Cell Biol.* **2007**, *19*, 628–633. [CrossRef]
8. Yin, H.; Price, F.; Rudnicki, M.A. Satellite cells and the muscle stem cell niche. *Physiol. Rev.* **2013**, *93*, 23–67. [CrossRef]
9. Rochlin, K.; Yu, S.; Roy, S.; Baylies, M.K. Myoblast fusion: When it takes more to make one. *Dev. Biol.* **2010**, *341*, 66–83. [CrossRef]
10. Kim, J.H.; Jin, P.; Duan, R.; Chen, E.H. Mechanisms of myoblast fusion during muscle development. *Curr. Opin. Genet. Dev.* **2015**, *32*, 162–170. [CrossRef]
11. Agrawal, G.; Aung, A.; Varghese, S. Skeletal muscle-on-a-chip: An in vitro model to evaluate tissue formation and injury. *Lab Chip* **2017**, *17*, 3447–3461. [CrossRef]
12. Pittenger, M.F.; Mackay, A.M.; Beck, S.C.; Jaiswal, R.K.; Douglas, R.; Mosca, J.D.; Moorman, M.A.; Simonetti, D.W.; Craig, S.; Marshak, D.R. Multilineage potential of adult human mesenchymal stem cells. *Science* **1999**, *284*, 143–147. [CrossRef] [PubMed]
13. Wilson, A.; Trumpp, A. Bone-marrow haematopoietic-stem-cell niches. *Nat. Rev. Immunol.* **2006**, *6*, 93–106. [CrossRef]
14. Gilbert, P.M.; Havenstrite, K.L.; Magnusson, K.E.; Sacco, A.; Leonardi, N.A.; Kraft, P.; Nguyen, N.K.; Thrun, S.; Lutolf, M.P.; Blau, H.M. Substrate elasticity regulates skeletal muscle stem cell self-renewal in culture. *Science* **2010**, *329*, 1078–1081. [CrossRef]
15. Dupont, S.; Morsut, L.; Aragona, M.; Enzo, E.; Giulitti, S.; Cordenonsi, M.; Zanconato, F.; Le Digabel, J.; Forcato, M.; Bicciato, S.; et al. Role of YAP/TAZ in mechanotransduction. *Nature* **2011**, *474*, 179–183. [CrossRef]
16. Wozniak, M.A.; Chen, C.S. Mechanotransduction in development: A growing role for contractility. *Nat. Rev. Mol. Cell Biol.* **2009**, *10*, 34–43. [CrossRef] [PubMed]
17. Hwang, Y.; Seo, T.; Hariri, S.; Choi, C.; Varghese, S. Matrix Topographical Cue-Mediated Myogenic Differentiation of Human Embryonic Stem Cell Derivatives. *Polymers* **2017**, *9*, 580. [CrossRef]
18. Hwang, Y.; Suk, S.; Shih, Y.R.; Seo, T.; Du, B.; Xie, Y.; Li, Z.; Varghese, S. WNT3A promotes myogenesis of human embryonic stem cells and enhances in vivo engraftment. *Sci. Rep.* **2014**, *4*, 5916. [CrossRef]
19. Berrier, A.L.; Yamada, K.M. Cell-matrix adhesion. *J. Cell. Physiol.* **2007**, *213*, 565–573. [CrossRef] [PubMed]
20. Cukierman, E.; Pankov, R.; Stevens, D.R.; Yamada, K.M. Taking cell-matrix adhesions to the third dimension. *Science* **2001**, *294*, 1708–1712. [CrossRef] [PubMed]
21. Park, J.S.; Chu, J.S.; Tsou, A.D.; Diop, R.; Tang, Z.; Wang, A.; Li, S. The effect of matrix stiffness on the differentiation of mesenchymal stem cells in response to TGF-beta. *Biomaterials* **2011**, *32*, 3921–3930. [CrossRef]
22. Reilly, G.C.; Engler, A.J. Intrinsic extracellular matrix properties regulate stem cell differentiation. *J. Biomech.* **2010**, *43*, 55–62. [CrossRef] [PubMed]
23. Teo, B.K.; Wong, S.T.; Lim, C.K.; Kung, T.Y.; Yap, C.H.; Ramagopal, Y.; Romer, L.H.; Yim, E.K. Nanotopography modulates mechanotransduction of stem cells and induces differentiation through focal adhesion kinase. *ACS Nano* **2013**, *7*, 4785–4798. [CrossRef]
24. Sah, J.P.; Hao, N.T.T.; Kim, Y.; Eigler, T.; Tzahor, E.; Kim, S.H.; Hwang, Y.; Yoon, J.K. MBP-FGF2-Immobilized Matrix Maintains Self-Renewal and Myogenic Differentiation Potential of Skeletal Muscle Stem Cells. *Int. J. Stem Cells* **2019**. [CrossRef]
25. Ghadiali, R.S.; Guimond, S.E.; Turnbull, J.E.; Pisconti, A. Dynamic changes in heparan sulfate during muscle differentiation and ageing regulate myoblast cell fate and FGF2 signalling. *Matrix Biol.* **2017**, *59*, 54–68. [CrossRef]
26. Nguyen, T.H.; Kim, S.H.; Decker, C.G.; Wong, D.Y.; Loo, J.A.; Maynard, H.D. A heparin-mimicking polymer conjugate stabilizes basic fibroblast growth factor. *Nat. Chem.* **2013**, *5*, 221–227. [CrossRef] [PubMed]
27. Nozaki, M.; Li, Y.; Zhu, J.; Ambrosio, F.; Uehara, K.; Fu, F.H.; Huard, J. Improved muscle healing after contusion injury by the inhibitory effect of suramin on myostatin, a negative regulator of muscle growth. *Am. J. Sports Med.* **2008**, *36*, 2354–2362. [CrossRef]
28. Sangaj, N.; Kyriakakis, P.; Yang, D.; Chang, C.W.; Arya, G.; Varghese, S. Heparin mimicking polymer promotes myogenic differentiation of muscle progenitor cells. *Biomacromolecules* **2010**, *11*, 3294–3300. [CrossRef] [PubMed]
29. Paluck, S.J.; Maynard, H.D. Structure Activity Relationship of Heparin Mimicking Polymer p(SS-co-PEGMA): Effect of Sulfonation and Polymer Size on FGF2-Receptor Binding. *Polym. Chem.* **2017**, *8*, 4548–4556. [CrossRef]
30. Chang, C.W.; Hwang, Y.; Brafman, D.; Hagan, T.; Phung, C.; Varghese, S. Engineering cell-material interfaces for long-term expansion of human pluripotent stem cells. *Biomaterials* **2013**, *34*, 912–921. [CrossRef]
31. Holle, A.W.; Tang, X.; Vijayraghavan, D.; Vincent, L.G.; Fuhrmann, A.; Choi, Y.S.; del Alamo, J.C.; Engler, A.J. In situ mechanotransduction via vinculin regulates stem cell differentiation. *Stem Cells* **2013**, *31*, 2467–2477. [CrossRef]
32. Bodine, S.C.; Stitt, T.N.; Gonzalez, M.; Kline, W.O.; Stover, G.L.; Bauerlein, R.; Zlotchenko, E.; Scrimgeour, A.; Lawrence, J.C.; Glass, D.J.; et al. Akt/mTOR pathway is a crucial regulator of skeletal muscle hypertrophy and can prevent muscle atrophy in vivo. *Nat. Cell Biol.* **2001**, *3*, 1014–1019. [CrossRef]
33. Clevers, H. Wnt/beta-catenin signaling in development and disease. *Cell* **2006**, *127*, 469–480. [CrossRef]
34. Han, X.H.; Jin, Y.R.; Seto, M.; Yoon, J.K. A WNT/beta-catenin signaling activator, R-spondin, plays positive regulatory roles during skeletal myogenesis. *J. Biol. Chem.* **2011**, *286*, 10649–10659. [CrossRef] [PubMed]

35. Sin, J.; Andres, A.M.; Taylor, D.J.; Weston, T.; Hiraumi, Y.; Stotland, A.; Kim, B.J.; Huang, C.; Doran, K.S.; Gottlieb, R.A. Mitophagy is required for mitochondrial biogenesis and myogenic differentiation of C2C12 myoblasts. *Autophagy* **2016**, *12*, 369–380. [CrossRef]
36. Quach, N.L.; Biressi, S.; Reichardt, L.F.; Keller, C.; Rando, T.A. Focal adhesion kinase signaling regulates the expression of caveolin 3 and beta1 integrin, genes essential for normal myoblast fusion. *Mol. Biol. Cell* **2009**, *20*, 3422–3435. [CrossRef] [PubMed]
37. Takebe, T.; Zhang, B.; Radisic, M. Synergistic Engineering: Organoids Meet Organs-on-a-Chip. *Cell Stem Cell* **2017**, *21*, 297–300. [CrossRef]
38. Nesmith, A.P.; Wagner, M.A.; Pasqualini, F.S.; O'Connor, B.B.; Pincus, M.J.; August, P.R.; Parker, K.K. A human in vitro model of Duchenne muscular dystrophy muscle formation and contractility. *J. Cell Biol.* **2016**, *215*, 47–56. [CrossRef] [PubMed]
39. Hannon, K.; Kudla, A.J.; McAvoy, M.J.; Clase, K.L.; Olwin, B.B. Differentially expressed fibroblast growth factors regulate skeletal muscle development through autocrine and paracrine mechanisms. *J. Cell Biol.* **1996**, *132*, 1151–1159. [CrossRef]
40. Vlodavsky, I.; Miao, H.Q.; Medalion, B.; Danagher, P.; Ron, D. Involvement of heparan sulfate and related molecules in sequestration and growth promoting activity of fibroblast growth factor. *Cancer Metastasis Rev.* **1996**, *15*, 177–186. [CrossRef]
41. Christman, K.L.; Vazquez-Dorbatt, V.; Schopf, E.; Kolodziej, C.M.; Li, R.C.; Broyer, R.M.; Chen, Y.; Maynard, H.D. Nanoscale growth factor patterns by immobilization on a heparin-mimicking polymer. *J. Am. Chem. Soc.* **2008**, *130*, 16585–16591. [CrossRef] [PubMed]
42. Larrain, J.; Carey, D.J.; Brandan, E. Syndecan-1 expression inhibits myoblast differentiation through a basic fibroblast growth factor-dependent mechanism. *J. Biol. Chem.* **1998**, *273*, 32288–32296. [CrossRef]
43. Rapraeger, A.C.; Krufka, A.; Olwin, B.B. Requirement of heparan sulfate for bFGF-mediated fibroblast growth and myoblast differentiation. *Science* **1991**, *252*, 1705–1708. [CrossRef]
44. Liekens, S.; Leali, D.; Neyts, J.; Esnouf, R.; Rusnati, M.; Dell'Era, P.; Maudgal, P.C.; De Clercq, E.; Presta, M. Modulation of fibroblast growth factor-2 receptor binding, signaling, and mitogenic activity by heparin-mimicking polysulfonated compounds. *Mol. Pharmacol.* **1999**, *56*, 204–213. [CrossRef]
45. Florini, J.R.; Ewton, D.Z.; Magri, K.A. Hormones, growth factors, and myogenic differentiation. *Annu. Rev. Physiol.* **1991**, *53*, 201–216. [CrossRef] [PubMed]
46. Perry, R.L.; Rudnick, M.A. Molecular mechanisms regulating myogenic determination and differentiation. *Front. Biosci.* **2000**, *5*, D750–D767. [CrossRef] [PubMed]
47. Yoon, M.S. mTOR as a Key Regulator in Maintaining Skeletal Muscle Mass. *Front. Physiol.* **2017**, *8*, 788. [CrossRef]
48. Risson, V.; Mazelin, L.; Roceri, M.; Sanchez, H.; Moncollin, V.; Corneloup, C.; Richard-Bulteau, H.; Vignaud, A.; Baas, D.; Defour, A.; et al. Muscle inactivation of mTOR causes metabolic and dystrophin defects leading to severe myopathy. *J. Cell Biol.* **2009**, *187*, 859–874. [CrossRef]
49. Cong, X.X.; Gao, X.K.; Rao, X.S.; Wen, J.; Liu, X.C.; Shi, Y.P.; He, M.Y.; Shen, W.L.; Shen, Y.; Ouyang, H.; et al. Rab5a activates IRS1 to coordinate IGF-AKT-mTOR signaling and myoblast differentiation during muscle regeneration. *Cell Death Differ.* **2020**, *27*, 2344–2362. [CrossRef]
50. Jin, Y.R.; Yoon, J.K. The R-spondin family of proteins: Emerging regulators of WNT signaling. *Int. J. Biochem. Cell Biol.* **2012**, *44*, 2278–2287. [CrossRef] [PubMed]
51. Abmayr, S.M.; Zhuang, S.; Geisbrecht, E.R. Myoblast fusion in Drosophila. *Methods Mol. Biol.* **2008**, *475*, 75–97. [CrossRef] [PubMed]
52. Fluck, M.; Carson, J.A.; Gordon, S.E.; Ziemiecki, A.; Booth, F.W. Focal adhesion proteins FAK and paxillin increase in hypertrophied skeletal muscle. *Am. J. Physiol.* **1999**, *277*, C152–C162. [CrossRef]
53. Madaro, L.; Marrocco, V.; Fiore, P.; Aulino, P.; Smeriglio, P.; Adamo, S.; Molinaro, M.; Bouche, M. PKCtheta signaling is required for myoblast fusion by regulating the expression of caveolin-3 and beta1D integrin upstream focal adhesion kinase. *Mol. Biol. Cell* **2011**, *22*, 1409–1419. [CrossRef]
54. Sunadome, K.; Yamamoto, T.; Ebisuya, M.; Kondoh, K.; Sehara-Fujisawa, A.; Nishida, E. ERK5 regulates muscle cell fusion through Klf transcription factors. *Dev. Cell* **2011**, *20*, 192–205. [CrossRef] [PubMed]
55. Kang, H.; Shih, Y.V.; Hwang, Y.; Wen, C.; Rao, V.; Seo, T.; Varghese, S. Mineralized gelatin methacrylate-based matrices induce osteogenic differentiation of human induced pluripotent stem cells. *Acta Biomater.* **2014**, *10*, 4961–4970. [CrossRef] [PubMed]
56. Weinman, C.J.; Gunari, N.; Krishnan, S.; Dong, R.; Paik, M.Y.; Sohn, K.E.; Walker, G.C.; Kramer, E.J.; Fischer, D.A.; Ober, C.K. Protein adsorption resistance of anti-biofouling block copolymers containing amphiphilic side chains. *Soft Matter* **2010**, *6*, 3237–3243. [CrossRef]
57. Hwang, Y.; Suk, S.; Lin, S.; Tierney, M.; Du, B.; Seo, T.; Mitchell, A.; Sacco, A.; Varghese, S. Directed in vitro myogenesis of human embryonic stem cells and their in vivo engraftment. *PLoS ONE* **2013**, *8*, e72023. [CrossRef]

International Journal of *Molecular Sciences*

MDPI

Article

RGDS-Modified Superporous Poly(2-Hydroxyethyl Methacrylate)-Based Scaffolds as 3D In Vitro Leukemia Model

Hana Svozilová [1,2], Zdeněk Plichta [3], Vladimír Proks [3], Radana Studená [1], Jiří Baloun [1], Michael Doubek [1,2], Šárka Pospíšilová [1,2] and Daniel Horák [3,*]

1 Center of Molecular Medicine, Central European Institute of Technology, Masaryk University, Kamenice 5, 625 00 Brno, Czech Republic; svozilova.hana@gmail.com (H.S.); radana.studena@gmail.com (R.S.); jiri.baloun@ceitec.muni.cz (J.B.); doubek.michael@fnbrno.cz (M.D.); sarka.pospisilova@ceitec.muni.cz (Š.P.)

2 Department of Internal Medicine—Hematology and Oncology, University Hospital Brno and Faculty of Medicine, Masaryk University, Jihlavská 20, 625 00 Brno, Czech Republic

3 Institute of Macromolecular Chemistry, Czech Academy of Sciences, Heyrovského nám. 2, 162 06 Prague, Czech Republic; plichta@imc.cas.cz (Z.P.); proks@imc.cas.cz (V.P.)

* Correspondence: horak@imc.cas.cz; Tel.: +420-296-809-260

Abstract: Superporous poly(2-hydroxyethyl methacrylate-*co*-2-aminoethyl methacrylate) (P(HEMA-AEMA)) hydrogel scaffolds are designed for in vitro 3D culturing of leukemic B cells. Hydrogel porosity, which influences cell functions and growth, is introduced by adding ammonium oxalate needle-like crystals in the polymerization mixture. To improve cell vitality, cell-adhesive Arg-Gly-Asp-Ser (RGDS) peptide is immobilized on the *N*-(γ-maleimidobutyryloxy)succinimide-activated P(HEMA-AEMA) hydrogels via reaction of SH with maleimide groups. This modification is especially suitable for the survival of primary chronic lymphocytic leukemia cells (B-CLLs) in 3D cell culture. No other tested stimuli (interleukin-4, CD40 ligand, or shaking) can further improve B-CLL survival or metabolic activity. Both unmodified and RGDS-modified P(HEMA-AEMA) scaffolds serve as a long-term (70 days) 3D culture platforms for HS-5 and M2-10B4 bone marrow stromal cell lines and MEC-1 and HG-3 B-CLL cell lines, although the adherent cells retain their physiological morphologies, preferably on RGDS-modified hydrogels. Moreover, the porosity of hydrogels allows direct cell lysis, followed by efficient DNA isolation from the 3D-cultured cells. P(HEMA-AEMA)-RGDS thus serves as a suitable 3D in vitro leukemia model that enables molecular and metabolic assays and allows imaging of cell morphology, interactions, and migration by confocal microscopy. Such applications can prospectively assist in testing of drugs to treat this frequently recurring or refractory cancer.

Keywords: poly(2-hydroxyethyl methacrylate); 3D scaffold; RGDS; chronic lymphocytic leukemia; B cell survival

Citation: Svozilová, H.; Plichta, Z.; Proks, V.; Studená, R.; Baloun, J.; Doubek, M.; Pospíšilová, Š.; Horák, D. RGDS-Modified Superporous Poly(2-Hydroxyethyl Methacrylate)-Based Scaffolds as 3D In Vitro Leukemia Model. *Int. J. Mol. Sci.* **2021**, *22*, 2376. https://doi.org/10.3390/ijms22052376

Academic Editors: Kyeongsoon Park and Peter John Jervis

Received: 29 January 2021
Accepted: 24 February 2021
Published: 27 February 2021

1. Introduction

Chronic lymphocytic leukemia (CLL) is the most common adult blood cancer in Caucasians. It is characterized by malignancy of morphologically mature CD5 positive B cells (B-CLLs), which results in the proliferation and accumulation of the neoplastic B cells in the blood, bone marrow, and secondary lymphoid tissues [1]. The nature of CLL is highly heterogeneous in terms of clinical course and overall survival. Some patients remain asymptomatic and survive for decades, whereas others can experience an aggressive course of the disease surviving for 3 to 5 years. Promisingly, novel drugs and their combinations have been introduced for CLL treatment [2]. While these treatments undeniably improve patient prognosis and can lead to long-lasting remissions, relapses still occur. Despite the worldwide effort to develop a cure for CLL, the disease is mostly incurable and the mechanisms describing CLL biology, relapses, and refractoriness have not been fully resolved [3,4]. Consequently, CLL research has been mostly focused on elucidating the molecular mechanism of the disease, while the search for the right treatment persists.

In current studies of cancer pathophysiology and drug development, the focus has been laid on 3D in vitro models [5–7]. 3D culture systems, rather than conventional ones (i.e., monolayers or suspension cultures), allow cells to grow, migrate, interact, and respond to various stimuli analogously to real in vivo behavior [8]. In vitro studies of CLL are also complicated by the fact that in the conventional cell culture, primary B-CLLs, which are dependent on their natural microenvironment, quickly undergo apoptosis and are not able to establish a stable cell line [9]. Consequently, novel bone marrow-like scaffolds that mimic the CLL microenvironment are being developed to reveal the mechanisms of CLL biology [10,11]. In order to fully understand the 3D in vitro cell interactions under different conditions, the development of novel 3D CLL models is thus of utmost importance.

Nowadays, 3D culture systems offer a vast number of options, including various scaffold-free or scaffold-based approaches [12,13]. A plethora of hydrogels or other materials based on both natural and synthetic polymers have been described for the scaffold design, including alginate, chitosan, collagen, hyaluronic acid, poly(lactic or glycolic acid), poly(hydroxybutyrate-*co*-hydroxyvalerate), poly(ester urea), poly(ester amide), poly(ethylene glycol), polyanhydrides, etc. [14–16]. Hydrogel scaffolds have an advantage in that they mimic an extracellular matrix, enabling the survival, proliferation, differentiation, and migration of the cells. This is due to biocompatibility (nontoxicity) of the hydrogels, the softness of which resembles living tissue, interconnecting pores, high water retention, efficient transport of oxygen and nutrients, etc. [17,18]. Among the hydrogels, poly(2-hydroxyethyl methacrylate) (PHEMA) plays a special role due to its tissue-like mechanical compliance, elasticity, mass transfer properties, and long history of successful applications in medicine and tissue engineering as soft contact lenses, surgical implants, and drug delivery vehicles [19,20]. Another advantage of PHEMA is that it can be easily fabricated into various configurations and shapes and conveniently chemically modified, e.g., by copolymerization of 2-hydroxyethyl methacrylate (HEMA), with various reactive comonomers. Moreover, PHEMA is partly transparent in water to light, enabling inspection of cell models with optical microscopic techniques.

When choosing artificial 3D matrices for B-CLLs, it is crucial to fulfill many requirements; most importantly, the scaffolds should be geometrically similar to the B-CLL microenvironment. In addition, the viability of B-CLLs strictly depends on the presence of essential interactions [9]. Thus, natural conditions have to be simulated by the addition of growth factors and/or by culture with cells which typically occur in the pores of the bone marrow, e.g., bone marrow stromal cells (BMSCs), T cells, or nurse-like cells [21]. While physical cell–cell contacts are necessary for BMSCs and B-CLLs [22], the presence of stimulating T cells can be advantageously replaced by the growth medium with CD40 ligand (CD40L) and interleukin 4 (IL-4), etc. Not only do the aforementioned interactions contribute to evading the B-CLLs from both in vivo and in vitro apoptoses, but they also stimulate the B-CLLs to boost their proliferation [23].

The aim of this study was to develop a 3D hydrogel scaffold that is biocompatible, partly transparent, and contains sufficiently large pores to accommodate the cells. The hydrogel was based on superporous poly(2-hydroxyethyl methacrylate-*co*-2-aminoethyl methacrylate) (P(HEMA-AEMA)), which was modified with the cell adhesion Ac-CGGGRG-DSGGGY-NH_2 (RGDS) peptide. The superporosity was reached by needle-shaped ammonium oxalate as a porogen and a final polymer was denoted as P(HEMA-AEMA)-RGDS. Another objective of the study was to optimize a suitable 3D system that resembles the natural environment of the primary B-CLLs, i.e., the trabecular bone, where hematopoiesis occurs and the leukemic B cells are present. Based on the results achieved with other 3D in vitro cancer models [24,25], our system should enable us to predict the sensitivity of B-CLLs to various treatments more accurately than the 2D cell culture. The final goal was to mimic cell–cell interactions by introducing the interaction partners, i.e., BMSCs or soluble factors produced by T cells, into the in vitro microenvironment.

2. Results

2.1. Poly(2-Hydroxyethyl Methacrylate-co-2-Aminoethyl Methacrylate) (P(HEMA-AEMA)) Scaffolds

P(HEMA-AEMA) scaffolds were obtained by 2,2′-azobis(2-methylpropionitrile) (AIBN)-initiated radical polymerization of 2-hydroxyethyl methacrylate (HEMA), ethylene dimethacrylate (EDMA), and 2-aminoethyl methacrylate (AEMA) in the presence of ammonium oxalate crystals as a porogen. The crystals were 30–60 μm thick, 0.3–2 mm long, and polydisperse in size (Figure 1a). In the synthesis, content of the ammonium oxalate crystals in the polymerization mixture during sedimentation amounted to 43 vol.%. After completion of the polymerization, the ammonium oxalate was removed from the polymer by washing with water, leaving interconnected pores in the hydrogel as reprints of the original crystals (Figure 1b–f); as a result, the pores were close to the structure of trabecular bone [26,27]. The interconnectivity of pores was documented in Figure 1c,f. P(HEMA-AEMA) hydrogels were fabricated in the form of cylinders by polymerization of the monomer mixture in injection syringes. The cylinders were then portioned to small cubes to facilitate future cell penetration (Figure 1d). The resulting P(HEMA-AEMA) contained 0.35 wt.% of N according to the elemental analysis, corresponding to 0.25 mmol of amino groups per g. This is more than the number of the groups in AEMA added in the polymerization feed. The discrepancy can be explained by the presence of residual nitrile groups originating from AIBN initiator. According to scanning electron microscopy (SEM) micrographs of the superporous P(HEMA-AEMA) hydrogels in the dry state, the diameters of the pores reached 15–60 μm (Figure 1b), which approximately corresponded to the size of original crystals. P(HEMA-AEMA) hydrogels swelled in water, where their pore sizes slightly increased (Figure 1f). It is worth noting that our earlier PHEMA-based scaffolds prepared with ammonium oxalate as a porogen reached water regain of 1.83 mL/g [28].

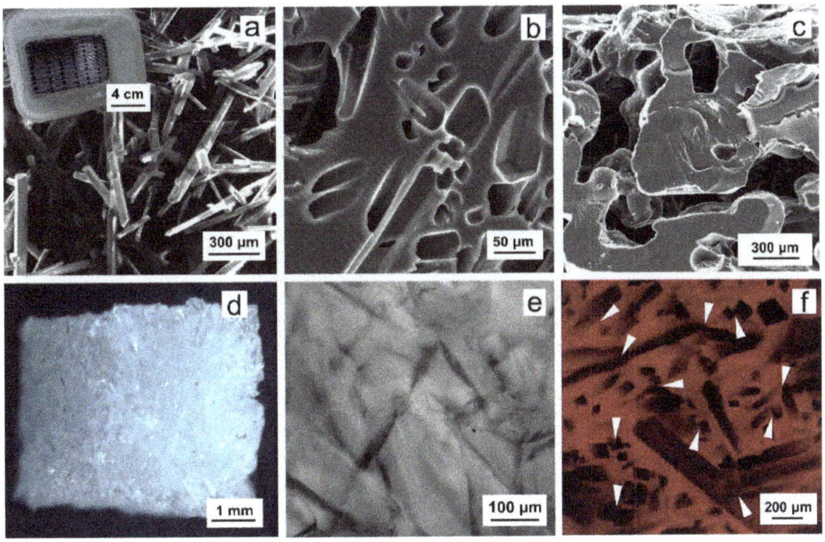

Figure 1. Scanning electron microscopy (SEM) micrographs of (**a**) ammonium oxalate crystals (inset: sieve for screening the needles), (**b**,**c**) superporous poly(2-hydroxyethyl methacrylate-*co*-2-aminoethyl methacrylate) (P(HEMA-AEMA)) hydrogel, and (**d**) light and (**e**,**f**) confocal micrograph of a hydrogel cube. (**f**) Single plane of hydrogel stained with AlamarBlue™, white arrows point to pore intersections. HEMA—2-hydroxyethyl methacrylate; AEMA—2-aminoethyl methacrylate.

2.2. Immobilization of RGDS Peptide on P(HEMA-AEMA)-RGDS Scaffolds

Since cell adhesion and motility strongly depend on the interactions between cells and scaffold, it is preferable to modify its surface with an adhesive peptide, although the cells alone can colonize the neat (unmodified) PHEMA scaffold used as cell support. P(HEMA-AEMA) properties were strengthened by *N*-(γ-maleimidobutyryloxy)succinimide (GMBS) activation and RGDS peptide to enhance cell adhesion and spreading on the scaffold. While the *N*-hydroxysuccinimide ester of GMBS was covalently conjugated with amino groups of the P(HEMA-AEMA) scaffold at pH 7.4 to form amide bonds, the maleimide reacted with thiol groups of RGDS at pH 6.8 to produce stable thioether bonds (Figure 2). The amount of immobilized RGDS was determined in our previous report, where 65 $\pm$ 10 μg of RGDS was found per P(HEMA-AEMA)-RGDS disc (i.e., 2.16 $\pm$ 0.9 $\times$ 10^{-13} mol of RGDS per cm^2) according to quantitative radioassay analysis using ^{125}I-labeled peptide [29].

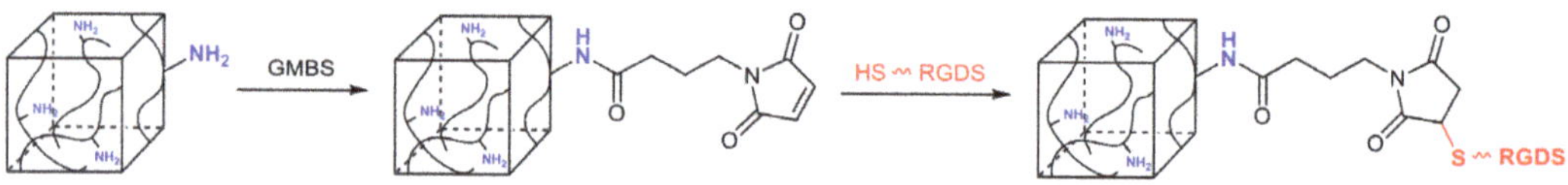

Figure 2. Scheme of immobilization of Ac-CGGGRGDSGGGY-NH_2 (RGDS) peptide on P(HEMA-AEMA) hydrogel via *N*-(γ-maleimidobutyryloxy)succinimide (GMBS) activation.

2.3. 3D Culture of Leukemia B Cells on P(HEMA-AEMA)-RGDS Scaffolds

The optimal cell seeding method, cell adhesion, viability, and metabolic activity on the P(HEMA-AEMA)-RGDS hydrogels were determined from confocal microscopy micrographs and AlamarBlue™ assay. Coculture experiments were only studied by confocal microscopy, as the AlamarBlue™ assay does not distinguish the contribution of each cell type to the overall metabolic activity. As the hydrogels were partly transparent in medium, they allowed the imaging of even the cells seeded deep inside of the matrix ($\leq$300 μm).

Each method of investigation and its conditions were first optimized in terms of the corresponding cell lines, which are immortal and resilient, compared to primary B-CLLs. Since BMSCs cultured with primary B cells mimic the natural microenvironment of the bone marrow, it was important to ensure that both the BMSCs and the B-CLLs penetrate the P(HEMA-AEMA)-RGDS scaffold and proliferate there. Consequently, the emphasis was laid on the cell culture conditions of both cell types. Coculture of BMSCs was further optimized preferably with murine BMSC lines since they resemble primary BMSCs in terms of phenotypic characters [30] and trigger more homogenous responses compared to primary or immortalized human BMSCs [31].

Cell morphology, distribution, and viability of the P(HEMA-AEMA)-RGDS scaffolds were determined by confocal microscopy (Figure 3). The HS-5 cells were mostly of polygonal shapes with fusiform morphologies in all culture types (Figure 3a–e). M2-10B4 also adhered to surface in flasks (Figure 3f) and RGDS-modified scaffolds (Figure 3i,j), but not in hydrogels without surface modifications (Figure 3g,h). The morphologies of both tested B-CLL lines MEC-1 (Figure 3k–o) and HG-3 (Figure 3p–t) corresponded with the supplier's description: round cells growing in suspension, MEC-1 being slightly adherent [32] and HG-3 partially forming clumps [33]. All studied cell lines manifested homogenous distribution within scaffold pores (Figure 3b,d,g,i,l,n,q,s).

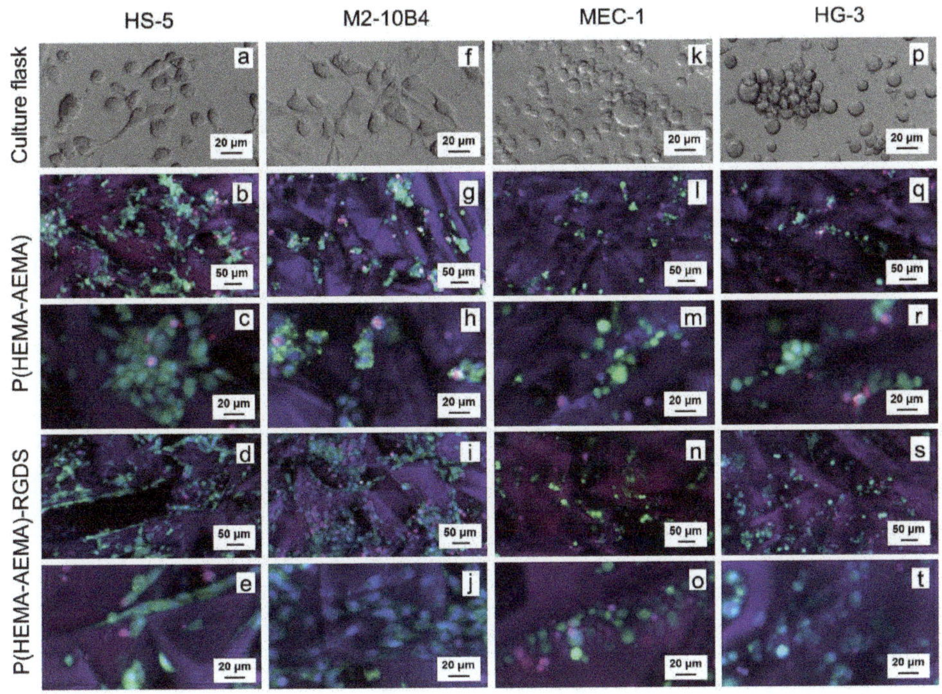

Figure 3. Confocal micrographs of P(HEMA-AEMA)-RGDS scaffolds seeded with (**a–e**) HS-5, (**f–j**) M2-10B4, (**k–o**) MEC-1, (**p–t**) HG-3 cell culture for 24 h. (**a**,**f**,**k**,**p**) Transmitted light; (**b–e**, **g–j**, **l–o**, **q–t**) live and dead cells and nuclei stained by calcein acetoxymethyl ester (AM) (green), propidium iodide (red), and Hoechst 33342 (blue), respectively.

Compared to primary B-CLL cells cultured on P(HEMA-AEMA) (Figure 4a), more cells, larger clumps and higher cell viability was observed on the P(HEMA-AEMA)-RGDS scaffolds (Figure 4b,c). While only 75% viability was calculated for P(HEMA-AEMA)-cultured cells, 95% (Figure 4b) or 98% (Figure 4c) live B-CLLs were seen in RGDS-modified hydrogels.

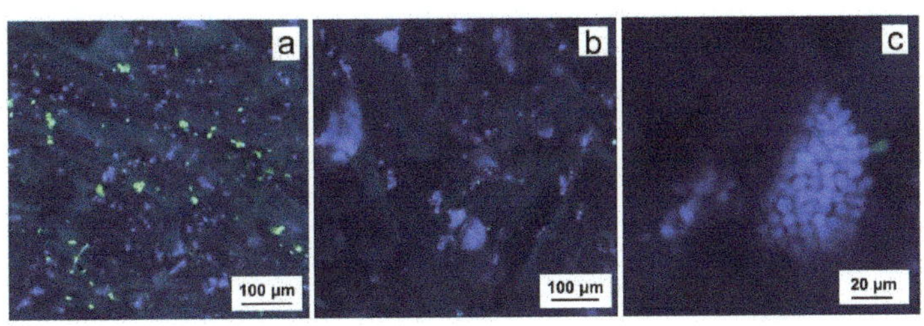

Figure 4. Confocal micrographs of (**a**) P(HEMA-AEMA) and (**b**,**c**) P(HEMA-AEMA)-RGDS scaffolds seeded with primary CD5 positive B cells (B-CLLs) for 24 h. B-CLLs and dead cells were stained by CellTrace™ violet (blue) and Sytox (green), respectively.

Optimal cell seeding concentration: The image analysis, i.e., cell counts and viability, assisted in the optimization of cell seeding concentration, which differed depending on the cell type. Multiple repetitions of the experiments showed highly heterogeneous results; therefore, the resulting optimal concentration is given in an interval in which the cell counts and viability were the highest. To enhance the probability of cell accommodation in the hydrogels, they were incubated with an excess of the cells. The adherent cell lines HS-5 and M2-10B4 generally required lower concentrations (3–4 million/mL) compared to the HG-3 and MEC-1 cell lines growing in the suspension (15–25 million/mL). Primary B-CLL cells were seeded in the highest concentration (~50 million/mL) to compensate for low proliferation and high rate of apoptosis, which is not seen in the immortalized cell lines. When coculture was performed, cell densities of 3 million/mL BMSCs and 50 million/mL B-CLLs were used.

Long-term cell line culture. Both adherent (Figure 5a,b) and suspension cells (Figure 5c–e) were cultured in the P(HEMA-AEMA) hydrogel to prove its capability of long-term cell accommodation, which is required for coculture of the primary B-CLLs and BMSCs. AlamarBlue™ assay was used to determine the metabolic activity of cells on the P(HEMA-AEMA) as well as P(HEMA-AEMA)-RGDS scaffolds during 70 days in media in both static and dynamic settings.

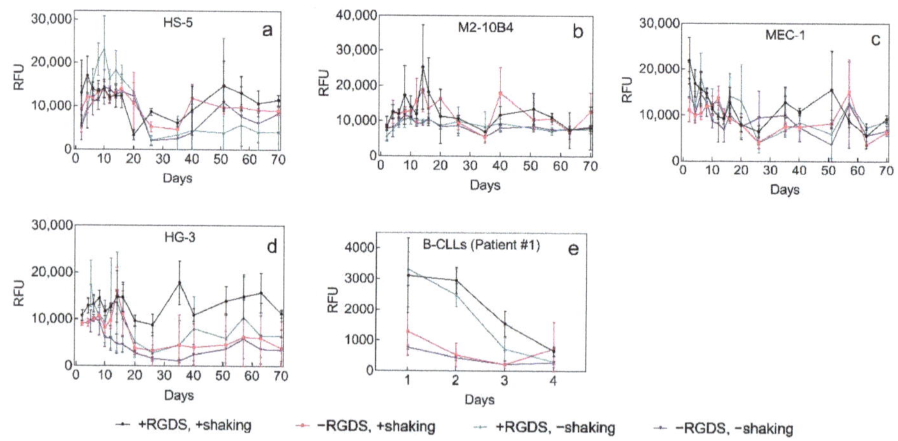

Figure 5. Time dependence of (**a**) HS-5, (**b**) M2-10B4, (**c**) MEC-1, (**d**) HG-3, and (**e**) primary B-CLL metabolic activity (patient No. 1, Table S1 in Supporting Materials (SMs)) in P(HEMA-AEMA) and P(HEMA-AEMA)-RGDS scaffolds without or with shaking to promote the medium flow. Metabolic activity was quantified by AlamarBlue™ assay for 70 days. RFU—relative fluorescence units; n = 3. Vertical bars in the graphs denote 95% confidence interval. Statistical analysis (Table S2) by Statistica revealed that metabolic activity was significantly affected by shaking in (**a**) HS-5 ($p < 0.001$) and (**b**) M2-10B4 ($p < 0.01$) cell lines and by RGDS surface modification in (**e**) primary B-CLLs ($p < 0.001$). No other significant differences were observed.

All studied cell lines demonstrated the ability to metabolize under the selected conditions for the whole period of 70 days. Even though their metabolic activities occasionally dropped, the cell lines always recovered their rates to a various nonzero extent. Irreversible drop in metabolic activity was only observed in primary B-CLLs, whose metabolic rates decreased after 4 days of seeding (Figure 5e). Even though B-CLLs were seeded at the same density, their initially observed metabolic activity differed. The most homogenous metabolic activity with the minimum number of fluctuations was seen in M2-10B4 cells, which makes them favorable for coculture with primary B-CLLs (Figure 5b).

Effect of medium flow and modification of P(HEMA-AEMA) with RGDS: A significant effect of medium flow was only seen in adhesion cell lines, where the motion of plates positively influenced cellular metabolic activity. In primary or immortalized B-CLL cells, the shaking did not stimulate metabolic activity of cells (Table S2). Interestingly, statistical analysis revealed a significant difference ($p < 0.001$) between the metabolic rate of B-CLLs cultured on the unmodified and RGDS-modified scaffolds; the latter hydrogel supported much higher metabolic rates. These results corresponded with higher adhesion and survival-supporting capacities of the RGDS-modified scaffolds observed by confocal microscopy (Figure 5d,e).

Prolonged survival of 3D-cultured primary B-CLLs was aimed for by introducing interaction partners into the microenvironment, i.e., either M2-10B4 cells (Figure 6), or soluble IL-4 (5 ng/μL), or CD40L (1 μg/μL) (Figure 7). The delivery of the nutrients was supported by shaking the plates and its effect on cell survival was studied as well. As cells of each individual might respond differently to specific external stimuli, cells from multiple patients were studied [34]. Patients selected for B-CLL culture had different levels of leukocytosis and carried the genetic burden of various severities resulting in adverse clinical implications (Table S1).

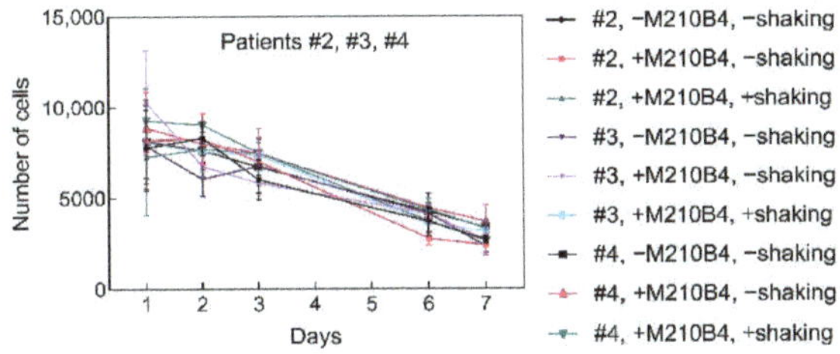

Figure 6. Time dependence of a number of viable B-CLLs of 3 different patients (No. 2–4; Table S1) seeded in the P(HEMA-AEMA)-RGDS hydrogel for 7 days and cultured in the presence or absence of bone marrow stromal cells (BMSCs) with or without supporting medium flow; n = 3. Vertical bars in the graphs denote 95% confidence interval. Statistical analysis (Table S3) by computing environment R; no significant differences were observed.

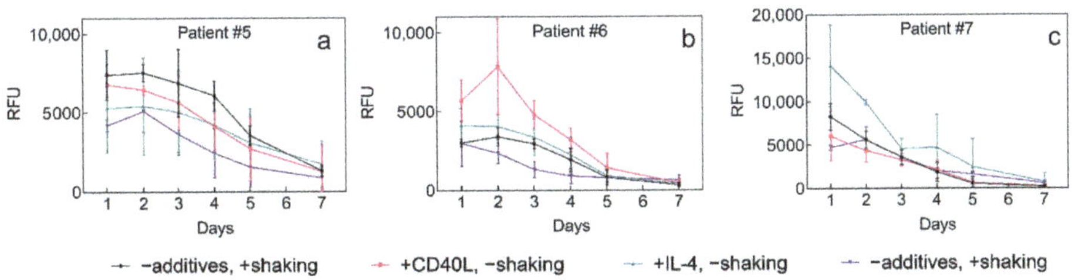

Figure 7. Time dependence of B-CLL metabolic activity of (**a–c**) three different patients (No. 5–7, respectively; Table S1) cultured with M2-10B4 cells in the P(HEMA-AEMA)-RGDS hydrogels with or without the addition of soluble IL-4 (10 ng/μL) and CD40L (1 μg/μL). RFU—relative fluorescence units; n = 3. Vertical bars in the graphs denote 95% confidence interval. Statistical analysis (Table S4) by computing environment R; no significant differences were observed.

The overall B-CLL survival did not differ among selected patients and did not correlate with selected patients' characteristics. Coculture with M2-10B4 in P(HEMA-AEMA)-RGDS scaffolds had no statistically significant impact on B-CLL survival, which was not improved even by shaking the plates (Figure 6; Table S3). The same applied for soluble additives IL-4, as well as for CD40L (Figure 7a–c; Table S4). On the other hand, shaking of the scaffolds reduced metabolic cell activity in all studied patients with probability $p = 0.052$.

Isolation of DNA: A protocol for isolation of DNA from 3D-cultured B-CLLs was optimized to introduce a methodology for prospective analysis of genetic mutations in time. Let us note that it is crucial to distinguish genetic information of the BMSCs and B-CLLs as both were cocultured. The coculture system of two cell types is inevitable since the protection by BMSCs can only be achieved via cell–cell contacts, not when the B-CLLs are separated from the BMSCs, e.g., by micropore filters [22,31]. Because of this, the cells were directly lysed on the 3D matrices, with the BMSCs and B-CLLs being of murine and human origin, respectively. DNA was successfully isolated within 24 h after B-CLL seeding on the superporous P(HEMA-AEMA)-RGDS scaffold using a FastDNA sample spin kit for soil. The amount of DNA isolated from three hydrogels reached 1.68 μg, with the DNA integrity being 6.7, which is comparable with the integrity achieved after DNA isolation from the whole blood by the automated protocol (Figure 8).

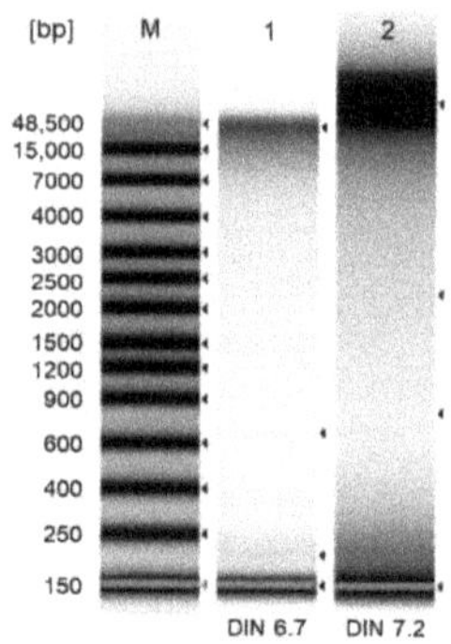

Figure 8. Electropherograms of ladder (1; standard), DNA isolated from B-CLLs cultured in P(HEMA-AEMA)-RGDS hydrogel (2), and DNA isolated from whole blood by MagCore genomic DNA whole blood kit (3; RBC Bioscience; Taipei, Taiwan; positive control). Electropherograms were obtained using a TapeStation 4200 system. DIN—DNA integrity number.

3. Discussion

PHEMA was selected for the preparation of superporous hydrogel scaffolds for long-term culture of leukemic cells. To increase the cell adhesive properties, the surface of PHEMA was modified with RGDS peptide via the GMBS coupling reaction [29]. In this report, PHEMA was crosslinked with a small amount of crosslinking agent (EDMA; 1 wt.%) to improve the mechanical properties and hinder the formation of microgels, which would compromise the cell experiment evaluation. In addition to EDMA, the limited amount of AEMA (1 wt.%) was used as a comonomer to introduce amino groups necessary for future reaction with GMBS and subsequent immobilization of RGDS peptide via SH groups (Figure 2). Because effective cell seeding into the hydrogel matrices requires the large pore size, inorganic crystals served as a porogen added in the polymerization feed; they are easily available and their size was regulated by controlling the crystallization conditions. Among various crystal shapes and sizes, in particular, ammonium oxalate needles (Figure 1a) have the advantage of producing continuous pores, in contrast to, e.g., sodium chloride, which is in the form of cubes. To achieve a ~1–5 mm length and prevent breaking of crystalline needles, a special sieve with longitudinal holes made on a 3D printer was used to classify the crystals (inset in Figure 1a). Ammonium oxalate has the additional advantage of introducing connectivity of the pores in places of contacts between the crystals, especially when the crystals are oriented in various directions (Figure 1c,f). These pore interconnections are important for prospective proper nutrient transport, cell adhesion, and possible ingrowth of blood vessels and/or bone tissue. It is worth mentioning that this uncontrolled pore orientation may lead to batch-to-batch variability influencing the consistency of results. In our experiments, a 15–60 μm pore diameter (Figure 1b) was chosen due to our previous successful experience with mesenchymal stem cells cultured in the pores of the same dimensions [29]. Such a diameter was also considered appropriate for the in vitro leukemia model since the size of cells ranged from 9 to 20 μm (Figure 3).

The superporous PHEMA hydrogel defined composition that was easy to control and replicable; the hydrogel was chemically and thermally stable, therefore sterilizable by ethanol or at high temperature, which made it suitable for cell culture. Moreover, nondegradability of PHEMA facilitated long-term incubations of cells in vitro. As the hydrogel porous structure did not collapse, it allowed the same support throughout the whole experiment. The hydrogel was easy to handle and could be transferred from well to well without any disruption.

All of the aforementioned aspects allowed long-term culture of HS-5, M2-10B4, MEC-1, and HG-3 cell lines in the P(HEMA-AEMA)-RGDS hydrogels for a minimum of 70 days (Figure 5a–d). The observed increase of metabolic activity of the cell line culture by day 20 (Figure 5a,b,d) could be explained by their log or lag phase of growth. On day 20, the cell lines probably reached their plateau, i.e., stationary phase, as well as excessive confluence, which temporarily led to the inhibition of contacts and increased death rate, hence, lower metabolic activity. Since the manipulation with 3D-cultured cell lines only involved medium exchange and no dilution, the cell lines had to develop their own mechanisms of coping with higher cell densities that exceeded medium capacity, e.g., by forming clumps that detached from the pores of scaffold and created space for the growth of new cells. Other adverse effects affecting the metabolic activity of cells might be following: (i) repeated exposure to AlamarBlue™ and its retention in the hydrogels, which could negatively influence cell viability, and metabolic activity; (ii) medium exchange that occurred every 2 to 3 days; thus, it depended on whether the measurement was performed after 2- or 3-day incubation. However, as the cancer cell line resilience dominated over the adverse effects, their metabolic activity mostly returned to the previous rate, which could be seen after 40–50 days, when the growth rate probably slightly exceeded the death rate.

Contrary to other cell lines, B-CLLs only have a short lifespan (Figures 5e, 6 and 7). However, the RGDS peptide proved to be essential for the cell-substrate adhesion of B-CLLs, since the initial number of seeded cells, their metabolic activity, and viability were higher in RGDS-modified scaffolds compared to unmodified ones (Figures 4a–c and 5e). Contrary to our expectations, the 3D culture itself did not affect prolongation of B-CLL survival, as their metabolic activity as well as viability dropped within 4–7 days (Figures 5e, 6 and 7), which can be explained by numerous theories suggested below.

The size of the pores, as well as their orientations and the process of cell seeding, led to the formation of cell clumps inside of the hydrogels (Figure 4b,c). The presence of the cell clumps can be beneficial, since the cells are in close contact, allowing efficient signal exchange and short-distance transport of growth factors. On the other hand, such proximity may later result in a deteriorated export of the cellular wastes from the pores, accounting for nutrient and oxygen deficiency. Although such hypoxia and malnutrition conditions are usually observed in the center of solid tumors or the cancer-cell spheroids in vitro, they induce abundant cell death, such as central necrosis [12], which may be one of the reasons why primary B-CLL metabolic activity dropped so fast. To avoid the diffusion-limited tumor growth, the B-CLLs could be dynamically cultured under direct perfusion flow [35], in capillary-like hollow fibers [36], or bioreactors using microgravity [10]. These might enhance the transport of medium components and also mechanically disrupt large clumps; nevertheless, such methods require special equipment.

Another explanation for reduced cell viability and metabolic activity is methodological. It should be noted that AlamarBlue™ is nontoxic for short periods, but cytotoxic in long time intervals (24–72 h) [37]. Thus, repeated exposure combined with resazurin retention in the hydrogels may have eventually damage the primary B-CLLs. The same applies to Sytox green, which was earlier considered as a nonpermeable and nontoxic fluorescent dye; however, its toxic effect was only evaluated during a 3-day incubation [38].

Surprisingly, coculture with M2-10B4, soluble IL4, or CD40L did not affect B-CLL survival, which is in disagreement with the existing literature [34,39]. The results suggest the distribution of the growth factors (either artificially supplemented or naturally produced by the cocultured cells) was insufficient under the tested culture conditions, or that the CLLs did not respond to the external stimuli provided by the BMSCs and the supplemented additives. To increase the probability of molecule delivery, peptides such as CD40L and IL-4 could be promising for direct immobilization on the scaffolds, since it has been already proven that, e.g., immobilized anti-immunoglobulin M provided a more potent CLL stimulus than a soluble one [40]. A boosted delivery of growth factors could also be achieved by transfecting BMSCs with vectors that encode human CD40L and IL-4, considering that B-CLLs and BMSCs tend to grow in proximity. Such a transfer might be preferable, assuming that membrane-bound CD40L shows a superior capacity to activate CD40 signaling and resembles natural interactions, where CD40L is probably delivered by cell–cell contacts [41].

Furthermore, patient variability (Table S1) did not cause differences between B-CLL in vitro survival (Figures 5e, 6 and 7), indicating that the CLL severity might have no impact on the B-CLL viability ex vivo, and/or there is a prevailing negative effect inducing B-CLL apoptosis before the differences can manifest. Let us emphasize that due to required high amount of cells used for the seeding (~50 million/scaffold), the results shown in Figures 5–7 only include a low number of replicates (n = 3). More research would be needed to identify and clarify any adverse effects.

Despite all the limitations, the RGDS-immobilized P(HEMA-AEMA) hydrogels can serve as a platform for studying B-CLL biology in vitro due to their partial transparency, which facilitates time-lapse studies of B-CLL migration or monitoring their interactions with adjacent cells in 3D. In addition, once the observed growth-specific vulnerabilities are clarified at the molecular level, they can potentially serve as therapeutic targets [42]. Last but not least, it was demonstrated that the direct DNA isolation from scaffold-cultured cells is efficient, which allows deep analysis, including sequencing. In the case of coculture with BMSCs, lines of nonhuman origin (e.g., M2-10B4) have to be used, to make the DNA from B-CLLs distinguishable by species-specific primers. Even though this adds one more step to the DNA analysis, the method is preferable to scaffold decellularization followed by cell sorting, since both these steps are also accompanied by high losses of cells, lowering thus DNA yield.

4. Materials and Methods

4.1. Materials

HEMA, EDMA, AEMA, AIBN, GMBS, ammonium oxalate, Ficoll® Paque Plus, penicillin-streptomycin (P/S), and sterile trypsin with 0.1% disodium ethylenediaminetetraacetate were obtained from Sigma-Aldrich (St. Luis, MO, USA). 1,4-Dioxane was from Fluka (Buchs, Switzerland). Phosphate buffer saline (PBS) was prepared from Na_2HPO_4 and KH_2PO_4. In-house prepared ammonium oxalate (LachNer; Neratovice, CR) crystals were obtained by repeated crystallization from aqueous solution at 50 °C with cooling to 23 °C; the crystals were then classified in saturated oxalate solution on a sieve (0.4 × 15 mm mesh). Calcein acetoxymethyl ester (AM), propidium iodide, Hoechst 33342, AlamarBlue™ cell viability reagent, CellTrace™ violet cell proliferation kit, and Sytox™ green nucleic acid stain were purchased from Thermo Fisher Scientific (Eugene, OR, USA). Reagents for peptide synthesis were from Iris Biotech (Marktredwitz, Germany). Dulbecco's Modified Eagle Medium (DMEM) with stable glutamine and sodium pyruvate, Iscove's Modified Dulbecco's Medium (IMDM) with stable glutamine and 25 mM 2-[4-(2-hydroxy-ethyl)piperazin-1-yl]ethanesulfonic acid (HEPES), RPMI 1640 medium with stable glutamine and 25 mM HEPES, and fetal bovine serum (FBS) were from Biosera (Nuaille, France). Recombinant human IL-4 and sCD40 ligand were purchased from Peprotech (Rocky Hill, NJ, USA). RosetteSep™ human B cell enrichment cocktail and CD3+ cell depletion cocktail were supplied by Stemcell™ Technologies (Vancouver, Canada). FastDNA sample spin kit for soil was obtained from MP Biomedicals (Santa Ana, CA, USA) and the genomic DNA screen tape was from Agilent (Santa Clara, CA, USA). An acridine orange/propidium iodide cell viability kit was supplied by Logos Biosystems (Anyang, Korea). Other chemicals were purchased from LachNer. Q-water was ultrafiltered on a Milli-Q Gradient A10 apparatus (Millipore; Molsheim, France).

Cell culture plates and flasks were manufactured by Techno Plastic Products (Trasadingen, Switzerland). The OptiPlate-96 White microplates were made by PerkinElmer (Waltham, MA, USA).

4.2. Synthesis of P(HEMA-AEMA) Scaffolds

HEMA (9.8 g), EDMA (0.1 g), AEMA (0.1 g), and AIBN (40 mg) were dissolved in 1,4-dioxane (5 mL) to obtain the monomer phase. Injection syringes (10 mL) were filled with ammonium oxalate crystals (4.33 mL) and the above-mentioned monomer phase (5.67 mL) and the mixture was polymerized at 60 °C for 16 h. Subsequently, the injection syringe was cut and the P(HEMA-AEMA) cylinder removed and washed with 10% aqueous ammonium chloride (60 mL) for three days to remove ammonium oxalate crystals. The cylinder was then transferred into water, washed with water ten times (50 mL each) for two days, and cut on cubes (4 × 4 mm), which were repeatedly washed with water until ammonium chloride was removed. The scaffolds were viewed with a Quanta FEG 200F scanning electron microscope (FEI; Brno, Czech Republic). The amount of nitrogen in the scaffolds was determined on a Perkin-Elmer 2400 CHN elemental analyzer (Norwalk, CT, USA).

4.3. Synthesis of Ac-CGGGRGDSGGGY-NH2 (RGDS) Peptide

The peptide was synthesized by a standard Fmoc/tBu solid-phase method on a TentaGel Rink Amide—R resin (0.18 mmol NH_2/g; Rapp Polymere; Tuebingen, Germany). An automatic CEM Liberty Blue microwave peptide synthesizer (Matthews, NC, USA) and software version 1.31.5252.26519 were used with default *N,N′*-diisopropylcarbodiimide/Oxyma Pure coupling and piperidine deprotection cycles. To avoid aspartimide formation, the pseudoproline Asp-Ser [Fmoc-Asp(OtBu)-Ser(Psi(Me,Me)pro)-OH] building block was used for incorporation of aspartic and serine amino acids. The peptide was cleaved from the resin with a CF_3COOH/triisopropylsilane/H_2O mixture (95/2.5/2.5 *v*/*v*/*v*) and isolated by the precipitation in diethyl ether. The purification was performed on a preparative Knauer HPLC system (Berlin, Germany) equipped with a Knauer diode-array detector and Kinetex LC column (5 μm, C18, 1000 Å, 250 × 21.2 mm) using a gradient elution with a water/acetonitrile mixture containing 0.1% formic acid. The identity of the peptide was confirmed on a Bruker Ultrafle Extreme MALDI-TOF mass spectrometer (Bremen, Germany) and Advion expression L CMS mass detector (Ithaca, NY, USA).

4.4. Activation of P(HEMA-AEMA) Scaffolds and Immobilization of RGDS Peptide

In a 200 mL Erlenmeyer flask, P(HEMA-AEMA) cubes were washed with water/ethanol (7:3 *v*/*v*) mixture (15 mL), the liquid was sucked away, and the cubes were immersed in a solution (5 mL) prepared from 0.07 M phosphate buffer (pH 7.4; 7 mL) and 1,4-dioxane (3 mL). GMBS (5 mg) in 1,4-dioxane (0.5 mL) was added and the mixture was shaken (50 rpm) at room temperature (RT) for 30 min. The cubes were washed twice with ethanol (10 mL each), twice with water (10 mL each), and with 0.1 M phosphate buffer (pH 6.8; 10 mL). The buffer was withdrawn, and a solution of RGDS peptide (2 mg) in 0.1 M phosphate buffer (pH 6.8; 6 mL) was added. The reaction continued at RT for 75 min with shaking (50 rpm), and the resulting P(HEMA-AEMA)-RGDS cubes were washed with water five times (10 mL each), sterilized in 60% aqueous ethanol (20 mL) for 16 h, washed with water to remove ethanol, and stored at 4 °C.

4.5. Primary B-CLL Cells and Cell Lines

Human B-CLL cell lines, MEC-1 (DSMZ ACC 497; Braunschweig, Germany) [32], and HG-3 (DSMZ ACC 765) [43] were used, being the only commercially available in vitro proliferating B-CLLs. Vital primary B-CLLs were obtained from patients diagnosed with CLL and treated at the Department of Internal Medicine, Hematology and Oncology, University Hospital Brno, Czech Republic. The patients signed their informed consent in accordance with the Declaration of Helsinki under protocols approved by the Ethical Committee of the University Hospital Brno (date of approval: 4 April 2018). Primary B-CLLs were separated from the peripheral blood using Ficoll® Paque Plus and Rosette Sep kits. The viability and number of cells directly after isolation (prior to cell culture) was evaluated in a Luna-FL™ dual fluorescence automated cell counter (Logos Biosystems; Anyang, South Korea) with an acridine orange/propidium iodide cell viability kit.

In order to mimic the microenvironment of the bone marrow, the B-CLL cells were cultured with adherent BMSCs, which support B-CLL survival in vitro [22,44]. BMSCs included human HS-5 cell line (CRL-11882™; ATCC; Manassas, VA, USA) [45] and murine M2-10B4 cell line (CRL-1972™; ATCC) [46].

For characterization of cell line sources, see Supporting Materials (SMs), Table S5. Patient's samples are described in Table S1.

4.6. Cell Culture Conditions

Cell cultures were incubated at 37 °C in a 5% CO_2 atmosphere in culture flasks in the medium recommended by the ATCC or DMSZ bioresource center and supplemented with 10% FBS and 1% P/S; medium was regularly exchanged two or three times a week. HS-5 or M210B4 were cultured in DMEM, MEC-1 in IMDM, and HG-3 or primary B-CLLs were incubated in RPMI 1640. Bone marrow stromal cells and primary B-CLLs were cocultured in RPMI 1640 medium supplemented with 20% FBS. In some cases, the cell suspensions were continuously shaken (30 rpm) on a Rocker 25 apparatus (Labnet International; Big Flats, NY, USA) at 37 °C under 5% CO_2 atmosphere. When the cell metabolic activity or viability was studied in the presence of CD40L and/or IL-4, the cells were stimulated continuously, as a single stimulation was insufficient to maintain cell division, survival, and differentiation [39].

4.7. Seeding of Cells in P(HEMA-AEMA) Scaffolds

Cells were seeded in the P(HEMA-AEMA) hydrogels according to a technique modified from an earlier published report [47]; schematic view of the seeding is shown in Figure 9. Prior to the seeding, the hydrogel cubes were placed in 2 mL round-bottom tubes filled with the corresponding medium (1 mL) and rotated (10 rpm; LabRoller II H5100; Labnet International; Big Flats, NY, USA) around the horizontal axis at RT for 45 min. Each scaffold was placed in the medium (250 µL) in the 96-well plate and cultured at 37 °C for 24 h under 5% CO_2 atmosphere. For BMSCs and B-CLLs coculture, the medium-imbibed cubes were placed in the BMSC suspension (1 mL) under the above-described rotation and culturing continued for 48 h. Finally, the freshly isolated B-CLL cell suspension (1 mL) with initial cell viability >98% was seeded in the scaffold with rotation and cultured in the medium in a 96-well plate for several days under continuous monitoring. The scaffolds were regularly (within one to three days) transferred to new wells containing fresh medium (250 µL each), always without any passaging, i.e., trypsinization.

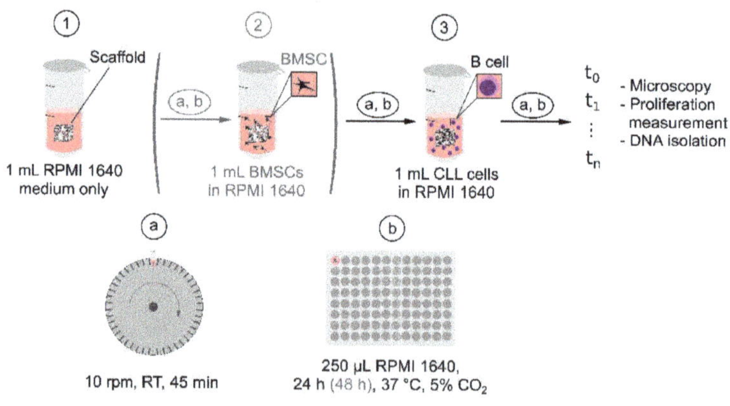

Figure 9. Schematic view of B cell (optionally BMSC—see the brackets) seeding in P(HEMA-AEMA) scaffold. Each imbibing (1) or seeding (2, 3) step was performed by placing the scaffold into the microtube with medium (1) or cell suspension (2, 3) followed by (**a**) rotation of the microtube around horizontal axis, and (**b**) transferring the scaffold into fresh medium in 96-well plate.

4.8. Cell Imaging and Image Analysis

For single time point measurements, including analysis of cell distribution and morphology, live and dead cells and nuclei of the 3D-cultured cells were stained with calcein AM (1:1000), propidium iodide (1:1000), and Hoechst 33342 (1:2000), respectively; these dyes are toxic, thus terminating the experiment [38,48,49]. To continuously monitor the 3D cell culture, the cells were initially stained with CellTrace violet and then with Sytox green (1:1000) each day, which was followed by washing with PBS (pH 7.4) three times. In both cases, the dyes were diluted in the corresponding medium with 1% P/S (without FBS) and the P(HEMA-AEMA)-RGDS cubes with incubated cells were kept at 37 °C for 45 min in the dark. The scaffolds were transferred in PBS (50 µL) and observed using a Zeiss LSM-800 confocal microscope (Jena, Germany) with z-axis stacking to scan multiple scaffold layers and tile scanning of adjacent positions in the x–y axis. At each time point, 10 square layers (area 8.64 mm^2) were scanned to the depth of 135 µm. After the acquisition, the images were postprocessed using ZEN and Fiji software (see Appendix A) [50].

4.9. Metabolic Activity

Metabolic activity was determined using AlamarBlue™ cell viability reagent diluted in 1:10 ratio (v/v) in complete medium with 10% FBS and 1% P/S. In a 96-well culture plate, diluted AlamarBlue™ reagent (150 µL) was added per well, each containing one P(HEMA-AEMA) cube. Incubation continued at 37 °C for 3 h in the dark, reagent (100 µL) surrounding the hydrogel was transferred into an OptiPlateTM 96-well plate without the lid, and fluorescence was measured using a Spark 10 M multimode microplate reader (Tecan; Männedorf, Switzerland) at 530 and 590 nm (excitation and emission, respectively). As AlamarBlue™ can interact with the cell culture medium, resulting in inconsistent fluorescence signal [51], P(HEMA-AEMA) cubes without the cells, but treated according to the same protocol, were used as a negative control. After the incubation with the reagent, the scaffolds were washed in PBS (pH 7.4) and put into a corresponding fresh complete medium.

4.10. DNA Isolation and Quality Control

DNA was isolated by FastDNA sample SPIN kit for soil according to the manufacturer's protocol 24 h after seeding the scaffolds with B-CLLs [52]. The following modifications were introduced to the protocol: homogenization was carried out for 8 min at 2000 rpm (Thermomixer C; Eppendorf; Hamburg, Germany); in the elution step, incubation continued at 55 °C for 5 min. The DNA quality was controlled by electrophoresis using a 4200 TapeStation system (Agilent; Santa Clara, CA, USA) with the genomic DNA screen tape as directed in the manual [53].

4.11. Statistical Analysis

Graphs were plotted using GraphPad Prism software, version 8.4.3 for Windows (San Diego, CA, USA). Vertical bars in the graphs denoted a 95% confidence interval. Statistical analysis was performed using Statistica, version 13.5.0.17 for Windows (Statsoft; Tulsa, OK, USA) with two-way ANOVA corrected for multiple comparisons, or using computing environment R [54] and ImerTest package, type III analysis of variance with Satterthwaite's method. p values < 0.05 were considered significant.

5. Conclusions

In the artificial scaffolds, pore size and type of pore structure play a key role in cell behavior. Interconnecting pores promote the removal of wastes, transport of nutrients, and facilitate proliferation and migration of the cells. This report demonstrates that the P(HEMA-AEMA)-RGDS hydrogel supported long-term viability and metabolic activity of B-CLL cell lines (MEC-1 and HG-3) and bone marrow stromal cell lines (HS-5 and M2-10B4), which mimic the microenvironment of bone marrow. This was confirmed by both confocal microscopy and AlamarBlue™ assay. Moreover, the culture of the primary

B-CLLs was performed, confirming that they survived in the 3D in vitro model for 4–6 days, independently on the cocultured cells or growth factors in the media. RGDS modification turned out to be beneficial for the survival of the B-CLLs in the culture, since it increased the probability of the cells to be seeded into the scaffold pores, promoting the cell–cell and cell–surface interactions. This is an important prerequisite for creating a 3D model of CLL, which could serve as a leukemia drug-testing platform that is much more realistic than the presently used 2D systems. As such, this model cannot generally replace phase II clinical trials, but it can supplement the use of animal in vivo protocols since it can exclude poor drug candidates, e.g., toxic and malfunctional ones, and thus lower financial and time costs. Our protocol for seeding the cells into the scaffolds does not apply for in vitro tests only, but it can be hypothetically extended to the xenograft models after transferring the scaffolds into immunodeficient animals (mice, rats, etc.). Additionally, the system can be subjected to thorough DNA analysis, as we introduced a novel methodology for DNA isolation of the 3D cocultured cells. Our new 3D model based on a RGDS-modified PHEMA hydrogel thus contributes to solving the etiology of chronic lymphocytic leukemia, a disease which remains incurable.

Supplementary Materials: The following are available online at https://www.mdpi.com/1422-0067/22/5/2376/s1: Table S1: Characterization of patients' samples; Table S2: Statistical evaluation of the effect of modification with RGDS peptide and plate shaking on cellular metabolic activity; Table S3: Statistical evaluation of coculture with M210B4 supported with medium flow on primary CLL cells (B-CLLs) survival; Table S4: Statistical evaluation of the effect of the interleukin 4 (IL-4) and CD40 ligand (CD-40L) added into shaken and unshaken medium on primary B-CLL metabolic activity; Table S5: Characterization of cell line sources.

Author Contributions: Conceptualization, H.S., V.P., R.S., J.B., M.D., Š.P., D.H. and V.P.; methodology, H.S., R.S., J.B., and Z.P.; validation, H.S. and Z.P.; formal analysis, H.S., V.P., and D.H.; investigation, H.S., Z.P., R.S., and J.B.; resources, V.P., M.D., Š.P., and D.H.; data curation, H.S., V.P., Z.P., R.S., J.B., and M.D.; writing—original draft preparation, H.S., V.P., and D.H.; writing—review and editing, H.S., Z.P., V.P., R.S., J.B., M.D., Š.P., and D.H.; visualization, H.S., Z.P., and D.H.; supervision, Š.P. and D.H.; project administration, H.S., R.S., Š.P., and D.H.; funding acquisition, Š.P., D.H., and V.P. All authors have read and agreed to the published version of the manuscript.

Funding: This research has been supported by the Ministry of Education, Youth and Sports of the Czech Republic (MEYS CR) under the projects CEITEC 2020 (LQ1601) and MUNI/A/1595/2020, the Czech Science Foundation (No. 20-07015S and 18-05510S) and the Ministry of Health of the Czech Republic under the research grant DRO (FNBr, 65269705). The first author is a Brno PhD. talent scholarship holder funded by the Brno city municipality.

Institutional Review Board Statement: The study was conducted according to the guidelines of the Declaration of Helsinki, and approved by the Ethics Committee of University Hospital Brno (date of approval: 4 April 2018, registration number SUp 8/18, University Hospital Brno).

Informed Consent Statement: Informed consent was obtained from all subjects involved in the study.

Data Availability Statement: All data presented in this study are contained within this article or its supplementary materials. Further details are available upon request from the corresponding author.

Acknowledgments: We would like to thank Daniela Kuruczova and Francesco Daniel Muto for statistical analysis and language correction, respectively. We acknowledge the core facility CELLIM supported by the Czech-BioImaging large RI project (LM2018129 funded by MEYS CR) for their support with obtaining scientific data presented in this paper.

Conflicts of Interest: The authors declare no conflict of interest. The funders had no role in the design of the study; in the collection, analyses, or interpretation of data; in the writing of the manuscript, or in the decision to publish the results.

Appendix A

Postprocessing of Microscopy Images

The images were postprocessed using ZEN build-in plug-ins. Multiple tiles in the x–y axis were combined with a stitching tool and the 3D image was converted into the 2D one by orthogonal projection, selecting the pixels with the maximum intensity from all the layers in the z-axis. The images were then processed using the open source Fiji software. Prior to counting, the image was converted to the mask using following plug-ins: (i) the background noise was removed and signal of overexposed areas was lowered by the "subtract background" plug-in, (ii) the image was converted to 8-bit by the "adjust threshold" plug-in, choosing the optimal threshold manually to avoid omission of any cell, (iii) the "watershed" plug-in drew a 1-pixel line between adjacent cells growing in the clusters to count each cell and not the clump, and (iv) the "analyze particle" plug-in counted the particles above the defined size (10-infinity).

References

1. Kipps, T.J.; Stevenson, F.K.; Wu, C.J.; Croce, C.M.; Packham, G.; Wierda, W.G.; O'Brien, S.; Gribben, J.; Rai, K. Chronic Lymphocytic Leukaemia. *Nat. Rev. Dis. Primer* **2017**, *3*, 16096. [CrossRef]
2. Hallek, M. Chronic Lymphocytic Leukemia: 2020 Update on Diagnosis, Risk Stratification and Treatment. *Am. J. Hematol.* **2019**, *94*, 1266–1287. [CrossRef]
3. Gribben, J.G. How I Treat CLL up Front. *Blood* **2010**, *115*, 187–197. [CrossRef] [PubMed]
4. Bosch, F.; Dalla-Favera, R. Chronic Lymphocytic Leukaemia: From Genetics to Treatment. *Nat. Rev. Clin. Oncol.* **2019**, *16*, 684–701. [CrossRef]
5. Satpathy, A.; Datta, P.; Wu, Y.; Ayan, B.; Bayram, E.; Ozbolat, I.T. Developments with 3D Bioprinting for Novel Drug Discovery. *Expert Opin. Drug Discov.* **2018**, *13*, 1115–1129. [CrossRef] [PubMed]
6. Chew, S.A.; Moscato, S.; George, S.; Azimi, B.; Danti, S. Liver Cancer: Current and Future Trends Using Biomaterials. *Cancers* **2019**, *11*, 2026. [CrossRef] [PubMed]
7. Li, D.; Lin, T.L.; Lipe, B.; Hopkins, R.A.; Shinogle, H.; Aljitawi, O.S. A Novel Extracellular Matrix-Based Leukemia Model Supports Leukemia Cells with Stem Cell-like Characteristics. *Leuk. Res.* **2018**, *72*, 105–112. [CrossRef]
8. Zhang, C.; Yang, Z.; Dong, D.-L.; Jang, T.-S.; Knowles, J.C.; Kim, H.-W.; Jin, G.-Z.; Xuan, Y. 3D Culture Technologies of Cancer Stem Cells: Promising Ex Vivo Tumor Models. *J. Tissue Eng.* **2020**, *11*. [CrossRef]
9. Burger, J.A.; Gribben, J.G. The Microenvironment in Chronic Lymphocytic Leukemia (CLL) and Other B Cell Malignancies: Insight into Disease Biology and New Targeted Therapies. *Semin. Cancer Biol.* **2014**, *24*, 71–81. [CrossRef]
10. Barbaglio, F.; Belloni, D.; Scarfò, L.; Sbrana, F.V.; Ponzoni, M.; Bongiovanni, L.; Pavesi, L.; Zambroni, D.; Stamatopoulos, K.; Caiolfa, V.R.; et al. 3D Co-Culture Model of Chronic Lymphocytic Leukemia Bone Marrow Microenvironment Predicts Patient-Specific Response to Mobilizing Agents. *Haematologica* **2020**. [CrossRef]
11. Dos Santos, J.; Enfield, L.; Dos Santos, S.B.; Allenby, M.C.; Zemenides, S.; Mantalaris, A.; Panoskaltsis, N. Primary Chronic Lymphocytic Leukemia Cells Can Be Maintained Long-Term in Serum-Free, Cytokine-Free 3D Culture. *Blood* **2017**, *130*, 2989. [CrossRef]
12. Verjans, E.-T.; Doijen, J.; Luyten, W.; Landuyt, B.; Schoofs, L. Three-Dimensional Cell Culture Models for Anticancer Drug Screening: Worth the Effort? *J. Cell. Physiol.* **2018**, *233*, 2993–3003. [CrossRef] [PubMed]
13. Datta, P.; Dey, M.; Ataie, Z.; Unutmaz, D.; Ozbolat, I.T. 3D Bioprinting for Reconstituting the Cancer Microenvironment. *Npj Precis. Oncol.* **2020**, *4*, 1–13. [CrossRef]
14. El-Sherbiny, I.M.; Yacoub, M.H. Hydrogel Scaffolds for Tissue Engineering: Progress and Challenges. *Glob. Cardiol. Sci. Pract.* **2013**, *2013*, 316–342. [CrossRef]
15. Dhandayuthapani, B.; Yoshida, Y.; Maekawa, T.; Kumar, D.S. Polymeric Scaffolds in Tissue Engineering Application: A Review. *Int. J. Polym. Sci.* **2011**. [CrossRef]
16. Lee, J.; Cuddihy, M.J.; Kotov, N.A. Three-Dimensional Cell Culture Matrices: State of the Art. *Tissue Eng. Part B Rev.* **2008**, *14*, 61–86. [CrossRef] [PubMed]
17. Drury, J.L.; Mooney, D.J. Hydrogels for Tissue Engineering: Scaffold Design Variables and Applications. *Biomaterials* **2003**, *24*, 4337–4351. [CrossRef]
18. Zhu, J.; Marchant, R.E. Design Properties of Hydrogel Tissue-Engineering Scaffolds. *Expert Rev. Med. Devices* **2011**, *8*, 607–626. [CrossRef] [PubMed]
19. Atzet, S.; Curtin, S.; Trinh, P.; Bryant, S.; Ratner, B. Degradable Poly(2-Hydroxyethyl Methacrylate)-*co*-Polycaprolactone Hydrogels for Tissue Engineering Scaffolds. *Biomacromolecules* **2008**, *9*, 3370–3377. [CrossRef] [PubMed]
20. Kůdela, J. Hydrogels. In *Encyclopedia of Polymer Science and Technology*; Wiley: New York, 1987; Volume 7, pp. 783–807.
21. ten Hacken, E.; Burger, J.A. Microenvironment Dependency in Chronic Lymphocytic Leukemia: The Basis for New Targeted Therapies. *Pharmacol. Ther.* **2014**, *144*, 338–348. [CrossRef]

22. Lagneaux, L.; Delforge, A.; Bron, D.; De Bruyn, C.; Stryckmans, P. Chronic Lymphocytic Leukemic B Cells but Not Normal B Cells Are Rescued from Apoptosis by Contact with Normal Bone Marrow Stromal Cells. *Blood* **1998**, *91*, 2387–2396. [CrossRef] [PubMed]
23. Crassini, K.; Shen, Y.; Mulligan, S.; Giles Best, O. Modeling the Chronic Lymphocytic Leukemia Microenvironment in Vitro. *Leuk. Lymphoma* **2017**, *58*, 266–279. [CrossRef] [PubMed]
24. Jabs, J.; Zickgraf, F.M.; Park, J.; Wagner, S.; Jiang, X.; Jechow, K.; Kleinheinz, K.; Toprak, U.H.; Schneider, M.A.; Meister, M.; et al. Screening Drug Effects in Patient-Derived Cancer Cells Links Organoid Responses to Genome Alterations. *Mol. Syst. Biol.* **2017**, *13*, 955. [CrossRef]
25. Sommerová, L.; Michalová, E.; Hrstka, R. New approaches for chemosensitivity testing in malignant diseases. *Klin. Onkol. Cas. Ceske Slov. Onkol. Spolecnosti* **2018**, *31*, 117–124. [CrossRef]
26. Lee, J.; Li, M.; Milwid, J.; Dunham, J.; Vinegoni, C.; Gorbatov, R.; Iwamoto, Y.; Wang, F.; Shen, K.; Hatfield, K.; et al. Implantable Microenvironments to Attract Hematopoietic Stem/Cancer Cells. *Proc. Natl. Acad. Sci. USA* **2012**, *109*, 19638–19643. [CrossRef] [PubMed]
27. Turnbull, G.; Clarke, J.; Picard, F.; Riches, P.; Jia, L.; Han, F.; Li, B.; Shu, W. 3D Bioactive Composite Scaffolds for Bone Tissue Engineering. *Bioact. Mater.* **2018**, *3*, 278–314. [CrossRef]
28. Kubinová, Š.; Horák, D.; Syková, E. Cholesterol-Modified Superporous Poly(2-Hydroxyethyl Methacrylate) Scaffolds for Tissue Engineering. *Biomaterials* **2009**, *30*, 4601–4609. [CrossRef]
29. Macková, H.; Plichta, Z.; Proks, V.; Kotelnikov, I.; Kučka, J.; Hlídková, H.; Horák, D.; Kubinová, Š.; Jiráková, K. RGDS- and SIKVAVS-Modified Superporous Poly(2-Hydroxyethyl Methacrylate) Scaffolds for Tissue Engineering Applications. *Macromol. Biosci.* **2016**, *16*, 1621–1631. [CrossRef]
30. Singh, S.; Ghode, S.; Devi, M.R.; Limaye, L.; Kale, V. Phenotypic and Functional Characterization of a Marrow-Derived Stromal Cell Line, M210B4 and Its Comparison with Primary Marrow Stromal Cells. *Biomed. Res. J.* **2015**, *2*, 120. [CrossRef]
31. Kurtova, A.V.; Balakrishnan, K.; Chen, R.; Ding, W.; Schnabl, S.; Quiroga, M.P.; Sivina, M.; Wierda, W.G.; Estrov, Z.; Keating, M.J.; et al. Diverse Marrow Stromal Cells Protect CLL Cells from Spontaneous and Drug-Induced Apoptosis: Development of a Reliable and Reproducible System to Assess Stromal Cell Adhesion-Mediated Drug Resistance. *Blood* **2009**, *114*, 4441–4450. [CrossRef]
32. Stacchini, A.; Aragno, M.; Vallario, A.; Alfarano, A.; Circosta, P.; Gottardi, D.; Faldella, A.; Rege-Cambrin, G.; Thunberg, U.; Nilsson, K.; et al. MEC1 and MEC2: Two New Cell Lines Derived from B-Chronic Lymphocytic Leukaemia in Prolymphocytoid Transformation. *Leuk. Res.* **1999**, *23*, 127–136. [CrossRef]
33. German Collection of Microorganisms and Cell Cultures GmbH: Details. Available online: https://www.dsmz.de/collection/catalogue/details/culture/ACC-765 (accessed on 22 October 2019).
34. Ghia, P.; Circosta, P.; Scielzo, C.; Vallario, A.; Camporeale, A.; Granziero, L.; Caligaris-Cappio, F. Differential effects on CLL cell survival exerted by different microenvironmental elements. In *Chronic Lymphocytic Leukemia*; Current Topics in Microbiology and Immunology; Springer: Berlin, Heidelberg, 2005; pp. 135–145. ISBN 978-3-540-29933-2.
35. Bourgine, P.E.; Klein, T.; Paczulla, A.M.; Shimizu, T.; Kunz, L.; Kokkaliaris, K.D.; Coutu, D.L.; Lengerke, C.; Skoda, R.; Schroeder, T.; et al. In Vitro Biomimetic Engineering of a Human Hematopoietic Niche with Functional Properties. *Proc. Natl. Acad. Sci. USA* **2018**, *115*, E5688–E5695. [CrossRef]
36. Walsby, E.; Buggins, A.; Devereux, S.; Jones, C.; Pratt, G.; Brennan, P.; Fegan, C.; Pepper, C. Development and Characterization of a Physiologically Relevant Model of Lymphocyte Migration in Chronic Lymphocytic Leukemia. *Blood* **2014**, *123*, 3607–3617. [CrossRef]
37. Nakayama, G.R.; Caton, M.C.; Nova, M.P.; Parandoosh, Z. Assessment of the Alamar Blue Assay for Cellular Growth and Viability in Vitro. *J. Immunol. Methods* **1997**, *204*, 205–208. [CrossRef]
38. Chiaraviglio, L.; Kirby, J.E. Evaluation of Impermeant, DNA-Binding Dye Fluorescence as a Real-Time Readout of Eukaryotic Cell Toxicity in a High Throughput Screening Format. *Assay Drug Dev. Technol.* **2014**, *12*, 219–228. [CrossRef]
39. Rush, J.S.; Hodgkin, P.D. B Cells Activated via CD40 and IL-4 Undergo a Division Burst but Require Continued Stimulation to Maintain Division, Survival and Differentiation. *Eur. J. Immunol.* **2001**, *31*, 1150–1159. [CrossRef]
40. Rombout, A.; Lust, S.; Offner, F.; Naessens, E.; Verhasselt, B.; Philippé, J. Mimicking the Tumour Microenvironment of Chronic Lymphocytic Leukaemia in Vitro Critically Depends on the Type of B-Cell Receptor Stimulation. *Br. J. Cancer* **2016**, *114*, 704–712. [CrossRef] [PubMed]
41. Natoni, A.; O'Dwyer, M.; Santocanale, C. A Cell Culture System That Mimics Chronic Lymphocytic Leukemia Cells Microenvironment for Drug Screening and Characterization. *Methods Mol. Biol. Clifton NJ* **2013**, *986*, 217–226. [CrossRef]
42. Han, K.; Pierce, S.E.; Li, A.; Spees, K.; Anderson, G.R.; Seoane, J.A.; Lo, Y.-H.; Dubreuil, M.; Olivas, M.; Kamber, R.A.; et al. CRISPR Screens in Cancer Spheroids Identify 3D Growth-Specific Vulnerabilities. *Nature* **2020**, *580*, 136–141. [CrossRef]
43. Rosén, A.; Bergh, A.-C.; Gogok, P.; Evaldsson, C.; Myhrinder, A.L.; Hellqvist, E.; Rasul, A.; Björkholm, M.; Jansson, M.; Mansouri, L.; et al. Lymphoblastoid Cell Line with B1 Cell Characteristics Established from a Chronic Lymphocytic Leukemia Clone by in Vitro EBV Infection. *Oncoimmunology* **2012**, *1*, 18–27. [CrossRef]
44. Crompot, E.; Van Damme, M.; Pieters, K.; Vermeersch, M.; Perez-Morga, D.; Mineur, P.; Maerevoet, M.; Meuleman, N.; Bron, D.; Lagneaux, L.; et al. Extracellular Vesicles of Bone Marrow Stromal Cells Rescue Chronic Lymphocytic Leukemia B Cells from Apoptosis, Enhance Their Migration and Induce Gene Expression Modifications. *Haematologica* **2017**, *102*, 1594–1604. [CrossRef]

45. Roecklein, B.A.; Torok-Storb, B. Functionally Distinct Human Marrow Stromal Cell Lines Immortalized by Transduction with the Human Papilloma Virus E6/E7 Genes. *Blood* **1995**, *85*, 997–1005. [CrossRef]
46. Lemoine, F.M.; Humphries, R.K.; Abraham, S.D.; Krystal, G.; Eaves, C.J. Partial Characterization of a Novel Stromal Cell-Derived Pre-B-Cell Growth Factor Active on Normal and Immortalized Pre-B Cells. *Exp. Hematol.* **1988**, *16*, 718–726.
47. Thevenot, P.; Nair, A.; Dey, J.; Yang, J.; Tang, L. Method to Analyze Three-Dimensional Cell Distribution and Infiltration in Degradable Scaffolds. *Tissue Eng. Part C Methods* **2008**, *14*, 319–331. [CrossRef]
48. Jonsson, B.; Liminga, G.; Csoka, K.; Fridborg, H.; Dhar, S.; Nygren, P.; Larsson, R. Cytotoxic Activity of Calcein Acetoxymethyl Ester (Calcein/AM) on Primary Cultures of Human Haematological and Solid Tumours. *Eur. J. Cancer Oxf. Engl. 1990* **1996**, *32A*, 883–887. [CrossRef]
49. Durand, R.E.; Olive, P.L. Cytotoxicity, Mutagenicity and DNA Damage by Hoechst 33342. *J. Histochem. Cytochem. Off. J. Histochem. Soc.* **1982**, *30*, 111–116. [CrossRef]
50. Schindelin, J.; Arganda-Carreras, I.; Frise, E.; Kaynig, V.; Longair, M.; Pietzsch, T.; Preibisch, S.; Rueden, C.; Saalfeld, S.; Schmid, B.; et al. Fiji: An Open-Source Platform for Biological-Image Analysis. *Nat. Methods* **2012**, *9*, 676–682. [CrossRef] [PubMed]
51. Munshi, S.; Twining, R.C.; Dahl, R. Alamar Blue Reagent Interacts with Cell-Culture Media Giving Different Fluorescence over Time: Potential for False Positives. *J. Pharmacol. Toxicol. Methods* **2014**, *70*, 195–198. [CrossRef] [PubMed]
52. FastDNATM SPIN Kit for Soil, MP Biomedicals—Instruction Manual. Available online: https://media.mpbio.com/productattachment/LS082019-EN-FastDNA-SPIN-Kit-for-Soil-116560200-Manual.pdf (accessed on 22 January 2021).
53. Agilent Genomic DNA Screentape—Quick Guide for TapeStation Systems. Available online: https://www.agilent.com/cs/library/usermanuals/public/gDNA_QuickGuide.pdf (accessed on 3 December 2019).
54. R Core Team. *R: A Language and Environment for Statistical Computing*; R Foundation for Statistical Computing: Vienna, Austria, 2020.

International Journal of *Molecular Sciences*

MDPI

Article

β-Cyclodextrin/Triclosan Complex-Grafted Methacrylated Glycol Chitosan Hydorgel by Photocrosslinking via Visible Light Irradiation for a Tissue Bio-Adhesive

Young Jae Moon [1,†], Sun-Jung Yoon [2,†], Jeung-Hyun Koo [1], Yihyun Yoon [3], Hye Jun Byun [3], Hyeon Soo Kim [3], Gilson Khang [4], Heung Jae Chun [3,5] and Dae Hyeok Yang [3,*]

1. Department of Biochemistry & Molecular Biology & Orthopaedic Surgery, Research Institute for Endocrine Sciences, Jeonbuk National University Hospital, Jeonbuk National University Medical School, Jeonju 54896, Korea; yjmoonos@jbnu.ac.kr (Y.J.M.); jhyuni81@jbnu.ac.kr (J.-H.K.)
2. Department of Orthopedic Surgery, Research Institute of Clinical Medicine of Jeonbuk National University, Biomedical Research Institute of Jeonbuk National University Hospital, Jeonbuk National University Medical School, Jeonju 54896, Korea; sunjungyoon@jbnu.ac.kr
3. Institute of Cell and Tissue Engineering, College of Medicine, The Catholic University of Korea, Seoul 06591, Korea; 21903676@cmcnu.or.kr (Y.Y.); 21904860@cmcnu.or.kr (H.J.B.); 21905010@cmcnu.or.kr (H.S.K.); chunhj@catholic.ac.kr (H.J.C.)
4. Department of BIN Convergence Technology & Polymer Nano Science & Technology and Polymer BIN Research Center, Jeonbuk National University, Jeonju 54896, Korea; gskhang@jbnu.ac.kr
5. Department of Biomedical & Health Sciences, College of Medicine, The Catholic University of Korea, Seoul 06591, Korea

* Correspondence: yangdh@catholic.ac.kr; Tel.: +82-2-2258-7497

† These authors contributed equally to this study.

Abstract: Accelerating wound healing with minimized bacterial infection has become a topic of interest in the development of the new generation of tissue bio-adhesives. In this study, we fabricated a hydrogel system (MGC-g-CD-ic-TCS) consisting of triclosan (TCS)-complexed beta-cyclodextrin (β-CD)-conjugated methacrylated glycol chitosan (MGC) as an antibacterial tissue adhesive. Proton nuclear magnetic resonance (^{1}H NMR) and differential scanning calorimetry (DSC) results showed the inclusion complex formation between MGC-g-CD and TCS. The increase of storage modulus (G′) of MGC-g-CD-ic-TCS after visible light irradiation for 200 s indicated its hydrogelation. The swollen hydrogel in aqueous solution resulted in two release behaviors of an initial burst and sustained release. Importantly, in vitro and in vivo results indicated that MGC-g-CD-ic-TCS inhibited bacterial infection and improved wound healing, suggesting its high potential application as an antibacterial tissue bio-adhesive.

Keywords: methacrylate glycol chitosan; beta-cyclodextrin; triclosan; antibacterial; tissue bio-adhesive

Citation: Moon, Y.J.; Yoon, S.-J.; Koo, J.-H.; Yoon, Y.; Byun, H.J.; Kim, H.S.; Khang, G.; Chun, H.J.; Yang, D.H. β-Cyclodextrin/Triclosan Complex-Grafted Methacrylated Glycol Chitosan Hydorgel by Photocrosslinking via Visible Light Irradiation for a Tissue Bio-Adhesive. *Int. J. Mol. Sci.* **2021**, *22*, 700. https://doi.org/10.3390/ijms22020700

Received: 15 December 2020
Accepted: 8 January 2021
Published: 12 January 2021

1. Introduction

Self-healing tissue adhesives have been regarded as attractive sealants for wound closure in surgery and trauma handling because they perform several important functions including promoting wound healing acceleration by preferably binding to tissues, as well as sealing leaks and stopping tissue adhesion [1]. However, cyanoacrylate-based adhesives are very toxic due to the formaldehyde that result from their degradation, and fibrin glues have weak tensile and adhesive strength, and problems with viral contamination [2,3]. Therefore, new types of tissue adhesives that overcome the drawbacks of the existing adhesives should be developed.

Hydrogels are promising candidates as self-healing tissue adhesives, since damaged wounds can be regenerated by their hydrophilic and crosslinked structures, such as soft tissues, and high permeability to oxygen, nutrients, and water-soluble metabolites [1]. In addition, hydrogels that can load drugs and proteins allow for the improvement in tissue regeneration [1,4]. In particular, antibacterial agent-containing hydrogels are currently a main focus in biomedical research and applications such as a surgical-site infection, because infection is a critical issue for successful wound healing [5,6]. Among the available antibacterial agents, triclosan (TCS) is a commonly used drug because of its broad spectrum of antibacterial activity, and TCS is widely used from industry to medical fields [7]. In medical fields, a TCS-coated suture is being used clinically.

Among the basic materials for hydrogels, water-soluble chitosan derivatives are coming into the spotlight as self-healing tissue adhesives [8,9]. These chitosan adhesives are potentiated by hydrogen bonding and electrostatic interaction between the polymer and collagen fibers [10]. The chitosan derivatives can be easily designed in the hydrogel form under an aqueous condition because amine groups in the polymer backbones have functional groups by various modification techniques, followed by hydrogel formation by various stimuli such as light, chemical, and temperature stimuli [11–16]. Ultraviolet/visible light-curing systems are efficient tools for preparing hydrogel tissue adhesives because they can be controlled by manipulating the presence and dose of light, and independence of spatial or temporal control of different biological processes in a wavelength-specific manner [17–19]. Notably, a visible light-curing system is safer than a system using ultraviolet light, because ultraviolet light can lead to melanoma and suppress the function of the immune system [20].

In this work, we prepared a self-healing hydrogel tissue bio-adhesive (MGC-g-CD-ic-TCS) with antibacterial activity based on visible light-cured methacrylated glycol chitosan (MGC) and succinyl-beta-cyclodextrin/triclosan (β-CD-ic-TCS) for wound closure. The efficacy was evaluated using a rat model of skin incision (Figure 1). A methacrylic group and succinyl-β-CD (β-CD-COOH) were conjugated to water-soluble GC for a visible light-curing system (MGC-g-CD). The water solubility of TCS was improved by host-guest inclusion complex with MGC-g-CD, which was characterized by proton nuclear magnetic resonance (^{1}H NMR) and differential scanning calorimetry (DSC) analyses. The MGC bio-adhesive was prepared using a blue visible light irradiation with riboflavin as a photoinitiator, which were characterized using rheology, in vitro drug release, in vitro cytotoxicity, and antibacterial activity assays. The tissue adhesion properties of MGC-g-CD-ic-TCS were compared with those of commercially used adhesives such as fibrin glue and cyanoacrylate. In addition, the self-healing capacity of MGC-g-CD-ic-TCS was compared with that of untreated incision, suture, cyanoacrylate, and MGC hydrogel.

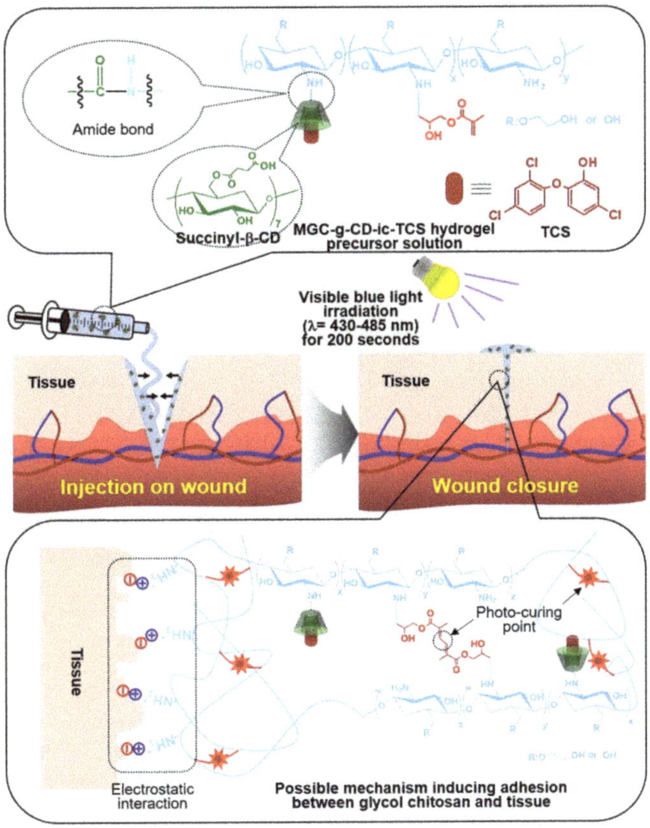

Figure 1. Schematic illustration on the application of β-cyclodextrin/triclosan complex-grafted methacrylated glycol chitosan (MGC-g-CD-ic-TCS) as a tissue bio-adhesive and possible mechanism of adhesion between MGC-g-CD-ic-TCS and soft tissue.

2. Results and Discussion

2.1. ^{1}H NMR Spectra of MGC-g-CD and MGC-g-CD-ic-TCS

Conjugation of succinyl-β-CD to the MGC backbone and the inclusion complex between MGC-g-CD and TCS were confirmed by ^{1}H NMR analysis in D_2O (Figure 2). We previously reported the successful preparation of MGC formed by reaction between GC and glycidyl methacrylate through ^{1}H NMR analysis using D_2O [11–14]. The ^{1}H NMR spectrum of MGC exhibited newly formed peaks at 2.02 ppm, 5.69 ppm, and 6.11 ppm assigned for the methyl and vinyl groups of methacrylation. Consistent with our previous reports [11–14], the ^{1}H NMR spectrum of MGC-g-CD also showed peaks related to methacrylation at the same locations, indicating formation of MGC. The degree of substitution (DS) of the methyl group was calculated in consideration of the integration ratio between the glucopyranosyl ring peak area of GC at 3.23–4.21 ppm and the -CH_3 peak area of the methacrylic group at 1.89 ppm, resulting in a methacrylate of 70%. MGC-g-CD produced new peaks at 2.42 ppm and 2.59 ppm, and 5.00 ppm, which were assigned for -CH_2CH_2- of the succinyl group and H-1 of the glucopyranosyl group, respectively [21,22]. In addition, the DS of β-CD calculated from the integration ratio between the glucopyranosyl ring peak at 3.23-4.21 ppm and the -CH_2- peak area of succinyl group at 2.34 ppm was 24%. Inclusion complex formation between MGC-g-CD and TCS made the appearance of new peaks possible, because ^{1}H NMR analysis of poorly water-soluble TCS is impossible in D_2O. The peaks of complexed TCS were observed at 6.45 ppm, 6.64 ppm, 6.88 ppm, 6.96 ppm, 7.30 ppm, and 7.56 ppm. In comparison with the -CH_2CH_2- peak of the succinyl group, the inclusion complex percentage of TCS was almost 100%.

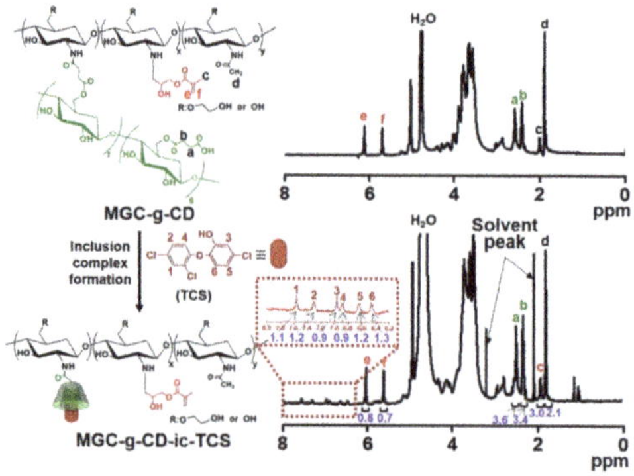

Figure 2. ^{1}H NMR spectra of MGC-g-CD and MGC-g-CD-ic-TCS analyzed using D_2O. a and b—CH2CH2- peak of succinyl β-CD; c, e, and f-methyl and vinyl groups of glycidyl methacrylate; d-methyl group of glycol chitosan. The numbers from 1 to 6 in the dotted line indicated triclosan.

2.2. DSC Curve and Storage/Loss Moduli of MGC-g-CD-ic-TCS

To further investigate the inclusion complex formation between MGC-g-CD and TCS, DSC was employed and compared with the results of MGC-g-CD and TCS (Figure 3A). TCS possesses a strong endothermic peak around 59 °C [23]. In a previous study, if an inclusion complex between cyclodextrins and TCS was successfully formed, the endothermic peak of TCS disappeared [23]. As shown in Figure 3A, MGC-g-CD-ic-TCS did not show the endothermic peak around 59 °C. Based on the previous study, we suggest that TCS is included into the ring molecules of MGC-g-CD-ic-TCS without uncomplexed TCS.

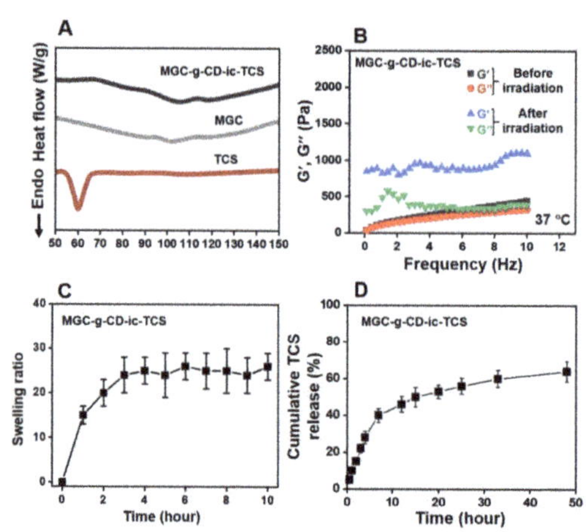

Figure 3. (**A**) DSC curves of triclosan (TCS), methacrylated glycol chitosan (MGC), and MGC-g-CD-ic-TCS monitored from 50 °C to 150 °C, (**B**) storage/loss moduli of MGC-g-CD-ic-TCS hydrogel precursor solution measured from 0 Hz to 10 Hz before and after visible light irradiation for 200 s, (**C**) swelling ratio of MGC-g-CD-ic-TCS measured at determined time intervals (0, 1, 2, 3, 4, 5, 6, 7, 8, 9, and 10 h), (**D**) Cumulative TCS release percentage calculated at determined time intervals (0.5, 1, 2, 3, 4, 7, 12, 15, 20, 25, 33, and 48 h). Swelling ratio and release behavior tests were performed in triplicate (n = 3).

The storage modulus of MGC-g-CD-ic-TCS hydrogel precursor solution before and after visible light irradiation for 120 s was monitored from 0 Hz to 10 Hz at 37 °C (Figure 3B). Visible light irradiation enhanced the storage modulus of the precursor solution. Before irradiation, the storage/loss moduli at 0 Hz and 10 Hz were 0/0 Pa and 440/326 Pa, respectively, but the two frequencies of the sample after irradiation were 850/300 Pa and 1098/392 Pa, respectively. This improved storage modulus may be from a three-dimensional (3-D) network formed from photocuring among the methacrylic groups of MGC-g-CD. This 3-D structure has a significant influence on the release behavior of hydrophobic triclosan [24]. Hydrogels exhibit controlled and sustained release of hydrophobic drugs [24]; this release behavior is closely related to the swelling property of hydrogels. Because of the open spaces in the polymer chains of hydrogels, the matrices allow for the absorption of solute and drugs and the diffusion of the drugs [25]. In addition, β-CD conjugated to polymer backbones in a hydrogel matrix makes it easier to release hydrophobic drugs [26]. Here, we conducted the inclusion complex formation between MGC-g-CD and TCS, and the swelling ratio and release test were evaluated (Figure 3C,D). The results were in accordance with the correlation between the swelling ratio and release behavior of hydrogels. MGC-g-CD-ic-TCS hydrogel was swollen by 3 h due to the migration of water molecules. Since then, the hydrogel remained maintained an equilibrium state for 24 h (Figure 3C). Figure 3D shows the release behavior of TCS from MGC-g-CD-ic-TCS over 7 days. The release behavior exhibited two patterns: an initial burst of 40% for 7 h and the sustained release of 64% for 2 days.

2.3. *In Vitro Cell Proliferation Ratio*

The cell proliferation test of TCS, MGC and MGC-g-CD-ic-TCS was performed at 37 °C for 7 days using a cell counting kit-8 (CCK-8) assay (Figure 4). A gradual increase in cell proliferation was observed in all samples for 7 days. TCS is known to have cytocompatibility below the concentration of 5 μg/mL, but above the concentration, it induces cytotoxicity by stimulating apoptosis [27,28]. Considering this fact, the concentration (1.8 μg/mL) of TCS used in this study is within biocompatible range. Although there was no marked difference of the cell proliferation ratio among the samples, MGC-g-CD-ic-TCS exhibited a higher cell proliferation ratio than TCS and MGC. In addition, the cell proliferation ratios of TCS, MGC and MGC-g-CD-ic-TCS at day 7 were 11.1-, 11.3-, and 10.4-fold higher than those at day 0. Some previous studies reported that photocured MGC hydrogels initiated by riboflavin can be used as biocompatible platforms for tissue engineering applications [11–14,29,30]. TCS has been reported to have toxicity against fish, algae, and aquatic organisms [24]. Hence, to reduce the side effects of TCS, hydrogels were introduced because they do not have direct contact with cells. Our results showed that MGC-g-CD-ic-TCS has a higher cell proliferation rate than TCS and MGC, indicating that the hydrogel is a safe material for controlling the cytotoxicity of TCS.

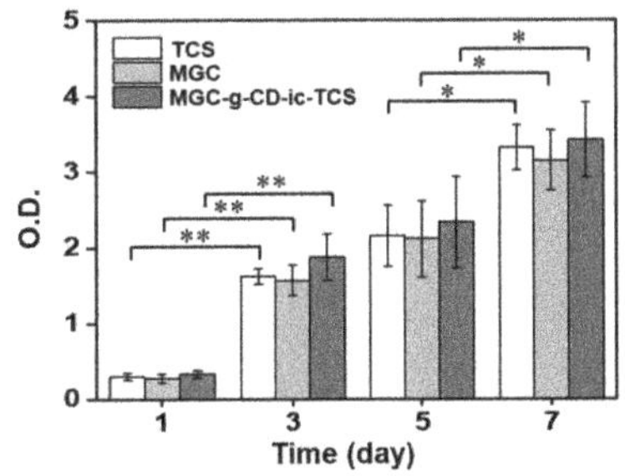

Figure 4. In vitro cell proliferation of L-929 cells cultured on TCS, MGC, and MGC-g-CD-ic-TCS for 1, 3, 5, and 7 days. This test was performed in triplicate ($n = 3$; * $p < 0.05$, ** $p < 0.01$).

2.4. Antibacterial Activity

The antibacterial activity of MGC-g-CD-ic-TCS was evaluated using *Escherichia coli* and *Staphylococcus aureus* (Figure 5). The *E. coli* and *S. aureus* are widely used as Gram-negative food-borne pathogens and Gram-positive pathogen, respectively [31]. The antibacterial activity of materials is deeply related to their surface charges and surface hydrophilic/hydrophobic natures [32]. We confirmed that MGC and MGC-g-CD-ic-TCS improve antibacterial activity against *E. coli* and *S. aureus*, as compared to PBS (Figure 5). It is well known that chitosan and its derivatives have good interaction with bacteria due to ionic interaction between the cationic charge of the polymers and the negative charge of bacteria [32,33]. The amine group of glycol chitosan (GC) is protonated when the bacterial medium is lower than pKa of 6.0 [34]. In the case of our study, the bacterial medium was almost 7.4, which indicated that MGC was not protonated. Therefore, the result of this study may be affected by the surface hydrophilic/hydrophobic nature of MGC. It has been reported that bacteria are well adhered to hydrophobic surface [33]. Based on this fact, the antibacterial activity of MGC is mainly dependent on its surface hydrophilicity. In addition to MGC, TCS improved antibacterial activity because it has a broad antimicrobial spectrum including bacteriostatic property as low concentration and induce cell death by disrupting bacteria membrane at high concentration [7,35]. The hydrogel with almost 40% of TCS release for 5 hours was found to have improved antibacterial activity [24]. From our result, together with the antibacterial activity of MGC, almost 30% of TCS release showed improved antibacterial activity. In addition, to accelerate antibacterial activity, TCS must have a concentration of above 5 μg/mL [28]. From this result, 1.8 μg of TCS containing MGC-g-CD exhibited excellent antibacterial activity. This reason can well explain that MGC-g-CD-ic-TCS is more effective against antibacterial activity than free TCS, due to the synergistic effect of MGC and TCS.

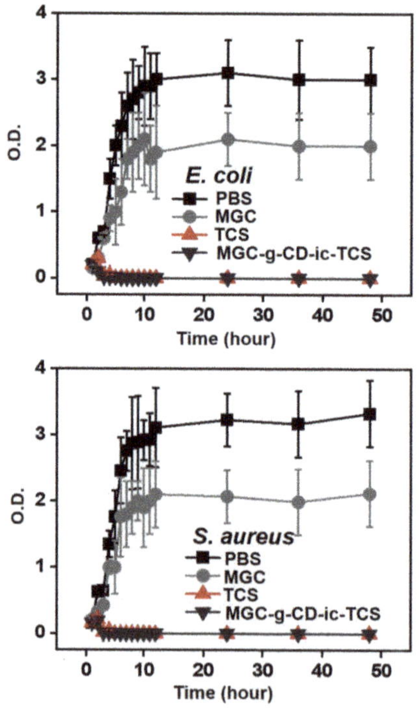

Figure 5. Antibacterial activity of MGC-g-CD-ic-TCS on *E. coli* and *S. aureus*, compared to PBS, MGC, and TCS. This test was performed in triplicate ($n = 3$).

2.5. Adhesion Strength of Hydrogels to Tissues

The adhesion strength of MGC-g-CD-ic-TCS was measured using a standard test method (ASTM F2255-05) for evaluating the tissue adhesive strength and the result was compared to those of fibrin glue, cyanoacrylate, and MGC, as shown in Figure 6. In hydrogel systems with non-surface-active groups, interactions including van der Waals forces and hydrogen bonding to soft tissues occur, which leads to weak binding ability. Meanwhile, positive-charged chitosan can contribute to stronger binding to tissue through multiple ionic interactions than neutral-charged polymers [36,37]. The results demonstrated that fibrin glue, cyanoacrylate, MGC, and MGC-g-CD-ic-TCS have 6.5 kPa, 530 kPa, 36 kPa, and 33 kPa of adhesive ability, respectively. The cyanoacrylate has a superior adhesiveness, but it is not suitable for tissue adhesion due to its toxicity, whereas fibrin glue is used as a control bio-adhesive due to its good biocompatibility. MGC, and MGC-g-CD-ic-TCS, had a stronger adhesion to tissue than fibrin glue due to its amine group can induce ionic interaction with tissues, resulting in adhesive strength more than five times.

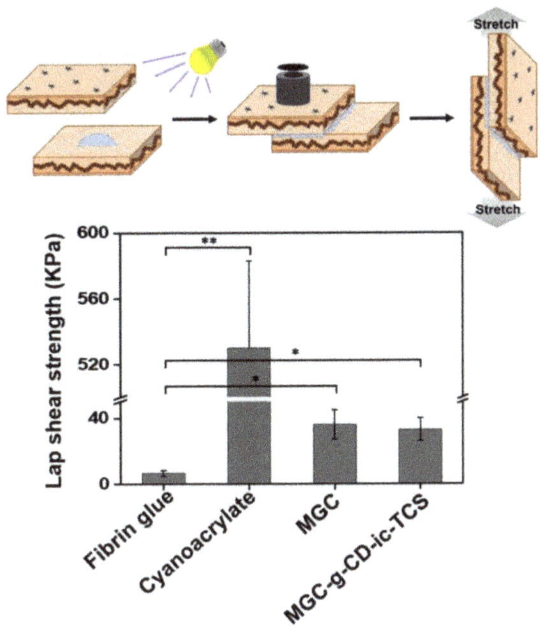

Figure 6. Lap shear strength of fibrin glue, cyanoacrylate, MGC, and MGC-g-CD-TCS. This test was performed in triplicate ($n = 3$; * $p < 0.05$, ** $p < 0.01$).

2.6. Observation of Gross Appearances of Skin Incisions

For improving the bioavailability of TCS in wound healing, its water solubility and release behavior must be advanced. β-CD-grafted MGC-based hydrogel can meet the characteristics by inclusion of complex formation between the ring molecule and the drug as shown in Figures 2 and 3. In addition, MGC-g-CD-ic-TCS was found to be cytocompatibility and has a superior tissue adhesive compared to fibrin glue, as shown in Figures 4–6. Figure 7 shows the gross appearances of skin incisions treated with suture, cyanoacrylate adhesive, MGC, and MGC-g-CD-ic-TCS for 14 days, as compared with that of sham. On day 7, the incisions in all experimental groups appeared to heal. Among the groups, more rapid healing was observed in the incisions treated with MGC and MGC-g-CD-ic-TCS. On day 14, scab was still observed in the sham, and suture-, and cyanoacrylate-treated incisions. On the other hand, MGC and MGC-g-CD-ic-TCS seemed to accelerate wound healing. On day 14, the incisions in the other groups remained scar except MGC-g-CD-ic-TCS-treated incision. Among incision, incision + suture, incision + adhesive, and incision + MGC, the MGC-treated sample showed the smallest scar. The incision + adhesive sample still exhibited scab due to the toxicity of cyanoacrylate.

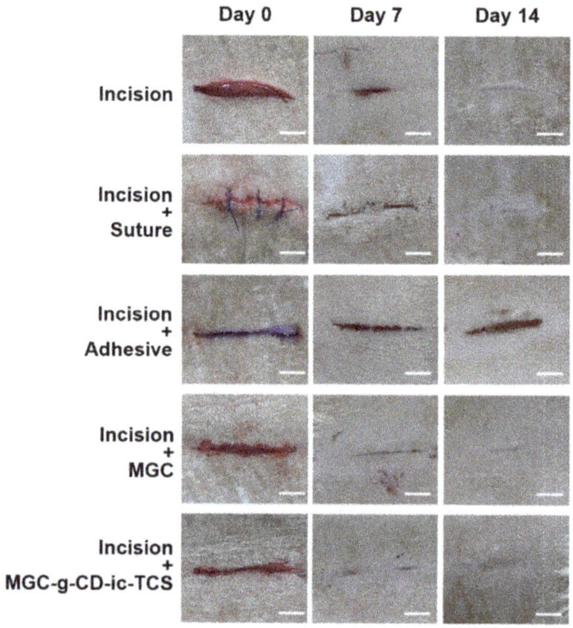

Figure 7. Gross appearances of untreated incision, and incisions treated with suture (incision + suture), cyanoacrylate (incision + adhesive), MGC (incision + MGC), and MGC-g-CD-ic-TCS (incision + MGC-g-CD-ic-TCS). The incisions were observed at day 0, 7, and 14. The white scale bars indicate 1 cm.

2.7. Histological Evaluations of Skin Incisions

To further investigate the histological changes in the incisions of all groups, hematoxylin and eosin (H&E) and Masson's trichrome (MT) staining were performed at days 7 and 14, respectively (Figure 8). From the H&E results (Figure 8A), at day 7, granulation tissue was newly formed in the incisions of all groups together with the infiltration of inflammatory cells. In addition, hair follicles and sebaceous glands began to develop in the MGC-g-CD-ic-TCS-treated incision. At 14 days, accelerated wound healing was observed in the incision + suture sample. However, incision, incision + suture, and incision + adhesive samples exhibited a thick epidermal layer, indicating abnormal wound healing. In particular, the proliferative epithelium was still observed in the incision + adhesive sample. In the incision + MGC and incision + MGC-g-CD-ic-TCS samples, an epidermis of normal thickness was formed. Hair follicles and sebaceous glands were first observed in the incision + MGC samples. In addition, hair follicles and sebaceous glands were mostly found in the MGC-g-CD-ic-TCS-treated incision.

Collagen deposition in the incisions was evaluated by MT staining (Figure 8B). Compared with incision + adhesive and incision + MGC samples, the incision, incision + suture, and incision + MGC-g-CD-ic-TCS showed active collagen production. However, on day 14, the incision, incision + suture, incision + adhesive, and MGC samples still seemed to stay in the proliferative stage. However, normal wound healing was observed in the incision + MGC-g-CD-ic-TCS sample. These results can be ascribed by the cocktail effect of MGC and TCS, because MGC and TCS can provide appropriate tissue junction and antibacterial/antifungal effect, respectively [7,35,38]. An animal study demonstrated that MGC-g-CD-ic-TCS accelerates re-epithelialization and wound remodeling in incision, compared with suture, cyanoacrylate, and MGC. Moreover, MGC induced a greater wound healing process than suture and cyanoacrylate. These results attribute to the adjustment of M1 and M2 polarization.

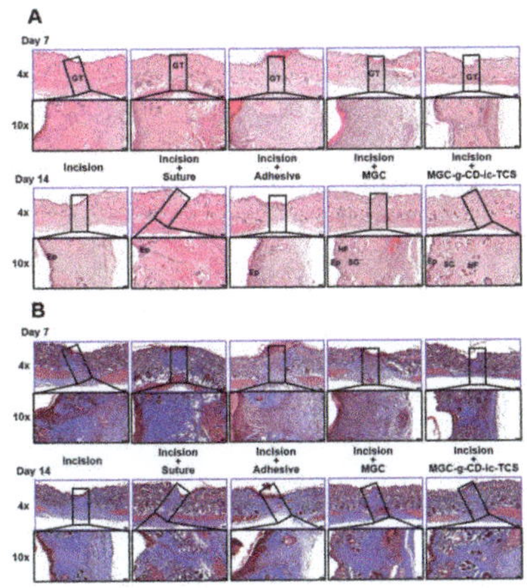

Figure 8. (**A**) H&E-stained and (**B**) MT-stained images of untreated incision, and incisions treated with suture (incision + suture), cyanoacrylate (incision + adhesive), MGC (incision + MGC) and MGC-g-CD-ic-TCS (incision + MGC-g-CD-ic-TCS) after 7 and 14 days. The slides were observed at 4× (200 μm) and 10× (100 μm), respectively. At 4× images, the rectangle boxes indicate incised part. GT-granulation tissue; EP-epithelial layer; HF-hair follicles; SC-sebaceous glands.

To investigate the infiltration of macrophages on days 7 and 14, we performed M1-specific anti-iNOS antibody and M2-specific anti-mannose receptor antibody immunostaining, as shown in Figure 9A. Over the time, MGC and MGC-g-CD-ic-TCS decreased the number of iNOS-positive cells and increased mannose receptor-positive cells. On day 7, the numbers of iNOS-positive cells in the incision, incision + suture, incision + adhesive, incision + MGC, and incision + MGC-g-CD-ic-TCS samples were 20, 16, 25, 15, and 14 cells/mm^2, respectively, and the numbers of mannose receptor-positive cells were 280, 550, 260, 500, and 600 cells/mm^2, respectively. At day 14, the M1/M2-related cells increased in the five groups, which exhibited 28, 18, 32, 20, and 19 cells/mm^2 in the case of iNOS-positive cells, and 310, 650, 290, 750, and 820 cells/mm^2 in the case of mannose receptor-positive cells. To further investigate the effect of M1 and M2 macrophages on wound healing, M1, M2, and M1/M2 ratio were calculated (Figure 9B). As shown in Figure 10A, incision + suture, incision + MGC, and incision + MGC-g-CD-ic-TCS samples exhibited a significantly fewer M1 cell numbers than incision and incision + cyanoacrylate. On the contrary to this, incision and incision + cyanoacrylate samples exhibited a significantly larger M2 cell number than the other samples (Figure 10B). Among all samples, the fewest M1 and the largest M2 cells were observed in incision + MGC-g-CD-ic-TCS. In addition, M1/M2 ratio tended to be similar in comparison with the M1 result (Figure 10C). However, on days 7 and 14, the increase of M1 and M2 cells in all samples may demonstrate the co-existence of inflammation, proliferation, and remodeling stages. These results imply that Incision + MGC-g-CD-ic-TCS provides a better wound healing process from inflammation to proliferative/remodeling stages.

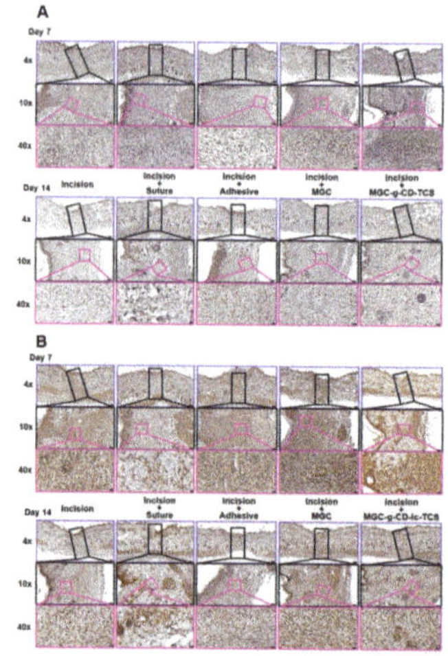

Figure 9. (**A**) Performed M1-specific anti-iNOS antibody and (**B**) M2-specific anti-mannose receptor antibody immunostained images of untreated incision, and incisions treated with suture (incision + suture), cyanoacrylate (incision + adhesive), MGC (incision + MGC), and MGC-g-CD-ic-TCS (incision + MGC-g-CD-ic-TCS) after 7 and 14 days. The slides were observed at 4× (200 μm), 10× (100 μm) and 40× (20 μm), respectively. At 4× images, the black rectangle boxes indicate incised part. At 10× images, the pink rectangle boxes indicate M1-specific anti-iNOS antibody and M2-specific anti-mannose receptor antibody immunostained images.

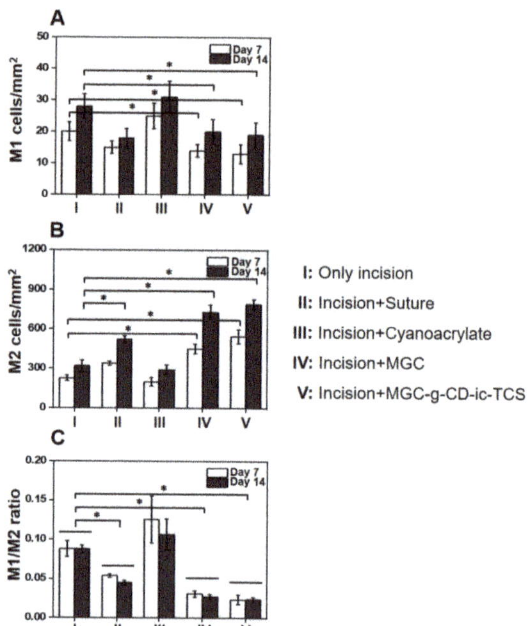

Figure 10. (**A**) M1, (**B**) M2, and (**C**) M1/M2 ratio of untreated incision, and incisions treated with suture (incision + suture), cyanoacrylate (incision + adhesive), MGC (incision + MGC), and MGC-g-CD-ic-TCS (incision + MGC-g-CD-ic-TCS) after 7 and 14 days ($n = 3$, * $p < 0.05$).

The immunostaining results of M1 and M2 suggest that MGC-g-CD-ic-TCS more reasonably modulates the polarization of macrophages than suture, cyanoacrylate, and MGC. As noted in a previous study, chitosan modulates macrophage polarization toward M2 phenotype at late differentiation stages in in vitro cell experiment [38]. A quick phenotypic switching from M1 to M2 in wound healing process leads to a fast resolution of inflammatory response and an improved local tissue regeneration [39]. In addition, chitosan hydrogel applied in the wounds of mice, rats, and pigs accelerates collagen synthesis, re-epithelialization, and angiogenesis [40–42]. In addition, TCS is known to possess a broad spectrum of antibacterial properties because it can inhibit the synthesis of fatty acid required for the membrane and wall of bacteria at low concentration and disrupt the membrane of bacteria at high concentration [7,35,38]. The stage of wound healing is a complex process involving numerous cells and cytokines in the temporospatial relationship [43]. Therefore, it is natural that it is not coordinated into a single molecule. The purpose of this study is not tissue-engineered skin regeneration or soft tissue regeneration, but hydrogel-based skin adhesive, designed for closing skin wounds. TCS has already been impregnated on absorbable sutures due to its antibacterial effect and is widely used [44]. Therefore, the purpose of this study was to perform a basic release test to evaluate whether triclosan contained in hydrogel-based polymeric adhesive exhibits antibacterial or even antibacterial induced anti-inflammatory effect. Our histologic evaluation is also focused on managing the inflammatory process rather than the entire soft tissue regeneration. From the results obtained by us, it is believed that hydrogel-based chitosan itself made the microenvironment of epidermis and partial dermal tissue favorable for wound healing, and that triclosan suppressed bacterial infection and maintained optimal inflammation throughout the wound. In this study, we used CD-ic-TCS-grafted glycol chitosan as a main material for bio-adhesive with normal wound healing in incision. Glycol chitosan and TCS have the same characteristics of chitosan as a type of chitosan derivative and help to steadily progress wound healing; therefore, the positive wound healing effect of MGC-g-CD-ic-TCS is affected by glycol chitosan and TCS.

3. Materials and Methods

3.1. Materials

GC ($\geq$60%, crystalline, Mw $\cong$ 585,000 g/mol) and glycidyl methacrylate (GM) were purchased from Sigma-Aldrich (St. Louis, MO, USA) for MGC preparation. β-CD-COOH (St. Louis, MO, USA) was used for MGC-g-CD preparation. Riboflavin 5′-monophosphate sodium salt (riboflavin; Santa Cruz, CA, USA) was used as a photoinitiator. 4-(4,6-Dimethoxy-1,3,5-triazin-2-yl)-4-methyl-morphoninium chloride (DMT-MM) as a condensing agent was supplied by Wako Pure Chemical Industries (Osaka, Japan). TCS obtained from Sigma-Aldrich (St. Louis, MO, USA) was used as an antibacterial agent. As a cyanoacrylate adhesive, Liquiband® Surgical S (Advanced Medical Solutions Group Plc, Cheshire, UK) was used. MGC-g-CD was purified using a dialysis membrane tube (cut-off: 25 kDa; Spectrum Laboratories Inc., Rancho Dominguez, CA, USA). Mouse fibroblast L-929 cell line was supplied by the Korean Cell Line Bank (Seoul, Korea). Chemicals were used as received.

3.2. Preparation of MGC-g-CD-ic-TCS

MGC was prepared according to procedures from our previous studies [11–14]. Next, β-CD-COOH was conjugated to the amine group of MGC through amide bond using DMT-MM as a condensation agent. To a solution of MGC (1.70 μmol, 1 g) in water, aqueous β-CD-COOH solution (1.15 mmol, 2.11 g, 500 mL) was slowly dropped and reacted at room temperature for 3 days. After dialysis for 7 days, the cleaned solution was lyophilized at −90 °C until dry. TCS (1.15 mmol, 333 mg) was dissolved in acetone (2 mL) and gently dropped in aqueous MGC-g-CD solution (1.5 g, 500 mL) filled in a beaker (1 L) with a continuous stirring (350 rpm). To evaporate acetone, the cap of the vial was removed and kimwipes® was covered to the entrance. The mixture was continuously stirred until acetone was perfectly evaporated and precipitate was eliminated by centrifugation at 1200 rpm. After lyophilization of the supernatant, MGC-g-CD-ic-TCS was characterized by analyses of ^{1}H NMR, DSC, and storage/loss moduli. The ^{1}H NMR analysis was performed using D_2O. The DSC (TA Instruments; DSC Q2000; New Castle, DE, USA) was employed for analyzing the thermal behavior of MGC-g-CD-ic-TCS, which was monitored from 50 °C to 170 °C at 10 °C/min by heating the hydrogel (50 mg) onto an aluminum pan. The storage/loss moduli of MGC-g-CD-ic-TCS were measured using an AR 2000ex rheometer (TA instruments, New Castle, DE, USA) at 37 °C from 0 Hz to 10 Hz.

3.3. *Preparation of MGC-g-CD-ic-TCS Hydrogel*

The hydrogel was prepared on a clean bench. Aqueous MGC-g-CD-ic-TCS solution was filtered using a 0.22 μm syringe filter for the removal of microorganisms, and then lyophilized. To a solution of MGC-g-CD-ic-TCS (200 mg) in DPBS (3 mL), aqueous riboflavin (12 μM; DPBS) was added, and the sample was vortexed gently. MGC-g-CD-ic-TCS was stabilized in a refrigerator (4 °C) for stabilization before use. For hydrogelation, the hydrogel precursor solution was irradiated for 200 s using a blue light (430–485 nm, 2100 mW/cm^2, light-emitting diode curing light, Foshan Keyuan Medical Equipment Co., Ltd., Foshan, China). The behavior was calculated by the ratio of swollen weight to initial weight at each time interval.

3.4. *In Vitro Release Test*

The release test was carried out according to a previously reported method [45]. A specific volume of MGC-g-CD-ic-TCS hydrogel (1 mL) was added in a dialysis bag (cut-off: 50,000 g/mol), and the drug-loaded sample was then immersed in PBS (10 mL, pH 7.4) containing 1% Tween-80 at 37 °C. At predetermined time intervals, 500 μL of PBS was extracted, and the same volume of fresh PBS was added. All extracted PBS solutions were examined by a UV-Vis spectroscopy (Multiskan® Spectrum; Thermo Fisher Scientific; Waltham, MA, USA) at a wavelength of 282 nm.

3.5. *In Vitro Cell Proliferation Assay*

L-929 cell line was cultured with Eagle's Minimal Essential Medium supplemented with FBS (5%), penicillin (100 units/mL) and streptomycin (100 μg/mL). For this experiment, cells cultured for 10 passages were used. The cell proliferation assay was performed on two hydrogel samples: TCS (TCS: 1.8 μg), MGC, and MGC-g-CD-ic-TCS (TCS: 1.8 μg). The hydrogel precursor solutions were immersed in a 96-well plate and photocured using a visible light irradiation for 200 s. After seeding the cells (1×10^4 cells/well) on the hydrogels, they were incubated for 1, 3, 5, and 7 days in an incubator set at 37 °C and 5% CO_2. In the case of TCS, the cells (1×10^4 cells/well) were attached on the wells of 96-well plate for 3 h and TCS-containing media was treated. At each time interval, CCK-8 (100 μL) was added in wells filled with the hydrogels and incubated for additional 2 hours. The optical density of supernatants extracted from CCK-8-treated wells was measured at 450 nm using a microplate reader (SpectraMax® i3; Molecular Devices, Sunnyvale, CA, USA).

3.6. *Tissue Adhesive Strength Measurement*

This test was performed using procedures based on ASTM F2255-05. The adhesive strength of MGC-g-CD-ic-TCS hydrogel was measured using a universal testing machine (Instron Model 3343, Norwood, MA, USA). Prior to the test, the fat layer of the skin, which was used for this study, was removed. MGC-g-CD-ic-TCS hydrogel precursor solution was applied on the surface of the cleaned skin with 10×10 mm^2 and another skin was covered to the sample-coated skin during visible light irradiation for 200 s. The adhesive strength of the bonded skins was measured at a crosshead speed of 10 mm/min with a 100 *n* load cell. This test was performed on 10 samples. As a control adhesive, the adhesive strength of fibrin glue was measured under the same conditions.

3.7. Evaluation of Antibacterial Activity

Escherichia coli ATCC25922 (*E. coli*) and *Staphylococcus aureus* ATCC6538 (*S. aureus*) were used for evaluating the antibacterial activity of MGC and MGC-g-CD-ic-TCS (TCS: 1.8 μg) hydrogels. It was estimated by monitoring the optical density of the two bacteria at 600 nm. *E. coli* or *S. aureus* (10 μL mid-long phase; O.D. = 0.3 units) were inoculated in fresh nutrient LB broth medium containing the two hydrogels (60 mg/L per each hydrogel) in flasks. LB medium alone was used as a negative control. Afterward, the flasks were incubated at 37 °C with a continuous stirring at 150 rpm. The optical density was measured for 48 h at 1-h intervals.

3.8. In Vivo Animal Study

This study was approved by the Institutional Animal Care and Use Committee of Chonbuk National University (23 March 2020; No: CBNU 2020-031). Male Sprague Dawley rats (three-month-old, 300–350 g) was used and randomized into five groups on each time interval (7 and 14 days; $n = 5$). A constant length of incision was produced on the back of each rat and the incision depth was made including the dermis. A specific volume (150 μL) of cyanoacrylate, MGC, and MGC-g-CD-ic-TCS (TCS: 1.8 μg). were used for three groups, respectively. Other two groups include incision + no treatment and incision + suture. After treating MGC and MGC-g-CD-ic-TCS hydrogel precursor solutions into the incisions, visible light (blue light; 430–485 nm, 2100 mW/cm^2, light-emitting diode curing light, Foshan Keyuan Medical Equipment Co., Ltd., Foshan, China) was irradiated for 120 s. At the predetermined time intervals, the mice of each group were anesthetized, the skins including the incisions were removed, and the mice were sacrificed. All skin tissue samples were used for histological evaluations.

3.9. Histological Evaluations

Removed skin tissues including incisions were immediately placed in 10% formalin solution. The fixed tissues were dehydrated using a series of ethanol solutions, and the dehydrated tissues were then embedded in paraffin for block preparation. A constant thickness of slide was sectioned (3 μm). As a control, the same thickness of paraffin slide using normal skin tissue was made. The tissue sections were treated with H&E and MT staining (Abcam: ab150686, Cambridge, UK). For immunostaining, deparaffinized tissue sections were treated using a microwave antigen-retrieval procedure with 0.01 M sodium citrate buffer. After blocking, the sections incubated with antibodies against iNOS (N-20, Santa Cruz Biochemical, Dallas, TX, USA) and mannose receptor (Abcam). For visualization, an HRP/DAB detection IHC kit (Abcam) was used according to the manufacturer's instructions. These sections were observed using a slide scanner (Pannoramic MIDI; 3DHISTECH Ltd., Budapest, Hungary) and a panoramic viewer (Version 1.15.3; Pannoramic MIDI; 3DHISTECH Ltd., Budapest, Hungary) program.

3.10. Statistical Analysis

All quantitative data were expressed as the mean ± standard deviation. Statistical analysis was performed with one-way analysis of variance (ANOVA) using SPSS software (SPSS Inc., Chicago, IL, USA). A value of * $p < 0.05$ was considered statistically significant.

4. Conclusions

Here, we designed and prepared a MGC-g-CD-ic-TCS hydrogel system as an antibacterial tissue bio-adhesive. Inclusion complexes between MGC-g-CD and TCS were confirmed by ^{1}H NMR and DSC analyses. Due to the migration of water molecules into its 3-D structure, the hydrogel was swollen, and this influenced the release behavior of TCS. MGC-g-CD-ic-TCS exhibited better tissue bonding strength than commercially used fibrin glue. In addition, the hydrogel showed high antibacterial activity against *E. coli* and *S. aureus* because of the cocktail effect of MGC and TCS. More importantly, MGC-g-CD-ic-TCS was associated with accelerated wound healing in vivo, suggesting its great potential as a tissue bio-adhesive.

Author Contributions: Conceptualization, Y.J.M., S.-J.Y., and D.H.Y.; formal analysis, J.-H.K., Y.Y., H.J.B. and H.S.K.; data curation, Y.J.M., S.-J.Y., and D.H.Y.; Writing—original draft preparation, Y.J.M., S.-J.Y., and D.H.Y.; supervision, G.K., H.J.C. and D.H.Y.; funding acquisition, S.-J.Y. and D.H.Y. Writing—review and editing, Y.J.M., S.-J.Y., H.J.C., and D.H.Y. All authors have read and agreed to the published version of the manuscript.

Funding: This research was supported by the Ministry of Trade, Industry & Energy (MOTIE, Korea) (20003560 and 20004627; D.H.Y.), the Technology development program (S2782460; D.H.Y.) funded by the Ministry of SMEs and Startups (MSS; Republic of Korea), funding from Biomedical Research Institute, Jeonbuk National University Hospital, and research funds for newly appointed professors of Jeonbuk National University in 2020.

Institutional Review Board Statement: The study was approved by the Institutional Animal Care and Use Committee of Chonbuk National University (23 March 2020; No: CBNU 2020-031).

Informed Consent Statement: Not applicable.

Data Availability Statement: Data is contained within the article.

Conflicts of Interest: The authors declare no conflict of interest.

Abbreviations

TCS	Triclosan
β-CD-COOH	Succinyl beta-cyclodextrin
MGC	Methacrylated glycol chitosan
^{1}H NMR	Proton nuclear magnetic resonance
DSC	Differential scanning calorimetry
MGC-g-CD-ic-TCS	TCS-complexed β-CD-conjugated MGC
3-D	Three-dimensional
CCK-8	Cell counting kit-8
GC	Glycol chitosan
H&E	Hematoxylin and eosin
MT	Masson's trichrome
GM	Glycidyl methacrylate
Riboflavin	Riboflavin 5′-monophosphate sodium salt
DMT-MM	4-(4,6-Dimethoxy-1,3,5-triazin-2-yl)-4-methyl-morphoninium chloride

References

1. Rahimnejad, M.; Zhong, W. Mussel-inspired hydrogel tissue adhesives for wound closure. *RSC Adv.* **2017**, *7*, 47380. [CrossRef]
2. Pascual, G.; Sotomayor, S.; Rodriguez, M.; Pérez-Köhler, B.; Kühnhardt, A.; Fernández-Gutiérrez, M.; San Román, J.; Bellón, J.M. Cytotoxicity of cyanoacrylate-based tissue adhesives and short-term preclinical in vivo biocompatibility in abdominal hernia repair. *PLoS ONE* **2016**, *11*, e0157920. [CrossRef] [PubMed]
3. Ge, L.; Chen, S. Recent advances in tissue adhesives for clinical medicine. *Polymers* **2020**, *12*, 939. [CrossRef] [PubMed]
4. Zhu, F.; Wang, C.; Yang, S.; Wang, Q.; Liang, F.; Liu, C.; Qiu, D.; Qu, X.; Hu, Z.; Yang, Z. Injectable tissue adhesive composite hydrogel with fibroblasts for treating skin defects. *J. Mater. Chem. B* **2017**, *5*, 2416. [CrossRef] [PubMed]
5. Li, S.; Dong, S.; Xu, W.; Tu, S.; Yan, L.; Zhao, C.; Ding, J.; Chen, X. Antibacterial hydrogels. *Adv. Sci.* **2018**, *5*, 1700527. [CrossRef]
6. Dai, T.; Tanaka, M.; Huang, Y.-Y.; Hamblin, M.R. Chitosan preparation for wounds and burns: Antimicrobial and wound-healing effects. *Expert Rev. Anti. Infect. Ther.* **2011**, *9*, 857–879. [CrossRef]
7. Puckett, M.M.; Morris, A.N.; Kuehn, J.L. The effect of triclosan on the proliferation of peritoneal tissue explants. *Bios* **2015**, *86*, 104–108. [CrossRef]
8. Du, X.; Liu, Y.; Rafique, M.; Li, S.; Shan, X.; Wu, L.; Qiao, M.; King, D.; Wang, L. Anti-infective and pro-coagulant chitosan-based hydrogel tissue adhesive for sutureless wound closure. *Biomacromolecules* **2020**, *21*, 1243–1253. [CrossRef]
9. Bhagat, V.; Becker, M.L. Degradable adhesives for surgery and tissue engineering. *Biomacromolecules* **2017**, *18*, 3009–3039. [CrossRef]
10. Wei, X.; Ma, K.; Cheng, Y.; Sun, L.; Chen, D.; Zhao, X.; Lu, H.; Song, B.; Yang, K.; Jia, P. Adhesive, conductive, self-healing, and antibacterial hydrogel based on chitosan-polyoxometalate complexes for wearable strain sensor. *ACS Appl. Polym. Mater.* **2020**, *2*, 2541–2549. [CrossRef]

11. Yoon, S.-J.; Yoo, Y.; Nam, S.E.; Hyun, H.; Lee, D.-W.; Um, S.; Kim, S.Y.; Hong, S.O.; Yang, D.H.; Chun, H.J. The cocktail effect of BMP-2 and TGF-β1 loaded in visible light-cured glycol chitosan hydrogels for the enhancement of bone formation in a rat tibial defect model. *Mar. Drugs* **2018**, *16*, 351. [CrossRef]
12. Yoo, Y.; Hyun, H.; Yoon, S.-J.; Kim, S.Y.; Lee, D.-W.; Um, S.; Hong, S.O.; Yang, D.H. Visible light-cured glycol chitosan hydrogel dressing containing endothelial growth factor and basic fibroblast growth factor accelerates wound healing in vivo. *J. Ind. Eng. Chem.* **2018**, *67*, 365–372. [CrossRef]
13. Yang, D.H.; Seo, D.I.; Lee, D.-W.; Bhang, S.H.; Park, K.; Jang, G.; Kim, C.H.; Chun, H.J. Preparation and evaluation of visible-light cured glycol chitosan hydrogel dressing containing dual growth factors for accelerated wound healing. *J. Ind. Eng. Chem.* **2017**, *53*, 360–370. [CrossRef]
14. Yoon, S.-J.; Hyun, H.; Lee, D.-W.; Yang, D.H. Visible light-cured glycol chitosan hydrogel containing a beta-cyclodextrin-curcumin inclusion complex improves wound healing in vivo. *Molecules* **2017**, *22*, 1513. [CrossRef] [PubMed]
15. Ahmadi, F.; Oveisi, Z.; Mohammadi Samani, S.; Amoozgar, Z. Chitosan based hydrogels: Characteristics and pharmaceutical applications. *Res. Pharm. Sci.* **2015**, *10*, 1–16. [PubMed]
16. Goycoolea, F.M.; Fernández-Valle, M.E.; Aranaz, I.; Heras, Á. pH- and temperature-sensitive chitosan hydrogels: Swelling and MRI studies. *Macromol. Chem. Phys.* **2011**, *212*, 887–895. [CrossRef]
17. Zhang, H.; Zhao, T.; Newland, B.; Duffy, P.; Annaidh, A.N.; O'Cearbhaill, E.D.; Wang, W. On-demand and negative-thermo-swelling tissue adhesive based on highly branched ambivalent PEG-catechol copolymers. *J. Mater. Chem. B* **2015**, *3*, 6420. [CrossRef]
18. Wang, R.; Yang, Z.; Lou, J.; Hsing, I.-M.; Sun, F. B12-dependent photoresponsive protein hydrogels for controlled stem cell/protein release. *Proc. Natl. Acad. Sci. USA* **2017**, *114*, 5912–5917. [CrossRef]
19. Zhao, X.; Lang, Q.; Yildirimer, L.; Lin, Z.Y.; Cui, W.; Annabi, N.; Ng, K.W.; Dokmeci, M.R.; Ghaemmaghami, A.M.; Khademhosseini, A. Photocrosslinkabel gelatin hydrogel for epidermal tissue engineering. *Adv. Healthc. Mater.* **2016**, *5*, 108–118. [CrossRef]
20. Ullrich, S.E. Sunlight and skin cancer: Lessons from the immune system. *Mol. Carcinog.* **2007**, *46*, 629–633. [CrossRef]
21. Song, M.; Li, L.; Zhang, Y.; Chen, K.; Wang, H.; Gong, R. Carboxymethyl-β-cyclodextrin grafted chitosan nanoparticles as oral delivery carrier of protein drugs. *React. Funct. Polym.* **2017**, *117*, 10–15. [CrossRef]
22. Jiang, R.-J.; Yang, B.; Yi, D.; Wang, F.; Han, B.; Zhao, Y.-L.; Liao, X.-L.; Yang, J.; Gao, C.-Z. Synthesis and characterization of a series of novel amino β-cyclodextrin-conjugated poly(ε-lysine) derivatives. *J. Polym. Eng.* **2014**, *34*, 133–139. [CrossRef]
23. Lee, J.H.; Park, S.H.; Kim, S.H. Fabrication of bio-based polyurethane nanofibers incorporated with a triclosan/cyclodextrin complex for antibacterial applications. *RSC Adv.* **2020**, *10*, 3450. [CrossRef]
24. Hu, Y.; Ren, G.; Deng, L.; Zhang, J.; Liu, H.; Mu, S.; Wu, T. Degradable UV-crosslinked hydrogel for the controlled release of triclosan with reduced cytotoxicity. *Mater. Sci. Eng. C Mater. Biol. Appl.* **2016**, *67*, 151–158. [CrossRef]
25. Li, J.; Mooney, D.J. Designing hydrogels for controlled drug delivery. *Nat. Rev. Mater.* **2016**, *1*, 16071. [CrossRef]
26. Lou, C.; Tian, X.; Deng, H.; Wang, Y.; Jiang, X. Dialdehyde-β-cyclodextrin-crosslinked carboxymethyl chitosan hydrogel for drug release. *Carbohy. Polym.* **2020**, *231*, 115678. [CrossRef]
27. Zuckerbraun, H.L.; Babich, H.; May, R.; Sinensky, M.C. Triclosan: Cytotoxicity, mode of action, and induction of apoptosis in human gingival cells in vitro. *Eur. J. Oral. Sci.* **1998**, *106*, 628–636. [CrossRef]
28. Zu, G.; Steinmüller, M.; Keskin, D.; van der Mei, H.C.; Mergel, O.; van Rijn, P. Antimicrobial nanogels with nanoinjection capabilities for delivery of the hydrophobic antibacterial agent triclosan. *ACS Appl. Polym. Mater.* **2020**, *2*, 5779–5789. [CrossRef]
29. Cui, Z.-K.; Kim, S.; Baljon, J.J.; Wu, B.M.; Aghaloo, T.; Lee, M. Microporous methacrylated glycol chitosan-montmorillonite nanocomposite hydrogel for bone tissue engineering. *Nat. Comm.* **2019**, *10*, 3523. [CrossRef]
30. Pereira, P.; Pedrosa, S.S.; Correia, A.; Lima, C.F.; Olmedo, M.P.; González-Fernández, Á.; Vilanova, M.; Gama, F.M. Biocompatibility of a self-assembled glycol chitosan nanogel. *Toxicol. Vitr.* **2015**, *29*, 638–646. [CrossRef]
31. Mori, T.; Akamatsu, M.; Okamoto, K.; Sumita, M.; Tateyama, Y.; Sakai, H. Micrometer-level naked-eye detection of caesium particulates in the solid state. *Sci. Technol. Adv. Mater.* **2012**, *13*, 015002. [CrossRef] [PubMed]
32. Rabea, E.I.; Badawy, M.E.-T.; Stevens, C.V.; Smagghe, G.; Steurbaut, W. Chitosan as antimicrobial agents: Applications and mode of action. *Biomacromolecules* **2003**, *4*, 1457–1465. [CrossRef] [PubMed]
33. Raafat, D.; von Bargen, K.; Haas, A.; Sahl, H.-G. Insights into the mode of action of chitosan as an antibacterial compound. *Appl. Environ. Microbiol.* **2008**, *74*, 3764–3773. [CrossRef] [PubMed]
34. Katsuro, S.; Akiko, N. Complex formation of heparin or sulfated cellulose with glycol chitosan. *Bull. Inst. Chem. Res.* **1975**, *53*, 392.
35. Kampf, G.; Kramer, A. Epidemiologic background of hand hygiene and evaluation of the most important agents for scrubs and rubs. *Clin. Microbiol. Rev.* **2004**, *17*, 863–893. [CrossRef] [PubMed]
36. Mizrahi, B.; Weldon, C.; Kohane, D.S. Tissue adhesives as active implants. In *Active Implants and Scaffolds for Tissue Regeneration*; Meital, Z., Ed.; Springer: Berlin/Heidelberg, Germany, 2001; Volume 8, pp. 39–56.
37. Duarte, A.P.; Coelho, J.F.; Bordado, J.C.; Cidada, M.T.; Gil, M.H. Surgical adhesives: Systematic review of the main types and development forecast. *Prog. Polym. Sci.* **2012**, *37*, 1031–1050. [CrossRef]
38. Oliveira, M.I.; Santos, S.G.; Oliveira, M.J.; Torres, A.L.; Barbosa, M.A. Chitosan drives anti-inflammatory macrophage polarization and pro-inflammatory dendritic cell stimulation. *Eur. Cell Mater.* **2012**, *24*, 136–152. [CrossRef]
39. Delavary, B.M.; van der Veer, W.M.; van Egmond, M.; Niessen, F.B.; Beelen, R.H.J. Macrophages in skin injury and repair. *Immunobiology* **2011**, *216*, 753–762. [CrossRef]

40. Chen, X.; Zhang, M.; Chen, S.; Wang, X.; Tian, Z.; Chen, Y.; Xu, P.; Zhang, L.; Zhang, L.; Zhang, L. Peptide-modified chitosan hydrogels accelerate skin wound healing by promoting fibroblast proliferation, migration, and secretion. *Cell Transpl.* **2017**, *26*, 1331–1340. [CrossRef]
41. Patil, P.S.; Evancho-Chapman, M.; Li, H.; Huang, H.; George, R.L.; Shriver, L.P.; Leipzig, N.D. Fluorinated methacrylamide chitosan hydrogel dressings enhance healing in an acute porcine wound model. *PLoS ONE* **2018**, *13*, e0203371. [CrossRef]
42. Hilmi, A.B.M.; Halim, A.S.; Jaafar, H.; Asiah, A.B.; Hassan, A. Chitosan dermal substitute and chitosan skin substitute contribute to accelerated full-thickness wound healing in irradiated rats. *Biomed. Res. Int.* **2013**, *2013*, 795458.
43. Maxon, S.; Lopez, E.A.; Yoo, D.; Danilkovitch-Miagkova, A.; Leroux, M.A. Concise review: Role of mesenchymal stem cells in wound repair. *Stem Cells Transl. Med.* **2012**, *1*, 142–149. [CrossRef] [PubMed]
44. Nakamura, T.; Kashimura, N.; Noji, T.; Suzuki, O.; Ambo, Y.; Nakamura, F.; Kishida, A. Triclosan-coated sutures reduce the incidence of wound infections and the costs after colorectal surgery: A randomized controlled trial. *Surgery* **2013**, *153*, 576–583. [CrossRef] [PubMed]
45. Ma, D.; Wu, T.; Zhang, J.; Lin, M.; Mai, W.; Tan, S.; Xue, W.; Cai, X. Supramolecular hydrogels sustained release triclosan with controlled antibacterial activity and limited cytotoxicity. *Sci. Adv. Mater.* **2013**, *5*, 1400–1409. [CrossRef]

International Journal of
Molecular Sciences

Review

Chitosan and Cellulose-Based Hydrogels for Wound Management

Sibusiso Alven and Blessing Atim Aderibigbe *

Department of Chemistry, University of Fort Hare, Alice Campus, Eastern Cape 5700, South Africa; 201214199@ufh.ac.za
* Correspondence: baderibigbe@ufh.ac.za; Tel.: +27-0406-022-266

Received: 18 August 2020; Accepted: 30 September 2020; Published: 18 December 2020

Abstract: Wound management remains a challenge worldwide, although there are several developed wound dressing materials for the management of acute and chronic wounds. The wound dressings that are currently used include hydrogels, films, wafers, nanofibers, foams, topical formulations, transdermal patches, sponges, and bandages. Hydrogels exhibit unique features which make them suitable wound dressings such as providing a moist environment for wound healing, exhibiting high moisture content, or creating a barrier against bacterial infections, and are suitable for the management of exuding and granulating wounds. Biopolymers have been utilized for their development due to their non-toxic, biodegradable, and biocompatible properties. Hydrogels have been prepared from biopolymers such as cellulose and chitosan by crosslinking with selected synthetic polymers resulting in improved mechanical, biological, and physicochemical properties. They were useful by accelerating wound re-epithelialization and also mimic skin structure, inducing skin regeneration. Loading antibacterial agents into them prevented bacterial invasion of wounds. This review article is focused on hydrogels formulated from two biopolymers—chitosan and cellulose—for improved wound management.

Keywords: chitosan; cellulose; hydrogel; wound dressings; burn wounds; diabetic wounds; chronic wounds; acute wounds

1. Introduction

Wounds are injuries on the skin [1]. They are classified as acute and chronic wounds. An acute wound is an injury to the skin that is sudden. It can heal within the time frame of 2–3 months, depending on its depth and size in the skin epidermis or dermis layers [2]. Chronic wounds are life-threatening because they fail to heal in a timely manner and examples of chronic wounds are burns, decubitus ulcers, infections, leg ulcers, etc. [1]. Wound management is expensive and there is an urgent need to design wound dressings that are affordable for the growing population. The United States of America spends about USD 20 billion annually for the management of chronic wounds, while the United Kingdom spent approximately GBP 184 million in 2012 for the management of chronic wounds [3–5]. Some of the currently used wound dressing materials suffer from several limitations such as poor antimicrobial effects, weak mechanical performance, and inability to provide moisture for acceleration of the wound healing process [1]. There is a pressing need for researchers to develop more advanced wound dressings that are cost-effective.

Wound dressings are mostly developed from biopolymers and synthetic polymers. The natural polymers that are commonly utilized include chitosan, cellulose, fibrin, elastin, hyaluronic acid, dextran, elastin, alginate, collagen, and gelatin [6]. These polymers possess interesting properties suitable for wound management such as good biocompatibility, non-toxicity, biodegradability, readily availability, and non-immunogenicity [7]. These polymers are usually cross-linked with synthetic polymers because of their poor mechanical properties. The synthetic polymers include poly(vinyl pyrrolidone) (PVP), poly(ethylene oxide)(PEO)/poly(ethylene glycol) (PEG), Poly(hydroxyethyl methacrylate) (PHEMA), poly(vinyl alcohol) (PVA), polyurethanes (PUs) and polyesters such as (polyglycolic acid (PGA), poly(lactic-co-glycolic acid) (PLGA) and polylactide (PLA)) [6,8]. In the design of an ideal wound dressing, factors that are usually considered are their ability to stop bleeding, prevention of microbial infections, good absorption of wound exudates, easy sterilization, their ability to promote wound debridement, good gas permeability, easy to use, biodegradability, and non-toxicity [9].

Wound dressing materials are formulated in various forms such as hydrogels [10–12], films [13–17], nanofibers [18,19], foams, topical formulations, wafers [20], transdermal patches [21–23], sponges [24–26], and bandages [27–29]. The limitations that are associated with some of the currently employed wound dressings include poor gaseous exchange between the wound and the surroundings, difficulty in the removal of the wound dressing, inability to protect the wound against microbial infection, poor mechanical properties, lack of sterility, induction of allergic reactions, poor absorption of wound exudates, and their inability to sustain a moist environment for accelerated wound healing [30]. Most of the aforementioned features that are lacking in some of the currently designed wound dressings are exhibited by hydrogels developed from biopolymers such as chitosan and cellulose. There are various interesting unique properties of chitosan and cellulose that are attractive for the development of wound dressings. The properties include excellent hemostatic capability and antibacterial effects, availability of various types of functional groups that can be used for their modification, bio-adhesiveness, highly biocompatible, and outstanding wound healing characteristics (Figure 1) [6]. Some of the biopolymers are not presenting these properties. Combining the aforementioned features with the properties of hydrogels can result in an effective wound care. This review article is focused on the in vitro and in vivo therapeutic outcomes and properties of hydrogels that were recently developed (2019–2020) from chitosan and its derivatives, and cellulose and its derivatives for wound management.

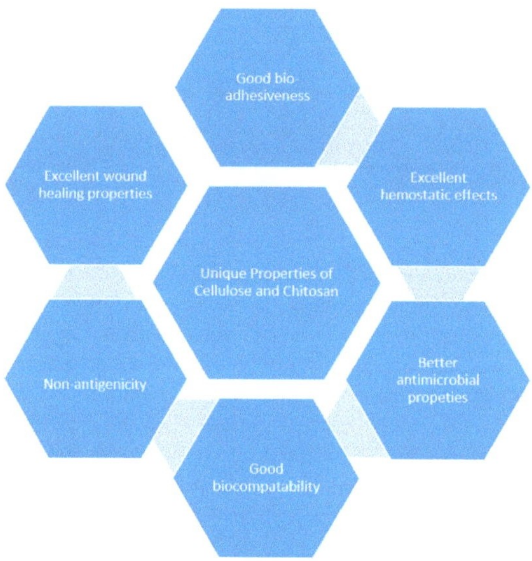

Figure 1. Properties of chitosan and cellulose.

2. Phases of Wound Healing

The wound healing process is a complex mechanism that leads to the repair of injured skin tissue. The phases of wound healing determine the type of wound dressing to be used for the proper management of the wound. The wound healing process is composed of four sequential phases that can overlap: hemostasis, inflammation, proliferation, and remodeling phase (Figure 2) [31]. The hemostasis phase takes place immediately after an injury, and this phase occurs rapidly. The blood vessel becomes narrow, restricting the blood flow. There is a platelet accumulation forming a clot that seals the ruptured wall of the blood vessel, thereby terminating the bleeding [32]. The inflammatory phase involves the cleansing of debris and bacteria removal by neutrophils, reactive oxygen species, and proteases which are released by phagocytic cells to provide a suitable environment for the healing process [33]. The hemostasis and inflammatory phase concurrently take place for 3 days [34]. The injured blood vessels produce transudate resulting in swelling. Inflammation is useful at this phase to prevent infection and control bleeding. However, prolonged inflammation can be problematic.

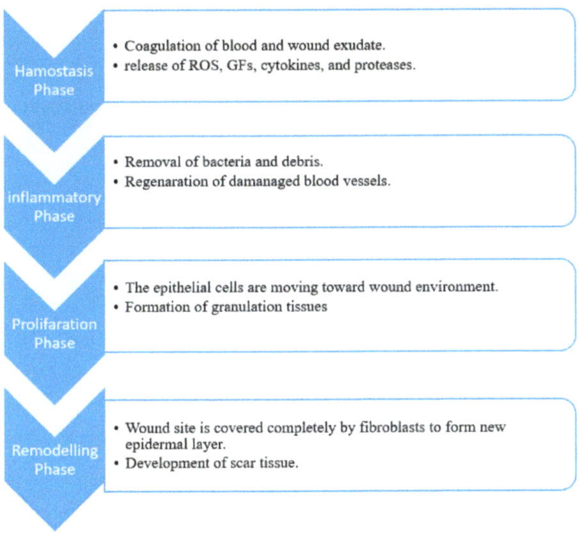

Figure 2. Summarized sequential phases of the wound healing process.

The proliferative phase involves the formation of connective tissue such as blood vessels, granulation tissues in the injury location to replace dead cells [35]. Extracellular matrix, including proteoglycans, elastin, hyaluronic acid, and collagen, produces a granulation tissue for the replacement of clot original formation. There are several types of cytokines and growth factors that participate in this phase, such as the transforming growth factor-β family (TGF-β, including TGF-β3, TGF-β1, and TGF-β2), angiogenesis factors, and interleukin (IL) family [35,36]. This phase normally takes place for days or weeks. The final step of the healing process of the wound is a remodeling phase. The collagen is remodeled by crosslinking to reduce the thickness of the scar with complete wound closure. Apoptosis occurs at this phase in which cells which were involved in the repair of the wound that are no longer useful are removed. This phase is very fragile and the inability of a wound to progress to this stage, results in a wound becoming chronic [37]. This significant phase can remain for months or even years [38,39].

3. Properties of Biopolymers

Biopolymers are derived from natural sources such as plant or animal and microbial sources [40]. They are biodegradable and it influences their physicochemical properties and their behavior (such as interaction with surrounding tissues, drug-releasing patterns, etc.) in the physiological environment [41]. The risks such as the transmission of infectious diseases from polymeric materials of natural origin can limit the use of biopolymer-based wound dressings in the wider applications of wound care due to their allogenic or xenogenic nature [42]. Other shortcomings of biopolymer-based systems include their poor stability and weak mechanical performance, which limits their biomedical applications [43]. Excitingly, there are numerous chemical modifications that have been advanced to overcome these shortcomings [44,45].

Some examples of biopolymers that are usually utilized for the preparation of wound dressings include chitosan **1**, cellulose **2**, hyaluronic acid **3**, alginate **4**, Elastin **5**, dextran **6**, fibrin **7**, pectin **8**, and collagen (Figure 3) [6]. Chitosan is a linear copolymer isolated from chitin, which is the key constituent of the exoskeletons of crustaceans such as crab and shrimp. This biopolymer and its derivatives are well-acknowledged for their heterogenic functionalities such as their non-toxicity, inertness, non-antigenicity, bioadhesiveness, biocompatibility, biodegradability, hemostatic effects, antimicrobial properties, and wound healing characteristics [46,47]. Additionally, chitosan is greatly versatile and possesses the ability to make a diversity of functionalized derivatives through the chemical modification of hydroxyl and amino groups. These derivatives include N,O-(carboxymethyl), N-carboxymethyl, N-succinyl, N-acyl, N-carboxybutyl, N-carboxyethyl, 5-methylpyrrolidinone, N-N-dicarboxymethyl, O-succinyl, and O-carboxymethyl chitosan derivatives etc.

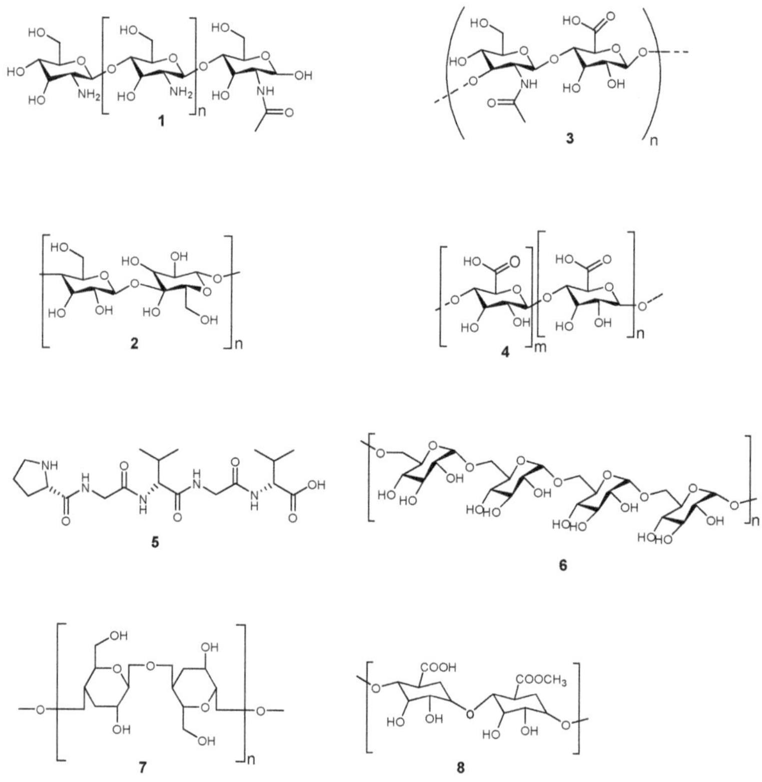

Figure 3. Molecular structures of biopolymers: chitosan **1**, cellulose **2**, hyaluronic acid **3**, alginate **4**, Elastin **5**, dextran **6**, fibrin **7**, pectin **8**.

On the other hand, cellulose is the major structural constituent of the cell walls of plants and is well-known as the most abundant biological polymer in the universe. This biopolymer is readily accessible and affordable. It is a linear organic polymer composed of β-1,4 combined D-glucose units which are linked to produce cellobiose repeating parts [48]. Cellulose and its derivatives display good biocompatibility [49]. Furthermore, the resorption of cellulose in cells does not happen since cells are unable to produce enzymes, cellulases [50]. The wound healing efficacy of cellulose results from its capability to accelerate the wound healing process through the maintenance and release of several growth factors at the injury site such as basic fibroblast growth factor, phosphodiesterase growth factor, and epidermal growth factor. These growth factors promote the migration and proliferation of dermal fibroblasts and inhibit the proliferation of bacteria in the wound [51].

4. Preparation Techniques and Properties of Hydrogels

There are four techniques that are adopted for the preparation of hydrogels: chemical crosslinking, physical crosslinking, radiation cross-linking, and grafting polymerization. These techniques can enhance their viscoelasticity and mechanical properties for their application in pharmaceutical and biomedical fields [52]. Chemical cross-linking can be defined as the incorporation of monomers on the polymers or link of two polymer chains using cross-linking agents. The cross-linking of synthetic and biopolymers can be performed through the reaction of their functionalities such as carboxylic, amine, and alcohol groups with the help of cross-linkers such as aldehydes (e.g., adipic acid dihydrazide and glutaraldehyde). The major methods under chemical cross-linking include grafting, crosslinker, and radiation in aqueous and/or liquid state which are employed to formulate hydrogels from a range of biopolymers [52].

Physical crosslinking has been of great interest in gel development because of the ease of formulation and the benefit of not utilizing cross-linking agents. The careful selection of pH, concentration, and hydrocolloid type can result in the development of a wide range of gel textures and is presently an area of great attention. The methods used to develop physically cross-linked hydrogels include cooling/heating a polymer solution, ionic interaction, complex coacervation, hydrogen-bonding and maturation freeze–thawing [52]. Radiation crosslinking is a broadly employed technique that does not require the use of chemical additives and therefore maintains the biocompatibility of the natural polymer. In addition, sterilization and modification can be performed in a single step, and hence it is an affordable process for natural polymer modification. This technique mainly depends on generating free radicals followed by exposure to a high energy source such as electron beam, X-ray or gamma ray. The direct or indirect action of radiation will rely on the polymer environment (i.e., solid-state, dilute solution, concentration solution). The major methods under radiation cross-linking include aqueous state radiation, radiation in a paste, solid-state radiation [53]. Graft polymerization is a process whereby monomers are bonded covalently and polymerized as side chains into the main polymer backbone. Grafting is an interesting approach that imparts a diversity of functionalities to a polymer. Graft polymers are also known as graft copolymers and comprise of at least two different types of monomer parts, such as the grafted side chains that are structurally different from the main chain. The monomer to be grafted may be of one or more than one kind; thus, the graft chains in the grafted copolymer may be homopolymers or copolymers [54].

Hydrogels can be defined as hydrophilic wound dressing scaffolds with three-dimensional networks (Figure 4) that can be designed from biopolymers and synthetic polymers. These wound dressings possess the capability to absorb a huge volume of water and biological fluids [55]. Hydrogels have been extensively utilized for applications in wound dressings because of their good porosity, ability to load and release various drugs, to offer debriding and desloughing capacity on fibrotic and necrotic tissue, to provide a moist environment, good flexibility, biocompatibility, and high water content [56,57]. The advantages of hydrogels also include soothing outcomes that enhances patient compliance, effective in softening necrotic tissue on the wound surface, hydrate wound surfaces, and they can be used or removed without interfering with the wound bed and they are non-adherent.

Although there are several interesting properties of hydrogels in the field of wound management, they also suffer from some limitations such as dehydration if they are not covered, result in skin maceration, they need a secondary dressing, it is not easy to secure them, and they possess poor mechanical stability at swollen state [58]. Other disadvantages of hydrogels are their ineffectiveness in wounds with excessive wound exudates, which may involve secondary wound dressings and they are not suitable scaffolds for moderate-to-extreme exudative wounds [58].

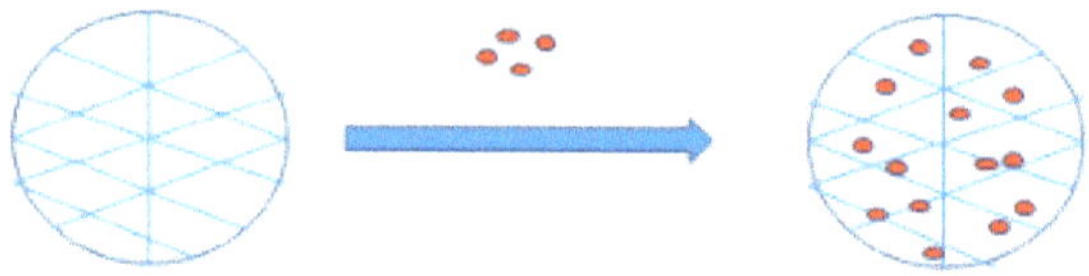

Figure 4. Schematic diagram of hydrogels.

4.1. Chitosan-Based Hydrogels

There are several reports of chitosan-based hydrogels for the management of wounds with good healing outcomes (Figure 5) (Table 1). Rasool and co-workers designed stimuli-responsive chitosan-based hydrogels by blending poly (N-vinyl-2-pyrrolidone) (PVP) and chitosan using solution casting technique for potential application in wound healing [59]. The hydrogels were thermally stable when compared to the free polymers. The hydrogels were biodegradable and pH-sensitive, with a maximum swelling degree of 10,220%. Their swelling capability increased with the increase in the pH and a decrease in swelling at neutral pH. The antimicrobial activity of the hydrogel samples was investigated using the disc diffusion method and all the chitosan-based hydrogels possessed antibacterial efficacy against Gram-negative *E. coli* strains which is attributed to the interaction of chitosan with the plasma membrane of the bacteria. Chitosan molecules have positively charged NH_3^+ groups and *E. coli* outer covering is composed of lipopolysaccharides that influence the significant negative charge of the bacterial surface. Furthermore, chitosan binds with bacterial DNA, thereby inhibiting transcription and translation. The in vitro drug release profile of the hydrogels loaded with antibiotics, silver sulfadiazine used in burn wound was sustained and controlled with over 91.2% of the drug released in 1 h. The results demonstrated that the chitosan-based hydrogels are potential scaffolds for burn wound management [59].

Xue and co-workers prepared chitosan–Matrigel–polyacrylamide hydrogels for wound healing and skin regeneration. These hydrogels showed good porous morphology with pore sizes ranging between 10 and 30 μm, which can significantly maintain a moist environment useful for cell migration, adhesion, and viability. These hydrogels showed good mechanical properties such as high fracture strain (1210%), compression, tensile strength (118.3 kPa), and resilience. The adhesive strength of the hydrogels on porcine skin was 1.3 kPa indicating that they can be easily removed from the wound surface without causing further trauma to the wound surface. The hydrogels swelling capability reached an equilibrium state of approximately 1200% after two days. The antibacterial activity of the hydrogels showed significantly better growth inhibition against *S. aureus*, *E. coli*, and *S. epidermidis* strains when compared to the control, kanamycin. The hydrogels reduced the amount of blood loss significantly in bleeding mouse liver. The wound healing capability of the hydrogels in vivo showed good wound contraction on the 5th day when compared to three control groups [60].

Khorasan and co-workers formulated chitosan/polyvinyl (alcohol)/zinc oxide hydrogels using the freeze–thaw method. The hydrogels were porous, with a mean pore size of 13.7 ± 5.9 μm. Their water absorption ranged between 680% and 850%. The mechanical properties of chitosan hydrogels showed decreased elastic modulus and tensile strength with increased elongation at the breakpoint. Their antibacterial efficacy result displayed the highest inhibition zone in *S. aureus* when compared to *E. coli* strain. The in vitro wound healing experiments of the prepared hydrogels displayed no toxicity and accelerated wound healing phase was significant [61]. Masood et al. designed chitosan-polyethylene glycol (PEG) hydrogel entrapped with silver nanoparticles for diabetic wound management. Their average particle size was 99.1 ± 2.3 nm. The hydrogel showed significant high porosity of 72.2%, which contributed to the accelerated wound healing, indicating a good oxygen penetration and absorption of exudates. The high swelling capacity of the plain hydrogels was significant when compared to nanoparticle-loaded hydrogels [62].

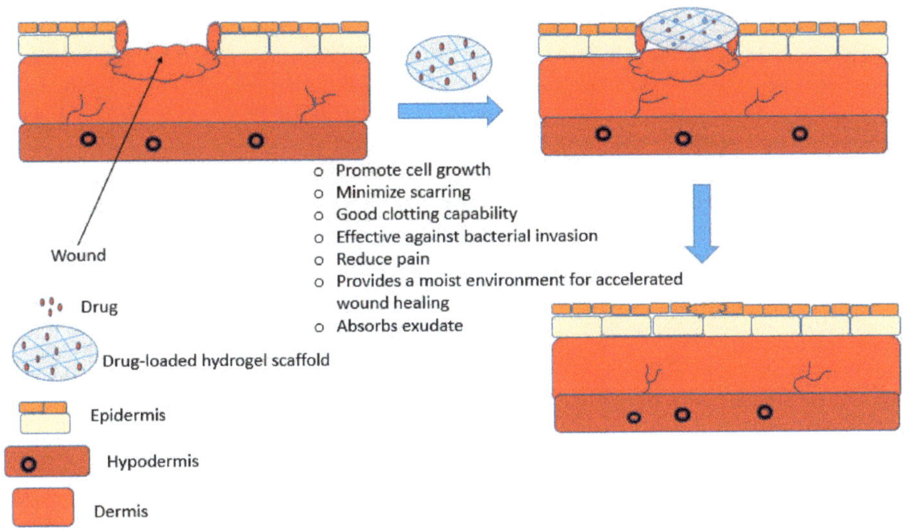

Figure 5. A schematic diagram illustrating the efficacy of chitosan-based hydrogel as an effective wound dressing.

The silver nanoparticle-loaded hydrogels demonstrated a water vapor transmission rate of 2104 $g{\cdot}m^{-2}{\cdot}24\ h^{-1}$ and 1391.3 $g{\cdot}m^{-2}{\cdot}24\ h^{-1}$ for the plain hydrogels indicating the suitability of nanoparticle-loaded hydrogels as a potent wound dressing for wound management. The in vitro release mechanism of nanoparticles from the chitosan-based hydrogels was slow and sustained at 37 °C. The antimicrobial activity evaluations demonstrated that chitosan-based hydrogels loaded with silver nanoparticle have good antimicrobial efficacy compared to the plain hydrogels and nanoparticles. The nanoparticle-loaded hydrogels exhibited higher zone inhibition of 21.5 ± 0.5, 15.5 ± 0.8, 21.8 ± 1.5, and 20.2 ± 1.0 mm against *S. aureus*, *B. subtilis*, *P. aeruginosa* and *E. coli*. The wound healing experiment exhibited that the loading of silver nanoparticles in chitosan-based hydrogels significantly improved the wound healing in diabetes-induced rabbits. The loaded hydrogels showed 47.7% wound contraction after 4 days compared to 12.6% in the negative control [62].

Wenbo et al. designed chitosan–heparin hydrogel for controlled release of SDF-1 α for endometrium wound injury healing in a mouse model. The results of Western blots assay, Masson trichrome staining, H&E staining, immunofluorescence staining and immunohistochemical staining demonstrated that the endogenous c-kit positive stem cells adhered to the wounded spot and stimulated the wound healing process. The in vitro drug release profile of DF-1α from the hydrogel was 5% in the 1st hour, and increased to 18 and 53% after 24 h and 102 h, respectively, indicating the slow and controlled release mechanism [63].

Zhang et al. prepared and evaluated carboxymethyl chitosan composite hydrogels. The results from the in vitro studies of the effects of composite hydrogels on the development of Human skin fibroblasts NS-FB and human hypertrophic scar fibroblasts (HS-FB) cells demonstrated the outstanding mechanical properties and good biological activity of chitosan-based hydrogels, and their potential application for the development of wound dressing. The highest swelling ratio of the hydrogel was over 581% with good water retention ability. The hemolysis rate of the hydrogel was 0.81–2.67% revealing good blood compatibility. The survival rate of NS-FB cells cultured by the hydrogel extract was greater than 93% showing good cytocompatibility and indicating their capability to induce cell proliferation. The survival rate of HS-FB cells was greater than 88% with no adverse side effects [64]. Chitosan-based hydrogels formulated by Zhang et al. showed modulating cationicity of chitosan with the capability to inhibit hypertrophic scar development during the wound healing process. The hydrogels were loaded with varying concentrations of genipin in the range of 2.5–15%. The hydrogel inhibited hypertrophic scar via suppression of the expression of a smooth muscle actin and induction of type I matrix metalloproteinases. The incorporation of 2% (*v*/*v*) Aloe vera gel further enhanced the inhibition of scar formation. The cationicity of the chitosan hydrogels is useful in the proliferation and differentiation of human skin fibroblasts, the formation of ECM and the production of growth factors [65].

Bagher et al. designed alginate/chitosan-based hydrogels loaded with hesperidin for wound healing in a rat model. These hydrogels displayed suitable porosity of 91.2% with interrelated pores, appropriate swelling capacity, and biodegradability confirmed by weight loss of over 80% after 2 weeks. The in vivo wound healing studies demonstrated that the formulated hydrogels had a better wound closure when compared to the gauze-treated wound, especially the hydrogels loaded with 10% of hesperidin [66]. Ehterami et al. reported similar findings as Baghe et al. for alginate/chitosan-based hydrogel loaded with Alpha-tocopherol for wound management in vivo [67]. The porosity of the hydrogel was 89.2% with interconnected pores. It was biodegradable with a weight loss of 80% in two weeks. The wound closure was accelerated when compared to the gauze-treated wound (the control). Neo-tissue and granulation tissue formation was visible in vivo when using the hydrogel on animal models [67].

He et al. prepared adhesive nanocomposite hydrogels via crosslinking of N-carboxyethyl chitosan and Pluronic F127 for wound healing. These hydrogels displayed remarkable photothermal antibacterial activity on *E. coli* and *S. aureus* infected wound in vivo [68]. These hydrogels displayed stable mechanical properties, suitable gelation time, good biodegradability, high water absorbency, and hemostatic properties. The hydrogels exhibited a pH-responsive drug release mechanism and good antimicrobial efficacy after they were loaded with antibiotic moxifloxacin. The hydrogel tissue adhesive property allowed these hydrogels to have an excellent hemostatic outcome in a mouse liver trauma model, mouse tale amputation model and mouse liver incision model. Furthermore, the in vivo assessment in a rat full-thickness skin wound-infected model showed that these formulations are excellent wound dressings for improved wound closure, angiogenesis, and collagen deposition [68]. Hamdi et al. formulated carotenoids enriched blue crab chitosan composite hydrogels. The release of the carotenoids was enhanced at pH 7.4 when compared to pH 4.0 and 2.0 showing that the hydrogels are pH-sensitive intelligent drug delivery systems. The in vivo wound healing assessments of the carotenoids enriched chitosan-based hydrogels in rat models demonstrated the acceleration of wound and complete healing. The hydrogels displayed low toxicity [69].

Du et al. prepared injectable chitosan-based hydrogels for application in wound healing. The hydrogels were active against *P. aeruginosa* and *S. aureus* and demonstrated killing efficiencies of 96.4% and 95.0%, respectively, at bacteria concentration of 108 CFU/mL. After two weeks, the wounds treated with the hydrogel displayed a 99.8% wound closure which was characterized by skin hair. The hemostatic activities of the hydrogels on a rat liver hemorrhaging model showed that the hydrogel adhered quickly to the tissue surrounding the bleeding site resulting in the formation of a barrier that prevents hemorrhaging. These results revealed that chitosan-based hydrogels are suitable for hemorrhagic healing and bacterially infected wound healing [70]. Zhang et al. formulated and evaluated novel chitosan–PVA–lignin composite hydrogels for wound management [71]. The hydrogels displayed good antibacterial activity and biocompatibility. The addition of lignin in these hydrogels effectively enhanced a wound moist environment for accelerated wound healing. The tensile strength was over 46.87 MPa. The hydrogels exhibited accelerated wound closure of 41% and 87% on the 5th and 10th day, respectively [71]. Pawar et al. designed chitosan-based hydrogels covalently incorporated with cefuroxime via ester bond for the treatment of infected wounds. The in vitro release profile of cefuroxime from the hydrogels was higher in alkaline medium of pH 10 and phosphate buffer of pH 7.4 with esterase enzyme when compared to the phosphate buffer. The drug release was via chemical and enzymatic hydrolysis. The hydrogel exhibited good biocompatibility on MG63 osteosarcoma and L929 fibroblast cell lines. The in vitro antibacterial activity against *S. aureus* revealed a high inhibition zone revealing that these hydrogels can be used for the treatment of infected wounds [72].

Pham et al. synthesized thermal-responsive chitosan-based hydrogels co-loaded with gelatin and curcumin for wound healing application. The in vitro drug release profile was sustained and controlled for curcumin and gelatin from the hydrogels. The in vivo evaluation demonstrated that these hybrid hydrogel displayed a synergistic effect of curcumin and gelatin, which stimulated the regeneration of the structure and the barrier's function of injured skin such as a wound [73]. Ferreira et al. prepared chitosan hydrogels loaded with 2 and 4% chlorhexidine. The formulation displayed 100% inhibition of *Staphylococcus aureus* growth. In vivo studies showed that on day 14, the animal models treated with the hydrogel containing 2% chlorhexidine displayed high contraction of the wound compared to the control groups [74]. Another study of hMSCs-seeded injectable chitosan hydrogel scaffolds conducted by Xu et al. demonstrated improved full-thickness cutaneous wound healing. The histological detection of these hydrogels further showed that the hydrogels encapsulated with hMSCs considerably accelerated wound closure, re-epithelialization, microcirculation, hair follicle regeneration, tissue remodeling, and inhibited over-inflammation in the surrounding and the central wounds [75]. Ragab et al. formulated soft hydrogels encapsulated with *P. granatum* peel extract that is based on chitosan for the treatment of chronic wound. These biopolymer hydrogels displayed tolerable cytotoxicity against human fibroblast cells. The in vivo wound healing study using a rat model showed good healing process [76].

Long et al. formulated 3D-printed chitosan–pectin hydrogels for drug delivery of lidocaine hydrochloride for wound treatment. The in vitro release profile at physiological conditions displayed burst release in 1 h, followed by sustained and controlled release over a period of 4 h. The burst release of lidocaine over 1 h from these hydrogels offer an effective pain relief [77]. Viezzer and co-workers formulated and evaluated chitosan-based polyurethane hydrogels loaded with transplant bone marrow mesenchymal cells for chronic wound treatment. These hydrogels demonstrated enhanced wound healing of ulcers in a diabetic rat model with a significant wound closure [78]. Moradi et al. synthesized and characterized chitosan-based hydrogels encapsulated with thymine oil cyclodextrin inclusion compound. The hydrogels displayed high elasticity and less stiffness. A significantly decreased number of bacteria was observed after inoculation with chitosan-based hydrogels loaded with thymine oil, indicating their potential use for the management of bacterially infected wounds [79]. The immersion of wounds in seawater containing salt and pathogenic bacteria, can result in serious infections. Wang et al. prepared composite hydrogel from a combination of hyaluronic acid and quaternized chitosan to accelerate healing of wounds immersed in seawater. The hydrogel inhibited bacterial growth. The in vivo study in a seawater-immersed wound defect model showed that the hydrogels decreased pro-inflammatory factors and promoted anti-inflammatory factors (TGF-β1) in the wound. The hydrogel promoted excellent re-epithelialization, increased the granulation tissue thickness and the density collagen deposition with good antibacterial activity [80].

Chen et al. developed citric-modified chitosan hydrogel loaded via a freezing and thawing treatment method and loaded with tetracycline hydrochloride. The hydrogel antimicrobial activity was significant against *E. coli* and *S. aureus*. The tensile strength and modulus of the hydrogel were 2.01 ± 0.01 MPa and 16.12 ± 0.11 MPa (3 wt%), respectively, when compared to the native skin with tensile strength of 3 ± 1.5 MPa and modulus in the range of 2.16 kPa–0.1 MPa. In vivo studies showed that granulation tissue was visible on the wounds covered with hydrogels after 12 days [81].

Huang et al. physically blended *Bletilla striata* and carboxymethyl chitosan with Carbomer 940 for good gel formation and improved water retention. The hydrogel exhibited good blood compatibility with 85% cell viability on M293T cells revealing the non-cytotoxicity. The full-thickness wound studies showed a 71.64 ± 6.64% and 83.80 ± 5.56% wound healing rates on day 7 and day 14, respectively. The connective tissues formed were dense with a well-developed sebaceous gland, granulation tissue and hair follicle [82]. Nguyen et al. reported hydrogel composed of chitosan, polyvinyl alcohol, and loaded with silver nanoparticles. The hydrogel excellent antibacterial activity was against *P. aeruginosa* and *S. aureus* [83]. Chen et al. developed carboxymethyl chitosan-based hydrogels loaded with melatonin. Evaluation in a full-thickness cutaneous wound model showed increased wound closure with enhanced proliferation of the granulation tissue, re-epithelialization, and rapid collagen deposition. The hydrogel-induced angiogenesis with the enhanced expression of vascular endothelial growth factor receptor protein and cyclooxygenase-2. The hydrogel was highly porous with interconnected interior structure and a pore size of 250 μm [84]. Nooshabadi et al. developed chitosan-based hydrogel loaded with exosomes. The hydrogel wound closure ability was 83.6% in full-thickness excisional wound model when compared to the sterile gauze that showed 51.5%. A high degree of re-epithelialization was observed with a reduction in wound size [85]. Ravishankar et al. crosslinked chitosan using alkali lignin. The electrostatic interaction between the phenoxide groups in lignin and the ammonium groups on the chitosan contributed to the ionotropic cross-linking. The gels were non-toxic to the *Mesenchymal* stem cells. The hydrogels supported good cell migration indicating their potential application for tissue regeneration [86]. Lin et al. prepared hydrogel from poly (vinyl alcohol) (PVA), dextran, and chitosan using glutaraldehyde as a cross-linker. Preparing the hydrogel from 6% PVA and 0.25% chitosan enhanced the antimicrobial capability. Combining the hydrogel with 4% dextran induced high cell proliferation. The thermostability, water retention capability, mechanical properties, and moisturizing ability of the hydrogel was influenced by the addition of chitosan [87]. Heimbuck et al. reported genipin-cross-linked chitosan hydrogels. The average water uptake was ~230% with good bacterial activity hindering 70% *E. coli*. It was biocompatible on fibroblast and keratinocyte cells in vitro. It induced high immune response and cellular proliferation in pressure wounds model used [88]. Ternullo et al. incorporated liposomes into hydrogel for dermal delivery of curcumin. The system displayed sustained delivery of curcumin with increased retention at the skin. The hydrogel bioadhesiveness was significant. Incorporation of neutral deformable liposomes in the hydrogel promoted the high bioadhesive nature of the hydrogel. The positively charged deformable liposomes promoted the bioadhesiveness and the sustained delivery of curcumin in ex vivo full human skin [89]. Qu et al. prepared hydrogel from N-carboxyethyl chitosan (CEC) and oxidized hyaluronic acid-graft-aniline tetramer. The hydrogels exhibited high swelling capability, good biodegradation property, and free radical scavenging capacity. The loading of amoxicillin prevented wound infection. It significantly accelerated wound healing rate with high granulation tissue thickness and more angiogenesis in a full-thickness skin defect model [90]. Cardoso et al. loaded phenytoin, an antiepileptic drug used for the treatment of epilepsy into chitosan hydrogels for the treatment of diabetic and burn wounds. The release of phenytoin was controlled from the hydrogel with good adhesion to the skin. Phenytoin promoted the formation of collagen fibers and fibroblasts in the wound site in vivo. The results reveal that phenytoin can be repurposed for wound healing [91]. Zahid et al. prepared hydrogel from chitosan, polyvinyl alcohol and S-nitroso-*N*-acetyl-DL-penicillamine. It induced significant angiogenesis in chronic wounds with sustained production of NO. The production of NO promoted angiogenic activity and accelerated wound healing. It supported the proliferation of 3T3 and HaCaT cells in vitro [92].

Bano et al. reported chitosan–PVA soft membranes for the treatment of burn wounds. The hydrogel supported normal growth of epidermis and accelerated the formation of granule and fibrous connective tissues. Accelerated wound healing and re-epithelialization of burn wounds resulted in a decreased deposition of collagen thereby preventing severe scar formation [93]. Nešović et al. prepared chitosan-poly(vinyl alcohol) hydrogels via a freezing–thawing method and loaded it with silver nanoparticles. The hydrogels were non-cytotoxic with good antibacterial activity against *S. aureus* and *E. coli*. Their swelling capability was improved by increasing the content of chitosan and the addition of AgNPs. The release profile of the drug from the hydrogel was an initial burst release followed by a slow drug release from day 5 to day 28. The release profile was appropriate for the effective treatment of a bacterially infected wound followed by a slow drug release thereby providing a sterile wound environment for an extended period [94]. Du et al. reported chitosan-based hydrogel prepared from hydrocaffeic acid-modified chitosan with hydrophobically modified chitosan lactate for sutureless closure of surgical incisions. The hydrogel was effective against *S. aureus* and *P. aeruginosa* in vitro. It was non-cytotoxic on 3T3 fibroblast cells, biocompatible and biodegradable. It exhibited good In situ antibleeding efficacy in rat hemorrhaging liver and full-thickness wound models. It closed the wound in a sutureless way thereby promoting wound healing [95]. Mousavi et al. developed chitosan/gelatin hydrogels in the ratio of 1:5 and 1:1, respectively. The presence of collagen in the hydrogel reduced the swelling capability, biodegradation rate of the hydrogels and the mechanical strength when compared to the gelatin. Collagen loaded in the hydrogel promoted cell attachment [96].

Wang et al. reported hydrogel prepared from a combination of hyaluronic acid and quaternized chitosan to promote wound healing and prevent bacterial infection. The biocompatibility of the hydrogels in vitro on fibroblast L929 cell was significant. The hydrogels displayed high repair in a seawater-immersed wound defect model. A decreased expression of pro-inflammatory factors such as (TNF-α, IL-1β, and IL-6) and increased expression of TGF-β1, an anti-inflammatory factors was significant in the wound. The hydrogel was effective against *S. aureus* [97]. The wound healing was accelerated and characterized by excellent reepithelialization, high deposition of collagen, good thickness of the granulation tissue and low level of endotoxin. Martínez-Ibarra et al. prepared hydrogels from a combination of xyloglucan and chitosan. The pore sizes of the hydrogels were in the range of 32.8–101.6 µm. The hydrogels were biodegradable with a significant weight loss over a period of two weeks. Their antibacterial activity was influenced by the biopolymers used for their preparation [97]. Soriano-Ruiz et al. prepared chitosan hydrogel for epidermal regeneration. The porous microstructure of the scaffold enhanced adequate oxygen and nutrient diffusion to the wound. Furthermore, hMSCs encapsulation provided an appropriate microenvironment that supported their viability for 7 days. The viability of hMSCs was above 75% [98]. Patil et al. investigated the benefit of chitosan hydrogels with or without oxygen. Wound healing processes are either directly or indirectly dependent on oxygen such as inflammation, angiogenesis, phagocytosis, cell proliferation, collagen synthesis etc. In vivo studies on an acute porcine wound model of the hydrogel was compared with a control, Derma-Gel™ hydrogel dressings. The combination of the hydrogel with oxygen accelerated the wound closure when compared to the control. The combination of the hydrogel with oxygen enhanced the formation of new blood vessel formation with maturation of keratinocyte [99]. Li et al. prepared collagen/chitosan gel composite loaded with cell-penetrating peptide (Oligoarginine, R8). The hydrogel composite inhibited *Staphylococcus aureus* growth. In vivo studies on animal models showed a complete wound surface healing rate of 98% on day 14. The formulation promoted the enhanced formation of granulation tissue, increased deposition of collagen and angiogenesis was visible in the wound tissue [100].

Soares et al. developed chitosan-based hydrogel loaded with a mixture of flavonoids obtained from Passiflora edulis Sims leaves for the treatment of diabetic wounds. In vivo wound healing studies on male Wistar rats showed that the treatment of the wound with the hydrogel stimulated the antioxidant defense systems. The release of the flavonoid from the hydrogel was rapid. On the 14th day of treatment, a high oxidative stress was visible which is attributed to hypoxia caused at the wound site due to the presence of the film. The hydrogel promoted increased lipid peroxidation at the injured tissue [101]. Movaffagh et al. reported chitosan hydrogel loaded with honey. The hydrogels were effective against *Staphylococcus aureus*, *Bacillus cereus*, *Escherichia coli*, *Pseudomonas aeruginosa*, and *Candida albicans*. The combination of honey with chitosan accelerated the wound healing process and enhanced the antibacterial activity [102]. Djekic et al. prepared chitosan hydrogel by ionic gelation technique. It was loaded with ibuprofen and the in vitro release profile of the drug was sustained over a period of 12 h followed by a zero-order release mechanism. The hydrogel displayed good cohesiveness and adhesiveness [103]. Lim et al. studied the wound healing efficacy of thermosensitive hydroxybutyl chitosan hydrogel loaded with human platelet lysate. The release of human platelet-derived growth factor from the hydrogel was sustained. The hydrogel promoted wound healing in vivo with the formation of new collagen. The hydrogel promoted a high level of Human Umbilical Vein Endothelial Cells proliferation and tube formation in vitro [104]. Li et al. reported IGF-1C domain-modified chitosan hydrogel for skin regeneration. The hydrogel had excellent proangiogenic effects with an accelerated wound closure and elevated remodeling of the extracellular matrix. It accelerated the cutaneous wound healing via stimulating angiogenesis [105].

Yan et al. prepared chitosan-gentamicin conjugate hydrogel with good water solubility and excellent antimicrobial activity. The hydrogel was effective against *P. aeruginosa* and *S. aureus* with the diameter zone of inhibition of 20.3 ± 0.06 mm and 20.0 ± 1.0 mm, respectively. The cell viability studies showed that the hydrogel co-incubation with L929 cells did not induce toxic effect (at a concentration 200 μg/mL) with cell viability of 101.68%, 100.44%, 86.73%, respectively, on day 1, 2 and 3. The hemocompatibility of the hydrogel was superior when compared to the free drug. On day 14 and 21, the wound healing rate was 89.18% and 99.61%, respectively, in vivo on the animal model treated with the hydrogel. On day 14, complete epidermal regeneration was visible [106]. Cifuentes et al. reported the efficacy of chitosan hydrogels functionalized with either bemiparin or unfractionated heparin for diabetic wound healing. The hydrogels accelerated the inflammation process and enhanced the re-epithelialization. The thickness of the neodermal in the animal model after treatment with the hydrogel was similar to the non-diabetic animal models. The hydrogels healing effect was similar. However, treating the animal model with chitosan hydrogel loaded with bemiparin resulted in good quality of tissue in the neoformed dermal tissue when compared to the chitosan hydrogels functionalized with unfractionated heparin [107].

Zhang et al. developed a temperature-responsive hydroxybutyl chitosan-based hydrogel from chitosan and dopamine at different concentrations. The hydrogels were non-toxic to mouse fibroblast cells (L929). The hydrogel was effective against *S. aureus* with good blood clotting capability [108]. Kalantari et al. prepared polyvinyl alcohol/chitosan hydrogel loaded with cerium oxide nanoparticles. The hydrogel with 0.5% CeO_2-NPs was effective against the growth of MRSA. It was non-toxic to human dermal fibroblasts healthy human dermal fibroblast with high viabilities of 90% over a period of 5 days [109]. Leonhardt et al. reported chitosan-loaded hydrogels that degrade at the site of the injury. The diameters of the hydrogel was 9.2 ± 3.7 nm resulting from an increase in the surface area of chitosan. The good hemostatic capability of the hydrogel revealed their promising application to control blood loss in skin injury [110]. Wang et al. developed chitosan/oxidized konjac glucomannan hydrogel loaded with silver nanoparticles for the treatment of irregular wounds. A comparative in vivo study of the hydrogel with a commercial hydrogel dressing (AquacelAg™) showed that the hydrogel displayed self-healing ability, good tissue adhesiveness, and antibacterial activity, making it a self-adapting wound dressing for the treatment of irregular wounds [111].

Yang et al. developed chitosan hydrogels loaded with LL-37 peptide for the treatment of pressure ulcer. The hydrogel at a concentration of 5 μg/mL was effective against *S. aureus* [112]. In vivo studies on animal models revealed high epithelial thickness and newly-formed capillary at day 15 and 21. The hydrogel's increased capacity to promote wound closure and re-epithelialization by keratinocytes was significant. The hydrogels exhibited excellent cytocompatibility and were non-cytotoxic in vitro. The expression of VEGF in the wound treated with the hydrogel was high at day 14, showing the correlation of HIF-1α with VEGF protein expression which indicates that HIF-1α promotes angiogenesis in deep tissue injuries [112].

4.2. Chitosan-Based In Situ Forming Hydrogels

In situ gels have been reported to be useful as wound dressings (Table 1). They display several advantages such as ease of administration, extended contact time of drug at the site (thereby improving drug bioavailability), reduced frequency of administration, improved patient compliance, etc. [113,114]. They are smart systems that represent a promising means of delivering drugs and they undergo sol–gel transition after administration. In situ gel formation occurs due to stimuli such as change in pH, temperature, etc. [115,116]. The application of in situ gels as wound dressing results from their capability to fit the irregular shape of the wound without causing any form of wrinkles [117].

Lin and co-workers formulated Histatin1 (Hst1)-modified thiolated chitosan-based hydrogels for wound dressing [118]. These hydrogels exhibited gelation time that ranged between 5 and 7 min at 37 °C in vitro while it was 8 min for the in situ injectable hydrogels in vivo. These results demonstrated that these hydrogels can quickly form in situ and can be utilized for skin defects repair in vivo. The drug release studies of the hydrogels in vitro at physiological conditions (pH 7.4 and 37 °C) using Rhodamine B (Rh B) and Rhodamine B–Hst1to stimulate the release mechanism of Hst1 peptide from the hydrogels showed that the release of Rh B increased within 72 h. The release of Rh B was higher than Rh B–Hst1 within 72 h. Furthermore, the Rh B–Hst1 release rate was decreased with an increase in the concentration of Rh B and Rh B–Hst1. The Hst1 was released in vivo via the degradation and the diffusion of chitosan. The wound healing studies of hydrogels in vivo at 0, 2, 5 and 7 days showed that the wound site contracted and the wound healing mechanism of Hst 1-loaded hydrogels were faster when compared to the free hydrogels [118].

Huang et al. formulated and evaluated antibacterial PEG diacrylate (DA)/chitosan hydrogels [119]. These hydrogels demonstrated in situ forming properties, adhesiveness, good mechanical strength, improved biocompatibility, and higher antibacterial activity. The PEG DA/chitosan hydrogels composed of 15 weight% of PEGDA and 2 weight% of chitosan displayed outstanding mechanical adhesiveness, sustained release of plasmid DNA and antibacterial peptides. Furthermore, the wound healing study in vivo displayed important acceleration in the wound healing mechanism in full-thickness skin defect mode by stimulating the angiogenesis and reducing the inflammation [119].

Lv et al. formulated in situ injectable hydrogels based on two biopolymers, carboxymethyl chitosan and alginate, for wound healing. The hydrogels containing 0.5% of chitosan oligosaccharide stimulated human umbilical cord mesenchymal stem cells proliferation and remarkably enhanced the wound healing mechanism in a mouse skin defect model. In addition, the microscopic wound analysis of the injectable hydrogels displayed an increase in the integrity and thickness of the epidermal tissue and increased the development of collagen fibers [120].

Song et al. designed a chitosan complex hydrogel for wound management. The rheological analysis revealed that chitosan-based hydrogels displayed a great anti-shear ability and broad linear viscoelastic region. The hydrogels exhibited moderate bactericidal efficacy against *S. aureus* and *E. coli*. The in vivo wound healing examination of chitosan-based hydrogels showed that the hydrogels significantly reduce wound size by 44% and 82.34% after 5th and 10th day, respectively [121].

Xu et al. designed in situ alginate-chitosan hydrogel for corneal wound healing via periodate-mediated sodium alginate oxidization. In vivo studies on alkali burn wounds in vivo showed improved epithelial reconstruction. Limbal stem cells were transplanted into the in situ hydrogel. The hydrogels provided favorable microenvironment for the cell differentiation and proliferation of the transplanted cells. The restoration of the integrity of the corneal after alkali burn was rapid. The in situ hydrogel displayed good biocompatibility, transparency, and biodegradability suitable for tissue regeneration [122].

Gholizadeh et al. developed chitosan-based in situ gels for the treatment of nasal wounds. Injury of nasal epithelium can lead to nose bleedings. The thermosensitive in situ gel was loaded with tranexamic acid. The formulation rapid liquid-to-gel phase change occurred in 5 min at the human nasal cavity temperature. Its liquid-to-gel capability retained the formulation in the anterior part of the nose with no lung deposition and also improved the drug bioavailability thereby prolonging the drug release on the wounded epithelium. The rate of wound closure was higher in cells treated with the in situ gel formulation after 1 h (6.89% ± 0.03) and 3 h (8.38% ± 0.03), when compared to treatment using the drug solution which was 1.16% ± 0.01 and 4.87% ± 0.01. The combination of factors such as the hemostatic properties of chitosan, the formulation of in situ gelation capability, sustained drug release profile and improved drug bioavailability at the wound site promoted the healing efficacy [123]. He et al. developed injectable hydrogel from catechol- and methacrylate-modified chitosan/gelatin. The formulation was effective against *P. aeruginosa* and *S. aureus*, killing more than 80% of these bacteria in a period of 2 h. It also displayed tissue adhesive that adhered to the skin at body temperature and it rapidly closed open wounds on the dorsum of a rat [124].

4.3. Cellulose-Based Hydrogels

There are also several hydrogels that are designed from cellulose and its derivatives for wound application (Table 1). Joorabloo formulated carboxymethyl cellulose-based hydrogels loaded with heparinized zinc oxide nanoparticle via a freeze–thaw method for application in wound dressing [125]. The average particle size of the loaded ZnO nanoparticles ranged between 16 and 36 nm. SEM images showed increased pore sizes after the addition of the nanoparticle with decreased pore density. The in vitro drug release profile at physiological conditions was an initial burst release of the heparinized nanoparticles from hydrogels followed by a sustained drug release. The mechanical properties of hydrogels increased with an increase in the amount of nanoparticles loaded into the hydrogel. The mechanical strength and Young's modulus increased while the elongation at breakpoint decreased [125]. Cell viability study of hydrogels loaded with nanoparticles demonstrated non-toxicity on HDF and L-929 cells after 2 days. The wound healing assessment studies revealed good wound healing capability on the created artificial wound after 24 h. The antibacterial activity of the hydrogels using disc diffusion method showed significant bactericidal efficacy of over 70% for ZnO-loaded hydrogels against *S. aureus* and *E. coli* [125].

Gupta et al. prepared bacterial cellulose hydrogels encapsulated with curcumin for wound dressing application [126]. The water vapor transmission rate of the hydrogels was in the range of 2526.32–3137.68 g/m^2/24 h suggesting that the hydrogels can provide a moist environment at the wound site. The in vitro drug release studies of curcumin from the hydrogels showed that 76.99 ± 4.46% of the loaded drug was released after 6 h followed by a slow and sustained drug release mechanism. The high bioavailability of the drug from the hydrogels at the wounded site was effective at controlling bacterial-associated infection. The curcumin-loaded hydrogels reduce oxidative stress at the wound site [126]. Fan and co-workers formulated and evaluated pH-sensitive dual drug-loaded cellulose-based hydrogels as potential wound dressings. In vitro degradation evaluation showed that the hydrogels degraded under slightly acidic conditions and the loaded drugs were released. The in vivo wound healing assessment of the hydrogels showed that during the wound healing process, the weight of the rats was maintained indicating that rats were in good health. Furthermore, these hydrogels promoted a high percentage of wound closure [127].

Erdagi et al. prepared gelatin–diosgenin–nanocellulose hydrogels [128]. Their morphology was interconnected with good porosity indicating that they are suitable for cell proliferation and adhesion. The gel yield of the hydrogels ranged between 83.67 ± 2.18% and 90.17 ± 3.51%. The hydrogels demonstrated good water uptake efficiency and reached equilibrium swelling within one day. The mechanical properties of the hydrogels showed compression modulus that ranged between 3.04 ± 0.15 kPa and 8.04 ± 0.31 kPa demonstrating good strength. These results from mechanical properties revealed these hydrogels as suitable scaffolds for application in wound healing. The in vitro antibacterial assessment of neomycin-loaded hydrogels showed higher bacterial inhibition against *S. aureus* when compared to *E. coli* [128]. Liu et al. formulated bacterial cellulose hydrogels with tailored crystallinity from *Enterobacter* sp. FY-07 by the controlled expression of colanic acid synthetic genes. These hydrogels exhibited significant water-holding capacity of 25,643% with good properties such as stability, tensile strength, purity, porosity, low immunogenicity, and biocompatibility. These properties are very important in dressings for wound management [129].

Table 1. Summary of hydrogels that are formulated from chitosan and cellulose.

Biopolymer Used for Hydrogel Formulation	Other Polymers Used	Loaded Bioactive Agents	Outcomes	References
Chitosan	PVP (Polyvinylpyrrolidone)	Silver sulfadiazine	Maximum swelling capacity, good antibacterial efficacy against Gram-negative *E. coli* strains with sustained and controlled drug release profile	[59]
Chitosan	Matrigel and polyacrylamide	–	Good mechanical properties and antibacterial activity	[60]
Chitosan	PVA Poly(vinyl alcohol)	ZnO nanoparticles	Good antibacterial efficacy	[61]
Chitosan	PEG Poly(ethylene glycol)	Ag nanoparticles	High porosity, good antimicrobial efficacy, and improved wound healing mechanism in diabetic wounds	[62]
Chitosan	–	SDF-1 α	Slow and controlled drug release kinetics	[63]
Carboxymethyl Chitosan	–	–	Excellent mechanical properties	[64]
Chitosan	–	–	Preventing hypertrophic scar development during the wound healing process	[65]
Chitosan	Alginate	Hesperidin	Good wound closure	[66]
Chitosan	Alginate	Alpha-tocopherol	Good wound contraction	[67]
N-carboxyethyl Chitosan	Pluronic F127	Moxifloxacin	pH-responsive drug release mechanism and good antimicrobial efficacy. Improved wound closure	[68]
Chitosan	–	Carotenoids	Accelerated wound healing process	[69]
Chitosan	–	–	Good hemostatic activity in a rat hemorrhaging and good antibacterial activity.	[70]
Chitosan	PVA and lignin	–	Accelerated wound healing with wound closure, good antibacterial activity, and good biocompatibility	[71]
Chitosan	–	Cefuroxime	Good biocompatibility on MG63 osteosarcoma and L929 fibroblast cell lines and excellent antibacterial efficacy	[72]
Chitosan	–	Gelatin and curcumin	Sustained and controlled drug release profiles	[73]
Chitosan	–	Human mesenchymal stem cells	Good cell viability, spreadability, and adhesive nature	[74]
Chitosan	–	Human mesenchymal stem cells	Improved full-thickness cutaneous wound healing	[75]
Chitosan	–	*P. granatum* peel extract	Good wound healing process	[76]
Chitosan	Pectin	Lidocaine hydrochloride	Sustained and controlled drug release	[77]
Chitosan	Polyurethane	Transplant bone marrow mesenchymal cells	Enhanced wound healing of diabetic wounds	[78]
Chitosan	–	Thymine oil cyclodextrin inclusion compound	Significantly decrease the number of bacteria	[79]
Chitosan	–	Silver nanoparticles	Non-cytotoxicity and high antibacterial activity against *E. coli* and *S. aureus*	[80]

Table 1. *Cont.*

Biopolymer Used for Hydrogel Formulation	Other Polymers Used	Loaded Bioactive Agents	Outcomes	References
Chitosan	–	Tetracycline hydrochloride	Excellent mechanical performance with superior antibacterial effects against *E. coli* and *S. aureus*	[81]
Carboxymethyl chitosan	Carbomer 940	*Bletilla striata*	Accelerated wound healing process and good blood compatibility	[82]
Chitosan	PVA	Silver nanoparticles	Excellent antibacterial activity against *P. aeruginosa* and *S. aureus*	[83]
Carboxymethyl chitosan	–	Melatonin	Increased wound closure with enhanced proliferation of the granulation tissue	[84]
Chitosan	–	Exosomes	Excellent wound healing effects	[85]
Chitosan	Lignin	–	Good biocompatibility	[86]
Chitosan	PVA and dextran	–	High cell proliferation	[87]
Chitosan	–	–	High water uptake with good antibacterial activity against *E. coli*.	[88]
Chitosan	–	Curcumin	Sustained drug release kinetics	[89]
N-carboxyethyl chitosan	Hyaluronic acid	Amoxicillin	High swelling capacity and wound healing process acceleration	[90]
Chitosan	–	Phenytoin	Good wound healing mechanism for diabetic and burnt wounds	[91]
Chitosan	PVA	S-nitroso-N-acetyl-DL-penicillamine	Good angiogenesis effects for wound healing process	[92]
Chitosan	PVA	–	Accelerated of wound healing process and re-epithelialization of burn wounds	[93]
Chitosan	PVA	Silver nanoparticles	Non-cytotoxicity with high antibacterial activity against *S. aureus* and *E. coli*	[94]
Chitosan	–	–	Effective bactericidal activity against *S. aureus* and *P. aeruginosa* with good wound healing effects	[95]
Chitosan	Gelatin	–	Good mechanical performance	[96]
Chitosan	Hyaluronic acid	–	Good biocompatibility with bactericidal efficacy against *S. aureus*	[97]
Chitosan	Xyloglucan	–	High porosity	[97]
Chitosan	–	–	Good biocompatibility	[98]
Chitosan	–	–	Good wound healing mechanism	[99]
Chitosan	–	Histatin1	High rate of wound healing process	[100]
Chitosan	–	Flavonoids	Increased lipid peroxidation at the injured tissue	[101]
Chitosan	–	Honey	High growth inhibition against various bacterial strains with accelerated wound healing process	[102]
Chitosan	–	Ibuprofen	Sustained drug release and good cohesiveness and adhesiveness	[103]
Hydroxybutyl chitosan	–	Human platelet lysate	Stimulated wound healing	[104]
Chitosan	–	–	Accelerated wound closure and high remodeling of extracellular matrix	[105]
Chitosan	–	Gentamicin	High antibacterial effects against *P. aeruginosa* and *S. aureus* with high cell viability and accelerated wound closure	[106]

Table 1. *Cont.*

Biopolymer Used for Hydrogel Formulation	Other Polymers Used	Loaded Bioactive Agents	Outcomes	References
Chitosan	–	Bemiparin	Accelerated the inflammation process and enhanced the re-epithelialization	[107]
Hydroxybutyl chitosan	–	–	Excellent biocompatibility and effective against *S. aureus* with good blood clotting capability	[108]
Chitosan	PVA	Cerium oxide nanoparticles	Bactericidal effective against MRSA with good biocompatibility	[109]
Chitosan	Cyclodextrin	–	Good hemostatic capability	[110]
Chitosan	Oxidized konjac glucomannan	Silver nanoparticles	Self-healing ability and good tissue adhesiveness	[111]
Chitosan	–	LL-37 peptide	High antibacterial efficacy against *S. aureus*	[112]
Chitosan	–	Histatin 1	Faster wound healing process	[118]
Chitosan	PEG diacrylate	Antibacterial peptide and plasmid DNA,	Important acceleration in wound healing mechanism on full-thickness skin defect model	[119]
Carboxymethyl chitosan	Alginate	Human umbilical cord mesenchymal stem cells	Remarkably speeded the wound healing mechanism in a mouse skin defect model	[120]
Chitosan	–	–	Significantly reduced wound size	[121]
Chitosan	Alginate	–	Rapid restoration of the integrity of the corneal after alkali burn	[122]
Chitosan		Tranexamic acid	The formulation was safe on human nasal epithelial cells with an efficient wound closure (six times faster than the control drug solution)	[123]
Cellulose	–	Curcumin	A potential antioxidant that can significantly reduce oxidative stress at the wound site	[124]
Cellulose	–	–	The high percentage of wound closure	[127]
Cellulose	Gelatin	Diosgenin	Good mechanical properties and higher bacterial inhibition	[128]
Cellulose	–	*Enterobacter* sp. FY-07	Significant water-holding capacity	[129]
Cellulose	PVA	Curcumin	100% cell viability, the sustained drug released, and significant wound closure	[130]
Cellulose	Alginate	–	Enhanced water retention properties	[131]
Carboxymethyl cellulose	Collagen	–	The better wound healing process	[132]
Cellulose	–	–	Enhanced wound healing process	[133]
Cellulose	PVA	–	Good mechanical properties and high wound healing efficiency	[134]
Cellulose	–	Thymol	Low cytotoxicity, increased cell viability, and reduced burn wound area	[135]
Carboxymethyl cellulose	–	Clindamycin	Good bactericidal activity	[136]
Cellulose	Fenugreek gum	–	Good biocompatibility and non-toxic	[137]
Cellulose	Acrylic acid	–	Accelerated burn wound healing	[138]
Cellulose	Hydroxypropyl-β-cyclodextrin	Silver nanoparticles and curcumin	Good antibacterial activity against various bacterial strains	[139]
cellulose	Polylactide	–	Self-detachable and did not require changing	[140]

Table 1. *Cont.*

Biopolymer Used for Hydrogel Formulation	Other Polymers Used	Loaded Bioactive Agents	Outcomes	References
Cellulose	Flaxseed gum	_	High swelling capacity with promoted hemostatic and wound healing process	[141]
Cellulose	Gelatin	Curcumin	Accelerated cell proliferation and the controlled release of curcumin from the hydrogels	[142]
Cellulose	_	Linezolid	Sustained drug release with accelerated wound closure and good cytocompatibility	[143]
Cellulose	PVA	_	High porosity, high drug-loading capacity, good mechanical properties	[144]
Hydroxyethyl cellulose	mesocellular silica	_	Enhanced wound healing	[145]
Carboxymethyl cellulose	K-carrageenan/graphene oxide/konjac glucomannan	Silver nanoparticles	Good antibacterial activity, biocompatible and accelerated wound recuperation	[146]

"_" means "No other polymer was used".

Shefa et al. prepared and evaluated oxidized cellulose nanofiber-PVA hydrogels incorporated with curcumin for wound healing acceleration [130]. The hydrogels displayed interconnected microscopic pores. The viscosity assessment showed that the viscosity of hydrogels was increased with increased in the polyvinylalcohol concentration and decreased by the addition of curcumin. The cell viability study of hydrogels using MTT assay showed over 100% cell viability on L929 cells while proliferation study in these cells displayed increased cell proliferation with the span of time (from day 1 to 7). The in vitro drug release profile of the hydrogels containing 7.5% and 10% polyvinylalcohol at pH 7.4 phosphate buffer saline solution revealed a sustained release of curcumin. In all of the hydrogels, curcumin release increased after 1 day of incubation. The wound healing analysis demonstrated the significant wound closure between 28.8 ± 1.3% and 29.9 ± 1.7% when compared to 8.3 ± 1.13% of the control [130]. Sulaeva et al. formulated bacterial cellulose-based hydrogels incorporated with alginate to improve their performance for wound management. These hydrogels demonstrated enhanced water retention properties. These scaffolds were moreover demonstrated to be biocompatible and useful for the treatment of bacterial colonized wounds [131].

Yang et al. formulated biocompatible and stable hydrogel composites that are based on dialdehyde carboxylmethyl cellulose cross-linked with collagen [132]. The high degree of crosslinking in these hydrogels was promoted by a high concentration of collagen because the ε-amino groups on the collagen is suitable for covalent crosslinking with aldehyde groups of dialdehyde carboxylmethyl cellulose. The high cross-linking capacity significantly contributed to the strong elastic behavior of the hydrogel. The stability of the hydrogel was enhanced by the formation of more covalent bonds, resulting in improved viscoelastic properties, resistance to enzymatic degradation and thermal stability. The cell viability studies on L929 fibroblasts using CCK-8 tests demonstrated that the dialdehyde carboxylmethyl cellulose hydrogels cross-linked with collagen did not promote any toxic cellular responses [132]. Fontes et al. designed and evaluated bacterial cellulose-based hydrogels for wound management. It was prepared from the combination of cellulose with either Calendula officinalis or Jacaranda caroba. The hydrogel provided a humid environment, promoted rapid tissue repair, reduced severe inflammation and did not damage granulation. The hydrogels showed faster re-epithelialization over a period of 7 days. The hydrogel was composed of an interconnected pores which promoted cell migration and rapid re-epithelialization. The hydrogels were suitable for dry wounds and their permeability to metabolites and decreased temperature of the wound bed can lead to reduced pain [133].

Liu and co-workers formulated cellulose nanocrystal reinforced nanocomposite hydrogels cross-linked with PVA with self-healing properties via a freeze–thaw cycle method for wound healing [134]. The SEM images of the hydrogels exhibited porous morphology and the pore size of the hydrogels decreased as the PVA amount increased. The mechanical properties analysis of the hydrogels displayed the compressive stress that ranged between 95 and 1056 kPa, with increased elasticity. Moreover, the Young's modulus and tensile stress of the hydrogels increased from 0.52 to 9.9 MPa and 17 to 33 kPa, respectively, when the PVA amount increased. These results showed that PVA can improve the poor mechanical properties of cellulose. The self-healing efficiency of the hydrogels was improved by adjusting the quantity of amount of PVA and a PVA content of 3 g increased the healing efficiency of the hydrogels to 37.03% [134].

Jiji et al. prepared and evaluated thymol incorporated bacterial cellulose hydrogels for the management of burn wounds [135]. The thermal and chemical changes of the hydrogels were successfully evaluated. The hydrogel water vapor transmission rate decreased slightly after the loading of thymol into the polymer matrix because of thymol hydrophobic nature. The in vitro biocompatibility assessments using MTT assay showed that the hydrogels facilitated the growth of mouse 3T3 fibroblast cells, demonstrating low cytotoxicity, and increased cell viability. The in vitro antibacterial studies revealed that thymol incorporated hydrogels possessed an excellent zone of inhibition in both Gram-positive and Gram-negative bacteria strains, especially in the case of *S. aureus* and *K. pneumonia* when compared to the standard antibiotic, gentamycin. The in vivo wound healing studies of the hydrogels using female Wister rats with burn wound demonstrated that the wound was enlarged at the 5th day. On day 15, the burn wound area was significantly reduced by 55%. The wound treated with plain cellulose-based hydrogels and thymol-loaded cellulose hydrogels were reduced by 74.5% and 90.7%, respectively, on day 20 [135].

Sadegh and co-workers formulated carboxymethyl cellulose-human hair keratin hydrogels for the controlled release of clindamycin [136]. A higher keratin content in the hydrogel reduced the water vapor transmission rate of the hydrogels from 3200 ± 196 to 1921 ± 92 g/m^2/day, which is suitable for wound management. The in vitro drug release mechanism at physiological conditions demonstrated an initial burst release of clindamycin from hydrogels during the first 4 h followed by a slow drug release profile. The total amount of released clindamycin after 7 days of incubating in phosphate buffer saline solution was 91.5 ± 3.1%, and it was decreased by the addition of keratin. This could be due to the hydrophilicity with increasing the keratin and lowering of water uptake value. These results suggest that the water uptake of the drug delivery system affects the released kinetics of hydrophilic drug (clindamycin). The in vitro antimicrobial activity of the hydrogel was 99.66% against *S. aureus* and this value was reduced by the addition of keratin because of the slow release of the antibiotic [136].

Deng et al. reported hydrogel prepared from fenugreek gum and cellulose. The composite hydrogel displayed a porous structure with good thermal stability and water absorption. The sustained release mechanism of the hydrogel was significant. The hydrogel also exhibited good biocompatibility and non-toxic properties. The hydrogels reduce blood loss and accelerated wound closure, with increased neovascularization and tissue repair [137]. Mohamad et al. reported bacterial cellulose/acrylic acid hydrogel loaded with human epidermal keratinocytes and human dermal fibroblasts for the treatment of burn wound. This in vivo study using athymic mice showed that the hydrogel loaded with cells accelerated burn wound healing. The % wound reduction on day 13 was (77.34 ± 6.21%) for the hydrogel when compared to the control which was (64.79 ± 6.84%). A high deposition of collagen was observed in the mice treated with hydrogel formulation which revealed their potential application for wound dressing [138]. Gupta et al. synthesized silver nanoparticles using curcumin:hydroxypropyl-β-cyclodextrin complex followed by loading into bacterial cellulose hydrogel. The hydrogels provided a moist environment with high cytocompatibility. They were effective against wound-infecting bacterial such as *Candida auris, Staphylococcus aureus* and *Pseudomonas aeruginosa* [139]. Koivuniemi et al. reported a single-center clinical trial evaluation of nanofibrillar cellulose wound dressing, (FibDex® by UPM-Kymmene Corporation) and comparing it with a polylactide-based copolymer dressing. Twenty-four patients requiring skin grafting were enrolled for the study. The skin elasticity was improved with nanofibrillar cellulose wound dressing. The performance of the cellulose wound dressing was comparable to the polylactide when evaluated for the treatment of skin graft donor sites. It was self-detachable and did not require changing. It also reduced pain [140]. Deng et al. prepared hydrogel from cellulose and flaxseed gum for the treatment of bleeding wounds. The hydrogel exhibited a high thermal stability. Its swelling capability was high with a moisture uptake of 200%. The % drug adsorption was 7.27 ± 0.15 mg/g and was biocompatible; it promoted a hemostatic and wound healing process [141]. Khamrai et al. prepared a biocompatible gelatin–cellulose-based polyelectrolyte hydrogel patch with self-healing capability. It was prepared from bacterial cellulose, obtained from *Glucanoacetobacter xylinus* (MTCC7795) bacterial strain. Bacterial cellulose in the hydrogel enhanced the self-healing activity of the hydrogel. Incorporation of curcumin in the hydrogel enhanced their wound healing activity. The curcumin-loaded hydrogel inhibited the bacterial growth with a zone of inhibition value of 15 ± 0.5 mm and 19 ± 1.0 mm against *E. coli* and *S. aureus*, respectively. The presence of curcumin accelerated cell proliferation and the release of curcumin from the hydrogel was controlled [142].

Forero-Doria et al. prepared cellulose hydrogel which was conjugated with carbon nanotubes for high uptake of bioactive compounds (resveratrol, dexpanthenol, allantoin, and linezolid). The wound closure in vivo was accelerated. The release of linezolid was sustained. The hydrogel had the capability to protect the antibiotic from degradation. The hydrogel displayed good biocompatibility with L929 mouse connective tissue fibroblasts with cell viability of 95% [143]. Muchová et al. prepared poly(vinyl alcohol)-2,3-Dialdehyde cellulose hydrogel as wound dressings. The hydrogels displayed unique features such as high porosity, high drug-loading capacity, and good mechanical properties making them easy to handle with good adherence to the skin. They were also biocompatible and non-cytotoxic [144].

Wang et al. reported quaternized hydroxyethyl cellulose/mesocellular silica hydrogel via one-pot radical graft copolymerization for the treatment of bleeding wounds. The hydrogel loaded with mesocellular silica foam (9.82 *w*/*w*%) activated the blood coagulation factors. The quaternized hydroxyethyl cellulose reduced the plasma clotting time to 59% in vitro resulting in less blood loss. It also enhanced wound healing in a full-thickness skin defect model in vivo [145]. Li et al. prepared carboxymethyl cellulose/K-carrageenan/graphene oxide/konjac glucomannan hydrogel loaded with silver nanoparticles synthesized with green deoxidizer mango peel extracts. The hydrogel exhibited good antibacterial activity, biocompatibility and accelerated wound recuperating with the development of fibroblasts and rapid epithelialization [146].

5. Hydrogels Currently in Clinical Trials

Currently, some cellulose and chitosan are under selected phases of clinical trials or are marketed product as shown in Table 2. They were found to be effective for the management of different types of wounds.

Table 2. Hydrogels in clinical trials/marketed products.

Hydrogels	Polymer	Clinical Trial Outcomes/Marketed Products	References
Nanofibrillar cellulose wound dressing (FibDex®)	Cellulose	Efficient wound healing at skin graft donor sites, required no dressing changes, self-detaches after re-epithelialization, it did not degrade into tissue and it reduced pain.	[140]
Nanofibrillar cellulose wound dressing	Cellulose	Detachment of the wound dressing from epithelialized skin graft donor site is presented in occurred in average of 18 days.	[147]

Table 2. *Cont.*

Hydrogels	Polymer	Clinical Trial Outcomes/Marketed Products	References
Bacterial nanocellulose	Cellulose	Most of the patched skins did not show any symptoms of edema, vesicle and bullae. It was non-irritant and safe for the further evaluation.	[148]
Polyuhexanide-containing cellulose dressing	Cellulose	Clinical trials was carried out on patients with pressure ulcers infected with Methicillin-Resistant *Staphylococcus aureus*. There was a 100% eradication of the bacteria.	[149]
Nanoderm™ Ag	Cellulose	Displayed increased flexibility and sustained antimicrobial properties. Effective for the management of infected wounds.	[150]
Nanoskin®	Bacterial cellulose	It is 100% natural, non-allergenic, biocompatible, effective for the management of burns, surgical wounds, diabetic ulcer wounds, dermal abrasions, Skin grafting sites etc.	[150]
CelMat®	Bacterial cellulose	Useful for the treatment of burns, ulcers and chronic wounds. It promotes pain relief, excellent gases exchange, absorption and desorption of fluids.	[150]
EpiProtect®	Cellulose	Effective for the management of pediatric burn wounds after enzymatic debridement.	[151]
HemCon™	Chitosan	Good hemostasis activity and antibacterial barrier against some strains of bacteria.	[152]
KA01 chitosan wound dressing	Chitosan	It enhanced wound healing by facilitating wound re-epithelialization and reducing pain level. It was safe and effective for the management of chronic wounds.	[153]
Chitosan Mesh Membrane	Chitosan	The mesh chitosan membrane promoted good adherence, excellent hemostasis, re-epithelialization of the wound, reduced itching and pain.	[154]
Chitosan gel	Chitosan	The gel is displayed rapid hemostatic activity and prevents adhesion formation. It is suitable for the management of common complications of sinus surgery.	[155]
Axiostat®	Chitosan	It exhibits rapid hemostasis and is suitable for emergency trauma and accidents. It is easily removed from the wound site without leaving any residue.	[156]
ChitoRhino	Chitosan gel	Good hemostasis activity and effective for wound healing after endoscopic sinus surgery.	[157]
ChitoHeal	Chitosan	Accelerates the rate of healing, scar reduction, biocompatible, effective for burns, cuts, scratches and diabetic foot ulcers	[158]
KytoCel	Chitosan fibers	It is a highly absorbent dressing suitable for the management of moderate and heavily exuding wounds.	[159]
PosiSep®	Chitosan fibers	A nasal dressing, easy to use, and display rapid expansion upon hydration for minimal bleeding procedures.	[160]
ExcelArrest® XT	Chitosan-based patch	It accelerates the clotting process to control bleeding from the skin.	[161]
ChitoClot Pad	Chitosan	It undergoes gelation after absorbing blood and prevents exudation of absorbed blood.	[162]

Table 2. *Cont.*

Hydrogels	Polymer	Clinical Trial Outcomes/Marketed Products	References
XSTAT	Chitosan	It is used to treat gunshot wounds.	[163]
Chitoderm® plus	Chitosan	Good absorbent properties.	[157]
ChitoClear®	Chitosan	A good hemostatic agent resulting in its capability to attract negatively charged red blood cells.	[164]
Celox™	Chitosan	Rapid hemostatic property and reduces blood loss.	[157]

6. Conclusions and Future Perspectives

The biopolymer hydrogels that are based on cellulose and chitosan demonstrated excellent features for the management of wounds. The hydrogels cross-linked with synthetic polymers displayed interesting mechanical properties that are useful in wound management, including good elasticity, flexibility, compressive stresses, Young's modulus, and tensile stresses. The microbial infected wounds can delay the wound healing mechanism and increase wound exudate resulting in chronic wounds. These wound dressings possess antibacterial efficacy, especially those loaded with antibiotics or nanoparticles against several Gram-negative and Gram-positive bacteria strains that infect wounds. Furthermore, the hydrogels containing antimicrobial agents promoted the healing process by limiting the number of pathogens on the wound site. The use of biopolymer-based hydrogels in these wound dressings were characterized by increased porosity, high water uptake, non-immunogenic effects with sustained and controlled drug release. The aforementioned properties improved rapid re-epithelialization, granulation tissue development, and wound healing in vitro and in vivo. The combination of nanotechnology with medicine offers great opportunities to the currently available wound dressings. Nanoparticles are beneficial to improve the therapeutic efficacy of synthetic compounds. Nanoparticles can act on the cellular and subcellular events during the wound healing process. The incorporation of metallic nanoparticles and antibacterial agents into the chitosan or cellulose-based matrix promoted sustained antimicrobial activity without affecting the normal cell viability. The mechanism of action of metal-based nanoparticles is influenced by the concentration and is via the formation of free radicals. Despite the efficacy of nanoparticles, there is a pressing need to thoroughly investigate their toxicology and biocompatibility over long term application. Although biopolymer-based hydrogels present interesting features in the series of in vitro and in vivo studies reported for wound management, very few of them have reached clinical trials. Therefore, there is an increasing demand for the formulation of advanced wound dressing materials with enhanced properties and they must be preclinically tested to ensure safety. There is a great promise that many hydrogels that are based on cellulose and chitosan will enter the clinical trials and market in the near future.

Author Contributions: Conceptualization, methodology, S.A. and B.A.A.; investigation, S.A. and B.A.A.; resources, S.A. and B.A.A.; data curation, S.A. and B.A.A.; writing-original draft preparation, S.A. and B.A.A.; writing—review and editing, S.A. and B.A.A.; supervision, B.A.A.; funding acquisition, B.A.A. All authors have read and agreed to the published version of the manuscript.

Funding: This research was funded by South African Medical Research Council, National Research Foundation, South African and Govern Mbeki Research and Development Council, University of Fort Hare, South Africa.

Conflicts of Interest: The authors declare no conflict of interest.

References

1. Dhivya, S.; Padma, V.V.; Santhini, E. Wound dressings—A review. *BioMedicine* **2015**, *5*, 24–28. [CrossRef]
2. Robson, M.; Steed, D.; Franz, M. Wound healing: Biological features and approaches to maximize healing trajectories. *Curr. Probl. Surger* **2001**, *38*, 77–89. [CrossRef]

3. Frykberg, R.G.; Banks, J. Challenges in the treatment of chronic wounds. *Adv. Wound Care (New Rochelle)* **2015**, *4*, 560–582. [CrossRef]
4. Schmidtchen, A.; Pang, C.; Ni, G.; Sönnergren, H.; Car, J.; Järbrink, K.; Bajpai, R. The humanistic and economic burden of chronic wounds: A protocol for a systematic review. *Syst. Rev.* **2017**, *6*, 7.
5. Aderibigbe, B.A.; Buyana, B. Alginate in Wound Dressings. *Pharmaceutics* **2018**, *10*, 42. [CrossRef]
6. Hussain, Z.; Thu, H.E.; Shuid, A.N.; Katas, H.; Hussain, F. Recent Advances in Polymer-based Wound Dressings for the Treatment of Diabetic Foot Ulcer: An Overview of State-of-the-art. *Curr. Drug Targets* **2017**, *19*, 527–550. [CrossRef]
7. Sudarsan, S.; Franklin, D.S.; Guhanathan, S. Imbibed salts and pH-responsive behaviours of sodium-alginate based eco-friendly biopolymeric hydrogels-A solventless approach. *MMAIJ* **2015**, *11*, 24–29.
8. Ulery, D.B.; Nair, L.S.; Laurencin, C.T. Biomedical applications of biodegradable polymers. *J. Polym. Sci. Pol. Phys.* **2011**, *49*, 832–864. [CrossRef]
9. Chittleborough, C.R.; Grant, J.F.; Phillips, P.J.; Taylor, A.W. The increasing prevalence of diabetes in South Australia: The relationship with population ageing and obesity. *Public Health* **2007**, *121*, 92–99. [CrossRef]
10. Wang, P.; Huang, S.; Hu, Z.; Yang, W.; Lan, Y.; Zhu, J.; Hancharou, A.; Guo, R.; Tang, B. In situ formed anti-inflammatory hydrogel loading plasmid DNA encoding VEGF for burn wound healing. *Acta Biomater.* **2019**, *100*, 191–201.
11. Dong, Y.; Cui, M.; Qu, J.; Wang, X.; Hyung, S.; Barrera, J.; Elvassore, N.; Gurtner, G.C. Conformable hyaluronic acid hydrogel delivers adipose-derived stem cells and promotes regeneration of burn injury. *Acta Biomater.* **2020**, *108*, 56–66. [CrossRef]
12. Rao, K.M.; Suneetha, M.; Zo, S.; Duck, K.H.; Han, S.S. One-pot synthesis of ZnO nanobelt-like structures in hyaluronan hydrogels for wound dressing applications. *Carbohydr. Polym.* **2019**, *223*, 115124. [CrossRef]
13. Rezvanian, M.; Ahmad, N.; Cairul, M.; Mohd, I.; Ng, S. Optimization, characterization, and in vitro assessment of alginate-pectin ionic cross-linked hydrogel film for wound dressing applications. *Int. J. Biol. Macromol.* **2017**, *97*, 131–140. [CrossRef]
14. Kaygusuz, H.; Torlak, E.; Akim-Evingur, G.; Ozen, I.; von Kitzing, R.; Erim, F.B. Antimicrobial cerium ion-chitosan crosslinked alginate biopolymer films: A novel and potential wound dressing. *Int. J. Biol. Macromol.* **2017**, *105*, 1161–1165. [CrossRef]
15. Li, S.; Li, L.; Guo, C.; Qin, H.; Yu, X. A promising wound dressing material with excellent cytocompatibility and proangiogenesis action for wound healing: Strontium loaded Silk fibroin/Sodium alginate (SF/SA) blend films. *Int. J. Biol. Macromol.* **2017**, *104*, 969–978. [CrossRef]
16. Türe, H. Characterization of hydroxyapatite-containing alginate—Gelatin composite films as a potential wound dressing. *Int. J. Biol. Macromol.* **2019**, *123*, 878–888. [CrossRef]
17. Shahzad, A.; Khan, A.; Afzal, Z.; Farooq, M.; Khan, J.; Khan, G.M. Formulation development and characterization of cefazolin nanoparticles-loaded cross-linked films of sodium alginate and pectin as wound dressings. *Int. J. Biol. Macromol.* **2019**, *124*, 255–269. [CrossRef]
18. El-aassar, M.R.; Ibrahim, O.M.; Fouda, M.M.G.; El-beheri, N.G.; Agwa, M.M. Wound healing of nanofiber comprising Polygalacturonic/Hyaluronic acid embedded silver nanoparticles: In-vitro and in-vivo studies. *Carbohydr. Polym.* **2020**, *238*, 116175. [CrossRef]
19. Mutlu, G.; Calamak, S.; Ulubayram, K.; Guven, E. Curcumin-loaded electrospun PHBV nanofibers as potential wound-dressing material. *J. Drug Deliv. Sci. Technol.* **2018**, *43*, 185–193. [CrossRef]
20. Labovitiadi, O.; Driscoll, N.H.O.; Lamb, A.J.; Matthews, K.H. Rheological properties of gamma-irradiated antimicrobial wafers and in vitro efficacy against Pseudomonas aeruginosa. *Int. J. Pharm.* **2013**, *453*, 462–472. [CrossRef]
21. Kataria, K.; Gupta, A.; Rath, G.; Mathur, R.B.; Dhakate, S.R. In vivo wound healing performance of drug loaded electrospun composite nanofibers transdermal patch. *Int. J. Pharm.* **2014**, *469*, 102–110. [CrossRef]
22. Sheth, N.S.; Mistry, R.B. Formulation and evaluation of transdermal patches and to study permeation enhancement effect of eugenol. *J. Appl. Pharm. Sci.* **2011**, *1*, 96–101.
23. Nilani, P.; Pranavi, A.; Duraisamy, B.; Damodaran, P.; Subhashini, V.; Elango, K. Formulation and evaluation of wound healing dermal patch. *Afr. J. Pharm. Pharmacol.* **2011**, *5*, 1252–1257.
24. Rossi, S.; Faccendini, A.; Bonferoni, M.C.; Ferrari, F.; Sandri, G.; Fante, C.D.; Perotti, C.; Caramella, C.M. "Sponge-like" dressings based on biopolymers for the delivery of platelet lysate to skin chronic wounds. *Int. J. Pharm.* **2013**, *440*, 207–215. [CrossRef] [PubMed]

25. Mohandas, A.; Anisha, B.S.; Chennazhi, K.P.; Jayakumar, R. Chitosan—Hyaluronic acid/VEGF loaded fibrin nanoparticles composite sponges for enhancing angiogenesis in wounds. *Coll. Surf. B Biointerfaces* **2015**, *127*, 105–113. [CrossRef] [PubMed]
26. Anisha, B.S.; Sankar, D.; Mohandas, A.; Chennazhi, K.P.; Nair, S.; Jayakumar, R. Chitosan—Hyaluronan/nano chondroitin sulfate ternary composite sponges for medical use. *Carbohydr. Polym.* **2013**, *92*, 1470–1476. [CrossRef]
27. Mohandas, A.; Sudheesh Kumar, P.T.; Raja, B.; Lakshmanan, V.K.; Jayakumar, R. Exploration of alginate hydrogel/nano zinc oxide composite bandages for infected wounds. *Int. J. Nanomed.* **2015**, *10*, 53–66. [CrossRef]
28. Kuddushi, M.; Patel, N.K.; Gawali, S.L.; Mata, J.P.; Montes-campos, H.; Varela, L.M.; Hassan, P.A.; Malek, N.I. Thermo-switchable de novo ionogel as metal absorbing and curcumin loaded smart bandage material. *J. Mol. Liq.* **2020**, *306*, 112922. [CrossRef]
29. Shteyer, E.; Ben, A.; Zolotaryova, L.; Sinai, A.; Lichtenstein, Y.; Pappo, O.; Kryukov, O.; Elkayam, T.; Cohen, S.; Ilan, Y. Reduced liver cell death using an alginate scaffold bandage: A novel approach for liver reconstruction after extended partial hepatectomy. *Acta Biomater.* **2014**, *10*, 3209–3216. [CrossRef]
30. Mehrabani, G.M.; Karimian, R.; Rakhshaei, R.; Pakdel, F.; Eslami, H.; Fakhrzadeh, V.; Rahimi, M.; Selehi, R.; Kafil, H.S. Chitin/silk fi broin/TiO2 bio-nanocomposite as a biocompatible wound dressing bandage with strong antimicrobial activity. *Int. J. Biol. Macromol.* **2018**, *116*, 966–976. [CrossRef]
31. Velnar, T.; Bailey, T.; Smrkolj, V. The Wound Healing Process: An Overview of the Cellular and Molecular Mechanisms. *J. Int. Med. Res.* **2009**, *37*, 1528–1542. [CrossRef]
32. Gauglitz, G.; Korting, H.C.; Pavicic, T.; Ruzicka, T.; Jeschke, M.G. Hyper-trophic scarring and keloids: Pathomechanisms and current and emerging treatment strategies. *Mol. Med.* **2011**, *17*, 113–125. [CrossRef]
33. Patel, S.; Srivastava, S.; Singh, M.R.; Singh, D. Mechanistic insight into diabetic wounds: Pathogenesis, molecular targets and treatment strategies to pace wound healing. *Biomed. Pharmacother.* **2019**, *112*, 108615. [CrossRef]
34. Wang, P.; Huang, B.; Horng, H.; Yeh, C.; Chen, Y.-J. Wound healing. *J. Chin. Med. Assoc.* **2018**, *81*, 94–101. [CrossRef]
35. Su, W.H.; Cheng, M.H.; Lee, W.L.; Tsou, T.S.; Chang, W.H.; Chen, C.S.; Wang, P.H. Nonsteroidal anti-inflammatory drugs for wounds: Pain relief or excessive scar formation? *Med. Inflamm.* **2010**, *2010*, 413238. [CrossRef]
36. Babaei, S.; Bayat, N.; Nouruzian, M.; Bayat, M. Pentoxifylline improves cutaneous wound healing in streptozotocin-induced diabetic rats. *Eur. J. Pharmacol.* **2013**, *700*, 165–172. [CrossRef]
37. Plikus, M.V.; Guerrero-Juarez, C.F.; Ito, M.; Li, E.R.; Dedhia, P.H.; Zheng, Y.; Shao, M.; Gay, D.L.; Ramos, R.; His, T.-C.; et al. Regeneration of fat cells from myofibroblasts during wound healing. *Science* **2017**, *355*, 748–752. [CrossRef]
38. Hart, J. Inflammation. 1: Its role in the healing of acute wounds. *J. Wound Care* **2002**, *11*, 205–209. [CrossRef]
39. Diegelmann, R.F.; Evans, M.C. Wound healing: An overview of acute, fibrotic and delayed healing. *Front. Biosci.* **2004**, *1*, 283–289. [CrossRef]
40. Tabata, Y. Biomaterial technology for tissue engineering applications. *J. R. Soc. Interface* **2009**, *6*, S311–S324. [CrossRef]
41. Malafaya, P.B.; Silva, G.A.; Reis, R.L. Natural-origin polymers as carriers and scaffolds for biomolecules and cell delivery in tissue engineering applications. *Adv. Drug Deliv. Rev.* **2007**, *207*, 207–233. [CrossRef]
42. Sell, S.; Wolfe, P.; Garg, K.; McCool, J.; Rodriguez, I.; Bowlin, G. The use of natural polymers in tissue engineering: A focus on electrospun extracellular matrix analogues. *Polym. Adv. Technol.* **2010**, *2*, 522–553. [CrossRef]
43. Huang, S.; Fu, X. Naturally derived materials-based cell and drug delivery systems in skin regeneration. *J. Control. Release* **2010**, *142*, 149–159. [CrossRef]
44. Jayakumar, R.; Prabaharan, M.; Kumar, P.T.S.; Nair, S.V.; Tamura, H. Biomaterials based on chitin and chitosan in wound dressing applications. *Biotechnol. Adv.* **2001**, *29*, 322–323. [CrossRef]
45. Tessmar, J.; Gopferich, A.M. Matrices and scaffolds for protein delivery in tissue engineering. *Adv. Drug Deliv. Rev.* **2007**, *58*, 274–291. [CrossRef]
46. Dai, T.; Tanaka, M.; Huang, Y.; Hamblin, M. Chitosan preparations for wounds and burns: Antimicrobial and wound-healing effects. *Expert Rev. Anti Infect. Ther.* **2011**, *9*, 857–879. [CrossRef]
47. Pérez, R.A.; Won, J.E.; Knowles, J.C.; Kim, H.W. Naturally and synthetic smart composite biomaterials for tissue regeneration. *Adv. Drug Deliv. Rev.* **2013**, *65*, 471–496. [CrossRef]
48. Mano, J.F.; Silva, J.; Asevedo, H.S.; Malafaya, P.B.; Sousa, R.A.; Silva, S.S.; Boesel, L.f.; Oliveira, J.M.; Santos, T.C.; Marques, A.P.; et al. Natural origin biodegradable systems in tissue engineering and regenerative medicine: Present status and some moving trends. *J. R. Soc. Interface* **2007**, *4*, 999–1030. [CrossRef]

49. Sannino, A.; Demitri, C.; Madaghiele, M. Biodegradable cellulosebased hydrogels: Design and applications. *Materials* **2009**, *2*, 353–373. [CrossRef]
50. Helenius, G.; Backdahl, H.; Bodin, A.; Nannmark, U.; Gatenholm, P.; Risberg, B. In vivo biocompatibility of bacterial cellulose. *J. Biomed. Mater. Res. Part A* **2006**, *76A*, 431–438. [CrossRef]
51. Cullen, M.B.; Silcock, D.W.; Boyle, C. Wound Dressing Comprising Oxidized Cellulose and Human Recombinat Collagen. U.S. Patent 7833790 B2, 16 November 2010.
52. Akhtar, M.F.; Hanif, M.; Ranjha, N.M. Methods of synthesis of hydrogels. A review. *Saudi Pharm. J.* **2016**, *24*, 554–559. [CrossRef] [PubMed]
53. Mohite, P.B.; Adhav, S.S. A hydrogels: Methods of preparation and applications. *Inter. J. Advanc. Pharm.* **2017**, *6*, 79–85.
54. Verma, C.; Negi, P.; Pathania, D.; Sethi, V.; Gupta, B. Preparation of pH-sensitive hydrogels by graft polymerization of itaconic acid on tragacanth gum. *Poly. Inter.* **2018**, *68*, 344–350. [CrossRef]
55. Almeida, J.F.; Ferreira, P.; Lopes, A.; Gil, M.H. Photocrosslinkable biodegradable responsive hydrogels as drug delivery systems. *Int. J. Biol. Macromol.* **2011**, *49*, 948–954. [CrossRef]
56. Sikareepaisan, P.; Ruktanonchai, U.; Supaphol, P. Preparation and characterization of asiaticoside-loaded alginate films and their potential for use as effectual wound dressings. *Carbohydr. Polym.* **2011**, *83*, 1457–1469. [CrossRef]
57. Jagur-Grodzinski, J. Polymeric gels and hydrogels for biomedical and pharmaceutical applications. *Polym. Adv. Technol.* **2011**, *21*, 27–47. [CrossRef]
58. Kamoun, E.A.; Kenawy, E.R.; Chen, X. A review on polymeric hydrogel membranes for wound dressing applications: PVA-based hydrogel dressings. *J. Adv. Res.* **2017**, *8*, 217–233. [CrossRef] [PubMed]
59. Rasool, A.; Ata, S.; Islam, A. Stimuli responsive biopolymer (chitosan) based blend hydrogels for wound healing application. *Carbohydr. Polym.* **2019**, *203*, 423–429. [CrossRef]
60. Xue, H.; Hu, L.; Xiong, Y.; Zhu, X.; Wei, C.; Cao, F.; Zhou, W.; Sun, Y.; Endo, Y.; Liu, M.; et al. Quaternized chitosan-Matrigel-polyacrylamide hydrogels as wound dressing for wound repair and regeneration. *Carbohydr. Polym.* **2019**, *226*, 115302. [CrossRef]
61. Khorasani, T.M.; Joorabloo, A.; Adeli, H.; Mansoori-Moghadam, Z.; Moghaddam, A. Design and optimization of process parameters of polyvinyl (alcohol)/chitosan/nano zinc oxide hydrogels as wound healing materials. *Carbohydr. Polym.* **2019**, *207*, 542–554. [CrossRef]
62. Masood, N.; Ahmed, R.; Tariq, M.; Ahmed, Z.; Masoud, M.S.; Ali, I.; Asghar, R.; Andleeb, A.; Hasan, A. Silver nanoparticle impregnated chitosan-PEG hydrogel enhances wound healing in diabetes induced rabbits. *Int. J. Pharm.* **2019**, *559*, 23–36. [CrossRef]
63. Qi, W.; Xu, L.; Zhao, S.; Zheng, J.; Tian, Y.; Qi, X.; Huang, X.; Zhang, J. Controlled releasing of SDF-1 α in chitosan-heparin hydrogel for endometrium injury healing in rat model. *Int. J. Biol. Macromol.* **2020**, *143*, 163–172. [CrossRef]
64. Zhang, C.; Yang, X.; Hu, W.; Han, X.; Fan, L.; Tao, S. Preparation and characterization of carboxymethyl chitosan/collagen peptide/oxidized konjac composite hydrogel. *Int. J. Biol. Macromol.* **2020**, *149*, 31–40. [CrossRef]
65. Zhang, N.; Gao, T.; Wang, Y.; Liu, J.; Zhang, J.; Yao, R.; Wu, F. Modulating cationicity of chitosan hydrogel to prevent hypertrophic scar formation during wound healing. *Int. J. Biol. Macromol.* **2020**, *154*, 835–843. [CrossRef] [PubMed]
66. Bagher, Z.; Ehterami, A.; Hossein, M.; Khastar, H.; emiari, H.; Asefnejan, A.; Davachi, S.M.; Mirzaii, M.; Salehi, M. Wound healing with alginate/chitosan hydrogel containing hesperidin in rat model. *J. Drug Deliv. Sci. Technol.* **2020**, *55*, 101379. [CrossRef]
67. Ehterami, A.; Salehi, M.; Farzamfar, S.; Samadian, H.; Vaez, A.; Ghorbani, S.; Ai, J.; Sahrapeyma, H. Chitosan/alginate hydrogels containing Alpha-tocopherol for wound healing in rat model. *J. Drug Deliv. Sci. Technol.* **2019**, *51*, 204–213. [CrossRef]
68. He, J.; Shi, M.; Liang, Y.; Guo, B. Conductive adhesive self-healing nanocomposite hydrogel wound dressing for photothermal therapy of infected full-thickness skin wounds. *Chem. Eng. J.* **2020**, *394*, 124888. [CrossRef]
69. Hamdi, M.; Feki, A.; Bardaa, S.; Li, S.; Nagarajan, S.; Mellouli, M.; Boudawara, T.; Sahnoun, Z.; Nasri, M.; Nasri, R. A novel blue crab chitosan/protein composite hydrogel enriched with carotenoids endowed with distinguished wound healing capability: In vitro characterization and in vivo assessment. *Mater. Sci. Eng. C* **2020**, *113*, 110978. [CrossRef] [PubMed]

70. Du, X.; Liu, Y.; Wang, X.; Yan, H.; Wang, L.; Qu, L.; Kong, D.; Qiao, M.; Wang, L. Injectable hydrogel composed of hydrophobically modi fied chitosan/oxidized-dextran for wound healing. *Mater. Sci. Eng. C* **2019**, *104*, 109930. [CrossRef]
71. Zhang, Y.; Jiang, M.; Zhang, Y.; Cao, Q.; Wang, X.; Han, Y.; Sun, G.; Li, Y.; Zhou, J. Novel lignin-chitosan—PVA composite hydrogel for wound dressing. *Mater. Sci. Eng. C* **2019**, *104*, 110002. [CrossRef]
72. Pawar, V.; Dhanka, M.; Srivastava, R. Cefuroxime conjugated chitosan hydrogel for treatment of wound infections. *Coll. Surf. B Biointer.* **2019**, *173*, 776–787. [CrossRef] [PubMed]
73. Pham, L.; Hang, L.; Dung, M.; Hiep, T.; Le, L.; Van Le, T.; Nam, N.D.; Bach, L.G.; Nguye, V.T.; Tran, N.Q. A dual synergistic of curcumin and gelatin on thermal-responsive hydrogel based on Chitosan-P123 in wound healing application. *Biomed. Pharmacother.* **2019**, *117*, 109183. [CrossRef]
74. Ferreira, M.O.; de Lima, I.S.; Morais, A.Í.; Silva, S.O.; de Carvalho, R.B.; Ribeiro, A.B.; Osajima, J.A.; Silva Filho, E.C. Chitosan associated with chlorhexidine in gel form: Synthesis, characterization and healing wounds applications. *J. Drug Deliv. Sci. Technol.* **2019**, *49*, 375–382. [CrossRef]
75. Xu, H.; Huang, S.; Wang, J.; Lan, Y.; Feng, L.; Zhu, M.; Xiao, Y.; Cheng, B.; Xue, W.; Guo, R. Enhanced cutaneous wound healing by functional injectable thermo-sensitive chitosan-based hydrogel encapsulated human umbilical cord- mesenchymal stem cells. *Int. J. Biol. Macromol.* **2019**, *137*, 433–441. [CrossRef]
76. Ragab, T.I.M.; Nada, A.A.; Ali, E.A.; Shimaa, A.l.; Shalaby, G.; Soliman, A.A.F.; Emam, M.; Raey, M.A.E. Soft hydrogel based on modified chitosan containing *P. granatum* peel extract and its nano-forms: Multiparticulate study on chronic wounds treatment. *Int. J. Biol. Macromol.* **2019**, *135*, 407–421.
77. Long, J.; Etxabide, A.; Nand, A.V.; Bunt, C.R.; Ray, S.; Seyfoddin, A. A 3D printed chitosan-pectin hydrogel wound dressing for lidocaine hydrochloride delivery. *Mater. Sci. Eng. C* **2019**, *104*, 109873. [CrossRef]
78. Viezzer, C.; Mazzuca, R.; Cantarelli, D.; Madalena, M.; Forte, D.C.; Ribelles, J.L.G. A new waterborne chitosan-based polyurethane hydrogel as a vehicle to transplant bone marrow mesenchymal cells improved wound healing of ulcers in a diabetic rat model. *Carbohydr. Polym.* **2020**, *231*, 115734. [CrossRef]
79. Moradi, S.; Barati, A.; Tonelli, A.E.; Hamedi, H. Chitosan-based hydrogels loading with thyme oil cyclodextrin inclusion compounds: From preparation to characterization. *Eur. Polym. J.* **2020**, *122*, 109303. [CrossRef]
80. Wang, X.; Xu, P.; Yao, Z.; Fang, Q.; Feng, L.; Guo, R.; Cheng, B. Preparation of antimicrobial hyaluronic acid/quaternized chitosan hydrogels for the promotion of seawater immersion wound healing. *Front. Bioeng. Biotechnol.* **2019**, *7*, 360. [CrossRef]
81. Chen, H.; Li, B.; Feng, B.; Wang, H.; Yuan, H.; Xu, Z. Tetracycline hydrochloride loaded citric acid functionalized chitosan hydrogel for wound healing. *RSC Adv.* **2019**, *9*, 19523–19530. [CrossRef]
82. Huang, Y.; Shi, F.; Wang, L.; Yang, Y.; Khan, B.M.; Cheong, K.L.; Liu, Y. Preparation and evaluation of Bletilla striata polysaccharide/carboxymethyl chitosan/Carbomer 940 hydrogel for wound healing. *Int. J. Biol. Macromol.* **2019**, *132*, 729–737. [CrossRef]
83. Nguyen, T.D.; Nguyen, T.T.; Ly Loan, K.; Tran Hien, A.; Nguyen, T.T.; Vo Minh, M.; Ho Minh, H.; Dang, N.T.; Vo Toi, V.; Nguyen, D.H.; et al. In vivo study of the antibacterial chitosan/polyvinyl alcohol loaded with silver nanoparticle hydrogel for wound healing applications. *Int. J. Polym. Sci.* **2019**. [CrossRef]
84. Chen, K.; Tong, C.; Cong, P.; Liu, Y.; Shi, X.; Liu, X.; Zhang, J.; Zou, R.; Xiao, K.; Ni, Y.; et al. Injectable melatonin-loaded carboxymethyl chitosan (CMCS)-based hydrogel accelerates wound healing by reducing inflammation and promoting angiogenesis and collagen deposition. *J. Mater. Sci. Technol.* **2020**. [CrossRef]
85. Nooshabadi, V.T.; Khanmohamadi, M.; Valipour, E.; Mahdipour, S.; Salati, A.; Malekshahi, Z.V.; Shafei, S.; Amini, E.; Farzamfar, S.; Ai, J. Impact of exosome loaded chitosan hydrogel in wound repair and layered dermal reconstitution in mice animal model. *J. Biomed. Mater. Res. Part A* **2020**. [CrossRef]
86. Ravishankar, K.; Venkatesan, M.; Desingh, R.; Mahalingam, A.; Sadhasivam, B.; Subramaniyam, R.; Dhamodharan, R. Biocompatible hydrogels of chitosan-alkali lignin for potential wound healing applications. *Mater. Sci. Eng. C* **2019**, *102*, 447–457. [CrossRef]
87. Lin, S.; Lo, K.; Tseng, T.; Liu, J.; Shih, T.; Cheng, K.C. Evaluation of PVA/dextran/chitosan hydrogel for wound dressing. *Cell. Polym.* **2019**, *38*, 15–30. [CrossRef]
88. Heimbuck, A.M.; Priddy-Arrington, T.R.; Padgett, M.L.; Llamas, C.B.; Barnett, H.H.; Bunnell, B.A.; Caldorera-Moore, M.E. Development of Responsive Chitosan–Genipin Hydrogels for the Treatment of Wounds. *ACS Appl. Bio Mater.* **2019**, *2*, 2879–2888. [CrossRef]
89. Ternullo, S.; Schulte, W.L.; Holsæter, A.; Škalko-Basnet, N. Curcumin-In-Deformable Liposomes-In-Chitosan-Hydrogel as a Novel Wound Dressing. *Pharmaceutics* **2020**, *12*, 8. [CrossRef] [PubMed]

90. Qu, J.; Zhao, X.; Liang, Y.; Xu, Y.; Ma, P.; Guo, B. Degradable conductive injectable hydrogels as novel antibacterial, anti-oxidant wound dressings for wound healing. *Chem. Eng. J.* **2019**, *15*, 548–560. [CrossRef]
91. Cardoso, A.M.; de Oliveira, E.G.; Coradini, K.; Bruinsmann, F.A.; Aguirre, T.; Lorenzoni, R.; Barcelos, R.C.; Roversi, K.; Rossato, D.R.; Pohlmann, A.R.; et al. Chitosan hydrogels containing nanoencapsulated phenytoin for cutaneous use: Skin permeation/penetration and efficacy in wound healing. *Mater. Sci. Eng. C* **2019**, *96AD*, 205–217. [CrossRef]
92. Zahid, A.; Ahmed, R.; ur Rehman, S.; Augustine, R.; Tariq, M.; Hazan, A. Nitric oxide releasing chitosan-poly (vinyl alcohol) hydrogel promotes angiogenesis in chick embryo model. *Int. J. Biol. Macromol.* **2019**, *136*, 901–910. [CrossRef]
93. Bano, I.; Arshad, M.; Yasin, T.; Ghauri, M.A. Preparation, characterization and evaluation of glycerol plasticized chitosan/PVA blends for burn wounds. *Int. J. Biol. Macromol.* **2019**, *124*, 155–162. [CrossRef]
94. Nešović, K.; Janković, A.; Radetić, T.; Vukašinović-Sekulić, M.; Kojić, V.; Živković, L.; Perić-Grujić, A.; Rhee, K.Y.; Mišković-Stanković, V. Chitosan-based hydrogel wound dressings with electrochemically incorporated silver nanoparticles—In vitro study. *Eur. Polym. J.* **2019**, *121*, 109257. [CrossRef]
95. Du, X.; Liu, Y.; Yan, H.; Rafique, M.; Li, S.; Shan, X.; Wu, L.; Qiao, M.; Kong, D.; Wang, L. Anti-Infective and Pro-Coagulant Chitosan-Based Hydrogel Tissue Adhesive for Sutureless Wound Closure. *Biomacromolecules* **2020**, *21*, 1243–1253. [CrossRef] [PubMed]
96. Mousavi, S.; Khoshfetrat, A.; Khatami, N.; Ahmadian, M.; Rahbarghazi, R. Comparative study of collagen and gelatin in chitosan-based hydrogels for effective wound dressing: Physical properties and fibroblastic cell behavior. *Biochem. Biophys. Res. Commun.* **2019**, *18*, 625–631. [CrossRef] [PubMed]
97. Martínez-Ibarra, D.; Sánchez-Machado, D.; López-Cervantes, J.; Campas-Baypoli, O.; Sanches-Silva, A.; Madera-Santana, T.J. Hydrogel wound dressings based on chitosan and xyloglucan: Development and characterization. *J. Appl. Polym. Sci.* **2019**, *136*, 47342. [CrossRef]
98. Soriano-Ruiz, J.L.; Gálvez-Martín, P.; López-Ruiz, E.; Suñer-Carbó, J.; Calpena-Campmany, A.C.; Marchal, J.A.; Clares-Naveros, B. Design and evaluation of mesenchymal stem cells seeded chitosan/glycosaminoglycans quaternary hydrogel scaffolds for wound healing applications. *Int. J. Pharm.* **2019**, *570*, 118632. [CrossRef]
99. Patil, P.S.; Fathollahipour, S.; Inmann, A.; Pant, A.; Amini, R.; Shriver, L.P.; Leipzig, N.D. Fluorinated Methacrylamide Chitosan Hydrogel Dressings Improve Regenerated Wound Tissue Quality in Diabetic Wound Healing. *Adv. Wound Care* **2019**, *8*, 374–385. [CrossRef]
100. Li, M.; Han, M.; Sun, Y.; Hua, Y.; Chen, G.; Zhang, L. Oligoarginine mediated collagen/chitosan gel composite for cutaneous wound healing. *Int. J. Biol. Macromol.* **2019**, *122*, 1120–1127. [CrossRef] [PubMed]
101. Soares, R.D.; Campos, M.G.; Ribeiro, G.P.; Salles, B.C.; Cardoso, N.S.; Ribeiro, J.R.; Souza, R.M.; Leme, K.C.; Soares, C.B.; de Oliveira, C.M.; et al. Development of a chitosan hydrogel containing flavonoids extracted from Passiflora edulis leaves and the evaluation of its antioxidant and wound healing properties for the treatment of skin lesions in diabetic mice. *J. Biomed. Mater. Res. Part A* **2020**, *108*, 654–662. [CrossRef] [PubMed]
102. Movaffagh, J.; Fazly-Bazzaz, B.S.; Yazdi, A.T.; Sajadi-Tabassi, A.; Azizzadeh, M.; Najafi, E.; Amiri, N.; Taghanaki, H.B.; Ebrahimzadeh, M.H.; Moradi, A. Wound Healing and Antimicrobial Effects of Chitosan-hydrogel/Honey Compounds in a Rat Full-thickness Wound Model. *Wounds A Compend. Clin. Res. Pract.* **2019**, *31*, 228–235.
103. Djekic, L.; Martinović, M.; Ćirić, A.; Fraj, J. Composite chitosan hydrogels as advanced wound dressings with sustained ibuprofen release and suitable application characteristics. *Pharm. Dev. Technol.* **2020**, *25*, 332–339. [CrossRef]
104. Lim, T.; Tang, Q.; Zhu, Z.; Wei, X.; Zhang, C. Sustained release of human platelet lysate growth factors by thermosensitive hydroxybutyl chitosan hydrogel promotes skin wound healing in rats. *J. Biomed. Mater. Res. Part A* **2020**, *108*, 2111–2122. [CrossRef]
105. Li, Q.; Cui, J.; Huang, H.; Yue, Z.; Chang, Y.; Li, N.; Han, Z.; Han, Z.C.; Guo, Z.; Li, Z. IGF-1C domain-modified chitosan hydrogel accelerates cutaneous wound healing by promoting angiogenesis. *Future Med. Chem.* **2020**, *12*. [CrossRef]
106. Yan, T.; Kong, S.; Ouyang, Q.; Li, C.; Hou, T.; Chen, Y.; Li, S. Chitosan-Gentamicin Conjugate Hydrogel Promoting Skin Scald Repair. *Mar. Drugs* **2020**, *18*, 233. [CrossRef] [PubMed]
107. Cifuentes, A.; Gómez-Gil, V.; Ortega, M.A.; Asúnsolo, Á.; Coca, S.; San, R.J.; Álvarez-Mon, M.; Buján, J.; García-Honduvilla, N. Chitosan hydrogels functionalized with either unfractionated heparin or bemiparin improve diabetic wound healing. *Biomed. Pharmacother.* **2020**, *129*, 110498. [CrossRef] [PubMed]

108. Zhang, X.; Sun, G.H.; Tian, M.P.; Wang, Y.N.; Qu, C.C.; Cheng, X.J.; Feng, C.; Chen, X.G. Mussel-inspired antibacterial polydopamine/chitosan/temperature-responsive hydrogels for rapid hemostasis. *Int. J. Biol. Macromol.* **2019**, *1383*, 321–333. [CrossRef]
109. Kalantari, K.; Mostafavi, E.; Saleh, B.; Soltantabar, P.; Webster, T.J. Chitosan/PVA Hydrogels Incorporated with Green Synthesized Cerium Oxide Nanoparticles for Wound Healing Applications. *Eur. Polym. J.* **2020**, *20*, 109853. [CrossRef]
110. Leonhardt, E.; Kang, N.; Hamad, M.; Wooley, K.; Elsabahy, M. Absorbable hemostatic hydrogels comprising composites of sacrificial templates and honeycomb-like nanofibrous mats of chitosan. *Nat. Commun.* **2019**, *10*, 1–9. [CrossRef]
111. Wang, Y.; Xie, R.; Li, Q.; Dai, F.; Lan, G.; Shang, S.; Lu, F. A self-adapting hydrogel based on chitosan/oxidized konjac glucomannan/AgNPs for repairing irregular wounds. *Biomater. Sci.* **2020**, *8*, 1910–1922. [CrossRef]
112. Yang, X.; Guo, J.L.; Han, J.; Si, R.J.; Liu, P.P.; Zhang, Z.R.; Wang, A.M.; Zhang, J. Chitosan hydrogel encapsulated with LL-37 peptide promotes deep tissue injury healing in a mouse model. *Mil. Med. Res.* **2020**, *7*. [CrossRef]
113. Sarada, K.; Firoz, S.; Padmini, K. In-situ gelling system: A review. *Int. J. Curr. Pharma Rev. Res.* **2014**, *15*, 76–90.
114. De Cicco, F.; Reverchon, E.; Adami, R.; Auriemma, G.; Russo, P.; Calabrese, E.C.; Porta, A.; Aquino, R.P.; Del Gaudio, P. In situ forming antibacterial dextran blend hydrogel for wound dressing: SAA technology vs. spray drying. *Carbohydr. Polym.* **2014**, *101*, 1216–1224. [CrossRef]
115. Singh, N.K.; Lee, D.S. In situ gelling pH-and temperature-sensitive biodegradable block copolymer hydrogels for drug delivery. *J. Control. Release* **2014**, *193*, 214–227. [CrossRef] [PubMed]
116. Guo, P.; Li, H.; Ren, W.; Zhu, J.; Xiao, F.; Xu, S.; Wang, J. Unusual Thermo-Responsive Behaviors of Poly (NIPAM-Co-AM)/PEG/PTA Composite Hydrogels. *Mater. Lett.* **2015**, *143*, 24–26. [CrossRef]
117. Thu, H.E.; Zulfakar, M.H.; Ng, S.F. Alginate based bilayer hydrocolloid films as potential slow-release modern wound dressing. *Int. J. Pharm.* **2012**, *434*, 375–383. [CrossRef] [PubMed]
118. Lin, Z.; Li, R.; Liu, Y.; Zhao, Y.; Ao, N.; Wang, J.; Li, L.; Wu, G. Histatin1-modi fi ed thiolated chitosan hydrogels enhance wound healing by accelerating cell adhesion, migration and angiogenesis. *Carbohydr. Polym.* **2020**, *230*, 115710. [CrossRef]
119. Huang, L.; Zhu, Z.; Wu, D.; Gan, W.; Zhu, S.; Li, W.; Tian, J.; Li, L.; Zhou, C.; Lu, L. Antibacterial poly (ethylene glycol) diacrylate/chitosan hydrogels enhance mechanical adhesiveness and promote skin regeneration. *Carbohydr. Polym.* **2019**, *225*, 115110. [CrossRef] [PubMed]
120. Lv, X.; Liu, Y.; Song, S.; Tong, C.; Shi, X.; Zhao, Y.; Zhang, J.; Hou, M. Influence of chitosan oligosaccharide on the gelling and wound healing properties of injectable hydrogels based on carboxymethyl chitosan/alginate polyelectrolyte complexes. *Carbohydr. Polym.* **2019**, *205*, 312–321. [CrossRef]
121. Song, R.; Zheng, J.; Liu, Y.; Tan, Y.; Yang, Z.; Song, X.; Yang, S.; Fan, R.; Zhang, Y.; Wang, Y. A natural cordycepin/chitosan complex hydrogel with outstanding self-healable and wound healing properties. *Int. J. Biol. Macromol.* **2019**, *134*, 91–99. [CrossRef] [PubMed]
122. Xu, W.; Liu, K.; Li, T.; Zhang, W.; Dong, Y.; Lv, J.; Wang, W.; Sun, J.; Li, M.; Wang, M.; et al. An in situ hydrogel based on carboxymethyl chitosan and sodium alginate dialdehyde for corneal wound healing after alkali burn. *J. Biomed. Mater. Res. Part A* **2019**, *7*, 742–754. [CrossRef] [PubMed]
123. Gholizadeh, H.; Messerotti, E.; Pozzoli, M.; Cheng, S.; Traini, D.; Young, P.; Kourmatzis, A.; Caramella, C.; Ong, H.X. Application of a Thermosensitive In Situ Gel of Chitosan-Based Nasal Spray Loaded with Tranexamic Acid for Localised Treatment of Nasal Wounds. *AAPS PharmSciTech* **2019**, *20*, 299. [CrossRef]
124. He, X.Y.; Sun, A.; Li, T.; Qian, Y.J.; Qian, H.; Ling, Y.F.; Zhang, L.H.; Liu, Q.Y.; Peng, T.; Qian, Z. Mussel-inspired antimicrobial gelatin/chitosan tissue adhesive rapidly activated in situ by H^2O^2/ascorbic acid for infected wound closure. *Carbohydr. Polym.* **2020**, *247*, 116692. [CrossRef]
125. Joorabloo, A.; Taghi, M.; Adeli, H.; Mansoori-Moghadam, Z.; Moghaddam, A. Fabrication of heparinized nano ZnO/poly (vinylalcohol)/ carboxymethyl cellulose bionanocomposite hydrogels using artificial neural network for wound dressing application. *J. Ind. Eng. Chem.* **2019**, *70*, 253–263. [CrossRef]
126. Gupta, A.; Keddie, D.J.; Kannappan, V.; Gibson, H.; Khalil, I.R.; Kowalczuk, M.; Martin, C.; Shuai, X.; Radecka, I. Production and characterisation of bacterial cellulose hydrogels loaded with curcumin encapsulated in cyclodextrins as wound dressings. *Eur. Polym. J.* **2019**, *118*, 437–450. [CrossRef]
127. Fan, X.; Yang, L.; Wang, T.; Sun, T.; Lu, S. pH-responsive cellulose-based dual drug-loaded hydrogel for wound dressing. *Eur. Polym. J.* **2019**, *121*, 109290. [CrossRef]

128. Erdagi, S.I.; Ngwabebhoh, F.A.; Yildiz, U. Genipin crosslinked gelatin-diosgenin-nanocellulose hydrogels for potential wound dressing and healing applications. *Int. J. Biol. Macromol.* **2020**, *149*, 651–663. [CrossRef]
129. Liu, D.; Cao, Y.; Qu, R.; Gao, G.; Chen, S.; Zhang, Y.; Wu, M.; Ma, T.; Li, G. Production of bacterial cellulose hydrogels with tailored crystallinity from Enterobacter sp. FY-07 by the controlled expression of colanic acid synthetic genes. *Carbohydr. Polym.* **2019**, *207*, 563–570.
130. Shefa, A.A.; Sultana, T.; Park, M.K.; Lee, Y.S.; Gwon, J.; Lee, B. Curcumin incorporation into an oxidized cellulose nanofiber-polyvinyl alcohol hydrogel system promotes wound healing. *Mater. Des.* **2020**, *186*, 108313. [CrossRef]
131. Sulaeva, I.; Hettegger, H.; Bergen, A.; Rohrer, C.; Kostic, M.; Konnerth, J.; Rosenau, T.; Potthast, A. Fabrication of bacterial cellulose-based wound dressings with improved performance by impregnation with alginate. *Mater. Sci. Eng. C* **2020**, *110*, 110619. [CrossRef]
132. Yang, H.; Shen, L.; Bu, H.; Li, G. Stable and biocompatible hydrogel composites based on collagen and dialdehyde carboxymethyl cellulose in a biphasic solvent system. *Carbohydr. Polym.* **2019**, *222*, 114974. [CrossRef] [PubMed]
133. Fontes, P.R.; Ribeiro, S.J.L.; Gaspar, A.M.M. Bacterial cellulose/phytotherapic hydrogels as dressings for wound healing. *Mater. Sci. Eng. Int. J.* **2019**, *3*, 162–173.
134. Liu, X.; Yang, K.; Chang, M.; Wang, X.; Ren, J. Fabrication of cellulose nanocrystal reinforced nanocomposite hydrogel with self-healing properties. *Carbohydr. Polym.* **2020**, *240*, 116289. [CrossRef] [PubMed]
135. Jiji, S.; Udhayakumar, S.; Rose, C.; Muralidharan, C.; Kadirvelu, K. Thymol enriched bacterial cellulose hydrogel as effective material for third degree burn wound repair. *Int. J. Biol. Macromol.* **2019**, *122*, 452–460. [CrossRef] [PubMed]
136. Sadeghi, S.; Nourmohammadi, J.; Ghaee, A.; Soleimani, N. Carboxymethyl cellulose-human hair keratin hydrogel with controlled clindamycin release as antibacterial wound dressing. *Int. J. Biol. Macromol.* **2020**, *147*, 1239–1247. [CrossRef]
137. Deng, Y.; Yang, X.; Zhang, X.; Cao, H.; Mao, L.; Yuan, M.; Liao, W. Novel fenugreek gum-cellulose composite hydrogel with wound healing synergism: Facile preparation, characterization and wound healing activity evaluation. *Int. J. Biol. Macromol.* **2020**, *160*, 1242–1251. [CrossRef] [PubMed]
138. Mohamad, N.; Loh, E.; Fauzi, M.B.; Ng, M.; Amin, M.C. In vivo evaluation of bacterial cellulose/acrylic acid wound dressing hydrogel containing keratinocytes and fibroblasts for burn wounds. *Drug Deliv. Transl. Res.* **2019**, *9*, 444–452. [CrossRef]
139. Gupta, A.; Briffa, S.M.; Swingler, S.; Gibson, H.; Kannappan, V.; Adamus, G.; Kowalczuk, M.; Martin, C.; Radecka, I. Synthesis of Silver Nanoparticles Using Curcumin-Cyclodextrins Loaded into Bacterial Cellulose-Based Hydrogels for Wound Dressing Applications. *Biomacromolecules* **2020**, *21*, 1802–1811. [CrossRef] [PubMed]
140. Koivuniemi, R.; Hakkarainen, T.; Kiiskinen, J.; Kosonen, M.; Vuola, J.; Valtonen, J.; Luukko, K.; Kavola, H.; Yliperttula, M. Clinical study of nanofibrillar cellulose hydrogel dressing for skin graft donor site treatment. *Adv. Wound Care.* **2020**, *9*, 199–210. [CrossRef]
141. Deng, Y.; Chen, J.; Huang, J.; Yang, X.; Zhang, X.; Yuan, S.; Liao, W. Preparation and characterization of cellulose/flaxseed gum composite hydrogel and its hemostatic and wound healing functions evaluation. *Cellulose* **2020**, *18*, 1–8. [CrossRef]
142. Khamrai, M.; Sanerjee, S.; Paul, S.; Samanta, S.; Kundu, P.P. Curcumin entrapped gelatin/ionically modified bacterial cellulose based self-healable hydrogel film: An eco-friendly sustainable synthesis method of wound healing patch. *Int. J. Biol. Macromol.* **2019**, *122*, 940–953. [CrossRef] [PubMed]
143. Forero-Doria, O.; Polo, E.; Marican, A.; Guzmán, L.; Venegas, B.; Vijayakumar, S.; Wehinger, S.; Guerrero, M.; Gallego, J.; Durán-Lara, E.F. Supramolecular Hydrogels Based on Cellulose for Sustained Release of Therapeutic Substances with Antimicrobial and Wound Healing Properties. *Carbohydr. Polym.* **2020**, *23*, 116383. [CrossRef] [PubMed]
144. Muchová, M.; Münster, L.; Capáková, Z.; Mikulcová, V.; Kuřitka, I.; Ícha, J. Design of dialdehyde cellulose crosslinked poly (vinyl alcohol) hydrogels for transdermal drug delivery and wound dressings. *Mater. Sci. Eng. C* **2020**, *26*, 111242. [CrossRef] [PubMed]
145. Wang, C.; Niu, H.; Ma, X.; Hong, H.; Yuan, Y.; Liu, C. Bioinspired, Injectable, Quaternized Hydroxyethyl Cellulose Composite Hydrogel Coordinated by Mesocellular Silica Foam for Rapid, Noncompressible Hemostasis and Wound Healing. *ACS Appl. Mater. Interfaces* **2019**, *11*, 34595–34608. [CrossRef]

146. Li, X.X.; Dong, J.Y.; Li, Y.H.; Zhong, J.; Yu, H.; Yu, Q.Q.; Lei, M. Fabrication of Ag–ZnO@ carboxymethyl cellulose/K-carrageenan/graphene oxide/konjac glucomannan hydrogel for effective wound dressing in nursing care for diabetic foot ulcers. *Appl. Nanosci.* **2020**, *10*, 729–738. [CrossRef]
147. Hakkarainen, T.; Koivuniemi, R.; Kosonen, M.; Escobedo-Lucea, C.; Sanz-Garcia, A.; Vuola, J.; Valtonen, J.; Tammela, P.; Mäkitie, A.; Luukko, K.; et al. Nanofibrillar cellulose wound dressing in skin graft donor site treatment. *J. Control. Release* **2016**, *244*, 292–301. [CrossRef] [PubMed]
148. Napavichayanun, S.; Yamdech, R.; Aramwit, P. The safety and efficacy of bacterial nanocellulose wound dressing incorporating sericin and polyhexamethylene biguanide: In vitro, in vivo and clinical studies. *Arch. Dermatol. Res.* **2016**, *308*, 123–132. [CrossRef]
149. Wild, T.; Bruckner, M.; Payrich, M.; Schwarz, C.; Eberlein, T.; Andriessen, A. Eradication of methicillin-resistant Staphylococcus aureus in pressure ulcers comparing a polyhexanide-containing cellulose dressing with polyhexanide swabs in a prospective randomized study. *Adv. Skin Wound Care* **2012**, *25*, 17–22. [CrossRef]
150. Portela, R.; Leal, C.R.; Almeida, P.L.; Sobral, R.G. Bacterial cellulose: A versatile biopolymer for wound dressing applications. *Microb. Biotechnol.* **2019**, *12*, 586–610. [CrossRef] [PubMed]
151. delli Santi, G.; Borgognone, A. The use of Epiprotect®, an advanced wound dressing, to heal paediatric patients with burns: A pilot study. *Burns Open.* **2019**, *3*, 103–107. [CrossRef]
152. Burkatovskaya, M.; Tegos, G.P.; Swietlik, E.; Demidova, T.N.; Castano, A.P.; Hamblin, M.R. Use of chitosan bandage to prevent fatal infections developing from highly contaminated wounds in mice. *Biomaterials* **2006**, *27*, 4157–4164. [CrossRef]
153. Mo, X.; Cen, J.; Gibson, E.; Wang, R.; Percival, S.L. An open multicenter comparative randomized clinical study on chitosan. *Wound Repair Regen.* **2015**, *23*, 518–524. [CrossRef] [PubMed]
154. Azad, A.K.; Sermsintham, N.; Chandrkrachang, S.; Stevens, W.F. Chitosan membrane as a wound-healing dressing: Characterization and clinical application. *J. Biomed. Mater. Res. B Appl. Biomater.* **2004**, *69*, 216–222. [CrossRef] [PubMed]
155. Valentine, R.; Athanasiadis, T.; Moratti, S.; Hanton, L.; Robinson, S.; Wormald, P.J. The efficacy of a novel chitosan gel on hemostasis and wound healing after endoscopic sinus surgery. *Am. J. Rhinol. Allergy* **2010**, *24*, 70–75. [CrossRef] [PubMed]
156. Kabeer, M.; Venugopalan, P.P.; Subhash, V.C. Pre-hospital Hemorrhagic Control Effectiveness of Axiostat® Dressing Versus Conventional Method in Acute Hemorrhage Due to Trauma. *Cureus* **2019**, *11*, e5527. [CrossRef] [PubMed]
157. Matica, M.A.; Aachmann, F.L.; Tøndervik, A.; Sletta, H.; Ostafe, V. Chitosan as a wound dressing starting material: Antimicrobial properties and mode of action. *Int. J. Mol. Sci.* **2019**, *20*, 5889. [CrossRef] [PubMed]
158. ChitoHeal Gel. Available online: https://chitotech.com/page/627/ChitoHeal-Gel (accessed on 20 September 2020).
159. KYTOCEL. Available online: https://www.wound-care.co.uk/dressings/kytocel.html (accessed on 20 September 2020).
160. PosiSep®. Available online: https://www.hemostasisllc.com/products/posisepc-c2/ (accessed on 20 September 2020).
161. ExcelArrest®XT. Available online: http://global.hemostasisllc.com/excelarrest.html (accessed on 20 September 2020).
162. ChitoClot Pad. Available online: http://www.anscare.co/english/products/detail.php?cpid=2&dpid=21 (accessed on 20 September 2020).
163. XSAT. Available online: https://www.revmedx.com/xstat/ (accessed on 20 September 2020).
164. A Natural Source of Chitosan. Available online: http://www.primex.is/products-services/chitoclear/ (accessed on 20 September 2020).

International Journal of
Molecular Sciences

MDPI

Article

Evaluation of Polyvinyl Alcohol/Cobalt Substituted Hydroxyapatite Nanocomposite as a Potential Wound Dressing for Diabetic Foot Ulcers

Wei-Chun Lin [1] and Cheng-Ming Tang [2,3,*]

1 School of Dental Technology, College of Oral Medicine, Taipei Medical University, Taipei 110, Taiwan; tukust94114wenny@gmail.com or Weichun1253@tmu.edu.tw

2 Graduate Institute of Oral Sciences, Chung Shan Medical University, Taichung City 40201, Taiwan

3 Department of Dentistry, Chung Shan Medical University Hospital, Taichung City 40201, Taiwan

* Correspondence: ranger@csmu.edu.tw; Tel.: +886-4-2471-8668 (ext. 55528); Fax: +886-4-2475-9065

Received: 21 September 2020; Accepted: 19 November 2020; Published: 22 November 2020

Abstract: Diabetic foot ulcers (DFUs) caused by diabetes are prone to serious and persistent infections. If not treated properly, it will cause tissue necrosis or septicemia due to peripheral blood vessel embolism. Therefore, it is an urgent challenge to accelerate wound healing and reduce the risk of bacterial infection in patients. In clinical practice, DFUs mostly use hydrogel dressing to cover the surface of the affected area as an auxiliary treatment. Polyvinyl alcohol (PVA) is a hydrophilic hydrogel polymer widely used in dressings, drug delivery, and medical applications. However, due to its weak bioactivity and antibacterial ability, leads to limited application. Filler adding is a useful way to enhance the biocompatibility of PVA. In our study, cobalt-substituted hydroxyapatite (CoHA) powder was prepared by the electrochemically-deposited method. PVA and PVA-CoHA nanocomposite were prepared by the solvent casting method. The bioactivity of the PVA and composite was evaluated by immersed in simulated body fluid for 7 days. In addition, L929 cells and *E. coli* were used to evaluate the cytotoxicity and antibacterial tests of PVA and PVA-CoHA nanocomposite. The results show that the addition of CoHA increases the mechanical properties and biological activity of PVA. Biocompatibility evaluation showed no significant cytotoxicity of PVA-CoHA composite. In addition, a small amount of cobalt ion was released to the culture medium from the nanocomposite in the cell culture period and enhanced cell growth. The addition of CoHA also confirmed that it could inhibit the growth of *E. coli*. PVA-CoHA composite may have potential applications in diabetic trauma healing and wound dressing.

Keywords: polyvinyl alcohol; cobalt-substituted hydroxyapatite; diabetic foot ulcers; hydrogels; antibacterial ability

---

1. Introduction

Diabetes is an endocrine disease caused by abnormal sugar and fat metabolism. It often causes other complications such as vascular disease, nephropathy, and diabetic foot ulcers (DFUs). Among them, DFUs are caused by poor long-term blood glucose control, which causes lower limb peripheral arterial disease and foot deformation [1]. When ulcers occur without proper treatment, it will cause tissue necrosis or amputation due to peripheral blood vessel embolism. In addition, bacteria can also spread through the blood circulation to cause septicemia and threaten the patient's life. Therefore, it is an urgent challenge to accelerate the speed of wound healing and reduce the risk of bacterial infection in patients. In clinical practice, dressing is often used to cover the surface of the affected area as one of the treatment methods for DFUs. The ideal wound dressing needs to have the following conditions:

(1) can provide gas and liquid exchange to keep the wound properly moist [2,3]. (2) It has favorable biocompatibility and does not cause any immune response. (3) Protects wounds to reduce infection and (4) the dressing is easily removed and does not cause secondary trauma [4,5]. Based on the above conditions, hydrogel-like materials are widely used in wound dressing.

Polyvinyl alcohol (PVA) has excellent film forming properties and outstanding chemical stability, widely used in industry applications such as food packaging [6]. In addition, PVA has non-toxic and low protein adsorption properties, resulting in low cell adhesion compared to other hydrogel-like materials [7]. Therefore, it is developed in artificial cartilage substitutes, corneal implants [8], and drug delivery [9]. However, PVA is relatively limited in terms of its bioactivity compared to other polymers. On the other hand, nano-scale particles widely added into polymer to enhance its mechanical properties and thermal properties is a common modification procedure [10], while the effect of nano-scale particle additives is much more remarkable than that observed in conventional composites because of the high surface to volume ratio of the nano-scale particles. For example, the mechanical and thermal properties of PVA were outstanding, improved upon the addition of a small amount (<5%) of bamboo charcoal nanoparticles (107.4 ± 9.8 nm) [11].

Furthermore, hydroxyapatite (HA) is the main component of human bones and the calcium atom in the apatite inner structure can be substituted by divalent metal cations [12–14], altering the original properties of HA such as the lattice, crystallinity, and grain shape [15]. In addition, cobalt ion is a hypoxia-mimicking agent, which can activate the hypoxia inducible factor-1 (HIF-1α) in bone marrow stromal stem cells and subsequently activate HIF-1α target genes including erythropoietin and vascular endothelial growth factor (VEGF) [16–19]. This indicates that cobalt ions have the property of inducing angiogenesis. For problem wounds caused by diabetes, in addition to reducing wound infections, the supply of nutrients will also affect wound repair. Therefore, cobalt ions can be used to promote angiogenesis to help transport nutrients to the wound to help repair. Previous literature shows that synthesized cobalt-substituted hydroxyapatite (CoHA) (Co^{2+} content 12 wt%) can enhance osteogenesis in osteoporosis-induced alveolar defects after 24 weeks [20]. Our previous studies have confirmed that the release of cobalt ions through CoHA can promote bone cell growth, reduce free radicals, and enhance antibacterial effects [21,22]. This shows that CoHA is a potential choice for wound dressing additives.

According to the above, this study involves the preparation of PVA/CoHA composites membranes. The main objectives are to analyze the effect of CoHA on the properties of PVA and on improving the biocompatibility, anti-bacterial ability, and anti-inflammatory response of the membrane. We hypothesized that the addition of CoHA is significantly effective for the bioactivity and anti-inflammatory properties of the membranes. Finally, we assess whether the composite membrane could be used for the application of diabetic traumas.

2. Results and Discussion

2.1. Characterization of CoHA

The surface morphology of CoHA shows a granular mixed structure (Figure 1A). It is evident that most of the nanoparticles are agglomerated. The surface elemental composition of CoHA was analyzed using energy dispersive X-ray spectroscopy (EDS) and cobalt elements were found (Figure 1B). The crystal structure of the powder was analyzed by XRD, diffraction peaks of HA crystals were observed at 26.01° (002), 32.04° (211), 39.68° (310), 46.81° (222), and 49.74° (123) and diffraction peak of Co_3O_4 was observed at 18.8° (111), which confirmed the presence of the CoHA structure (Figure 1C). The chemical compositions of HA and CoHA were obtained by FTIR. The absorption peaks at wavenumbers of 570.9 cm^{-1} represent PO_4^{3-}. The peak at 3383 cm^{-1}, representing OH^- is observed in HA and CoHA, among which CoHA showed the largest peak shift (Figure 1D). This could be caused by the addition of $CoCl_2$, which was combined with OH^- to observe the peak shift [21].

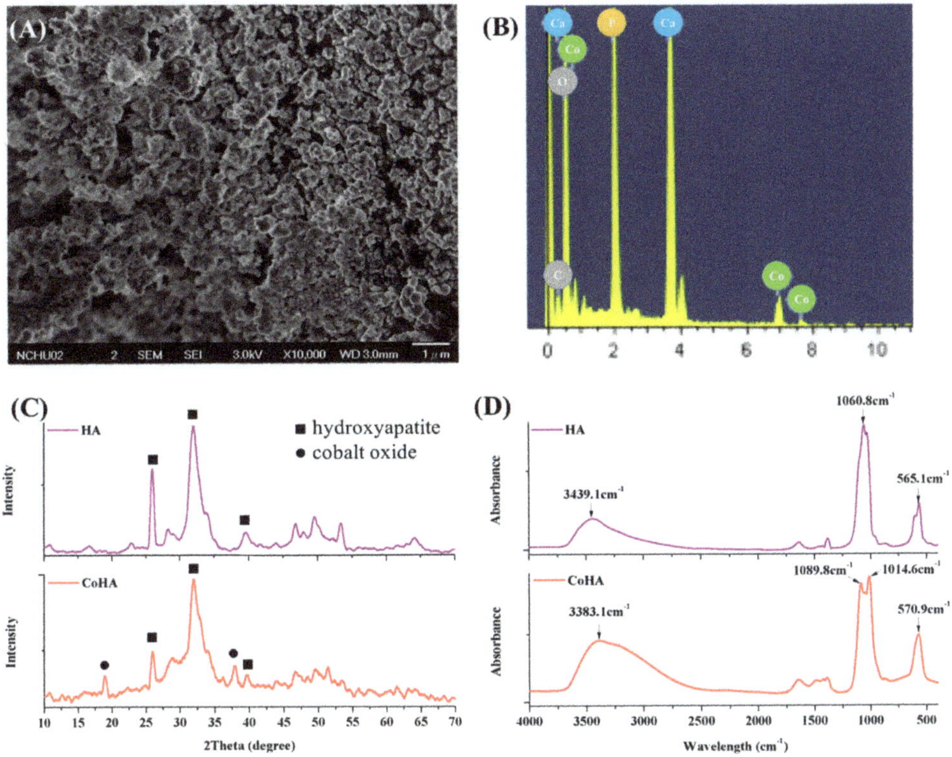

Figure 1. (**A**) Field emission scanning electron microscopy (FESEM) image of cobalt-substituted hydroxyapatite (CoHA) powder, (**B**) surface element analysis of CoHA by energy dispersive X-ray spectroscopy (EDS), (**C**) X-Ray diffraction (XRD) patterns of hydroxyapatite (HA) and CoHA, (**D**) IR spectra of HA and CoHA by FTIR.

2.2. Characterization of PVA Nanocomposite

Dried PVA and nanocomposite membranes had a smooth surface (data not shown). When the membranes were continuously immersed in distilled water at 37 °C for 24 h, the overall structure retains its integrity. However, swollen membranes were dehydrated and an interconnected porous structure on the surface was observed by FESEM (Figure 2A–C).

The crystal structure is one of the major factors that affect the mechanical properties of a material. The XRD patterns of the PVA and PVA/CoHA nanocomposites are presented in Figure 3A. The XRD pattern of the pure PVA membranes revealed strong crystalline reflections at around $2\theta = 19.88°$ and a shoulder at 22.74°. The two peaks are characteristic of PVA, representing reflections from (101) and (200) from a monoclinic unit cell [23]. In the XRD profile of PVA/Co-HA nanocomposite membranes, no peak from Co-HA was observed in the XRD curves of the nanocomposites [11,24].

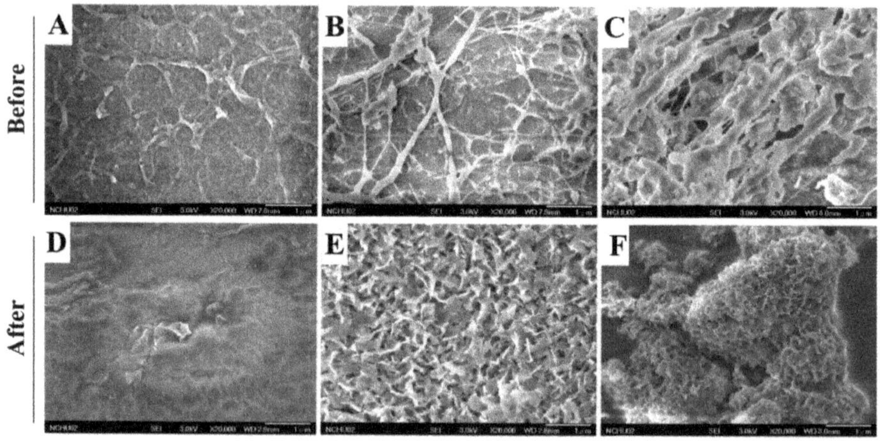

Figure 2. Surface morphology of (**A**,**D**) PVA, (**B**,**E**) PVA-HA nanocomposites, and (**C**,**F**) PVA-CoHA nanocomposites before and after immersion in SBF for 7 days.

ATR-FTIR analysis was conducted on the PVA and PVA nanocomposite membranes (Figure 3B,C). PVA absorption peaks appeared at 3255 cm^{-1} (stretching of OH), 2941, 2906 cm^{-1} (symmetric stretching of CH_2), 1569 cm^{-1} (O–H and C–H bending), 1416 cm^{-1} (bending of OH and wagging of CH_2), 1658 and 1330 cm^{-1} (C = O), 1142 cm^{-1} (CH wagging), 1086 cm^{-1} (stretching of CO and bending of OH from amorphous sequence of PVA), 920 cm^{-1} (CO symmetric stretching), and 837 cm^{-1} (CH_2 rocking) [11,24–28]. The small bands found at 854 cm^{-1} were generated by the stretching vibrations of C–C bonds [23]. The crystallinity of PVA was obtained from the peak at 1143 cm^{-1} in the ATR-FTIR spectrum. The peak symmetric stretching mode of the C-C or C-O stretch of the chain is related to the intramolecular hydrogen bonding between two adjacent OH groups on the same side of the carbon plane [29]. The peak at 1143 cm^{-1} indicates that C-O is stretched from the crystalline sequence of PVA, while the peak at 1086 cm^{-1} indicates C-O stretching from the amorphous sequence of PVA [30]. Therefore, the ratio of the two peaks reflects the crystallinity of PVA. The position strengths through 1143 cm^{-1}/1086 cm^{-1} is calculated as surface crystallinity of PVA (%) = [Absorbance at 1143 cm^{-1}/(Absorbance at 1143 cm^{-1} + Absorbance at 1086 cm^{-1})] × 100%. The results show that the addition of pure HA and CoHA slightly increases the crystallinity on the surface of the membranes (Table 1), but there is no significant difference.

Table 1. The relative intensities of 1143 cm^{-1} and 1086 cm^{-1} peak in ATR-FTIR spectra of PVA and PVA/CoHA nanocomposites.

Sample	Absorbance (1143 cm^{-1}/1086 cm^{-1})	Surface Crystallinity (%)
PVA	0.524	34.4%
PVA-HA	0.544	35.2%
PVA-CoHA	0.550	35.5%

Tensile testing showed that the addition of CoHA significantly increased the tensile strength and ductility of PVA (Figure 3D). This result is similar to the literature, adding a small number of nanoparticles can improve the mechanical properties of the polymer [11]. In addition, the OH group on the molecular chain of CoHA and PVA generates a hydrogen bond to restrict the movement of the molecular chain which is called the crystalline region. However, the unaffected molecular chains can move freely and are called amorphous regions. As shown in Figure 3E, the crystalline region acts to enhance mechanical strength, while the non-crystalline region imparts ductility to the film.

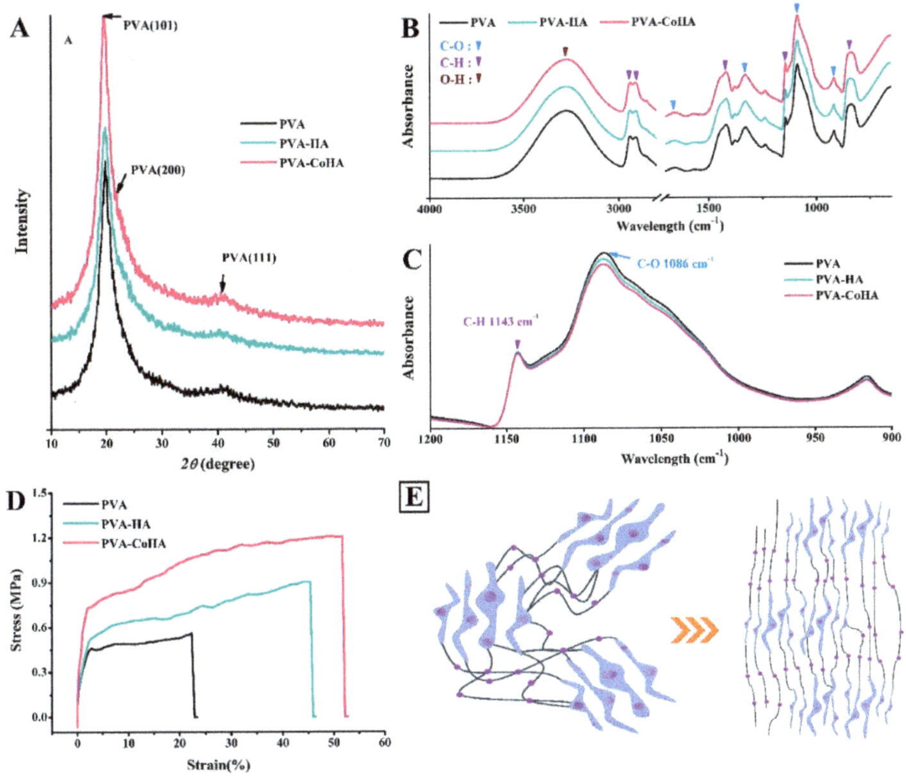

Figure 3. (**A**) XRD patterns, (**B**,**C**) ATR-FTIR spectra, and (**D**) stress–strain curve of PVA and PVA-CoHA nanocomposites. (**E**) Structure model for extended PVA-CoHA.

2.3. Thermal Properties

Thermal properties of the PVA and nanocomposites were assessed using thermo-gravimetric analyzer (TGA) and differential scanning calorimetry (DSC) data and are summarized in Table 2. Treating PVA at 100 °C completely removed any moisture present, [24] and heating it at 247–362 °C caused rapid decomposition (Figure 4A,B). The composite film changes at 380 °C. The addition of CoHA was found to slow down the weight loss of PVA. From the SEM (Figure 2A–C) speculation, this phenomenon can be explained by the fact that PVA uniformly coats CoHA particles and makes the structure denser. It also explains that PVA-CoHA has a higher crystallinity of PVA internally. PVA-CoHA heated to 400 °C led to 20.75% residual material, significantly higher than the case of pure PVA (13.32% residual, Table 2). The ash content of PVA-CoHA at the temperature end point (600 °C) is still higher than that of pure PVA (3.08% at 400 °C), denoting that CoHA content also increases the thermal properties of PVA. The results of Xc are also listed in Table 3. DSC results show that the addition of CoHA can significantly increase the crystallization of PVA, which is consistent with the XRD results where the typical PVA peak (101) in PVA-CoHA is the highest.

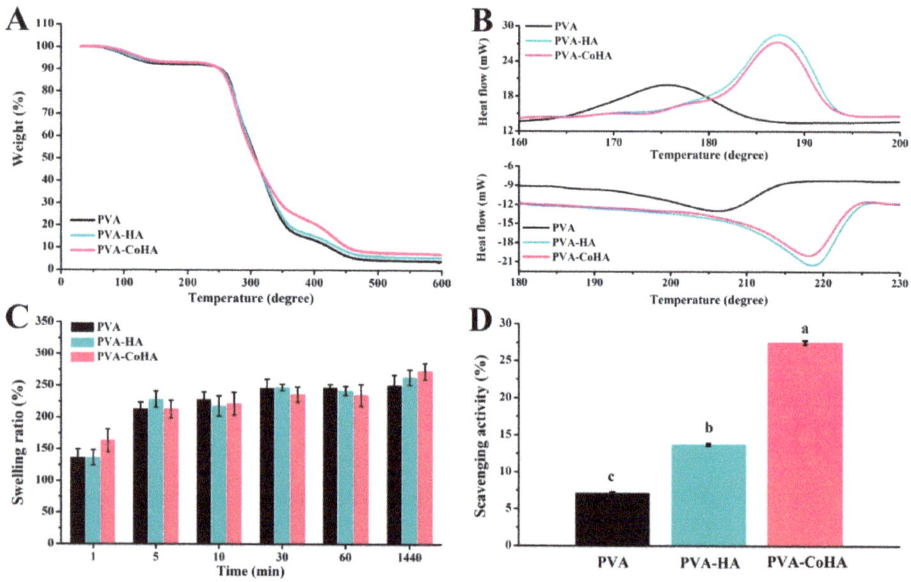

Figure 4. (**A**) TGA patterns, (**B**) DSC patterns, (**C**) swelling behaviors curve, and (**D**) free radical scavenging ability of PVA and PVA-CoHA nanocomposites. Means with different letters (a–c) were significantly different ($p < 0.05$, mean ± SD, $n = 4$).

Table 2. Thermal properties of PVA and PVA/CoHA nanocomposites.

Sample	T_{onset} (°C)	T_p (°C)	Ash (%)	T_c (°C)	T_m (°C)	ΔH_m (J/g)	X_c
PVA	113.0	308.3	0	175.5	206.2	40.7	27.1
PVA-HA	118.0	307.2	1.7	187.4	218.5	53.6	36.1
PVA-CoHA	127.8	306.2	3.1	187.2	218.0	61.6	41.5

T_{onset}: onset temperature of pyrolysis, obtained from TGA curves at 95% weight; T_p: peak pyrolytic temperature, obtained from TGA curves at 50% weight. Crystallization temperature (T_c), melting temperature (T_m) and heat of fusion of PVA (ΔH_m), obtained from DSC.

2.4. Swelling Behavior

Swelling property of wound dressing will affect drug release properties and interactions with tissue fluid on wound position. Therefore, swelling behavior is one of the important indicators for evaluating dressings. PVA and PVA-CoHA membranes were immersed in PBS as a model for human contact (pH = 7.42) at 37 °C, and the swelling behavior was calculated at different immersion times (Figure 4C). The results showed that PVA and PVA-CoHA nanocomposites membranes have rapid and good swelling properties (Figure 4C), where soaking in PBS for one minute leads to 150% expansion. After 24 h soaking in PBS, the membrane reached 260% expansion and an equilibrium state, and statistical analysis indicated no significant differences among the tested sample groups. Therefore, the addition of CoHA does not affect the swelling behavior of PVA and can ensure that the membrane maintains a good water absorption effect.

2.5. Free Radical Scavenging Ability

Anti-inflammatory effects of wound dressings have attracted more attention. Biological defense and the differentiation process will produce free radicals. However, excessive free radicals may cause excessive oxidation. Antioxidants can capture excessive harmful free radicals to maintain the normal growth of organisms. Therefore, they play an important role in this balance. Figure 4D shows the free radical scavenging ability of the PVA-Co-HA composite membrane. The capture capacity of PVA-CoHA was significantly higher than that of pure PVA or PVA-HA ($p < 0.05$). It is mainly the exposed CoHA on the surface that allows PVA-CoHA to combine with more free radicals. Therefore, it indirectly indicates that PVA-CoHA can reduce the anti-inflammatory effect of cells and reduce the inflammatory response of wounds.

2.6. Hydrophilicity Test

The hydrophilicity and hydrophobicity of the material surface is one of the important indicators of biomedical materials [31]. Use contact angle measurement to evaluate the hydrophilicity and hydrophobicity of the material surface to observe the effect of the material surface on cell adhesion. It is reported in the literature that increasing the hydrophilicity of the material can improve its biological activity and contribute to cell adhesion [32]. In order to evaluate the surface contact angle of the PVA membranes after contact with the skin. In this study, PBS was used as the test solution. The results show that both PVA and PVA-CoHA nanocomposites membranes are hydrophilic materials (Figure 5). In addition, the time that the solution stays on the surface is negatively correlated with the contact angle. This is similar to the result of swelling. The PVA membranes start to absorb the liquid on the surface after 1 min and reduced the contact angle. However, the surface of PVA-HA and PVA-CoHA is relatively hydrophilic due to the combination of the structural solution with OH on the surface and water molecules. Especially the SEM image can find that the exposure of CoHA in PVA-CoHA is more obvious, which leads to a significantly smaller contact angle than other groups ($p < 0.05$). The above addition of HA and CoHA can effectively improve the hydrophilicity of the PVA surface.

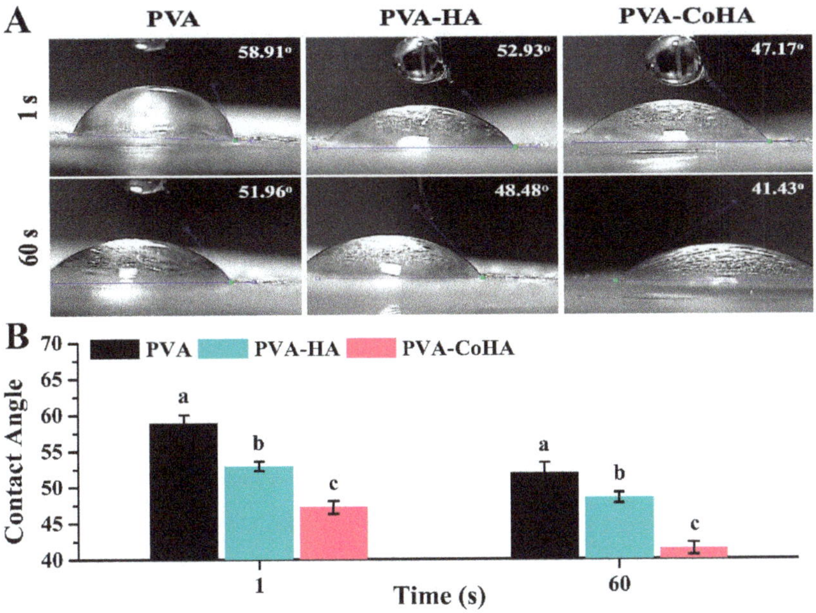

Figure 5. The surface wettability of PVA and PVA nanocomposites. (**A**) Image of surface contact angle. (**B**) Quantitative analysis of the contact angle was performed. Means with different letters (a–c) were significantly different ($p < 0.05$, mean ± SD, $n = 5$).

2.7. Bioactivity

Many studies use surface mineralization results to evaluate the bioactivity of materials. Surface mineralization results are given in Figure 2D–F for PVA and PVA-CoHA nanocomposites membranes placed in 37 °C SBF solution for 7 days. The pure PVA membrane surface did not produce HA at any exposure time, while PVA-HA and PVA-CoHA produced HA, especially after 7 days of SBF exposure.

Based on EDS data, the Ca/*p* ratios of different sample types were 1.29 (PVA-HA) and 1.39 (PVA-CoHA). The apatite crystal faces also differ between samples, where the PVA-HA crystalline morphology is a sheet of flat HA. PVA-CoHA feature lamellae divided into regions rather than a flat surface, which may be due to the impact of cobalt ion remineralization during crystallization. Notably, there were cobalt ions before immersion in the CoHA samples, and after 7 days of SBF immersion these ions disappeared due to apatite deposition (Table 3). This indicates that the membrane surface undergoes new HA deposition, covering the cobalt ions. XRD data show that the diffraction peaks of pure PVA do not produce new peaks after SBF soaking, indicating that the membrane does not undergo mineralization (Figure 6A). HA and CoHA were added to membranes, creating a new diffraction peak at 26° and 32°. The diffraction peaks for HA are (002) and (112), respectively [11]. The ATR-IR spectrum of the PVA film mineralization is shown in Figure 6B. In addition to the pure PVA membrane, the peak at 1027 cm^{-1} is generated from HA with a phosphate group [33]. As the apatite surface is exposed, the calcium and phosphorus ions in the SBF solution deposited on the membrane surface. These results confirm that the addition of HA and CoHA can significantly enhance the bioactivity effect of PVA membranes.

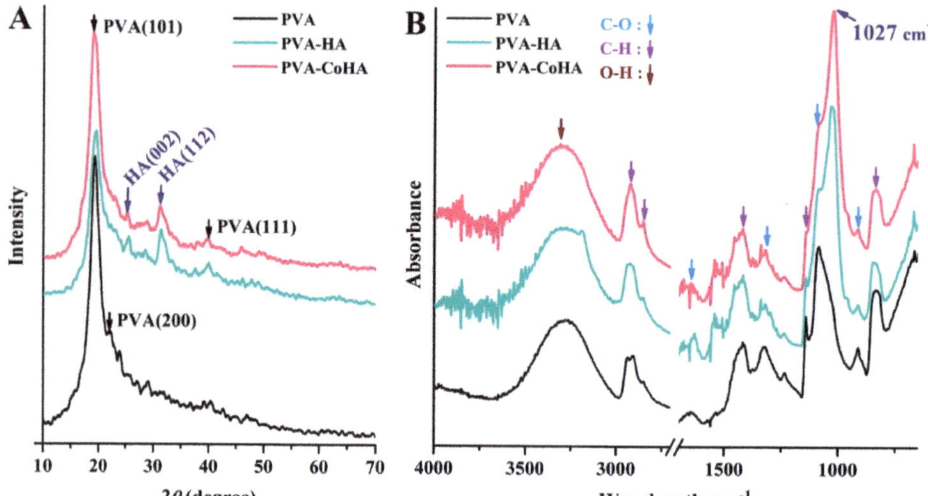

Figure 6. (**A**) XRD patterns and (**B**) ATR-FTIR spectra of PVA and PVA-CoHA nanocomposites after immersion in SBF for 7 days.

Table 3. Calcium to phosphate (Ca/P) ratio of PVA and PVA nanocomposites surface.

Group	Ratio	PVA	PVA-HA	PVA-CoHA
Before Immersion	Ca/P	N.D.	1.02	0.86
	Ca + Co/P	N.D.	1.02	0.94
Immersed in SBF for 7 Days	Ca/P	N.D.	1.29	1.40
	Ca + Co/P	N.D.	1.29	1.40

2.8. Biocompatibility

Membrane biocompatibility was evaluated using L929 fibroblasts cultured with PVA and PVA-CoHA nanocomposites in vitro. PVA is a hydrophilic material and has excellent swelling effects, so it will reduce the adhesion ability of cells after swelling, making it difficult to evaluate the cytotoxicity of surface cells.

This part of the study is divided into two parts: the membrane surfaces directly in contact with cultured cells, and the extraction of membrane fluid to test the cytotoxicity. As shown in Figure 7A, tissue culture plate (TCP) is significantly higher than the other groups, indicating that cells in the membrane have less desirable surface adhesion. After 24 h, cell attachment for PVA was significantly higher than for the other groups. Seventy-two hours later, the number of PVA-HA and PVA-CoHA cells was not significantly different than for pure PVA. Affixed experiments show that PVA membrane surface cells have poor attachment. For wound dressings however, this phenomenon means that the tissue surrounding the wound is less likely to stick with the dressing and will cause secondary damage.

As to whether the reduction of cells is caused by the release of cobalt ions, we do not know, so the second part uses the membrane material extraction solution for cytotoxicity testing according to the method of ISO 10993-12. The results of the cell culture with the membrane extraction solution are shown in Figure 7B. After the initial 24 h exposure, the results of PVA-HA were significantly higher than for all other groups, and the CoHA groups grow slowly. After 72 h, PVA-CoHA was significantly higher than TCP and pure PVA, but did not differ significantly from the HA groups. The extraction results confirm that the PVA composite membrane has no obvious cytotoxicity. Notably, the release of cobalt ions slows cell growth when cells are initially attached, but it aids cell growth in the latter part of cell proliferation.

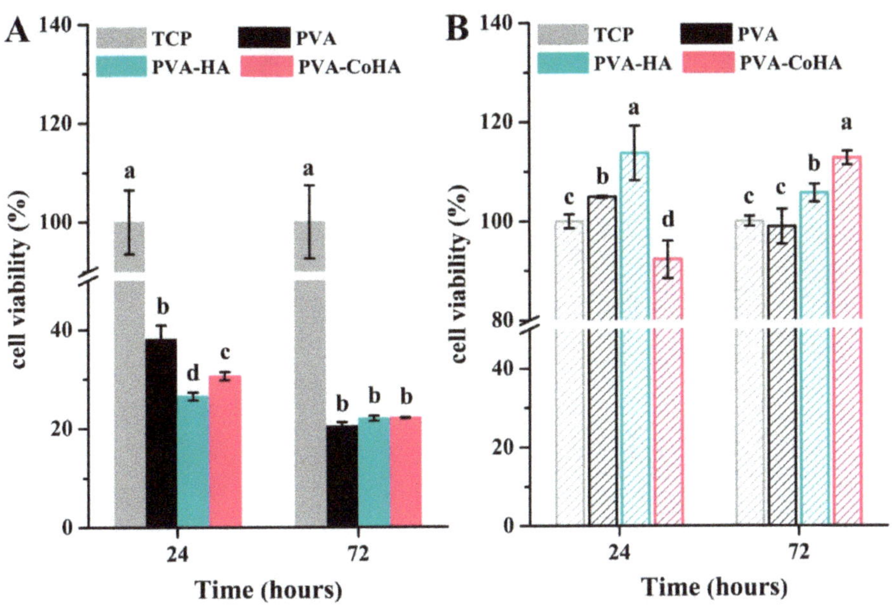

Figure 7. Biocompatibility of the PVA and PVA-CoHA nanocomposite. (**A**) Cells are grown on the surface. (**B**) The cells are cultured using extraction solutions. Means with different letters (a–d) were significantly different ($p < 0.01$, mean ± SD, $n = 4$).

2.9. Antibacterial Ability

Due to slower recovery in diabetic traumas, wounds are vulnerable to invasion by surrounding bacteria. Therefore, current wound dressings still add antibacterial agents or nanoparticles to achieve antibacterial effects. Through this method, the wound is protected from infection by bacteria on the periphery. In this study, *E. coli* was used to evaluate the antibacterial ability of the membrane. The sample surface of each *E. coli* test was added to a new medium, and 100 λ bacteria was extracted to a petri dish-coated plate for 17 h growth (Figure 8A–C). The number of colonies on a labelled petri dish (n = 3) and single factor analysis of variance (ANOVA) were used to assess the statistical significance of the results in Figure 8. That PVA-CoHA group eminently reduced the bacteria viability almost by 87.4% compared to the PVA group (control). The results show that PVA-CoHA has a good antibacterial effect (Figure 8D). Previous studies have also shown that CoHA particles can inhibit the growth of bacteria through the release of cobalt ions [34]. In this study, the CoHA content in PVA-CoHA was very small. Therefore, achieving 100% antibacterial is challenging. Perhaps in the future, the effect of antibacterial ability can be improved by increasing the content of CoHA. Moreover, literature indicates that using Co_3O_4 nanoparticles for *E. coli* treatment is effective, with potential for use as antimicrobial agents [35]. This is similar to the results of this experiment and also confirms that the antibacterial effect comes from the addition of cobalt ions. Therefore, PVA added CoHA has the opportunity to inhibit the bacteria around the wound and prevent wound infection.

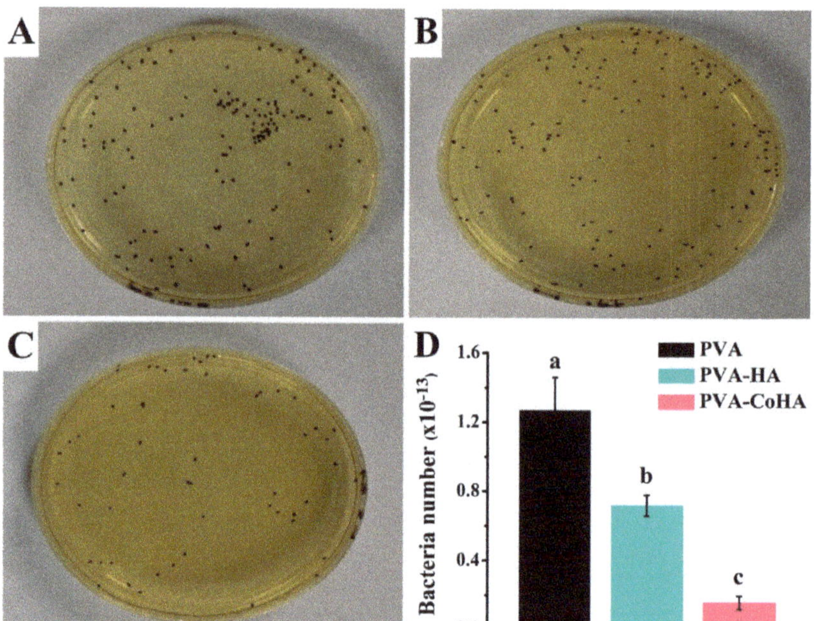

Figure 8. Evaluation of the antibacterial activity of (**A**) PVA, (**B**) PVA-HA and (**C**) PVA-CoHA nanocomposites against *E. coli*. (**D**) Bacteria adhesion on PVA and PVA-CoHA nanocomposites. Means with different letters (a–c) were significantly different ($p < 0.01$, mean ± SD, n = 4).

3. Materials and Methods

3.1. Synthesis of CoHA

The CoHA powder was prepared by electrochemical deposition [21]. The electrolyte solution was formulated with 42 mM calcium nitrate (Shimakyu's pure chemical, Osaka, Japan), 25 mM ammonium dihydrogen phosphate (Showa, Tokyo, Japan), and 7.98 mM cobalt chloride (Shimada chemical works, Tokyo, Japan) in de-ionized water. The titanium sheet was a cathode, the stainless-steel sheet was an anode. The electrodeposition procedure was carried out with a direct current power supply (GR-50H10, GICEK, Taipei, Taiwan) and at constant potential by 5.5 V at 328 K for 20 min. After the reaction was completed, rinsed with deionized water and dried, the powder was removed from the surface and collected.

3.2. Preparation of PVA and Nanocomposite

The PVA used in this study was obtained from Sigma-Aldrich (St. Louis, MO, USA). The PVA has a molecular weight of 70,000–110,000 g/mole and hydrolysis grade of 98.5%. The homogeneous PVA solution was prepared by adding 1.485 g of PVA in 10 mL of distilled water and stirring at 85 °C for 30 min. Then, 15 mg of CoHA (or HA) powder was then added to the PVA solution and stirred for 24 h at room temperature. The above solution was poured into a polytetrafluroethylene mold (90 mm in diameter) and placed in a fume hood until the solvent evaporated.

3.3. Characterization of PVA and Nanocomposite

The surface morphology of the PVA and PVA/CoHA nanocomposite was examined by field-emission scanning electron microscopy (FESEM) (JSM-6700F, JEOL, Tokyo, Japan). Phase indication in samples was confirmed by X-ray diffraction (XRD) (Miniflex II, Rigaku corporation, Tokyo, Japan), operating at 30 kV with Cu-Kα radiation within the scanning range of 10–70° (2θ) and a scanning speed of 4°/min. Infrared spectra of PVA and composite were recorded using a Fourier transform infrared (FTIR) system (FTIR-8000, Shimadzu, Tokyo, Japan) in the spectral range of 650–4000 cm^{-1} with a spectral resolution of 2 cm^{-1}, using the attenuated total reflection (ATR) mode. Tensile tests of the PVA and PVA nanocomposites were carried out with universal testing machine (GF-AI-7000M, GO TECH, Taichung, Taiwan) instruments by applying a 10 N load cell at a crosshead speed of 10 mm/min. All samples were cut with a dumbbell-shaped mold 40 mm long and 20 mm wide. The average value of tensile properties was obtained from the results of 3 tests.

3.4. Thermal Properties

The thermos-gravimetric analysis (TGA) was carried out on the PVA and PVA nanocomposites using a Q500, TA instrument analyzer (New castle, DE, USA). The samples were heated from 30–600 °C at a heating rate of 10 °C/min under nitrogen, with a nitrogen flow rate of 30 mL/min. The differential scanning calorimetry (DSC) of samples were measured under N_2 atmosphere with a Q10, TA instrument thermal analyzer (New castle, DE, USA). Samples for DSC measurements were prepared from several circular pieces cut from the polymer film to a mass of about 10 mg. The temperature range studied was 30–400 °C. The heating/cooling rate was 10 °C/min. The degree of inside crystallinity (Xc) of the PVA membrane was calculated by the following well Equation (1) [11]:

$$X_C = \frac{\Delta Hm}{W \times \Delta H_0} \times 100, \tag{1}$$

where ΔH_0 is the apparent enthalpy of crystallization, ΔH_m is the extrapolated value of the enthalpy corresponding to the melting of a 100% crystalline sample (138.6 J/g).

3.5. Swelling Behavior

The swelling was measured by immersing the samples in phosphate buffer saline (PBS). All samples were dried in an incubator which was maintained at 37 °C until no change of mass was observed. After, the excessive water on the surface was removed with filter paper. The fully swollen samples were again weighed. The swelling ratio can be calculated as a function of time as swelling ratio (%) = $[(W_w-W_d)/W_d] \times 100\%$, where W_d is the weight in the dry state of a sample and W_w is the weight in the swollen state of the sample.

3.6. Free Radical Scavenging Ability

The 2,2-diphenyl-1-picrylhydrazyl (DPPH) was used to evaluate the free radical scavenging ability of PVA and PVA nanocomposites [11]. A control of distilled water (1 mL) or 1 mL of deionized water containing PVA and PVA nanocomposites (30×10 mm^2) was added to 3 mL of DPPH in methanol and left to stand for 90 min. Absorbance of the reaction mixture was then measured at 515 nm with an ultraviolet–visible spectrophotometer (Helios Zeta, Thermo, Waltham, MA, USA). Free radical scavenging ability effect is determined by the following equation: scavenging ratio (%) = [1 − (absorbance of test sample/absorbance of control)] × 100%.

3.7. Hydrophilicity Test

The hydrophilicity and hydrophobicity were evaluated through the contact angle test of the material surface. The surface hydrophilicity of the PVA and PVA nanocomposites was evaluated by a contact angle meter (CA-D, Kyowa interface science, Tokyo, Japan). The static contact angle was determined at 25 °C by employing drops of PBS. The quantitative titration of the water drop used was 4 μL and the time was 1 and 60 s. After that, the image was analyzed using the Drop-analysis software attached to Image J for contact angle analysis.

3.8. Bioactivity

Simulated body fluid (SBF) was prepared with ionic concentrations nearly equal to those of human blood plasma (K^+ 5.0, Na^+ 142.0, Ca^{2+} 2.5, Mg^{2+} 1.5, Cl^- 103.0, HCO_3^- 4.2, HPO_4^{2-} 1.0, and SO_4^{2-} 0.5 mM; pH 7.42). Samples were pre-cut into square shapes (10 mm × 20 mm) and then immersed in 10 mL of SBF solution in a plastic container, which was tightly closed and kept at 37 °C. The samples were removed after seven days, rinsed with distilled water and lyophilized by a freeze-drying device (FDU-1200, EYELA, Tokyo, Japan) for 24 h [11].

3.9. Biocompatibility

To evaluate the biocompatibility, 100 μL polymer solution was coated on a circular glass substrate (15 mm in diameter) and dried. Mouse-derived fibroblasts cell line (L929) was maintained in Dulbecco's modified eagle medium (DMEM) and supplemented with 10% fetal bovine serum (FBS) at 37 °C in a 5% CO_2 incubator. Before cell seeding, all samples were sterilized with UV radiation for 1 h. The sterilized samples were placed into 24-well culture plates and seeded with a cell suspension with a cell density of 5×10^4 cells/mL, followed by culturing for 24 and 72 h at 37 °C in 5% CO_2 incubator. After incubation, samples were rinsed with PBS, followed by incubation in a culture medium containing 1 mL MTT reagent for 4 h. After removal of the medium, 0.5 mL of dimethyl sulfoxide (DMSO) was added to the wells. The 0.1 mL from each sample solution was transferred to 96-well plates and the optical density (O.D.) was measured at 563–650 nm. In addition, the membrane extraction in this study followed the ISO 10993-12 standard. PVA and PVA nanocomposite membranes were cut into square shapes (3 cm^2) and subjected to sterilization by UV light for 30 min. Then, they were immersed in 75% ethanol solution and followed by PBS replacement. The membrane was immersed in DMEM at 37 °C for 24 h. After, immersion in 1 mL DMEM with 10% FBS at 37 °C in a 5% CO_2 incubator for 24 h. The extraction solution acquired for each membrane was harvested and used in MTT to determine the cytotoxicity.

3.10. Antibacterial Ability

PVA and PVA nanocomposite were investigated against *E. coli* as a model Gram-negative bacteria by the colony plate count method in order to quantify the bacterial effect of our system. The *E. coli* were prepared from fresh brain heart infusion (BHI, Becton Drive, Franklin Lakes, NJ, USA) and incubated at 37 °C for 24 h. The BHI containing *E. coli* was dilute to 10^{11} times the original concentration. Disc shapes (5.5 mm in diameter) of sample was placed into a centrifuge plastic tube and 1 mL of bacteria liquid was extracted and co-cultured for 2 h. Then, the sample was removed and placed in the new BHI solution, then agitated for 10 min. The 100 μL was extracted and applied on the BHI Agar (Becton, Dickinson and Company, TX, USA) petri dish before being cultured for 17 h at 37 °C. Finally, the colonies were counted and the results were expressed as percentage reduction rates bacteria number = $[\alpha \times 10^{11}]$, where α is the number of bacterial colonies.

3.11. *Statistical Analysis*

All data were expressed as mean ± standard deviation (SD) from three repeat samples. The data were analyzed using JMP 13 software (Statistics Analysis System, NC, USA). A one-way ANOVA followed by a Tukey's HSD post hoc test was used to determine the level of significance, where $p < 0.05$ was considered to be significant.

4. Conclusions

In this study, PVA-CoHA nanocomposite membranes show good swelling behavior, antibacterial, and bioactivity. At the same time, increasing the extensibility of the PVA reduces the damage caused by the dressing when pulled. It is expected that the release of cobalt ions in PVA-CoHA will reduce wound infection and reduce the damage caused by epithelial tissue attachment, discharge the bad tissue fluid through the excellent swelling effect, and keep the wound moist to promote healing. The results support our hypothesis that the addition of CoHA is very effective for the biological activity and anti-inflammatory properties of the membrane. Based on the above, the PVA-CoHA composite membrane described in this study can be a potential choice for diabetic traumas dressings. In the future, we will continue to explore the performance of the PVA-CoHA composite membrane on diabetic wound animal models.

Author Contributions: W.-C.L. is the core author for contributions to the analysis and interpretation of data, drafting of the manuscript, critical revision of the manuscript and carried out the experiment; C.-M.T. is the corresponding authors and contributed to the design and implementation of the research, to the analysis of the results, and to the writing of the manuscript. All authors have read and agreed to the published version of the manuscript.

Funding: This research was funded by the ministry of science and technology of Taiwan, Republic of China, via grants NSC-100-2221-E-040-001.

Conflicts of Interest: The authors declare no conflict of interest.

References

1. Young, M.; Boulton, A.; MacLeod, A.; Williams, D.; Sonksen, P. A multicentre study of the prevalence of diabetic peripheral neuropathy in the united kingdom hospital clinic population. *Diabetologia* **1993**, *36*, 150–154. [CrossRef] [PubMed]
2. Krasner, D.; Sibbald, R. Diabetic foot ulcer care: Assessment and management. In *Levin and O'Neals The Diabetic Foot*, 6th ed.; Mosby: St. Louis, MO, USA, 2001.
3. Foster, A.; Greenhill, M.; Edmonds, M. Comparing two dressings in the treatment of diabetic foot ulcers. *J. Wound Care* **1994**, *3*, 224–228. [CrossRef] [PubMed]
4. Boateng, J.S.; Matthews, K.H.; Stevens, H.N.; Eccleston, G.M. Wound healing dressings and drug delivery systems: A review. *J. Pharm. Sci.* **2008**, *97*, 2892–2923. [CrossRef] [PubMed]
5. Fonder, M.A.; Lazarus, G.S.; Cowan, D.A.; Aronson-Cook, B.; Kohli, A.R.; Mamelak, A.J. Treating the chronic wound: A practical approach to the care of nonhealing wounds and wound care dressings. *J. Am. Acad. Dermatol.* **2008**, *58*, 185–206. [CrossRef]
6. Tănase, E.E.; Popa, V.I.; Popa, M.E.; Râpă, M.; Popa, O. Biodegradation study of some food packaging biopolymers based on pva. *Bull. Uasvm Anim. Sci. Biotechnol.* **2016**, *73*, 1–5. [CrossRef]

7. Baker, M.I.; Walsh, S.P.; Schwartz, Z.; Boyan, B.D. A review of polyvinyl alcohol and its uses in cartilage and orthopedic applications. *J. Biomed. Mater. Res. Part B Appl. Biomater.* **2012**, *100*, 1451–1457. [CrossRef]
8. Vijayasekaran, S.; Fitton, J.; Hicks, C.; Chirila, T.; Crawford, G.; Constable, I. Cell viability and inflammatory response in hydrogel sponges implanted in the rabbit cornea. *Biomaterials* **1998**, *19*, 2255–2267. [CrossRef]
9. Chen, W.; Hou, Y.; Tu, Z.; Gao, L.; Haag, R. Ph-degradable pva-based nanogels via photo-crosslinking of thermo-preinduced nanoaggregates for controlled drug delivery. *J. Control. Release* **2017**, *259*, 160–167. [CrossRef]
10. Ismail, H.; Freakley, P.; Sheng, E. The effect of carbon black particle size on multifunctional additive-carbon black interaction. *Eur. Polym. J.* **1995**, *31*, 1049–1056. [CrossRef]
11. Tang, C.-M.; Tian, Y.-H.; Hsu, S.-H. Poly(vinyl alcohol) nanocomposites reinforced with bamboo charcoal nanoparticles: Mineralization behavior and characterization. *Materials* **2015**, *8*, 4895–4911. [CrossRef]
12. Stojanović, Z.; Veselinović, L.; Marković, S.; Ignjatović, N.; Uskoković, D. Hydrothermal synthesis of nanosized pure and cobalt-exchanged hydroxyapatite. *Mater. Manuf. Process.* **2009**, *24*, 1096–1103. [CrossRef]
13. Veselinović, L.; Karanović, L.; Stojanović, Z.; Bračko, I.; Marković, S.; Ignjatović, N.; Uskoković, D. Crystal structure of cobalt-substituted calcium hydroxyapatite nanopowders prepared by hydrothermal processing. *J. Appl. Crystallogr.* **2010**, *43*, 320–327. [CrossRef]
14. Stanić, V.; Radosavljević-Mihajlović, A.S.; Živković-Radovanović, V.; Nastasijević, B.; Marinović-Cincović, M.; Marković, J.P.; Budimir, M.D. Synthesis, structural characterisation and antibacterial activity of ag+-doped fluorapatite nanomaterials prepared by neutralization method. *Appl. Surf. Sci.* **2015**, *337*, 72–80. [CrossRef]
15. Kramer, E.R.; Morey, A.M.; Staruch, M.; Suib, S.L.; Jain, M.; Budnick, J.I.; Wei, M. Synthesis and characterization of iron-substituted hydroxyapatite via a simple ion-exchange procedure. *J. Mater. Sci.* **2013**, *48*, 665–673. [CrossRef]
16. Fan, W.; Crawford, R.; Xiao, Y. Enhancing in vivo vascularized bone formation by cobalt chloride-treated bone marrow stromal cells in a tissue engineered periosteum model. *Biomaterials* **2010**, *31*, 3580–3589. [CrossRef] [PubMed]
17. Zhou, J.; Zhao, L. Hypoxia-mimicking co doped tio2 microporous coating on titanium with enhanced angiogenic and osteogenic activities. *Acta Biomater.* **2016**, *43*, 358–368. [CrossRef]
18. Kulanthaivel, S.; Mishra, U.; Agarwal, T.; Giri, S.; Pal, K.; Pramanik, K.; Banerjee, I. Improving the osteogenic and angiogenic properties of synthetic hydroxyapatite by dual doping of bivalent cobalt and magnesium ion. *Ceram. Int.* **2015**, *41*, 11323–11333. [CrossRef]
19. Tahmasebi Birgani, Z.; Fennema, E.; Gijbels, M.J.; de Boer, J.; van Blitterswijk, C.A.; Habibovic, P. Stimulatory effect of cobalt ions incorporated into calcium phosphate coatings on neovascularization in an in vivo intramuscular model in goats. *Acta Biomater.* **2016**, *36*, 267–276. [CrossRef]
20. Ignjatovic, N.; Ajdukovic, Z.; Rajkovic, J.; Najman, S.; Mihailovic, D.; Uskokovic, D. Enhanced osteogenesis of nanosized cobalt-substituted hydroxyapatite. *J. Bionic Eng.* **2015**, *12*, 604–612. [CrossRef]
21. Lin, W.-C.; Chuang, C.-C.; Wang, P.-T.; Tang, C.-M. A comparative study on the direct and pulsed current electrodeposition of cobalt-substituted hydroxyapatite for magnetic resonance imaging application. *Materials* **2019**, *12*, 116. [CrossRef]
22. Lin, W.C.; Yao, C.; Huang, T.Y.; Cheng, S.J.; Tang, C.M. Long-term in vitro degradation behavior and biocompatibility of polycaprolactone/cobalt-substituted hydroxyapatite composite for bone tissue engineering. *Dent. Mater.* **2019**, *35*, 751–762. [CrossRef] [PubMed]
23. Ionita, M.; Crica, L.E.; Tiainen, H.; Haugen, H.J.; Vasile, E.; Dinescu, S.; Costache, M.; Iovu, H. Gelatin–poly(vinyl alcohol) porous biocomposites reinforced with graphene oxide as biomaterials. *J. Mater. Chem. B* **2016**, *4*, 282–291. [CrossRef] [PubMed]
24. Vashisth, P.; Nikhil, K.; Roy, P.; Pruthi, P.A.; Singh, R.P.; Pruthi, V. A novel gellan-pva nanofibrous scaffold for skin tissue regeneration: Fabrication and characterization. *Carbohydr. Polym.* **2016**, *136*, 851–859. [CrossRef] [PubMed]
25. Bryaskova, R.; Pencheva, D.; Kale, G.M.; Lad, U.; Kantardjiev, T. Synthesis, characterisation and antibacterial activity of pva/teos/ag-np hybrid thin films. *J. Colloid Interface Sci.* **2010**, *349*, 77–85. [CrossRef]
26. Shi, Y.; Xiong, D.; Liu, Y.; Wang, N.; Zhao, X. Swelling, mechanical and friction properties of pva/pvp hydrogels after swelling in osmotic pressure solution. *Mater. Sci. Eng. C Mater. Biol. Appl.* **2016**, *65*, 172–180. [CrossRef]

27. Mishra, S.K.; Ferreira, J.M.; Kannan, S. Mechanically stable antimicrobial chitosan-pva-silver nanocomposite coatings deposited on titanium implants. *Carbohydr. Polym.* **2015**, *121*, 37–48. [CrossRef]
28. Abdelrazek, E.M.; Elashmawi, I.S.; Labeeb, S. Chitosan filler effects on the experimental characterization, spectroscopic investigation and thermal studies of pva/pvp blend films. *Phys. B Condens. Matter.* **2010**, *405*, 2021–2027. [CrossRef]
29. Adhikari, B.B.; Kanemitsu, M.; Kawakita, H.; Ohto, K. Synthesis and application of a highly efficient polyvinylcalix [4] arene tetraacetic acid resin for adsorptive removal of lead from aqueous solutions. *Chem. Eng. J.* **2011**, *172*, 341–353. [CrossRef]
30. Nie, L.; Chen, D.; Suo, J.; Zou, P.; Feng, S.; Yang, Q.; Yang, S.; Ye, S. Physicochemical characterization and biocompatibility in vitro of biphasic calcium phosphate/polyvinyl alcohol scaffolds prepared by freeze-drying method for bone tissue engineering applications. *Colloids Surf. B Biointerfaces* **2012**, *100*, 169–176. [CrossRef]
31. Lim, Y.J.; Oshida, Y.; Andres, C.J.; Barco, M.T. Surface characterizations of variously treated titanium materials. *Int. J. Oral Maxillofac. Implant.* **2001**, *16*, 333–342.
32. Buser, D.; Broggini, N.; Wieland, M.; Schenk, R.K.; Denzer, A.J.; Cochran, D.L.; Hoffmann, B.; Lussi, A.; Steinemann, S.G. Enhanced bone apposition to a chemically modified sla titanium surface. *J. Dent. Res.* **2004**, *83*, 529–533. [CrossRef] [PubMed]
33. Saeed, A.M.; Hassan, R.A.; Thajeel, K.M. Synthesis of calcium hydroxyapatite powder from hen's eggshell. *Iraqi J. Phys.* **2011**, *9*, 24–28.
34. Lin, W.C.; Chuang, C.C.; Chang, C.J.; Chiu, Y.H.; Tang, C.M. The effect of electrode topography on the magnetic properties and mri application of electrochemically-deposited, synthesized, cobalt-substituted hydroxyapatite. *Nanomaterials* **2019**, *9*, 200. [CrossRef] [PubMed]
35. Kavitha, T.; Haider, S.; Kamal, T.; Ul-Islam, M. Thermal decomposition of metal complex precursor as route to the synthesis of co 3 o 4 nanoparticles: Antibacterial activity and mechanism. *J. Alloy. Compd.* **2017**, *704*, 296–302. [CrossRef]

MDPI
St. Alban-Anlage 66
4052 Basel
Switzerland
Tel. +41 61 683 77 34
Fax +41 61 302 89 18
www.mdpi.com

International Journal of Molecular Sciences Editorial Office
E-mail: ijms@mdpi.com
www.mdpi.com/journal/ijms

www.ingramcontent.com/pod-product-compliance
Lightning Source LLC
Chambersburg PA
CBHW061601220926
51850CB00005B/396

* 9 7 8 3 0 3 6 5 7 2 5 1 2 *